Creating Agile Business Systems with Reusable Knowledge

Agility and innovation are necessary to acheive global excellence and customer value in twenty-first century business; yet most approaches to business process engineering in existence today sacrifice these in favor of operational efficiency and economics. Moreover, the IT systems used to automate and encapsulate business processes are inflexible and unable to respond to the constantly changing business environment. Mitra and Gupta provide insight to close this gap – they show how innovation can be systematized with normalized patterns of information, how business processes and information systems may be tightly aligned, and how these processes and systems can be designed to automatically adapt to change by re-configuring shared patterns of knowledge. The authors present a modular approach to building business systems that parallels that of object oriented software. They describe how business knowledge can be identified, encapsulated, and shared, as well as how reusable process modules can be developed to offer the systems flexibility. The book provides practical templates required for accelerating integration, analysis, and design. Mitra and Gupta lay the foundation of a new paradigm in which computers manipulate meanings, not blind symbols.

This book will appeal to consultants, analysts, and managers in IT firms looking to develop new, more flexible products for their clients. It will also be of interest to researchers and graduate students in business, management, and IT. By deeply integrating business knowledge and processes with IT systems design, this book is a valuable addition to both communities.

Amit Mitra is Managing Consultant at Headstrong LLC, in addition to President and Principal Consultant at Sprybiz LLC. He is an alumnus of KPMG and former Chief Methodologist of the American International Group.

Amar Gupta holds a number of positions at the University of Arizona, Tucson. He is Professor of Entrepreneurship and MIS; Thomas R. Brown Chair in Management and Technology; and Senior Director for Research and Business Development at the Eller College of Management; and he is Professor of Computer Science in the College of Science. In addition, he serves as Visiting Professor, Engineering Systems Division, College of Engineering at the Massachusetts Institute of Technology (MIT).

Creating Agile Business Systems with Reusable Knowledge

A. Mitra and A. Gupta

CAMBRIDGE
UNIVERSITY PRESS

CAMBRIDGE UNIVERSITY PRESS
Cambridge, New York, Melbourne, Madrid, Cape Town, Singapore, São Paulo

Cambridge University Press
The Edinburgh Building, Cambridge CB2 2RU, UK

Published in the United States of America by Cambridge University Press, New York

www.cambridge.org
Information on this title: www.cambridge.org/9780521851633

First published 2006

Printed in the United Kingdom at the University Press, Cambridge

A catalog record for this publication is available from the British Library

ISBN-13 978-0-521-85163-3 hardback
ISBN-10 0-521-85163-7 hardback

I dedicate this book to my father, Ajoy Mitra, my mother, Sevati Mitra, and my teachers, who helped shape my thoughts, to my wife Snigdha and my children Tanya and Trishna, who helped shape my life, and in turn shape this book.

Amit Mitra

I dedicate this book to my mother, my sister (Beena), my wife (Poonam), my children (Amrit and Amita), and to my teachers in India (especially at IIT, Kanpur and IIT, Delhi), the UK (at Birkbeck College and the Imperial College of Science and Technology), and the US (primarily at MIT).

Amar Gupta

Contents

Boxes

Figures

Tables

Foreword

The health and well-being, and today the very survival, of an enterprise depend on its ability to respond and adapt in *timely*, *innovative*, and *effective* manners. The relatively static behaviors of the past have been changed unalterably by the explosion of *telecommunication* and *information technologies/capabilities* as typified through the emergence of the World-Wide Web (W^3). Enterprises are learning to adapt to the challenges of the new global business and national security environment by exploiting the same capabilities that are driving the *dynamic* environment, telecommunications and IT. In essence, *information* and the *knowledge* derived therefrom have emerged as key *assets of the enterprise* in responding and adapting to the demands of the global environment.

The experiences over the past decade for a wide variety of enterprises, including both governmental and commercial entities, are reflected by *more failures* than successes in embracing successful strategies and solutions for creating, engineering, and evolving the *knowledge system* that serves the enterprise most effectively. Certainly, problems have arisen through failures of *leadership* and *management*, who have been unable to break the static behaviors and narrow organizational views that served them well in the past. On the other hand, *engineers*, given responsibility for exploiting information technologies to meet the information needs of leadership and management for knowledgeable decision making, have found the challenges of dealing with the *complex event-driven environments* and the *complex* array of enterprise *stakeholders* and *systems* vastly more difficult than the systems engineering problems of the past.

From both successful and failed efforts, there is an ever-growing body of knowledge about some of the keys to successfully reengineering the global enterprise as a flexible and adaptive entity. The concepts that are seeing increasing attention include enterprise architecting, service-oriented architectures, business process modeling, enterprise and e-business patterns, enterprise systems engineering, and agile development methods. All of these approaches and methods contribute to one or more of the fundamental advantages that are driving the developments. These advantages span a number of enterprise dimensions and can be summarized in the following way:

- *Strategic focus:* provides a basis for understanding the contributions of complex, large, distributed information systems in achieving enterprise goals and missions.
- *Broadened communications:* enhances communication across the enterprise community from leaders/managers/users to engineers/developers/testers.

- *Performance and QoS:* improves performance and quality of service (QoS) for decision support and knowledge-based decision making.
- *Timely and flexible response:* enables flexibility for timely and effective response to new and unexpected situations.
- *Integration and interoperable operations:* enables mechanisms for assuring integrated and interoperable applications, both among legacy and new systems.
- *Commercial technology evolution:* facilitates the introduction and effective use of rapidly changing commercial information systems and technology.
- *Cost-effective migration:* establishes a foundation for value-based thinking, analysis of alternatives, and investment planning to establish cost-effective system evolution.
- *Organizational efficiencies:* allows organizational efficiencies due to reduced staffing requirements, easier system evolution process, etc.

But from reviewing and assessing a myriad of enterprise developments, it is apparent that a much deeper understanding is needed to increase the probability of success for enterprises working to meet the challenges of the global environment. In this book, complemented by their earlier book, *Agile Systems with Reusable Patterns of Business Knowledge*, the authors provide a very comprehensive and integrated perspective on the range of topics mentioned above. Starting from basic principles, the book presents an approach to enterprise reengineering that merits careful attention and thoughtful application. As they say in the Preface, the book provides a description of a "hidden and elegant theoretical framework: a framework that is a direct bridge between business process engineering and systems engineering." The approach that is presented is ambitious and provocative, and commands thoughtful consideration from developers and researchers in this field of ever-increasing importance.

Harold W. Sorenson

Professor of Engineering Systems
Faculty Director
Graduate Program in Architecture-based
Engineering of Enterprise Systems
Jacobs School of Engineering
Rady School of Management
University of California, San Diego
La Jolla, CA 92037

[Former Chief Scientist, US Air Force;
Former Senior Vice President and General
Manager for Air Force Systems at MITRE;
Former Chief Engineer for AF Electronic
Systems Center]

Preface

Why this book? – Because it is a book *begging* to be written. The real world is chaotic and never stands still. Businesses constantly strive to re-invent themselves under continual, and often intense, pressures of competitive, regulatory, and technological change. The pivotal issue in business computing lies in incorporating new learning in automated information systems; adding to what is already known and adapting automatically as perspectives and priorities continually change. This is the challenge for which we have sought the answers presented in this book.

Where did it all start? It was 1992. One of the authors was the Chief Methodologist for the American International Group, an unusual global corporation that believed in turning on a dime. The firm needed a systems development discipline to facilitate nimble and innovative business practices. Thus a truly exciting and wonderful journey began – a journey we want to share with you. In this book, you will find readily usable patterns and models you can leverage to establish business requirements, object models, and knowledge bases to support the agile and exacting business needs of the twenty-first century. You will also find the exciting and simple beauty of a framework that is the direct bridge between business process engineering and systems engineering. Yes, it is a proven framework that works for every industry and business application we have tested it on – from telecommunications to insurance, from financial services to manufacturing.

In the chapters that follow, we will share with you not only how this framework works in practice, but also how it actually anticipates key requirements even before users articulate them, such as those that flowed from strategic shifts in the regulatory bedrock of the US telecommunications industry. The theoretical foundation of the approach is not only deep, but also elegant in its simplicity.

Where will it eventually lead? In the short term, it can make your business more agile. It can provide reusable models, processes, and business knowledge components to compress your time to market new or improved products, services, and processes. It can also show you how you can compress systems development and integration times. However, it is the vision at the end of this journey that is the most fascinating of all. The concepts in this book can provide the foundation of disciplines that can make business systems truly maintenance free – systems based on software that can automatically adapt to change and chaos. These systems can be supported by automated intelligent agents[1] that will, some day, maintain

[1] The "Intelligent agents" section of the Bibliography at the end of this book lists papers that describe agent technology and the-state-of-the-art.

software and respond to environmental change at the speed of thought, a vision we will share with you at the end of this book.

As practitioners, managers, and teachers in the field of information systems, we often talk of change control. Change plays havoc with our plans and products. However, the wealth in the knowledge economy will flow from global excellence, thriving on change and innovation. The only justification for technology will, and must be, change facilitation, not change control. Are we ready?

"Wouldst thou," – so the helmsman answered
"Learn the secret of the sea?
Only those who brave its dangers
Comprehend its Mystery!"
(Henry Wadsworth Longfellow,
The *Galley of Count Arnaldos*)

Acknowledgements

We deeply appreciate the time that Dr. E. C. Subbarao, our former teacher, spared from his busy schedule to help shape this book. We thank Srividhya Subramanian, Yan An Surendra Sarnikar, Shivram Mani, and Prithi Avanavadi for going through this entire manuscript carefully. We are grateful to Jayant Pal and Raghunandan Dhat of Setu Inc. for their enthusiasm in instantiating the concepts in software, and to Pat Fayad, for keeping the faith, and her unflagging encouragement as this book was being written. Finally, one of the authors (Amar Gupta) would like to thank the Engineering Systems Division of the Massachusetts Institute of Technology (MIT), especially Professor Daniel Hastings and Professor Joel Moses, for providing a Visiting Professorship and other support in 2005 and 2006 when this book was being finalized.

Introduction

1 What is this book about and who should read it?

This book is about facilitating change with component technology, but is different from most approaches to the topic. The components in the book are not traditional I/T components. Rather they are shared components of knowledge from which patterns of business knowledge are assembled. A fundamental premise of this paradigm is that meanings are patterns of information, abstract structures that may be derived from other components, which are also meanings. If we can identify and describe these components and their structures with precision, we can automate the process. Business processes and information systems configured from these components will be extremely flexible, configurable, and coordinated.

This book lays the foundation of a new computing paradigm – a paradigm in which computers manipulate meanings, not program code or blind symbols. Computers of the future, built on the principles described in the book, will operate on the plane of meanings – a little like we do.

Business meanings, patterns, and rules jointly constitute the substance of a business process. Without the business layer, technological standards have little meaning. The return on investment from reusing business knowledge can complement, and be orders of magnitude larger than by adherence to technical standards alone. This book establishes a framework for the transfer and reuse of business knowledge in different contexts. This is why we urge architects interested in service oriented architecture and business process management to read this book. This book is for information and process architects. It is also a book for architects of *languages* for specifying business processes (languages like BPMN and information sharing standards such as SBVR and BPDM from object Management Group (OMG).).[1]

This book is about automating the configuration of business processes from components of business knowledge. We urge software architects and technologists to read this book

[1] The Business Process Management Initiative (BPMI) consortium is a consortium of diverse firms. Its purpose is to standardize business process definitions "that span multiple applications, corporate departments and business partners, behind the firewall, and over the Internet." BPMI has published the BPML language in support of business process automation. See http://www.bpmi.org. XML is from W3C, another consortium for data standards.

because it is about the technical principles that drive information architecture and information sharing in the form of meanings and concepts. The principles and patterns in this book complement the work that has been already been done in developing technology and interfacing standards for information systems. The purpose of this book is not to propose yet another technical standard. It is to describe the business intelligence, in component form, that these standards must support and be joined to. It is the next step.

2 What will the information be used for?

The patterns in this book can address the following business issues:

1. *Agility and adaptability of businesses processes and systems*: facilitate designing of agile business processes and flexible systems based on the reusable patterns of information in this book. They will help automate the alignment between business processes and information systems, speed development of systems to support new products and distribution channels, and accelerate process and systems integration when businesses integrate or reinvent themselves in their product markets.
2. *Integration and coordination of information*: coordinate integration of information and processes across supply chains, enterprises, and databases.
3. *Reduction of time-to-market of new concepts*: accelerate formulation of functional requirements and process models based on the prefabricated reusable patterns in this book.
4. *Creation of automated tools for aligning of information systems with business*: facilitate the development of an integrated Computer Aided Process Engineering (CAPE) and Computer Aided Systems Engineering (CASE) tool; provide a framework for testing the completeness and validity of a language or methodology for business rule/process definition and modeling, and for evaluating CAPE and upper CASE tools.
5. *Compression of the time to develop prototypes*: the patterns delineated in this book can serve as the basis for early prototypes when iterative prototyping methodologies are used for developing or integrating information systems and business processes.

Ultimately this book is about change. It describes a technology for automating and facilitating change – a technology that will facilitate the innovation and adaptation so necessary for corporations to remain competitive in a fast-changing, diverse, and tumultuous business universe that will not forgive the tardy.

3 Technology's broken promise

Why is change so difficult? That is a question with an easy answer. We have all experienced how a seemingly simple change to a business process or information system has many impacts – often unanticipated, at multiple places, in multiple ways – on other different business processes and at many different layers of the information systems legacy that support these processes. Each impact has several other impacts in turn, which ricochet through our processes and systems until we are caught in an explosive cascade of change. Business sponsors requesting the change are then faced with a painful choice – either make

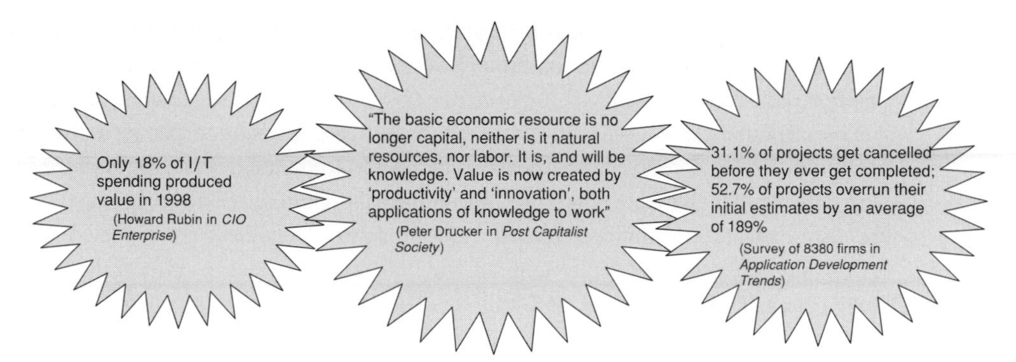

Figure 1 Information is a key resource, but investment in information systems is fraught with risk

the changes at a cost, both in time and money, and take a risk that might be excessive, or abandon the competitive benefits of innovation because the risk is too high and the change is not timely. Studies have indicated that information system projects are fraught with risk, and businesses do not realize the value they should from their investments in information systems (figure 1).

A recent example was the Y2K problem that seemed trivial to the layman, but the state-of-the-art of computer technology was such that it may have cost industry as much as $600 billion[2] and a significant part of the world's resource pool of professionals to merely express the year in four, instead of two, digits.[3] Many strategic benefits have been difficult to implement for similar reasons. Denial of strategic benefits to consumer and provider alike are so frequent that examples litter the industrial landscape in almost every direction. Examples of missing capabilities include:

- Straight through processing and "T+0" settlement in the financial securities industry (the ability to settle a trade immediately with almost no manual processing).
- Real time billing for telephone subscribers and personal telephone numbers in the telecommunications industry (where a unique contact phone number automatically follows an individual regardless of location or geography).
- Timely and reliable order fulfillment and innovative customer service for manufacturers and retailers.
- Risk assessment when providing insurance coverage to complex global clients in the insurance industry.

This is only a small slice of such wish lists – strategic innovations and improvements in almost every industry are often deemed too risky or impossible because supporting processes and information systems are deemed too complex and risky, if not impossible, to change.

Despite inventing new technology at a prodigious rate to make change easy (including technologies such as CASE tools, code generators, structured programming, relational databases, expert systems, object technology, and reusable components), systems still cannot

[2] Sources: Gartner Group & Congressional Research Service estimates quoted by Steve C. Yuen, Ph.D., University of Southern Mississippi and Jo Ann Mitchell, Jones Junior College on our website.

[3] For example, 1/1/2001 instead of 1/1/01. Computer calculations involving dates beyond 1999 had a very high risk of error if the year was not expressed in four digits.

change fast enough. Why are information systems a bottleneck? Can process re-engineering, business innovation, and time to market be accelerated? Why have these technologies not fulfilled their full potential?

The principal reason why the problem of change continues to persist is that we have not found a way of representing business rules and knowledge in a single place in such a way that we can change a rule once, and reflect the change wherever it impacts business processes and supporting automation. Instead, rules of business are repeated in different forms and formats in multiple, tangled ways in several systems at several places, which makes change complex, risky, and difficult. This has been a core problem.

It was not solved in the 1950s when we replaced the tangled code of machine language with assembler languages, or in the 1960s when we replaced the spaghetti code of assembly languages with that of third-generation languages like COBOL and FORTRAN; nor was it solved in the 1970s and 1980s with the coming of relational databases, expert systems, and CASE tools, or even in the 1990s when tangled object inheritance became so much of a problem that many advocated making multiple inheritance illegal in tools of the day. *More automation merely automated tangling of more business rules faster.*

For this reason, the authors asked a different question: what information do we need to model the stimulus response behavior of business processes and the organizations they support in the real world, and what is the *natural* real-world structure of information that can represent business knowledge in fully normalized,[4] and hence reusable form across the universe of diverse global business environments?

Why would the proposed approach work when so many others have failed? It works because it untangles business rules. It untangles business rules even if they were tangled in legacy models and systems. Thus it allows us to represent business knowledge *once* in a repository of knowledge, from where it can *naturally* manifest itself in different business contexts. Changes made in the right place will automatically impact business systems where they must. It is no surprise that many businessmen and professionals have intuitively felt that business knowledge acquired in one context might often be reused in another. We discovered in 1992 that this intuition is a fundamental truth that flows from the natural structure of business knowledge in the real world. However, we must explicitly recognize this structure and express it with mathematical precision to use it effectively. We will share this vision with you in the chapters that follow.

4 Component reuse – *the genesis*

The concept of using reusable components to compress application development time is almost as old as the software industry. Components have evolved from concepts such as copy libraries, common subroutines, and general purpose applications packages, in the early days of batch computing, to reusable GUI, network, and data services objects, based on standards such as CORBA, COM, and web services such as XML and WSOL which support distributed, interactive Web, and client–server computing.

[4] "Normalized" means represented uniquely in a single place once.

For historical reasons, software component engineering first focused on the back end of the process engineering value chain. Its first concern was program code and interfaces for communicating data in terms of streams of bits and bytes. The economic impact, however, is usually far larger at the front end of the re-engineering value chain – on reusing business knowledge to configure and innovate business processes, services, and products. It is not surprising that business had only very limited interest and no involvement in the kind of components that software engineers were interested in. Consequently, the business community's support for the software community's component technology was lukewarm at best. The focus has shifted in step with evolving technology. Now the time is ripe to look at the reuse of knowledge. This territory, long neglected by the software community is, and has always been, where the major benefits to business are found. Let us analyze these imperatives.

Box 1 Example of the process engineering value chain

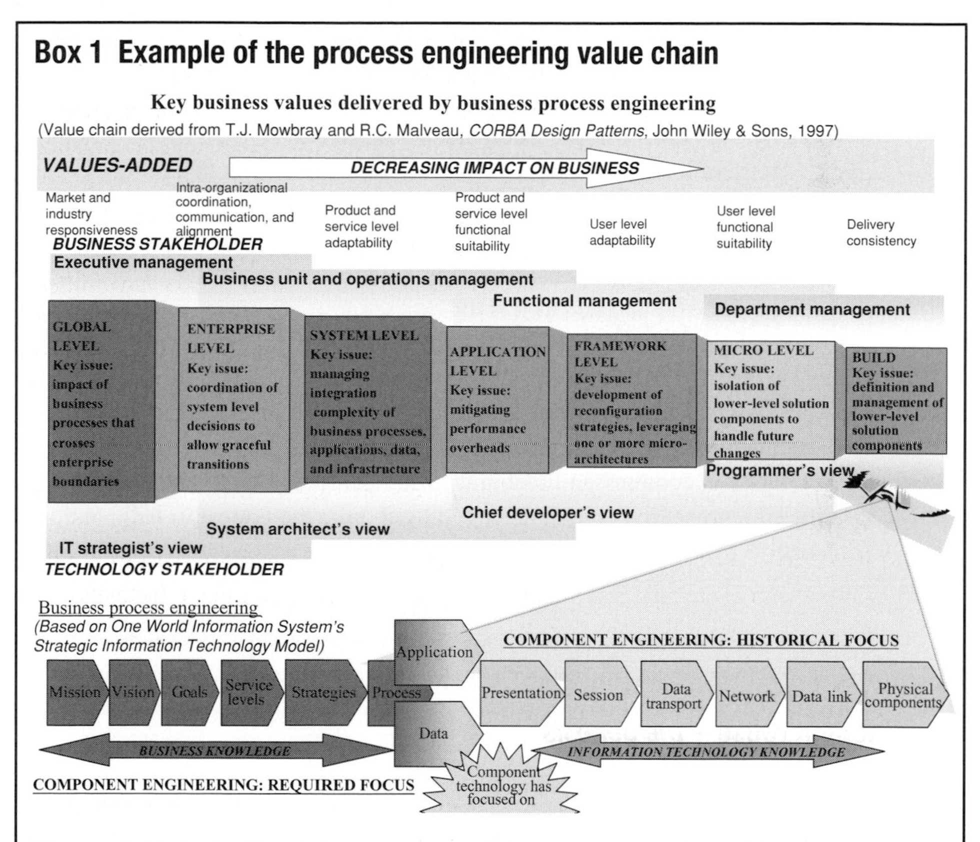

The supply chain for IT solutions represents the process of technology initiatives, application development, implementation and business use. For example, the supply chain for an ERP system may comprise of technology platform selection, systems specifications, systems development, packaging and documentation, implementation, and use in the end user business. The concept of the demand chain, which transfers demand from end users to technology suppliers, is less familiar. To give one example, the demand chain for an ERP may start with business users spotting new opportunities for using the system to support their business. The next link in the chain is the IT

> department of the user organization looking for potential solutions already in place in the business. In the demand chain of the ERP system, it is 'missing' process solutions that drives the next stage – a process innovation stage – where new processes and solutions are outlined. The last step in the demand chain, is demand for resources and skills needed for using, operating, and developing the ERP system in the user organization over time . . . What is needed is capabilities to capture an increasing number of business opportunities already in the use . . . the supply chain for IT solutions needs to be managed so that both the current and future applications architectures are scalable, flexible and modular. (Jan Holmström and Tiina Tissari, *IT Value Capture: Creating an Effective Demand–Supply Chain for IT Solutions*)

As businesses have become increasingly reliant on automation, the line between technology and the corporation's key business operations has started to blur. Industry has begun to recognize that the greatest benefit to business will flow from reusing business intelligence embedded in software. Consequently the software industry has been striving to craft software components to reuse this embedded business intelligence across the supply chain. The intent is to speed up business processes, to make corporations more agile, and, above all, to position the business at a competitive advantage.

However, this kind of reuse has remained elusive in spite of over 15 years of industry effort. The reason why the promise has not been fulfilled is that industry was not ready to leverage the technology – processes had not matured, technology was still groping for the right answers, software developers were loth to frontload effort on software projects, and, most of all, business sponsorship was weak because software architecture was not as critical to successful business as it is today.

E- (as well as M-) commerce has forced cross-enterprise transparency into business processes and driven the need for standards. The market is now ripe for a product offering software components that will encapsulate and reuse business knowledge to build software architected to facilitate business innovation, speed, and agility: software that must be developed in compressed timeframes. Competitors are few and it is a prime opportunity for entrepreneurial corporations willing to take the plunge to build and sell software components that reuse business intelligence.

THE NEW OPPORTUNITY

Components will reduce the need for large scale integration – the bread and butter of the Big Six and others. But even as they mourn the loss, a new practice will emerge: helping users pick from the rapidly growing set of component based options. (The Forrester Report, "Package Application Strategies," June 1, 1996)

5 Scope of this book

In the following chapters, we will examine how project managers, requirements analysts, process engineers, and information modelers can leverage the frameworks and patterns proposed in this book to do their jobs faster, better, and with fewer resources.

Balancing risk with reward is at the heart of every business. Therefore it might be ironic, but hardly surprising, that business operations are largely deterministic. They are designed to minimize uncertainty. The scope of this book is therefore limited to deterministic patterns and processes. This assumption will simplify our model of knowledge. However, we cannot do justice to business knowledge if we ignore uncertainty and risk altogether.[5] Therefore the model of knowledge in this book does provide some structures and components that partially, but adequately,[6] compensate for the purely deterministic nature of the model.

Unlike engineering systems, the vast majority of business processes deal with discrete, not continuous, change. For example, a business agreement might be negotiated in discrete steps that start with a first draft, progress through a series of reviews and revisions, and finally end with a binding contract. The scope of this book is limited to models of discrete, not continuous, change. This is adequate for almost all business processes. An engineering process, however, might contend with feeds and parameters that change or flow continuously. For example, when hydrocarbon-based resins are made in a chemical reactor, the density of the resin produced varies continuously with changing temperatures and pressures in the reactor. Instead of focusing on continuous technical processes, our focus, is on the discrete business process.

The book focuses on normalizing,[7] encapsulating, and reusing business knowledge across the value chain described in box 1. Business knowledge is technology independent. This knowledge may be embedded in processes that are supported by diverse technologies, both automated and manual. Often, in large organizations, the same business rules are expressed in different systems and procedures, on different technology platforms, in different countries or organizational units. The choice of the technology often depends on the organization's legacy, its local environment, and its infrastructure. Although business knowledge is independent of the technology that implements it, if an organization wants to reuse business knowledge explicitly, it must store this knowledge in an electronic repository. Thus business knowledge in such a repository is an item of information that is expressed in some physical format and medium and is an electronic artifact. For this reason, we have named these components of knowledge *business knowledge artifacts.*

Business knowledge artifacts complement, but are different from traditional software components. The following chapters will show you how to link business knowledge artifacts to software components.

Because this book focuses on the rules of business, *business knowledge artifact* has often been abbreviated to *knowledge artifact* in the material that follows. *Knowledge artifacts* encapsulate bits of formal business intelligence – meanings – that can be stored as reusable components in a repository of business knowledge. Standardized knowledge artifacts will be central to the evolving knowledge economy, especially to the global supply chains emerging in the new economy.

6 Foundation of knowledge reuse: three pillars

Business knowledge is not about files, data flows, formats, screens, or computers. Rather it is about processes, practices, norms, products, policies, regulations, infrastructure, and people, constrained by the physical, regulatory, and ethical contexts in which they function.

[5] Publications in the bibliography discuss stochastic processes (processes based on chance and uncertainty).
[6] Adequate for the vast majority of business processes. [7] See the endnote on normalization.

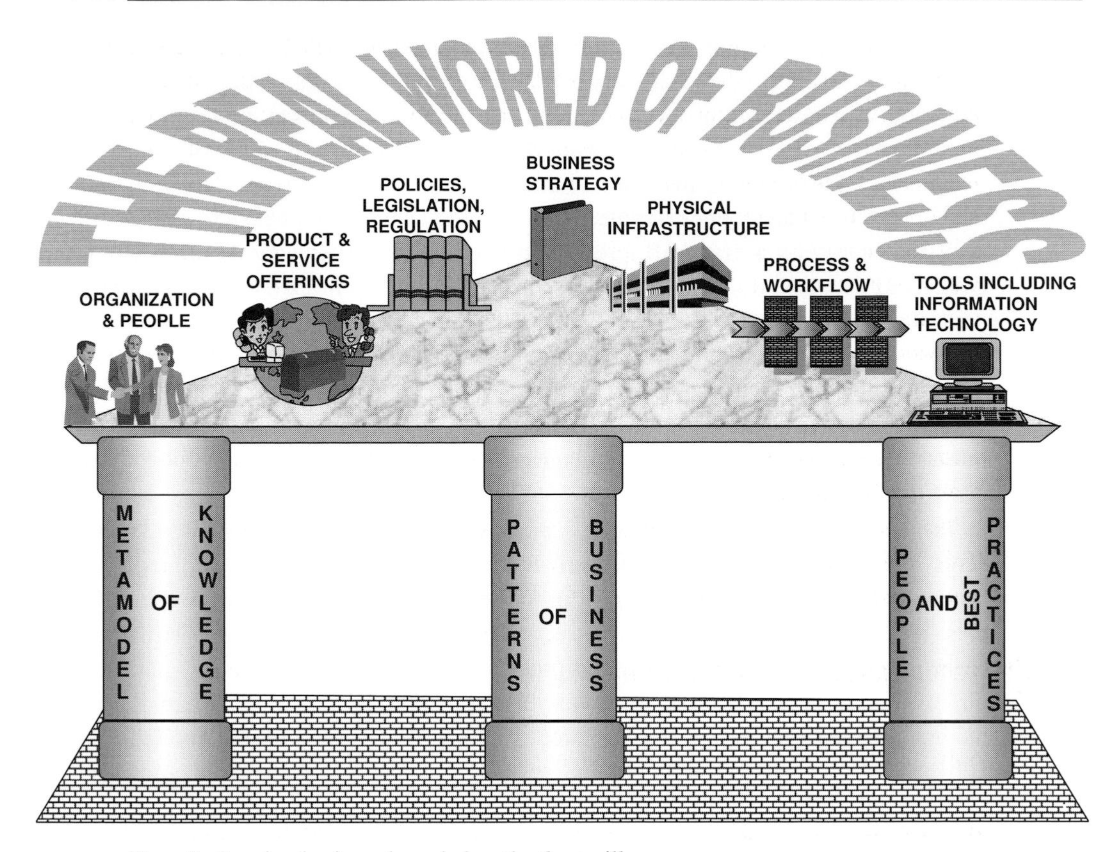

Figure 2 Reusing business knowledge: the three pillars

To recast this knowledge in the form of normalized[8] and reusable capsules of information that can be assembled into configurations of knowledge and innovative ideas, we must first know how knowledge can be normalized. We must also know which parts are reusable and how to organize people and business practices to leverage these reusable knowledge components. The three pillars in figure 2 are the pillars on which business knowledge components must stand.

6.1 The first pillar: metamodel of business knowledge

Knowledge can only be reused if it is extracted and stored as a single piece of information in the knowledge repository. This information can then be used in as many different contexts as necessary, whenever and wherever it is needed. Additionally, in order to track its impact, we must know the relationship this piece of information has with other similar bits of knowledge in our repository of business knowledge.

[8] Normalization is a structured method of representing information in a non-redundant way. The endnote on normalization describes it in more detail.

For example, if the exchange rate between the US dollar and British pound changes, it could impact several valuations such as amounts on invoices, credit limits, checks, payment amounts, cash on hand, and fixed assets overseas. In other words, we must know the *structure* of information in the real world – that there are interrelated entities such as processes, resources, work products, and units of measure.

This information about information is collectively called a *metamodel*. The metamodel will provide the scheme for storing business knowledge in a non-overlapping, non-redundant way. The abstract objects in the metamodel, such as process, resource, unit-of-measure, and their interrelationships, are containers of non-redundant (normalized) business knowledge. Individual business knowledge artifacts would be classified and stored in these containers provided in the knowledge repository.

Specific knowledge artifacts can then be extracted from these containers and assembled into complex business processes and bodies of knowledge around which information systems can be built, the metamodel is the schema of the knowledge repository. It is the first pillar on which knowledge reuse stands. Without it, there can be no knowledge components. This book develops the metamodel of business knowledge. A companion book *Knowledge Reuse and Agile Processes, Catalysts for Innovation* extends the metamodel.

6.2 The second pillar: business patterns

How many business rules does an enterprise need in order to do business? We know that only a small fraction of business knowledge is explicitly recorded and recognized by most operating businesses. Most business knowledge is implicit. Some are common sense rules that seem foolish to explicitly publish, such as "accept payment for goods sold," while others might be embedded in the experience or common understanding of the firm's employees, such as "breaking my budget will be a career limiting move." However, automation has no innate commonsense unless it is explicitly built in. Extracting and storing all rules of business, implicit and explicit, for even a small and simple business like a mom and pop corner store is not just a daunting task, it is an impossible task (figure 3): there are too many rules. There can be only one outcome if an analyst attempts to discover *every* rule of business for even the simplest enterprise: analysis paralysis.

Fortunately there is a solution. The knowledge repository is an electronic warehouse that holds the inventory of knowledge components and facts about how the business operates. Manufacturers and retailers who deal with large and diverse component and product inventories stored in brick and mortar warehouses are familiar with two fundamental laws of inventory management:

1 Only a few kinds of items account for the most frequent movement of inventory. Businesses need the vast majority of other items less frequently.[9]

2 Only a few items (not necessarily those with volatile inventories) are most critical to the business.[10]

[9] The analysis of which items in inventory are needed frequently and which infrequently is called ABC analysis. Category A item inventories are the most volatile and category C the least volatile.

[10] An analysis of the degree of criticality of items in inventory to business operations is called 123 analysis. Category 1 items are the most critical to business and category 3 items are the least important.

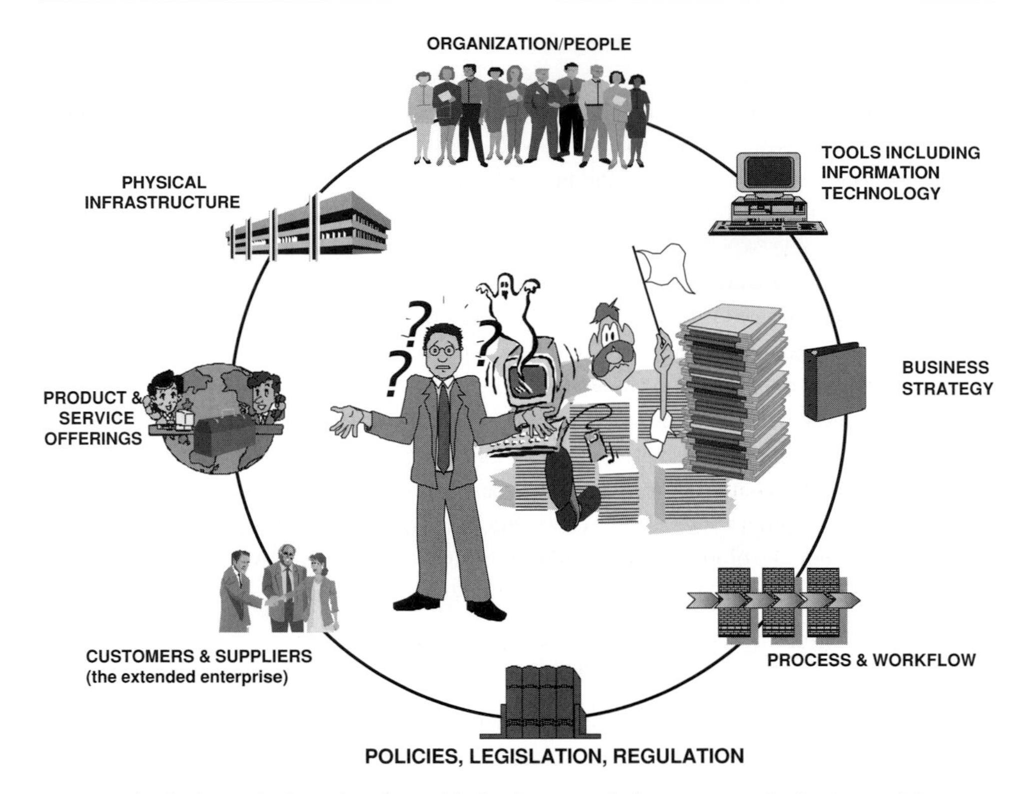

Figure 3 Analysis paralysis: only a few critical rules, reused often, connect the business of the enterprise but they are lost in a tangled web of minutiae

Knowledge inventories too follow these laws. Every rule of business need not be extracted and stored before the business can benefit from the concept of knowledge artifacts. Also there are only a few critical items of business knowledge that are reused most often. These items can be discovered in common business patterns that not only orchestrate the internal operations of the enterprise and its many diverse functions, but also connect the enterprise to stakeholders across supply and demand chains. This is also the knowledge that is of utmost value to the business and impacts it maximally! (See box 1.)

One of the authors served as the director of systems architecture for NYNEX at the time when it was one of US's largest telecommunications firms. NYNEX was then wrestling with the implications of the impending deregulation of the US telecom industry. Earlier, when he had worked for AIG, a large insurance firm, he had identified several fundamental patterns of business that were common to all businesses, regardless of what they produced, or where they were located. He was delighted to find that these common patterns (from his AIG experience) could be applied to the core processes, products, and services of the telecommunications industry as well. Indeed, they even anticipated key changes driven by deregulation before users articulated their requirements. We believe that the opportunity to leverage knowledge patterns and artifacts exists for many other industries and applications, and show how this may be done in the chapters that follow in this book and in the modules available on our website.

These patterns are not always obvious to the practitioner's eye; the practitioner's often has to focus too narrowly on a small part of the business. However, as e- and m-commerce begin to shape the business paradigm, it is becoming clear that not just individual corporations, but entire supply chains must compete as coordinated units in order to succeed in the marketplace. Recognizing the critical pieces of frequently used business knowledge that orchestrate not only the tasks of the enterprise but also operations across several enterprises will become increasingly critical.

This book identifies these common patterns. Today, systems analysts focus on translating the most critical components of business knowledge into requirements for the design of business processes and information systems. Tomorrow, businesses may be able to concentrate only on adding those few components that can truly distinguish their business from its competitors. This is how analysis paralysis can be circumvented, and, as we will see within these pages, much of the effort can even be automated.

(Common patterns are available in a supplementary module on our website. A companion book, *Agile Systems with Reusable Patterns of Business Knowledge – A Component Based Approach* [337], by the same authors, elaborates, and adds to the patterns at the website.)

6.3 The third pillar: people and best practices – managing change effectively

Technology can take quantum leaps, but to effectively utilize new technology, new methods or new processes, organizations, people, skills, and culture must also be realigned.

Change cannot be accomplished in quantum leaps where people and organizations are involved. Evolution is key, and the migration path determines risk. The optimal trajectory depends on environmental factors: business drivers, culture, available skills, risk tolerance, and others. This is where organizations usually stumble.

Random or improvised trajectories of change carry a high risk of failure. Change can become chaotic, credibility of the new technology can erode, and, unless the transition is managed carefully, the organization can even regress to become less capable than before.

Organizations often underestimate the risk of failure. The most common mistake is to try to mitigate risk through staff training or hiring and acquisition of tools alone.

For this reason, a significant part of a companion book by the same authors, *Agile Systems with Reusable Patterns of Business Knowledge – A Component Based Approach* [337], is devoted to managing change. It elaborates on change management themes and best-of-breed practices needed to facilitate effective use of the technology described in this book.

7 How this book is organized

To encapsulate business knowledge in common reusable themes, and then forge components of normalized business information from these themes, we must first understand the concept of knowledge itself – the themes, structures, and abstract information that define knowledge. Only then can we use these structures to describe common components of business knowledge to automate the design of agile business processes and information systems. Any kind of business knowledge has, at its root, the concept and understanding of knowledge. Therefore, it is the components in the metamodel of knowledge that will be used and reused as we forge business knowledge. For this reason, the primary focus of this book is on the metamodel of knowledge.

Supplementary modules on the Web summarize some of the key components of business knowledge that form the next tier of frequently used themes. The companion book, *Agile Systems with Reusable Patterns of Business Knowledge – A Component Based Approach* [337] elaborates on these themes of business that flow from the metamodel of knowledge.

This book and its supplementary materials are organized as follows:

The **Introduction** gives an outline of the book: its scope, principles, audience, structure, and utility.

Chapter 1 is an introduction to the definition and structure of knowledge and its reuse in diverse scopes. The purpose of the chapter is to help the reader develop an intuitive understanding of key concepts and semantics without getting lost in definitions and detail. It can serve as an introductory chapter for managers and non-technical readers who want a broad understanding of the topic.

Chapter 2 introduces the object paradigm and the state machine (without recourse to complex mathematics). It describes how the object paradigm can encapsulate knowledge and automate its propagation through mechanisms such as inheritance, and how this is the basis for sharing and reusing knowledge in component form. This chapter also describes the fluidity of knowledge and how systems could adapt to new learning and new behavior by reconfiguring components of old knowledge and/or adding new learning. The chapter goes on to discuss the problem of multiple perspectives and adaptation to shifting scopes. It introduces a solution, based on shared patterns of knowledge.

Chapter 3 elaborates on properties of objects and constraints. It introduces configurations and patterns that define constraints on business and object behaviors. It elaborates on the concept of "state" introduced in the previous chapter. Concepts are illustrated with real-world/business examples.

Chapter 4 formally describes the concepts of pattern, measurability, and, more importantly, the meaning of immeasurability. The chapter addresses the spectrum of meanings that range from those that precisely quantify and measure numerically versus those that are purely qualitative. It describes components, configurations, and patterns of information that derive these meanings from each other and eventually lead to the very concepts of existence, qualities, properties, patterns, states, languages, and meaning itself. The concepts are illustrated with real and hypothetical examples.

The contents of the book are supplemented by:

Web pages on our website at http://publishing.cambridge.org/resources/0521851637. The purpose of these web pages is to provide the reader with additional value in the form of more material than a single book can hold. The user name to access these pages is "Mitra" and the password is "Gupta." The supplementary material is organized into the following modules:

Modules I through IV on the Web supplement Chapters 1 through 4.

Modules V through VIII extend the metamodel of knowledge.

Module V focuses on interaction, configuration, innovation, causality, and process. It includes mathematical transforms that turn business meanings into rules of automation. It describes how processes automatically blossom from generic (non-process) knowledge.

Module VI describes how the concept called *constraint* turns inchoate information into choate meanings that become components of knowledge.

Module VII provides a broad overview of the model of knowledge and its components. It integrates the metamodels and patterns developed in previous chapters and modules into a single overarching pattern of coherent information that describes the generic concept of knowledge built with shared, reusable components.

Module VIII extends the metamodel into shared themes of business. It describes shared patterns of business knowledge derived by adding information to the patterns in the metamodel. With these patterns, business knowledge will neither have to be rediscovered, nor components redesigned, each time a business needs to rebuild or integrate its processes and systems.

Other supplementary material

In addition to the main text, the book provides supplementary materials in the form of:

Boxes which may be embedded in the text of the book and also found on our website. These boxes elaborate on concepts described in the book.

Endnotes which contain technical and mathematical details and comments. Sometimes they include suggestions for further reading. They are referenced in the text and are located on the website. The companion book from Artech House contains a hardcopy of the Endnotes [337].

Bibliography expresses business knowledge in component form and reuses these components to draw on a wide variety of areas of active research as well as business experience. The bibliography at the end of the book covers this. We have provided URLs wherever possible to make it easy for readers to access carefully chosen papers and publications on the Web. These URLs were valid at the time this book was being written. However, the Web is forever changing, and we cannot guarantee that these links will always exist.

1 On the nature of reality and the nature of business

Introduction to the MetaWorld

This chapter lays the basic foundation on which the metamodel of knowledge will be built. It introduces fundamental concepts that are at the bedrock of the metamodel. The purpose of the chapter is to help the reader develop an intuitive understanding of key concepts without getting lost in definitions and detail. The chapter:

- describes the emerging need for coordinated business knowledge in software;
- introduces the concept of normalized knowledge and its configuration from *atomic rules* or *irreducible facts*;
- introduces the concept of behavior, the concept of modeling it, and describes how knowledge gets replicated in analysis artifacts;
- introduces fundamental components of the metamodel and their roles in normalizing knowledge:
  - objects, relationships, processes, and events as repositories for behavior
  - domains and their role in measuring, normalizing, storing, and expressing information

Does an arcane discussion on the nature of reality really have a place in a book on information systems? True, the full tapestry of reality in all its richness is better left to philosophers, but we must understand how the real world structures meaning, for meaning is the foundation of information, and knowledge bereft of information is just an empty word. Meaning is the foundation of the metamodel of knowledge.

> And moving thro' the mirror clear
> That hangs before her all the year,
> shadows of the world appear . . .
> There sees the highway near
> winding down to Camelot
> (Lord Tennyson)

Even so, why do we, engineers and practitioners of information systems, need this arcane discussion when we have built systems, and built them well for over half a century? Why

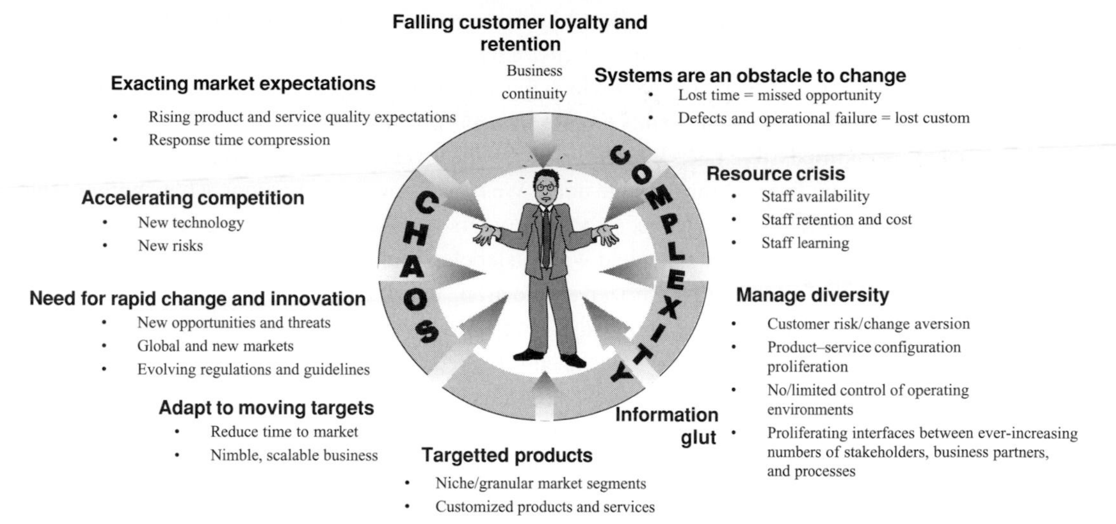

Figure 4 More and more, past experience is under pressure from the demands of scale, scope, and agility that businesses are placing on systems

do we need to step into uncharted waters when we have ready experience with tried and tested techniques that have served us well?

We must do this, dear reader, because we *must*. The world we have known is changing, and, with the coming of the information age, it will surely change at the speed of thought. Yesterday's paradigms are fading ever more rapidly. Our reach has become global and our businesses have become bigger and more complex by quantum leaps; technology is making yesterday's impossibility into today's imperative. Missed opportunities and lurking threats will annihilate businesses that do not move in step with the times.

As we have grown, our customers have become less forgiving and more fickle. Loyalty can only be bought with performance, and, even then, it may be lost as quickly as it was bought. Customers' expectations are high, and standards stringent, yet the scale, scope, and complexity of our systems have grown ever larger. Our employees and partners cannot deliver without automation. E- and m-commerce are here, and, even as customers' expectations rise, they interface more with automation than with people. The methods we have used in the past worked in a smaller, simpler age. Increasingly, the tried and true are giving way under the demands of scale, scope, and agility that are becoming the keynotes of business. In the Introduction, we described why it has become imperative for business to thrive on change, while systems have become the principal obstacle to the very change that is the life-blood of business. Businesses pay a price for this. The price is often much more than just the cost of maintaining and revamping systems. Real costs are measured by cost of opportunities lost or delayed, the revenues lost, market shares and competitive standing eroded, goodwill not realized, customers not satisfied, and much more.

Systems are an obstacle to agile and adaptive business practices, principally because change has a domino effect on systems. Changes explosively and chaotically ricochet through the system, each impact of which must be managed and resolved before the change can take effect. In many large and complex systems prevalent today, this is not just a difficult task, it is an impossible task. Defects are often discovered and resolved long after applications have gone live and the damage done. We have known of situations when, without the supplier's consent or knowledge, savvy customers changed product prices on e-commerce applications. Once a toxic chemical was mislabeled and shipped to the wrong destination because of a defect in a computerized application. This happens because business rules are not normalized and business knowledge is repeated in multiple ways in several places, all of which should be, but are not always, coordinated.

What is knowledge and how can we coordinate it? How can we adapt to moving targets as businesses constantly flex and maneuver for competitive advantage? The answer, paradoxically, lies in the real world we live in, not computer systems. The natural, or real world, frames all business opportunities, threats, goals, strategies, and operations. All businesses are bound by, not only the laws of nations, but also the immutable laws of nature. Therefore we must look at the structure of knowledge in the real world, where knowledge is naturally normalized. Knowledge gets fragmented and replicated only when we store *information* about the real world in our systems, designs, and artifacts.[1] The solution is to incorporate knowledge and business rules into systems *as they are in the real world.* To do so, we must first understand the nature of reality and the nature of knowledge.

1 The nature of knowledge

Knowledge is meaning. It is the meanings of goals, policies, and practices, and how they fit together into a cogent whole. *How* this was said has changed, but *what* was said has not. The meaning has endured the passage of thirteen hundred years across a sea change of time and a panorama of ages.

Knowledge conveys information about the business environment. Knowledge conveys information about how business goals and guidelines are coordinated with business opportunities and operations. Knowledge conveys information about how the business' products and processes are aligned with business mandates and markets. Knowledge conveys information about breach and recovery: which rules to follow and which to dilute; what can be safely ignored and what *must* be ignored. Knowledge conveys information about how practices and people coordinate resources and requirements. Knowledge is information about customers and competition, and about business constraints and configurations.

Thus knowledge is coordinated information about how rules of business, imposed by man or nature – expressed explicitly or understood implicitly, called policy, common sense, culture, or collective wisdom – can mutually orchestrate the business. Knowledge is how this symphony of information moves business towards its goals, helps the business achieve its minor successes and crowning glories, and, also, occasionally create minor embarrassment,

[1] See box 2.

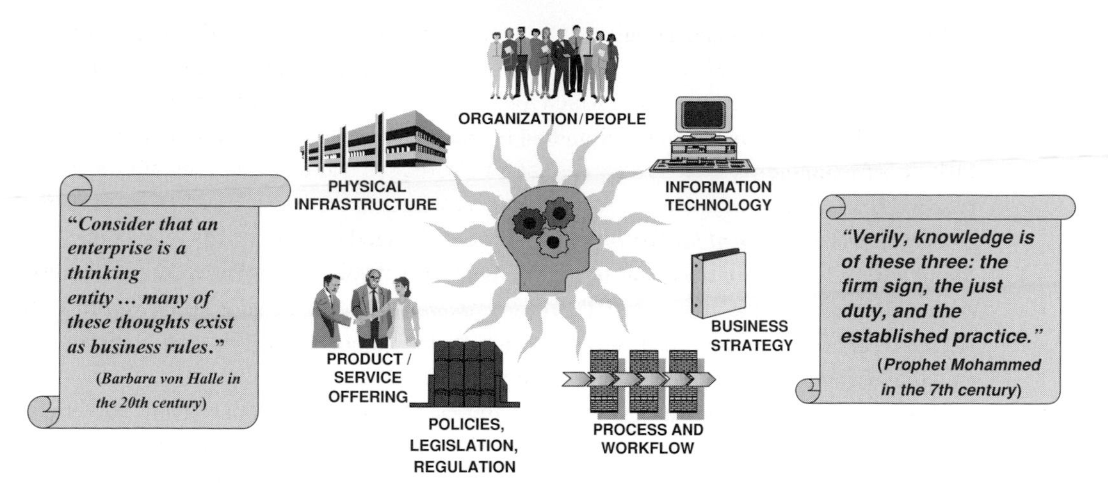

Figure 5 Knowledge is the meaning of business practices, rules, goals and guidelines and how they fit into an integrated whole

or even catastrophic failure. Yes, knowledge is not only about what to do, how to do it, and when to do it; but it is also about what not to do, how not to do it, and when not to do it. Thus, knowledge is an orchestra of rules in harmony, guided by meaning and rationale.

Rules are assertions. *Knowledge is the configuration of rules and reasons*. Rules may be simple or complex; they may stand alone, or might include several other rules and caveats. Rules carry information about the business. Together they orchestrate knowledge. Business rules convey the components of information that we can assemble into configurations of knowledge and best practices.

Engineers build complex machinery. Architects build complex facilities. Both build large and complex things from simple parts. They are familiar with techniques that divide and conquer complexity. They know that complex things must be made from simple ones. Small and simple components must be first tested and assembled into subassemblies, which in turn should be retested and assembled into even more complex components. These then might fit into yet larger components, and so on, until the end product is finally assembled.

Business knowledge too is complex and requires the same approach, but there is an added complication. Knowledge is intangible. Unlike buildings, bridges, and machine parts, components of knowledge cannot be obviously seen or felt.

The first step towards forging components of knowledge is understanding which assertions we can divide without losing information, and which we cannot divide without loss of meaning, or *knowledge*. If, by breaking an assertion into smaller parts, we lose information that we cannot recover by reassembling the pieces into a "subassembly of knowledge," then we have gone too far. We will call these lowest level indivisible rules *atomic rules* or *irreducible facts*.[2]

[2] These are called *atomic rules* (Ronald Ross [294]) or *irreducible facts* (G. M. Nijjsen in [297]) because they cannot be divided without losing information. [252] has an advanced discussion on coordination of rules.

Take for example, the assertion:

Jenny is a woman who has a son named Michael.

The truth of this statement can be expressed in two smaller and simpler statements without any loss of meaning:

1 *Jenny is a woman.*

2 *Jenny has a son named Michael.*

Taken together they can only, and uniquely, mean that "*Jenny is a woman who has a son named Michael*" and nothing else. Therefore "*Jenny is a woman who has a son named Michael*" is not an atomic (or irreducible) fact. Its meaning can be fully derived from the meanings of two other simpler assertions about Jenny.

Now let us go one step further and try to break the assertion about Jenny into three smaller, and even simpler, assertions.

1 *Jenny is a woman.*

2 *Jenny has a son.*

3 *A son is named Michael.*

At first glance it might seem that these three assertions together can only mean, "*Jenny is a woman who has a son named Michael*," but they do not – not really. The three bald assertions tell us only three things: (1) Jenny is a woman, (2) that she has a son, and (3) that somebody's son, not necessarily Jenny's, is named Michael. We have lost information: that Jenny's son's name is Michael. We lost Jenny's son's name when we tried to break "*Jenny has a son named Michael*" into smaller, simpler components because that assertion was an irreducible fact or *atomic rule*.

What do irreducible facts have to do with managing change? Irreducible facts have everything to do with managing change because they are at the root of coordinated requirements. Normalized knowledge will help coordinate requirements in today's complex corporations and cross company supply chains. Change has a domino effect that radiates chaotically through the system because the same irreducible facts are scattered chaotically, with little control or even awareness, through the software.

Knowledge (meaning) is not replicated in the natural world, i.e. it is normalized, whereas in today's information systems it may be fragmented and replicated. In present-day systems, irreducible facts, the basic building blocks of knowledge, may be replicated and unsynchronized in different applications, in requirements recorded in different forms, in design artifacts, in databases, in "help" files and deliverables to such an extent that it is sometimes unrecognizable as the same root knowledge. Therein lies the problem, as we shall see in the two examples that follow.

A customer orders voice mail services from a telephone company. The company adds the service to the customer's record and starts charging the customer. The firm must also reprogram telephone switches to activate the service. The software that instructs the switch does not recognize voice mail services. Now there is not just an unhappy customer who has been billed for services not provided, but also an unhappy phone company that is spending time and incurring the high cost of skilled human resources needed to service an irate customer.

Voice mail is a feature of telephone service is an irreducible fact. You cannot break it into simpler assertions without losing information. *Voice mail is a service offering* was recognized by the billing system, but not by the service provisioning system. Knowledge was not normalized, hence requirements were not coordinated in the phone company's systems. That was the root of the problem. (The companion book, *Agile Systems with Reusable Patterns of Business Knowledge – A Component Based Approach* [337], examines irreducible facts that describe products and services in more detail.)

Let us take a more complex example where consequences were less serious because customers were not directly impacted, but significant opportunity costs were incurred. John was a deliveryman. He worked for "Zippy" Courier Company. Zippy's scheduling system downloaded his delivery route to John's palm computer at the beginning of each work day. Sometimes delivery priorities changed, or Zippy's command central, which coordinated deliveries, got information about traffic congestion on parts of John's route. Depending on which delivery persons are where, they re-organized delivery routes and schedules, downloaded changes, and alerted their delivery persons by wireless link, alerted their customers to revised timing by telephone, and informed warehouse operations of these changes.

Zippy also had a sophisticated facility in the warehouse for sorting and loading packages on to delivery trucks. Which package was allocated to what truck depended on the final destination of the package and the route of the truck. Sometimes containers or trucks got full before all packages for that route were loaded. These were then loaded on other trucks that might cover similar routes. Warehouse operations informed command central when that happened.

Zippy used two very different systems for two very different applications. Yet both scheduled deliveries, one over roads to geographic addresses, and the other over conveyor belts, picking, packing, and staging systems, to trucks. Both could use multiple routes to deliver their shipments, and, in both, routes could sometimes be filled to capacity.

Many scheduling and routing requirements were common between the two systems, but when Zippy improved command central's scheduling algorithms, warehouse operations neither knew nor cared, let alone took advantage of the improvements. This was an opportunity cost that was completely hidden from Zippy's management.

Then Zippy's delivery scheduling system was enhanced to allow customers to specify special instructions that would facilitate coordinated delivery of two different packages from different pick-up points to a common destination. *Customers could ask that there be no more than a day's gap in the arrival of multiple packages.* Zippy's loading system could have added a new business rule that *deferred items must not wait more than a day to be loaded on to a truck that would cover the required delivery address.* It was the same *atomic rule* masquerading as a different requirement for a different system. This was not done and the new service commitments were harder to satisfy. Consistency and reliability of service suffered, while command central's operations became more complicated.

The state-of-the-art made it very difficult for Zippy to use the software and design artifacts of one system to change the other. This is true of most firms today. Changes come harder and improvements take longer, and both cost more than necessary.

The root of both Zippy's and the phone company's problem was not malfunctioning technology. The hardware, software, and networks performed according to design. The problem was uncoordinated requirements. The systems development process did not normalize, or leverage normalized knowledge, to coordinate requirements. Nor did the process seek to save time and development cost through knowledge reuse. Systems professionals could say, with some justification, that the requirements they were given were incomplete, but the bottom line was that the systems failed the customer and the company.

They failed because the firms were too large and their operations so automated that coordination of knowledge across the firm was complex. Systems failed because *knowledge was not reflected in systems as it was in the real world* where meaning is unique and its expression naturally coordinated at the root. Knowledge, like matter and energy, frames reality and is framed by it. The real world of immutable meaning automatically normalizes knowledge. Thus the real world becomes the yardstick for success and failure of automation. To reflect knowledge in our artifacts, as it exists in the real world, we must first understand reality, and how reality structures *meaning and information.* After all, reality frames the artifacts we create.

2 Modeling the real world

"Understand that as the mighty wind blowing everywhere, rests always in the sky, all created beings rest in Me." (Translated from the *Bhagvat Gita*, the holy book of Hinduism by Swami Prabhupad)

The nature of reality: we open the discussion with an extreme and radical assertion. *We assert that in the real world there is no such thing as data, and no such thing as process. There is only* ***behavior***. Data and process are mere *artifices* we have created in order to represent *information* about the manifest behavior of real-world objects.

What is behavior? You knew about behavior long before you even learned to read – long before you knew of process, data, or normalization. Hit a sheet of glass, it will shatter. Hit a sheet of metal, it will ring. Hit hard and it may bend.

Behavior is how an object in the real world responds to a stimulus (or an event). Behavior involves events, constraints, rules, location, and shape, but, most of all, it involves *change*, and change involves *time*. Step back to that time in your childhood when you knew only about objects you could see, events that influenced them, and the flow of time, and we will be ready to model reality.

It is important to remember that models are *not* reality. They only *represent* reality in a limited scope. The real world is too complex a tapestry to represent fully in all its richness and intricacy. A model represents limited information about reality in a repeatable, consistent, and accurate manner. The scope of the model is circumscribed by the real-world behaviors it targets. The reliability and accuracy of the model are circumscribed by the range of error or inconsistency we will tolerate – tolerances in terms of deviations from repeatedly consistent accurate predictions of target behaviors.

Box 2 Example of a model for baking a cookie

This model demonstrates:

1 How limited a model is compared to reality.
2 How easily knowledge becomes denormalized in artifacts which must then be coordinated.

The scope of this model is restricted to presenting information about a *sequence* of a select set of events involved in making cookies. The arrows show a succession of events. The event at the end of an arrowhead cannot occur until the event at the beginning of that arrow has happened. Thus we cannot bake dough unless we have put a glob of dough on the cookie sheet.

Events like starting the oven, acquiring the cookie dough, and eating the cookie are beyond the scope of this model. The behavior of the dough, such as shaping into globs, hardening under heat, its color and fragrance are also out of scope.

The information in the model could also have been expressed in a different syntax. For example, instead of a set of labeled boxes connected by arrows, the sequence and constraints could have been written in English sentences. That would not change the *model* or its meaning. It would only change the syntax, or *technique* of expressing information. The *information* and its meaning would be exactly the same in both expressions.

Although the meaning and information are identical in the two syntaxes, there are now two artifacts or deliverables with the same information, or *meaning*. To be consistent, the two must be coordinated. This is an example of how easily the information and *meaning* of a single real-world phenomenon can be replicated in our records. If one changes, the other too must change. By repeating information in two different artifacts, we have just denormalized real-world knowledge about baking cookies and made change more complex. We did not even try. It just happened.

3 Metaworld of information

He is distant in His nearness and near in His distance, He fashions 'how' so it is not said of Him, "How?" He determines the where so it is not said of Him 'Where?' He sunders "how" and "where" so He is "One- the Everlasting Refuge". *(Qur'an*, 112:1–2)

To normalize and reflect real-world knowledge in our systems as it is normalized in the real world, we must understand and model its structure.

Objects, relationships, processes, and events

Let us start by examining the nature of the model in box 2. We can almost hear you say that we just contradicted ourselves. We asserted there was no such thing as process in the real world, and almost in the same breath drew processes in box 2. You might contend that each box, connected by arrows in the model in box 2, actually represents a process. You are absolutely right! – but we are not being inconsistent and this is why:

We have already seen that objects and their observed behavior manifest reality. In the real world, objects can, and do, influence each other. The hammer can hit glass and break it. The dough, the oven, and the cook together bake the dough glob into a cookie. Buildings are located in geographies. One or more objects acting in concert with each other make the real world and orchestrate its behavior. In other words, objects relate to each other. Some of these relationships, such as baking the cookie, involve the passage of time, while others, such as the location of the building, are assertions that do not involve time.

These relationships are natural repositories for certain kinds of behaviors of real-world objects acting in concert. As such, they too are objects in their own right (see Module V on our website). For example, we could interrupt and stop "bake cookie" before the cookie is fully baked. This is a behavior of *bake cookie*, the object. Similarly, the *same* person may become an *employee* through an employment relationship with an organization, and a *spouse* via a marital relationship with another person. In addition to behaviors common to *persons* in general, such as breathing and growing older, *employees* and *spouses* can have special behaviors. For example, spouses may get divorced and employees may be promoted.

*Processes are artifacts for expressing **information** about the **behavior** of those relationships that involve the passage of time*, i.e. involve *before and after effects*. For example, the "bake cookie" object in figure 6 captures the information carried by the "bake cookie" relationship.

Not only does *bake cookie* relate six objects in the model: "dough," "oven," new and used cookie sheets, the cook, and the cookie, but it also *sequences* them. The object "bake cookie" tells us that the objects to the left in figure 6, namely the dough, the oven, a new cookie sheet, and the cook must precede the existence of objects to the right, namely, the cookie and the used cookie sheet. *Bake cookie* is a process only because it carries information about a temporal sequence. *Processes are thus special kinds of relationships that contain sequencing information besides being objects in their own right.*

What triggers behavior? What starts a process? We all know that events do.[3] Objects respond to events[166], and their response is behavior. The hammer hit the glass to break it. The hammer strike was an event. Something triggered the bake cookie process. It might have been that the chef asked the cook to start. Thus the chef's request may have been the trigger. In box 2, the end of the preceding process, *make cookie dough*, triggered the process, *arrange dough glob on cookie sheet*. These triggers are events.

[3] Objects may sometimes exhibit spontaneous behavior. Spontaneous behavior is not triggered by any obvious external event. For example, stock prices may move at random from minute to minute. Spontaneous changes are also events.

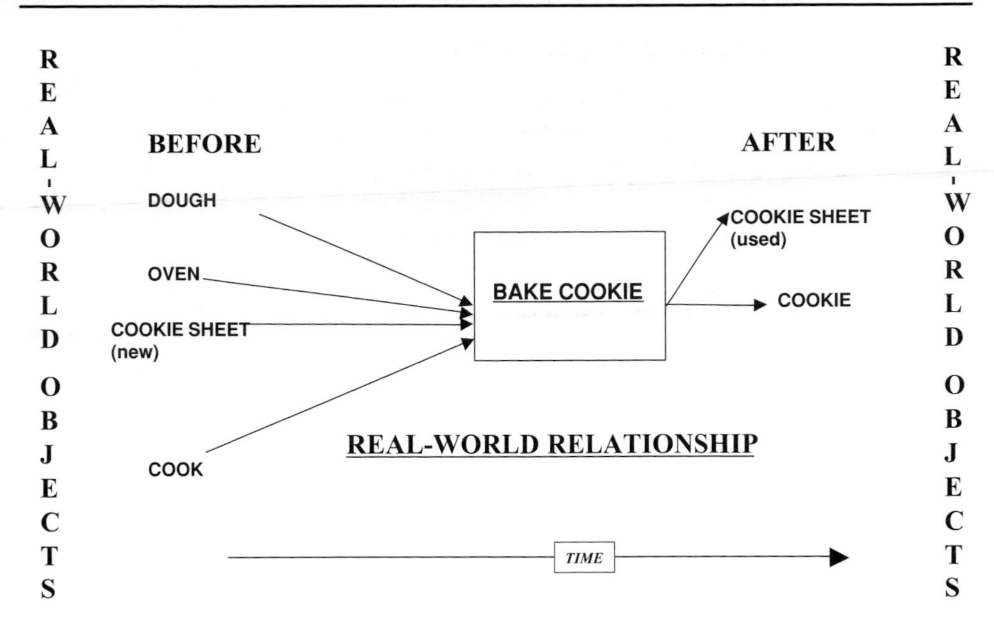

Figure 6 Processes are a special kind of relationship. They carry information on "before and after" effects between objects

An event is an occurrence that, unlike a process, may transform nothing. Processes, like events, occur in time, but all processes have a distinct beginning, a finite duration, and a distinct end. Events, on the other hand, may never end. Processes always make change or seek information. Events may not. Business process engineers often call the time interval from the beginning to the end of a process its *cycle time*. A process can even be instantaneous, but end it must. An event may go on forever. For instance, a deep space probe like the *Pioneer* will climb forever into interstellar space. A process may be considered to be a special kind of event – one that makes change in a finite time interval. Processes, of course, may also be instantaneous, like a blip in time with zero duration. The concept of event subsumes the concept of process, even an instantaneous process. We all know that anything that happens in the real world must take some time, even if the time taken is infinitesimally small. For example, the chef would take a few seconds to vocalize his or her instruction to start baking cookies. However, for modeling purposes, we can consider that the cook's request is a zero duration occurrence, or, in other words, an *event* and also a process.

Events are important because they trigger actions, processes, and behavior. For example, the cook might hit the stop button on the oven and interrupt the *bake cookie* process in box 2. Hitting the stop button would then be the event that suspended the *bake cookie* process. Remember that processes are special kinds of objects. As such, hitting the stop button was an event that triggered specific behavior of the *bake cookie* object. Two key events implicit in the model in box 2 are the start and end of a process. It is important to bear in mind that these two events, the start and the end of a process are implicit, intrinsic, and inalienably associated with the existence of every process. The importance of this concept will become evident in Module V on our website, where we discuss the behavior of processes.

Figure 7 How is information naturally manifest in the real world?

Perception and information naturally speaking: domains, units of measure, and formats

What is the nature of information manifest in the behavior of reality? What is the relationship between the information intrinsic in reality and its perception through our senses? – an exotic, abstruse, and arcane discussion? Perhaps, but critically important to normalizing business rules as we shall see. The example in box 2 shows us that meaning must be separated from its expression if we must normalize knowledge.

We need a home for *meaning* to ensure that business rules that involve meaning are normalized. For this we must look beyond physical objects and relationships. We must look at how reality structures the information it contains. To do this, we must augment our metamodel to represent additional entities beyond physical objects and relationships. In this section, we will add three new entities to our metamodel: *domain*, *unit of measure (*UOM *for short)*, and *format*.

Information exists in the real world, as do matter and energy, but the rules are different. Matter might be more tangible, but no one today would argue that energy is in any way less real or natural than matter is. This was not always true. It took humanity a thousand years to reach that conclusion,[4] and even longer to realize that matter and energy may be expressed in different forms, but cannot be created or destroyed. Information is even more abstract and its laws more complex, but information is no less real than matter or energy. Only, it is *manifested* in the behavior of real objects and physical energy.

Unlike matter or energy, *meaning* is not located at a particular place in space and time. Only its *expression* is.[5] Accordingly, in the example of box 2, the same meaning was found in two different artifacts that had no spatial or temporal relationship with each other. Their only relationship was in their shared meaning, or *information content*.[6] Although *meaning*

[4] See the endnote on how the twin concepts of matter and energy were developed.

[5] Shannon's information theory described in the endnotes measures the *quantum* of information. Meanings *structure* information. The two concepts complement each other.

[6] Physical phenomena linked purely by information that just *is*, as opposed to information transmitted spatially and temporally by messages, is illustrated by the aspect experiments described in the endnote on messages between objects.

in its true sense (and hence the *information* it conveys) does not occupy space and is immutable in time, it is ironic that we can only know meaning from information *expressed and observed* in the physical world framed by space, time, and real-world objects. A single meaning may have many expressions.[7] The same information may be stored on printed paper in a filing cabinet as well as on the hard disk of a computer; the Spanish and English versions of the owner's manual of your car (should) contain the same information; the Japanese Primeminister's speech at the UN should have the same meaning or *information* as its English translation.[8]

This then is a fundamental difference between information on the one hand, and matter and energy on the other. The same information can exist at many different places and times, whereas a specific material object or packet of energy can exist at only a single location at any given moment in time.[9]

Matter or energy mediates our observation of information. We can only *observe* the *behavior* of reality *manifested in the behavior of objects* located in space and time. This is a very important concept and we will repeat it again: the information carried by *meaning* is *non-local*, i.e. is independent of space and time, whereas specific physical objects such as documents, bits of energy, screw drivers, and people like you and I are *local*, i.e. they exist in a particular place at any given moment in time. To normalize business rules, it is critical that we understand the natural structures that connect information to its physical expression(s).

What mediates information and its expression in the physical world? There are two metaobjects that do. One is intangible. It deals with the quantum of information[10] that is intrinsic to the *meaning* being conveyed, and is closely tied to nature. We shall call it the *domain of information*, or *domain* for short. The other is more tangible – it is the format or the physical form of expression. It is easy to recognize the format, and many tools and techniques have done so explicitly. It is much harder to be aware of *domain*,[11,12] but nature does not care about what we know. Domain just *is*. If we did not know or care, and clubbed domain with format,[13] it would come back to haunt us in the form of replicated business rules and inflexible software. Let us see how.

> The curtains seem to part;
> A sound is on the stair,
> As if at the last. . I start;
> Only the wind is there.
> (Bliss Carmaon, *A Northern Vigil*)

[7] This concept can be confusing: Is the information or its *expression* the right meaning? In our metamodel, we will treat *meaning*, *expression*, and the *quantum of information* as separate objects.

[8] See the endnote on how information relates to physical objects.

[9] See the endnote on the locale of matter and energy.

[10] The endnote on Shannon's information theory discusses the measure of information.

[11] See the endnote on the mathematical theory of categories.

[12] Mathematical discussions on generic domains can be found in several mathematical and engineering texts, including [308], [232], [233], [234] and [235] also describe sets, domains, and functions.

[13] Many CASE tools and professional publications club *domain* and format together and call the composition domain. In this book, we will distinguish between the two. Readers will not be confused if they remember this.

A parable of Jim, Jane, Jugs of milk, and Robert in the United States of Information

The information content of reality manifests itself to us through the behavior, or properties, of objects we observe. For example, people have birthdays, they age, like some colors more than others on their cars, have a gender that determines certain physical attributes and the ability to bear children. Let us take a completely different object. Say a jug of milk. It stores milk. You can measure the amount of milk in the jug. You can quantify both your age and the volume of milk stored in the jug with numbers that describe their individual magnitudes.

Your intuition tells you that, in some sense, the values of both these very dissimilar qualities of very dissimilar objects (person's age and the volume of milk stored in a jug) are defined on a *domain of information* that contains some common behavior – not of the objects themselves, but of the information conveyed *in* (not *by*) the *act* of measurement – that each quality can be quantitatively measured. Another example of such a quality shared by disparate objects is temperature. We can measure the temperature of all three: the jug, the milk, and the person. Your intuition is right. Let us understand the kind of information, or behavior, that *domains* naturally normalize by comparing the *amount* of information *intrinsically* conveyed by each of these qualities of people and jugs of milk.

Nominal domains

Let us start with gender. We know that it conveys that men are different from women and nothing else. It has no information on how men and women can be arranged in any natural order, nor does gender carry any quantitative information on differences between men and women.

When we *store* this information on a physical medium, we could choose to arbitrarily represent "male" with a numeric code 1, and "female" with 2. If our friend, Robert, a professional and dedicated mad scientist devoted to divining the true nature of things, then claimed that men precede women because the number 1 precedes 2,[14] we would know that Robert's claim is meaningless because the *domain* on which gender is naturally defined has no information about sequence.

This will always be true regardless of how we *physically* express or code the information: it is also meaningless to subtract 1 from 2 to find the amount by which men and women differ, or to divide 1 by 2 to find the proportion of difference. *The domain just does not have that information*. It has nothing to do with how the information is physically expressed. *What does not intrinsically exist cannot be expressed*; you cannot squeeze blood from stone.

Domains of this type that contain just enough information to classify objects based on their properties (or relationships) are called *nominally scaled domains* or *nominal* domains[15] in short.

Ordinal domains

Next consider a person's color preference for cars. Say, if the cars are identical in every other way, Jane likes blue cars more than green and red, and cares even less for black cars. Between green and red cars, she really has no preference if all else is equal.

[14] This is called coercive polymorphism. See the endnotes on polymorphism and the mathematical theory of categories.

[15] See *discrete distance* in the endnote on metric space for more information.

Box 3 Objects, domains, and formats

Domains carry meaning. Formats are how information is *physically presented* to a person, system, or instrument.[16]

For example, gender may be formatted, with a numeric code such as "1" for "male" and "2" for "female", or "M" for "male" and "F" for "female"; it may be spelt out in written or spoken (where technology supports multimedia) English words – "male" and "female"– or in another language. It may even be graphic icons or pictures, static or moving, of a man or woman, or any other physical expression. All these are examples of FORMAT, the physical *expression*(s) of meaning, *not meaning* itself.

Objects and domains together convey *meaning*. Objects frame the context of the meaning conveyed by domains (to be described with more precision later). For example, the meaning of the fact that objects may be female (carry progeny), male (fertilize females to enable them to have offspring), or neuter (neither) is conveyed by the domain alone. A common domain normalizes the common meaning and behavior of gender across objects like people, plants, dogs, deer, and other living things. An object such as a person or an animal puts this generic behavior into context, giving it a specific *meaning*[17] for that kind or instance of animal.

Thus, there may be male and female people, parts of flowers, dogs, spiders, and cats. This conveys the fact that a property, "gender," of a class of objects called "persons" (or parts of flowers, dogs, spiders, cats etc.) maps to the gender domain, with the restriction that only a male *or* female gender is allowed for an instance of this object. It records an *irreducible fact*, that people must be either male or female. Similarly, other classes of objects such as dogs, spiders, and parts of flowers would map to the gender domain with the same restriction – an *irreducible fact* about these objects.

Each earthworm, on the other hand, must be *both* male and female because each earthworm may carry *and* fertilize earthworm eggs.[18] This too is an atomic rule or irreducible fact. Each object thereby provides the *context* of maleness and/or femaleness (or neither), whereas the domain is the bucket for recording the *common meaning* of maleness, femaleness, or neutrality.

Sometimes more than one property of an object may map to the same domain. Each will represent a distinct irreducible fact needed to represent the real world. For example, the length, breadth, and height of a room all map to the length domain. The domain normalizes the facts that these three properties of room can have the same units of measure, which have the same conversion factors. Thus they need not be repeated for each property. The same logic holds when different properties of very different kinds of objects map to the same domain. For example people's heights and the lengths of rooms both map to the length domain, which provides the common home for their units of measure and conversion rules between units of length.

[16] Called *actor* in the language of object technology, or *observer* in the parlance of physics. Readers interested in more information about actors may refer to books on UML, or the resources in the bibliography at the end of this book. UML, the acronym for universal modeling language, is becoming the *de facto* standard. The Object Management Group and Rational Corporation are strong advocates of UML. Also see Rational Corporation's Web resources in the Bibliography.

[17] See *polymorphism* in the endnote on the mathematical theory of categories for more information.

[18] See the endnote on the question of gender.

She agrees to participate in a consumer survey of preferred colors of cars. First, Jim, the researcher, asks her to rank the four car colors she likes most in order of preference, starting with the color she likes most. That is easy: blue first, followed by red and green at par, and finally black at the end. So far she has no trouble.

Next, Jim asks her to quantify how *much* she likes each color by assigning a number to each. Now Jane has a problem. She does not know how to respond. She knows that she should give blue the highest score, followed by an equal score for green and red, and a lower score for black, but what should these scores be? She has no idea. All she knows is that she likes blue more than green and red, green and red equally, and black the least, but cannot quantify her liking. The information is just not there.

The domain on which Jane defines her *preference* for color of cars intrinsically and naturally contains sequencing, or ranking information, but no information about *magnitudes*. Should Jim insist, she might quote some numbers, but these numbers will convey no information beyond Jane's ranking of color preferences for cars.[19] It does not matter how Jim codes her color preference – with numbers, letters, colors, or graphic icons.

Domains of this type that have no quantitative information, but do convey enough information to arrange objects in some sequence, or order, are called ***ordinal domains****.*[20]

Note that, because she can rank cars in order of color preference, Jane can *automatically* group cars into separate groups (in this case, green and red cars would be grouped together – the criterion is her color *preference*, not the actual *color* of the car). However, if she just groups, not ranks cars in order of color preference, she is *withholding information* from Jim. This shows that *ordinal domains* ***intrinsically*** *carry more information than nominal domains.*[21] They carry sequencing information as well as, by implication, classification information.

Now suppose Jim, frustrated by Jane's inability to quantify her preferences, assigns some kind of number to her preferences – say, for arguments sake, the rank Jane assigned to each color – 1 to blue, 2 to red and green, and 3 to black.

We know that it would be entirely incorrect for Jim to conclude on this basis that Jane likes blue cars three times more than she likes black cars. Nor can Jim conclude that the *gap*, or difference, in Jane's preference between blue and red cars is equal to her preference gap between red and black cars. The domain simply does not *have* this information. *You cannot squeeze blood from stone.*

Difference scaled domains

Let us consider Jim's and Jane's temperature next.

Jane liked Jim and asked him to stay for lunch. After lunch, they went to Domain's Metaphysical Diner for a cup of good Columbian coffee. Robert, the mad scientist, happened to be drinking coffee at the next table. Robert was researching the true meaning of temperature and had a superb collection of thermometers of every kind in his brief case.

[19] See coercive polymorphism in the endnotes.

[20] [211] has mathematical detail on ordinal measurement.

[21] Shannon's information theory in the endnotes describes the mathematical measure of information.

Mr. Domain took great pride in his special coffees and always served coffee with a separate warm jug of milk for each customer. Jane found that the new waitress had accidentally served her chilled milk in the jug. Robert overheard Jane, and sprang up with missionary zeal to ask if he might address any issues with Jane's and Jim's milk. Mistaking him for the new waiter, Jane graciously accepted.

Robert immediately flung open his brief case and extracted two high-tech digital thermometers, a scientific calculator, and an elegant notebook. Without further delay, he plunged a thermometer into each jug, did a quick calculation, and declared that Jim's milk is twice as warm as Jane's.

Jenny, the waitress, was piqued and asked Robert how he knew. "Simple," Robert explained, "look at the display of each thermometer. Jane's shows *40 degrees Fahrenheit*, and Jim's shows *80 degrees Fahrenheit*. Since 80 is twice as large as 40, Jim's milk is twice as hot as Jane's."

"Also," said Robert to show off his high-tech thermometers and impress Jenny with his erudition, "these thermometers can show you the temperature in either Fahrenheit or Celsius at the touch of a button! Always be sure that you use the same *units of measure* for both jugs, otherwise you will not be comparing like readings." He hit two identical buttons on the thermometers, and the temperature of Jane's jug of milk read "4.44 DEGREES CELSIUS" and Jim's read "26.67 DEGREES CELSIUS."

"Look what you have done now!" Jenny complained. You made Jim's milk more than six times hotter than Jane's, because 26.7 divided by 4.44 is more than 6!"

"I did *not*," Robert retorted, "I just changed my *unit of measurement*."

"Sir!" exclaimed Jim, "my temperature is rising as well! We want to drink our coffee in peace. All we need is a fresh jug of warm milk for Jane, or she may use some of mine."

"Impossible!" cried Robert. "Your temperature can only rise if you are sick, or the mechanism for regulating your temperature cannot cope with the extreme heat of summer! See, my remote sensing thermometer can even sense your temperature from a distance, and it shows that you are holding at a steady 98.4 degrees Fahrenheit."

Fortunately Mr. Domain arrived just then, before things got out of hand. "What's the fuss about?" he asked Jenny. Jenny, almost in tears by now, cried, "Robert just made Jim's milk six times warmer than Jane's by measuring its temperature in Celsius rather than Fahrenheit! It was only twice as hot before!"

"Now ladies and gentlemen, let us be civilized about this," said Mr. Domain, fixing Robert with a specially penetrating glare. "I happen to know all your birthdays. You, Jenny, were born on 1/1/1977. It is now 2001. That makes you 24 years old. Your daughter was born on 1/1/1995. That makes her six years old. I can also calculate that there was an 18-year gap between the date of your birth and that of your daughter's.

"Now, what would I learn by dividing the date on which you were born by the date on which your daughter was born?" They thought hard about it. No one had an answer.

"Well, there you are," said Mr. Domain triumphantly, "you would learn *nothing*. It is meaningless to divide one date by another."

"But why?" asked Robert, quite intrigued by Mr. Domain's question.

"Simple, Robert," answered Mr. Domain. "You, of all my guests here should know. The date domain has no information on ratios because it has no *natural* zero. The zero hour for the Gregorian calendar (the 'normal' calendar most used in the Western Hemisphere) was arbitrarily set. You can certainly measure the *difference* between any two dates as I just did – the unit of measurement is your choice, it could be days, years, minutes, hours, seconds, or any other measure of time – but ratios are meaningless. The domain just does not have the information if it has no natural zero."

"Ah!" exclaimed Robert, a new light dawning in his eager eyes, "so it is meaningless to take the ratio of temperatures as well! After all, zero degrees Celsius was arbitrarily set at the temperature at which water freezes, as was 32 degrees Fahrenheit."

"That is correct," Mr. Domain replied, addressing both Jenny and Robert, "so it was meaningless to say that Jim's milk is hotter than Jane's by any multiple, be it two, six, or anything else.[22] However, you can say that the difference in temperature is 40 degrees Fahrenheit, or 22.27 degrees Celsius." To make it less embarrassing for Robert, he added, "You were right about the unit of measure. When you talk of magnitudes of differences, you must express them in some unit of measure, otherwise they are meaningless. There can be a wide choice of units, but you must choose one.

"Domains of this type that convey information on classification, order (or sequence), and the ***magnitudes of gaps*** *(or differences) between points in sequence, but no information on ratios or proportions, are called* ***difference scaled domains****.*[23]

"All differences in a difference scaled domain must be expressed in at least one, but perhaps many, unit(s) of measure. Nominal and ordinal domains, on the other hand, need no unit of measure. All they need to express information in the world framed by space and time is *format*. Difference scaled domains are different. To *express* information (that already exists in the domain, regardless of whether it was actually expressed in space and time), all *difference scaled domains must be associated with at least one, and perhaps many,* ***unit(s) of measure***. Formats must then be linked to each unit of measure. I will tell you more about that in a bit."

"I see the truth of that," replied a much more contemplative Robert. "For example, the distance between the door and my table is 10 feet, or 120 inches. The unit of measure I use does not change the *actual distance* between the door and me, but it certainly changes the *number* I write down."

"You are right," Mr. Domain replied, "but your observation about *length* takes this discussion to an entirely different level altogether. It should be obvious by now that difference scaled domains are rich in information. They convey enough information to not only classify and sequence objects but also to measure the magnitude of gaps between objects in a sequence. However there is another kind of domain that carries even more information. We need to talk about *ratio scaled* domains."

To Jim and Jane, he added, "I did not mean to intrude, and I thank you for being so even tempered. Robert was right though. His remote sensing thermometers show that both your

[22] Chapter 4, section 3 addresses the information content of ratios. Also see the endnote on the natural zero of temperature and time.

[23] [211] describes the mathematical relationship between ordinal and difference scaled measurement.

temperatures are normal at 98.4 degrees Fahrenheit, and, if I may take the liberty of saying so, you are a well matched pair and very close to each other in the temperature domain. The gap between you is very close to zero."

Ratio scaled domains

"What about me!" exclaimed Jenny quite pleased with the thought of difference scaled domains, "Does that mean no one can say I am four times as old as my daughter?"

"Sorry to disappoint you Jenny," said Mr. Domain, "Date and age are defined on very different kinds of domains. Age is the *gap* between the date on which you were born, and today's date. It is not a *date*. Indeed, it *is* valid to talk about ratios and proportions of gaps between objects measured in difference scaled domains, but the ratios themselves must map to *ratio scaled domains*. For example, you can say that you are six times as old as your daughter and Jim will be perfectly right if he says that the *difference* in temperature between his jug of milk and Jane's was five times the difference in temperature between ice and Jane's jug of milk. That ratio will hold regardless of the units of measure you use to measure the *gap*, as long as you use the units consistently. Try it for yourself."

Box 4 Mr. Domain's calculations

Robert quickly verified Mr. Domain's calculations

In Fahrenheit:

Temperature of Jane's milk = 40 degrees Fahrenheit
Temperature of ice = 32 degrees Fahrenheit
Difference between temperature of ice and Jane's milk = 8 degrees Fahrenheit

Temperature of Jim's milk = 80 degrees Fahrenheit
Temperature of Jane's milk = 40 degrees Fahrenheit
Difference between temperature of ice and Jane's milk = 40 degrees Fahrenheit

Ratio of differences $= 40/8 = 5$

In Celsius:

Temperature of Jane's milk = 4.44 degrees Celsius
Temperature of ice = 0 degrees Celsius
Difference between temperature of Jim's and Jane's milk = 4.44 degrees Celsius

Temperature of Jim's milk = 26.67 degrees Celsius
Temperature of Jane's milk = 4.44 degrees Celsius
Difference between temperature of ice and Jane's milk = 22.26 degrees Celsius

Ratio of differences $= 22.26/4.44 = 5$

(The result of the actual calculation is 5.01, not exactly 5, because of rounding errors. The temperature in Celsius has been computed to only two decimal places. Had the temperatures not been rounded, the two ratios would match exactly.)

(Mr. Domain sighed quietly. "What a pity," he thought, "Jim and Jane are so close in the temperature domain, and I wish Jane's jug of milk was closer to Jim's as well.[24] Then none of this commotion would have happened." However, he continued.)

"*Gaps* between objects in a difference scaled domain will always map to a ratio scaled domain, but so do many other things. Almost everything physicists measure such as mass, length, area, volume, time (in the same sense as *age*, not date), probability, and many other things of interest to business, such as money or process defect densities, map to ratio scaled domains.

"Of nominal, ordinal, difference, and ratio scaled domains, ratio scaled domains are the richest in information.

"*Ratio scaled domains convey enough information to classify, sequence, and measure differences, as well as ratios* between objects that map to them.

"Like difference scaled domains, ratio scaled domains must also have at least one (and perhaps many) unit(s) of measure. *Units of measure are needed to express information* that these domains already have in the world framed by space and time."

"Like my distance from the door!" exclaimed Robert, "It maps to the *length* domain and I can certainly say that I am twice as far as Jim from the door because I am standing 10 feet away from the door, while Jim is sitting only 5 feet from the door. Even if I changed my unit of measure to inches, meters, or anything else, the individual *numbers* might change (although the *distance* would not), but their *ratio must stay the same* as long as I use the same unit to measure both our distances. Now it has all started making sense."

Mr. Domain's secret

Mr. Domain was beaming happily at Robert. "Now I am almost ready to share my secret with you. It is my secret map of knowledge. It is very old – as old as the universe we live in.

"But first we must pause to take stock of what we know. There are four kinds of domains:

- "Nominal domains contain only classification information. They have no information on sequencing, distances, or ratios of properties of objects. In order to physically express this information, it must be physically formatted and recorded on some medium. A single piece of information must be recorded in at least one format, and possibly many formats. For example, a person's gender may be coded as a number (say, 1 for 'female' and 2 for 'male') or letter (say F for 'female' and M for 'male') or a picture of a man for 'male' and a woman for 'female', or a hexadecimal code on magnetic disk that only computers can read, or almost any coding scheme you can think of.
- "Ordinal domains contain both classification and sequencing information. They have no information on the magnitudes of gaps or ratios of properties of objects. To physically express this information, we only need to choose a physical format and record it on some medium. A single piece of information must be recorded in at least one format, and possibly many formats. Moreover, regardless of format, we can compare which objects are greater or less than others in terms of properties that map to ordinal domains.

[24] Box 16 describes how domains extend conventional concepts of distance.

- "Difference scaled domains let us classify and arrange objects in a natural sequence and also let us measure the magnitude of point-to-point differences in the sequence, but carry no information on ratios. They have no natural zero.

  "To physically express this information, we not only need at least one physical format, but also a unit of measure (UOM). A single piece of information must be recorded in at least one format, and perhaps several units of measure. For example, the ambient temperature may be recorded in Fahrenheit or Celsius, and the date may be expressed in the Gregorian or Islamic calendars.

  "The UOM is not enough by itself to express the information. Each UOM must be expressed in at least one, but possibly several formats. For example, Fahrenheit may be spelt out as Fahrenheit, or printed as '°F' in different documents; it can be in different fonts or colors. It may even be spoken out aloud, displayed in a graph or icon, or recorded as a binary code on disk for computers to interpret.
- "Ratio scaled domains let us classify and arrange objects in a natural sequence, measure the magnitude of differences in properties of objects, and take their ratios. They always have a natural zero.

  "Like difference scaled domains, both UOMs and formats must be specified in order to physically express the information. A single piece of information must be recorded in at least one, and perhaps several units of measure. For example, lengths of rooms may be recorded in feet, inches, meters, or centimeters.

  "As in the case of difference scaled domains, the UOM is not enough by itself to express the information. Each UOM must be expressed in at least one format, but possibly several formats. For example, the US dollar may be printed as 'USD' or '$' in different documents."

Then, Mr. Domain opened a weathered and ancient book. With a twinkle in his eyes, he said, "Here is my secret map. It is not complete, and I regret I cannot let you into all my secrets just yet, but I promise I will. This map is only an introduction to the territory of domains. It summarizes only what I have just told you, but, let me show you how to read it." (In Chapter 4, domains will be examined in more detail.)

"Domains are objects in the metamodel of knowledge, as are units of measure and formats. For this reason we can call them metaobjects. Not all metaobjects are shown on this map – not even some you have been introduced to, like relationship and process. The hierarchy at the top of the map classifies different kinds of domains and arranges them in order of *intrinsic* information content. (In Chapter 4, you will see how these hierarchies will help you normalize business rules.)

"The lower half of the map (figure 8) shows the relationships between various meta-objects. The metaobjects are shown as rectangles, and the meta-relationships, as arrows. To understand the rules you must read along the arrows.

"For example, starting with 'quantitative domain', the full sentence along the arrow reads 'Quantitative Domain *is expressed by one, or many*, units of measure.' Note that the lower limit (1) on the occurrence of unit of measure shows that each quantitative domain *must* have at least one unit of measure, else it cannot be expressed at all. Similarly, the next sentence, starting with unit of measure reads, 'Unit of measure *is expressed by one, or many*, formats.'

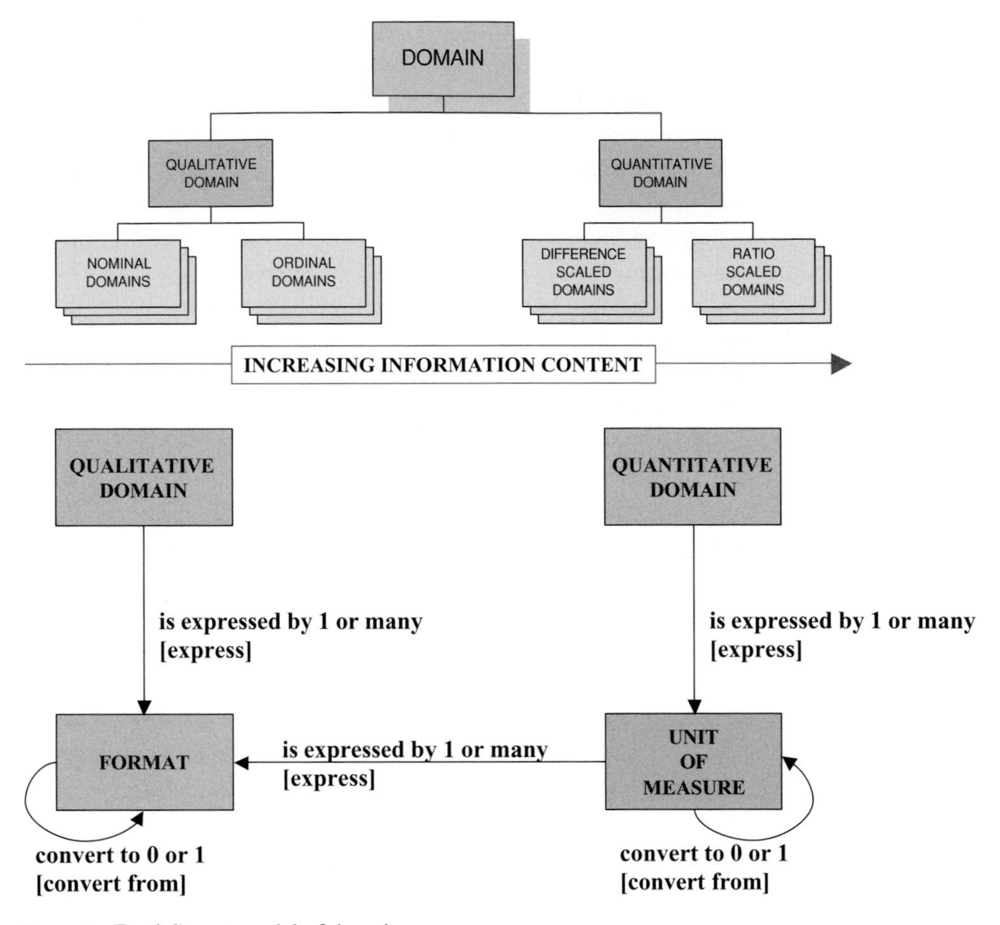

Figure 8 (Partial) metamodel of domain

"The arrow that starts from, and loops back to unit of measure, reads 'Unit of measure *converts to none, or at most one*, unit of measure.' This is the metaobject (remember relationships are objects too) where conversion rules, such as those for currency conversion or conversion from feet to meters, reside. This metaobject facilitates storage of the conversion rule in a single place."

"I understand why you have a lower limit of zero – if you had only one UOM, there is nothing else to convert to," interjected Robert, "but why did you restrict the conversion rule to only one other UOM? Cannot yards, for example, be converted to feet by multiplying by 3, or to inches by multiplying by 36? So right there, you have yard, a UOM for length related to two, not one, other UOM."

"A very perceptive question, Robert," said Mr. Domain. "Of course, you are right, each UOM of a quantitative domain can be converted to every other UOM, but remember the purpose of the metamodel is to avoid redundancy, and you need only one conversion rule per UOM. You could then navigate to any other UOM in the domain via a chain of conversion relationships.[25] One relationship per UOM is all you need. Another

[25] The chain must be *acyclic*, but more on that later.

Box 5 Conversion between UOMs

Measurements in any given difference or ratio scaled domain can be converted from one UOM to another by multiplying by a conversion ratio. If one or more UOMs (and conversion ratios) are already in use in a domain and a new UOM is introduced, we need to introduce only one new conversion ratio to enable us to convert measurements expressed in the new UOM to any, and every, other UOM already in use. We do not need individual ratios for conversion from the new UOM to each of the UOMs already in use. Indeed, we would denormalize knowledge if we were to specify each ratio individually; each of these ratios can be derived from just one conversion ratio.

The following example illustrates these real-world facts. In order to keep the example simple, we have based it on the length domain, but the same arguments will apply to UOMs in any ratio or difference scaled domain.

Let us assume that doctors in different countries decided that they would conduct a survey to find the average height of people. Soon after they started the project they realized their scales had different units of measure: inches in Inland, feet in Footland, and meters in Metland. They realized they would all have to agree on a single unit of measure to succeed. The conversion rules between inches, feet, and meters are in the following table:

	TO		
FROM	*Inches*	*Feet*	*Meters*
Inches			
Feet	$\times 12$		
Meters		$\times 3.2808$	

For example, to find the rule for converting from feet to inches in the table above, find "Feet" under the "From" column on the extreme left, and then look along the "Feet" row to find the cell under the "Inches" column. That cell contains the rule "$\times 12$", which means multiply by 12, i.e. to convert feet to inches multiply by 12. (Thus 5 feet $= 5 \times 12 = 60$ inches.) Similarly, the rule for converting meters to feet is "multiply by 3.2808." We also know that division is the inverse of multiplication. (We will revisit rules like these in more detail in Chapter 4.) Thus the table contains three atomic rules.

We need only these three rules to be able to convert between any units of measure in the table. For example, although the table contains no explicit rule for converting inches to feet, we can derive it by using the rule that division is the inverse of multiplication (to convert inches to feet, we divide by 12.) Similarly, although there is no explicit rule for converting meters to inches we can derive it from the information in the table. We can convert meters to feet by multiplying by 3.2808, and then multiply the result by 12 to convert to inches. Had we included the conversion ratio for explicitly converting meters to inches in the table, it would have been redundant, and knowledge would be denormalized. (We will revisit this issue in Chapter 4 in more detail.)

Had everyone used the same UOM, there would have been be no need to convert at all, and there would have been no need for conversion rules. (Note that the diagonal cells of the table are all blank.) *This is what Robert meant when he said that he understood why the lower limit of the conversion relationship was zero.* (We will revisit this relationship in more detail in Chapter 4.)

The countries were uncomfortable with UOMs that they were unfamiliar with and could not agree on which of the three UOMs (inches, feet, or meters) they would use for their project. They finally decided that to be fair to all, they would settle on centimeters, a UOM that none of them used. If we added centimeters to our list of UOMs for length, the conversion rule table would become:

	TO			
FROM	*Inches*	*Feet*	*Meters*	Centimeters
Inches				
Feet	×12			
Meters		×3.2808		
Centimeters			×0.01	

There is only one conversion rule we would add to the table: "*multiply by 0.01 to convert from centimeters to meters.*" With this single new rule we could convert centimeters to any of the other units of measure in the table. (For example, although there is no explicit rule in the table for converting centimeters to inches, we could multiply by 0.01 to convert centimeters to meters, multiply the result by 3.2808 to convert to feet, and then multiply that result by 12 to convert to inches. We leave it to the reader to try to convert centimeters to other UOMs in the table.) This is an example of what Robert meant when he said that he did not need to add a separate conversion rule for every UOM in the domain each time he added a new UOM to the table – that adding a single new conversion rule would be enough.

would *add* no information. It would be redundant. I once had a very interesting visitor – I remember his name was Claude Shannon[26] – who helped me understand this."

"It also implies that if you ever create a new UOM for the domain, you have to add just one conversion rule," said Robert excitedly, "you do not need a separate rule for ever other UOM in the domain. Boy, isn't that a nice saving! This object (conversion relationship) has both the conversion ratio and the rule that tells you to multiply the value as measured by the source UOM to convert to the target UOM."

"Also note, whenever you have two or more UOMs for any given domain, there is automatically and intrinsically a pair of conversion rules that will let you convert one UOM to the other. This is true whether business is currently interested in converting between UOMs or not. It just *is*, and these rules will apply to any objects, as widely disparate as they

[26] See the endnote on measure of information.

may seem to us that map properties to the domain. For example, the rule for converting feet to inches will apply to heights of people, dimensions of rooms, lengths of wire, and any other property that maps to the length domain.

"So you see," continued Mr. Domain, "even though we have just started, and our meta-model is still rudimentary, some kinds of reusable components are already becoming self evident." And looking at Jane who was shaking her head incredulously, he added, "Of course, you might argue that this is all common sense, and it could well be, but, bear with me, as the metamodel fills out, many other reusable components will emerge *naturally*. After all, we are modeling the nature of nature," he added with a smile.

"And I can see from the map that much of what we have discussed for UOMs has parallels with formatting issues as well!" exclaimed Robert. "Now I am beginning to see, even if it is still just a glimmer in my eye, how the metamodel of knowledge can help me normalize rules!"

"And now that you can read the map, understanding the verbs in parentheses are easy," said Mr. Domain. "They merely show the relationship in the reverse direction (i.e. read in the direction opposite to the arrow). It is called the *inverse* relationship,[27] but more on that later.

"And one more thing, if I may, before we move on," continued Mr. Domain. "Note that none of these meta-relationships involves time. They are *not* processes. The rules just *are*. There is no data flow, or conversion process in the real world. It is just knowledge. Later we will see how these can naturally map to computer implementation and still stay normalized. We will have to link each implementation to a single piece of immutable knowledge that just *is*."

The structure of domains – perception, five senses, and aliens in the lost worlds of metanesia

Jane was getting a little confused and felt it was high time she made her presence felt. "Whoa! Hold your horses there for a moment! I really don't understand. What is this fuss about information intrinsic to meaning that exists beyond any spatial or temporal frame? After all, we can only know about the existence of information through our five senses. We know about the behavior of objects that *are* framed by space and time only because we can see, hear, smell, touch, or feel them. How can we claim something exists when we cannot see, hear, smell, touch, or feel it?"

"Good point, Jane," said Mr. Domain with a delighted smile. "Indeed, you have raised questions that philosophers have long debated.[28] I will need to take you on a tour of my secret zoo in the lost world of Metanesia to explain.

"A word of warning though," he added a little anxiously, "it is a fantastic journey, but not everyone returns from Metanesia. Moreover, even if they do, they can get somewhat eccentric like my friend Robert here. Are you sure you want to go?"

[27] See the endnote on the mathematical theory of categories or publications in the bibliography [252], [253], and [308].

[28] See the endnote on positivism.

"I am not going to be scared off so easily," thought Jane to herself, "Mr. Domain, or whatever he likes to be called, is probably out of touch with reality. I am not going to let him off the hook so easily!" To Mr. Domain, she simply said, "Yes, I am sure I want to go."

"Come then, and remember that it was your choice!" Jane was suddenly plunged into a strange stygian darkness. She could not even see the tip of her nose in the dark. A complete silence, unlike anything she had ever experienced before, enveloped her totally. She felt strangely disembodied. She realized with some trepidation that she was completely cut off from her senses. She could neither see, nor hear, nor feel, nor smell. Even the taste of the air she used to breathe was gone. Only the core of her being was left. Gradually Jane sensed strange presences gathering around her, and as if from very far away, she heard Mr. Domain:

"You are in the presence of a very powerful alien being. It is not like anything you have ever known, or can ever imagine. Its senses are completely different from yours. It does not see, feel, smell, hear, or taste. Yet it *knows*. Even I do not know how; and it has a mind. Even I do not understand its thoughts or perceptions. Be extremely cautious."

Jane was beginning to wish that she had not accepted the challenge. Oh, wouldn't it have been so much nicer to go back to a warm conversation with Jim, even if her jug of milk was a tad cold! Anyway, here she was. Slowly she felt a thought forming at the fringes of her consciousness. It seemed like a question:

"What are you?"

"I am Jane."

"Do not comprehend response."

"I am human, 5 feet 6 inches tall."

"Do not comprehend response. Explain human. Explain tall."

"Tall is the same as high."

"Do not comprehend response."

She heard Robert's presence replying, "Humans are sets of properties mapped to space."

"Now *I* don't understand," thought Jane, "but that's okay if we can get out of here."

"Explain space."

"Boy, whatever this thing is, it must be dumb," thought Jane.

"Careful Jane!" – that seemed to be Mr. Domain.

"Let me try" (Jane was almost sure that it was Robert responding). "Space has three independent attributes (that we call three dimensions). Each is mapped to the same ratio scaled domain we call length. The three attributes of space are labeled length, breadth, and width. These attributes are uncorrelated."[29]

"Comprehended!"

"Yes Robert, as long as you can exchange pure information, any mind, however alien will have common ground to understand your message. This being senses the world in ways we cannot even begin to fathom, and our senses are equally alien to it. If you talk in terms of things you see, hear, touch, smell, or taste, it will be confused," Jane was sure that this was from Mr. Domain.

Then she sensed Robert's response, "Understood. However, whatever incomprehensible and unimaginable senses it has, it has to be aware of its environment. That can happen only if it gets

[29] [255] in the bibliography defines space mathematically, and has succinct descriptions of common spaces of various kinds.

information from reality. The same thing is true for us as well. The common structure of information is the only way we can understand it, and it can understand us."

"Oh get real!" thought Jane.

No sooner was the thought out, she found herself standing on a sidewalk with Jim, back in the good old United States of Information. Had it all been a dream? Jane was sure that Domain's Metaphysical Diner was somewhere in the neighborhood. She just couldn't see it. She has looked for Mr. Domain's diner ever since. She has never found it. After all, as she understood in a flash just before she found herself on the sidewalk, he is everywhere and every when. You can't really get away from him. Not now. Not ever.

> O World invisible, we view thee,
> O world intangible, we touch thee,
> O world unknowable, we know thee,
> Inapprehensible, we clutch thee.
> (Francis Thompson, *No Strange Land*)

4 Basic metaobject inventory

Let us pause here and make a list of the metaobjects we have discussed in this chapter. These metaobjects will help us normalize real-world behavior (or Nijssen's *irreducible facts* [297] or Ross' *atomic rules* [294]).

The concepts we have covered are:

- object
- property
- relationship
- process
- event
- domain
- unit of measure (UOM)
- format

The kinds of rules each metaobject normalizes are shown in figure 15. Moreover, we have seen that *behavior* and *irreducible facts* (or *atomic rules*) are merely different perspectives of the information content, or properties of objects. They are simple in and of themselves, but are the building blocks of knowledge of varying complexity.

5 Metaobjects and the natural repository of knowledge

Wisdom, we saw in section 1, is the symphony of collective knowledge that helps to move the firm towards its goals. Knowledge is configurations of rules. The most fundamental building block of knowledge, as well as the ultimate repository of information, is the atomic rule – a rule we cannot break into smaller, simpler parts without loss of meaning. The metaobjects

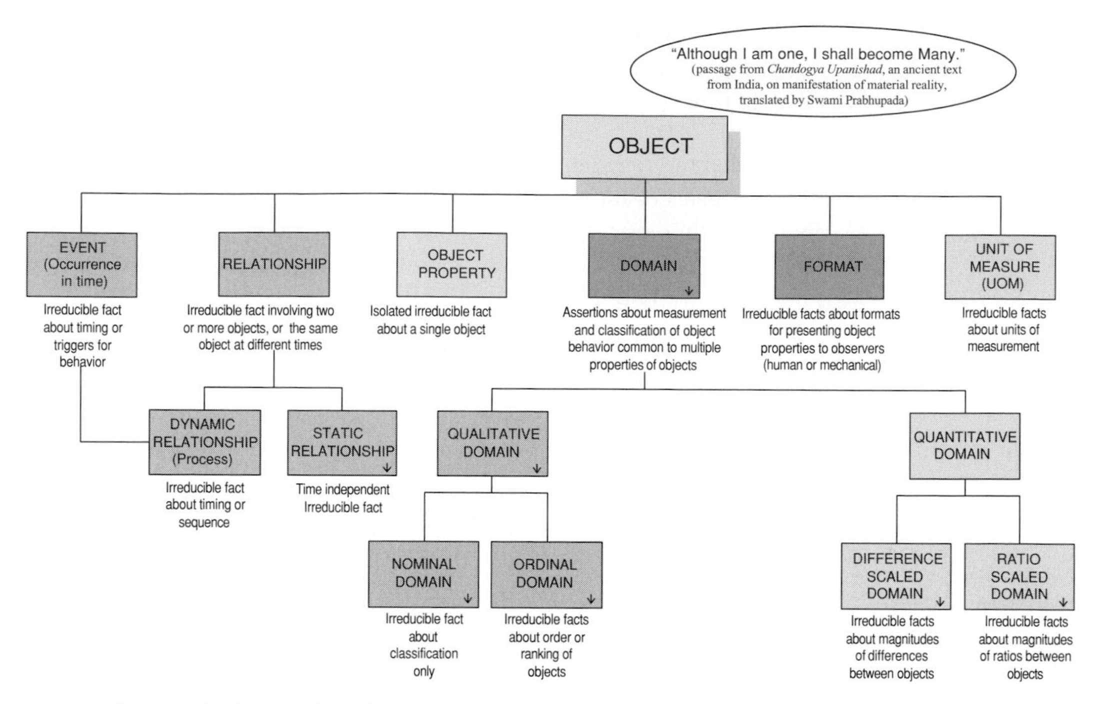

Figure 9 Basic metaobject inventory: *kinds of rules each metaobject normalizes*

of figure 9 are the natural wellspring of atomic rules and the repository of knowledge: they are the home, and the basis of real-world meaning.

In this section, we will understand how atomic rules are configured into knowledge, and how reusable components of knowledge *naturally* emerge from the metaobjects of figure 9. The intent is to develop a basic understanding. In the following chapters, we will revisit these issues again in depth and with greater precision.

A configuration of rules is not merely a loose collection of atomic rules. It possesses structure and patterns. It incorporates *atomic rules that are assembled from other atomic rules*. Some atomic rules are reused repeatedly as we assemble new rules to seek competitive advantage by specializing and implementing our business operations in new and innovative ways. These reusable rules constitute our reusable components of knowledge. Sometimes entire structures and configurations themselves may be reused to build other, more specialized domains of knowledge. This is analogous to manufacturers assembling reusable subassemblies from standard (reusable) parts. These reusable subassemblies may in turn be incorporated into a multitude of versions and variations of the end product. Reusability springs from the syntax of objects. Therefore let us first understand how the objects in figure 9 serve as repositories of atomic rules.

Let us start by revisiting the simple example in box 2. Each process in box 2 is an object. These processes are strung in a chain that shows which process must precede which other. These links are relationships, and therefore objects in their own right. These relationships carry irreducible facts about mutual dependencies between processes they connect. The

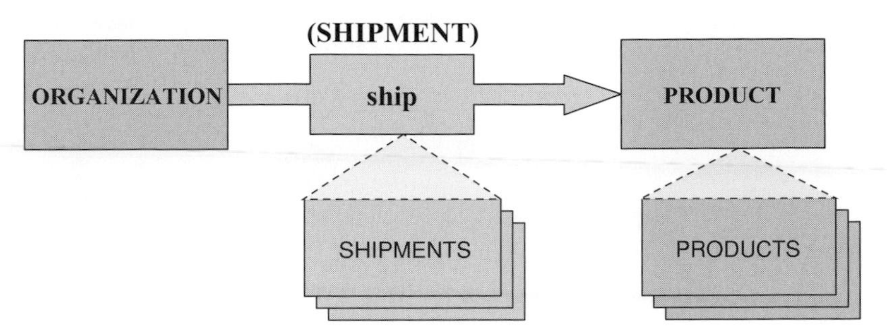

Figure 10 A rule: organization ships product assembled from objects

chain of processes is a *structure* assembled from atomic rules. It is a very simple configuration of atomic rules.

To understand how atomic rules may be assembled from other atomic rules, and to understand how subassemblies of rules may be reused, let us take another example – a simple atomic rule common to many businesses: *organization ships product.*

The shipment is a relationship between organization and product. It is also an object in its own right (just as all relationships are). The rule *organization ships product* is shown in figure 10. (This is an illustrative diagram, not one characterized by a rigorous syntax. Subsequent chapters will describe more precise ways of modeling atomic rules.) Read it as you would the diagram of figure 8; only remember that the arrows, i.e. relationships, are objects in their own right.

Figure 10 illustrates two atomic rules:

1 an organization may make many shipments and
2 each shipment may contain many products.

This is a simple configuration of knowledge. It is just a set of two atomic rules that are not mutually linked in a structure.

Now let us examine a scenario that forces change. In the following scenario, as rules of business change, we will add or alter rules, changing the simple configuration above step by step. As we do this, we will understand how knowledge, configured in the metaobjects of figure 9, is *naturally* normalized in the real world. We will step through the process of assembling knowledge from components, one step at a time.

Assume that the firm had negotiated a flat rate per shipment, but the contract is about to expire and shipping cost will depend on the gross weight of the shipment in the new contract. The scope of the shipping model must be expanded to include the gross weight.

Assume also that the firm has access to components of knowledge as a part of an inventory of knowledge artifacts that it has already built and stored in a repository. First we must look for the relevant knowledge in the repository.

We locate the weight domain. It is a ratio scaled domain. We understand (from figure 8 and box 5) that it must be associated with units of measure and conversion rules in the structure shown in figure 8. We also understand that weight can never be a negative number. It is a constraint, an atomic rule, associated with the domain (constraints are objects we will examine in depth later). As such, there is a *natural* structure of irreducible facts associated

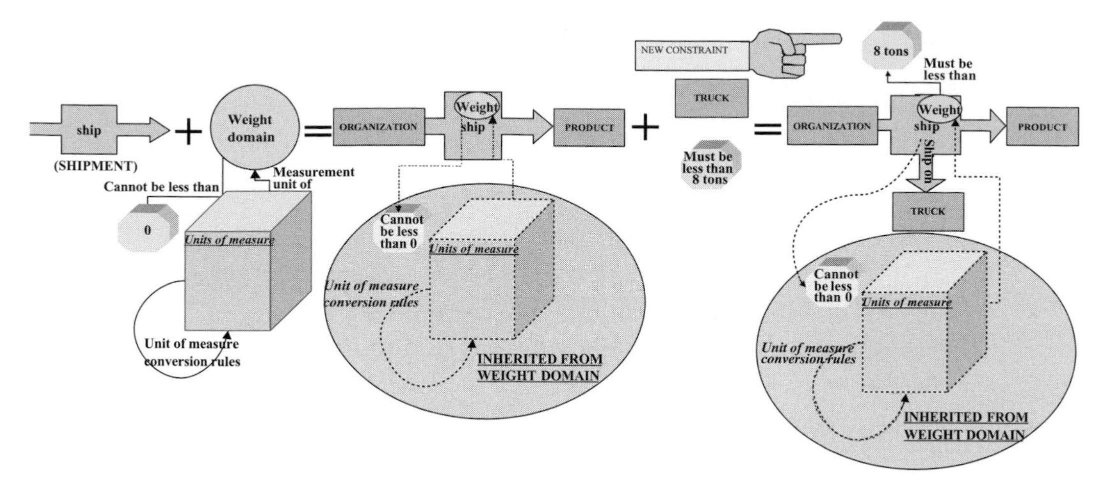

Figure 11 Adding components to assemble configurations of rules

with the weight domain. Assume the artifact in the repository reflects this. This natural structure may then be considered to be a subassembly of knowledge stored in the repository. Figure 11 shows this structure (second structure from the left).

When we assemble shipment with weight, it implicitly and naturally inherits the entire structure associated with the domain. The units in which shipment weight might be measured, the conversion rules between these units, and the fact that the shipment weight cannot be negative are all irreducible facts that flow from the subassembly. We might choose a preferred, default unit of measure to express the shipment weight when we design the business process. Our default unit of measure might depend on the context (default and initial states are described in detail in Chapter 2). For example, we might prefer kilograms when we deploy the process in the European Union, and tons when we deploy it in the US. In neither context will the *structure* change. Indeed, it will be common to both. This is an example of how knowledge is reused.

If the unit weight of the *product* were also needed, we would reuse the weight domain again. We would assemble the product object with the weight domain and inherit the same structures and rules. We would not need to redefine these constraints, the units of measure, and the conversion rules separately for shipment weight and product weight,. If a conversion rule was changed, or a new unit of measure was added to the weight domain, it would *automatically* be available to both shipment weight and product weight, *because knowledge was normalized.*

Now we will understand how irreducible facts may be reused to build other irreducible facts. Assume that the new contract with the shipping company specifies that all products must be shipped by truck. The atomic rule will read: *organization ships product by truck.*

First let us test this rule to validate that it *is* an atomic rule. Let us check if we lose information when we break the rule into smaller, simpler pieces:

1 *Organization ships product.*

2 *Organization ships by truck.*

Even if the two rules are taken together, they do not mean that the *product* will be shipped by truck. For example, both statements will be true even if the organization ships other items such as supplies and documents by truck, and ships products by air. Therefore we have lost information by dividing *organization ships product by truck*. It is an atomic rule. We obtained that atomic rule by turning *shipment* from a two-way relationship between *organization* and *product* into a three-way relationship between *organization*, *product*, and *truck*. We have created a new atomic rule from another. It is a special case of the more general atomic rule it was derived from.[30] This new structure, as well as its assembly from knowledge artifacts in the repository, is illustrated in figure 11.

Now we have another requirement: we find that trucks cannot carry more than 8 tons, i.e. the gross weight of each shipment by truck must be no more than 8 tons. It is another irreducible fact. This is not a generic constraint attached to the weight domain like the fact that weights must equal or exceed zero was; rather it is specific to shipment by truck. Therefore the constraint is attached to *shipment weight*, a property of *shipment* (an object in its own right as shown in figure 9), not to the weight domain. This constraint will not be automatically inherited by weights that are properties of other objects (for example product weight) because it is attached specifically to shipment weight, and not to the generic weight domain.

The effect of attaching this constraint of 8 tons to *shipment weight* implies that this property of *shipment* now has two constraints:

1 inherited automatically from the weight domain that no weight may be negative;
2 specific to shipment weight, that no shipment may exceed 8 tons.

The combined effect of both constraints is to restrict shipment weight to a 0–8 ton range. The structure on the extreme right of figure 11 shows how these rules have been configured to reflect knowledge about product shipment.

If we re-engineered the process to ship by air as well as truck, we would use the structure *organization ships product* in figure 11 again. Only, *airplane* would substitute *truck* in the structure on the extreme right-hand side of figure 11. (The weight limitation might also have to change.) This is another example of how knowledge can be *naturally* normalized and reused.

Since the weight limitations imply use of the weight domain, all conversion rules will also be automatically inherited from the domain and will apply to all constraints on weight. In our example, this might facilitate interoperability between European and US operations.

6 The architecture of knowledge and the scope of the metamodel in this book

The following chapters will show how meanings are components configured from other, more elementary meanings. The metamodel of knowledge starts with the broadest, most widely shared concepts first, and then, layer by layer, adds meaning to these to build

[30] The universal perspective and the metamodel of knowledge have the most frequently used generalized rules, a starting point for reusable components.

irreducible facts much like we did in the previous section. In other words, shared behavior is rooted in shared components of knowledge reused frequently.

Traditional analysis follows a different dogma. It uses simplistic stimulus–response models to describe behavior. In order to portray behavior accurately, traditional approaches, such as the black box and node branch methods, require detailed rules be completely known up front. This is why historically it has been hard to identify reusable components of knowledge with these methods. This has negatively impacted the resilience and agility of automation.

Box 6 The architecture of knowledge

(See the detailed supplementary discussion under "The architecture of knowledge" at our website. This section summarizes the information in Module I on our website.

The black box approach

Figures 12 and 13 describe the black box philosophy, and figure 14 shows the node branch approach.

In the black box approach, the focus is on the behavior of the "black box" that changes observed properties and values of inputs into outputs. It is analyzed by "process decomposition," which attempts to represent the behavior of the black box by analyzing interactions between smaller black boxes inside it (figure 13). The actual mechanisms inside the "black boxes" are ignored (details on our website).

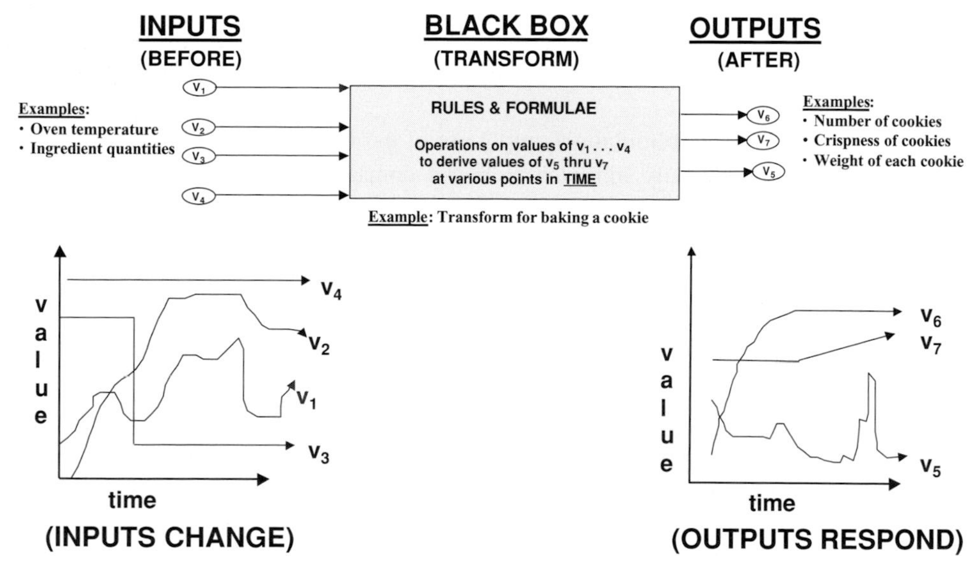

Figure 12 The black box perspective of behavior

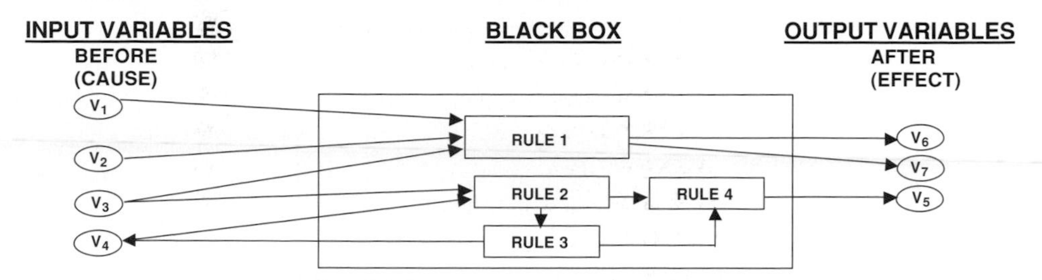

Figure 13 Process decomposition

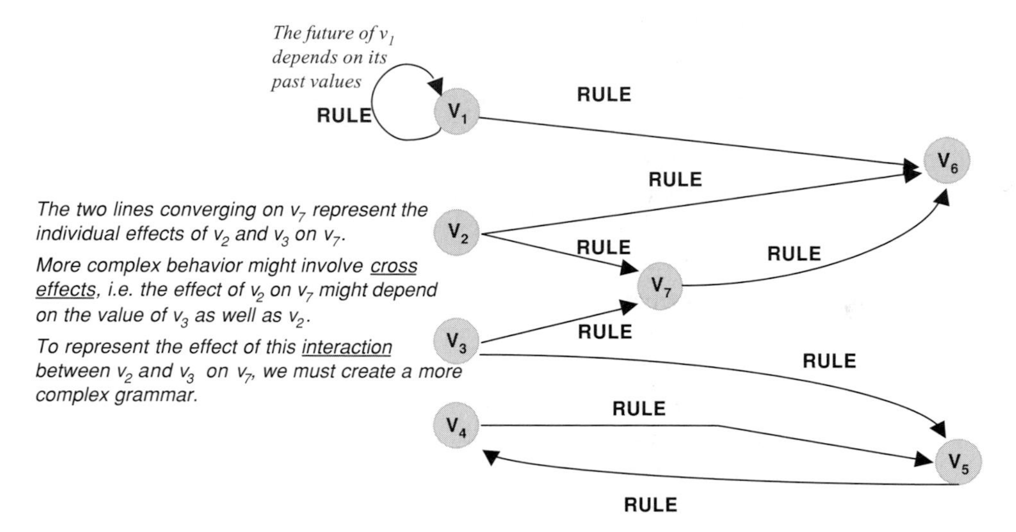

Figure 14 Node–branch representation

The node–branch approach

The node–branch approach uses a network of cause-and-effect rules to estimate changes in variables of interest and thus to model the behavior of a system over time (figure 14).

This technique is useful when large numbers of variables are involved. However, it has the same drawback as the black box method: all the rules and variables are needed up front. In neither approach, the focus is on incremental development and discovery, based on common rules that can be progressively specialized in step with new learning (details on our website).

The architecture of knowledge

The node branch and black box methods or their variants were adequate when business systems were smaller, simpler, and addressed small scopes. The problem with using these techniques today is that scopes are larger, ideation more rapid, and interactions more complex than they were in the past.

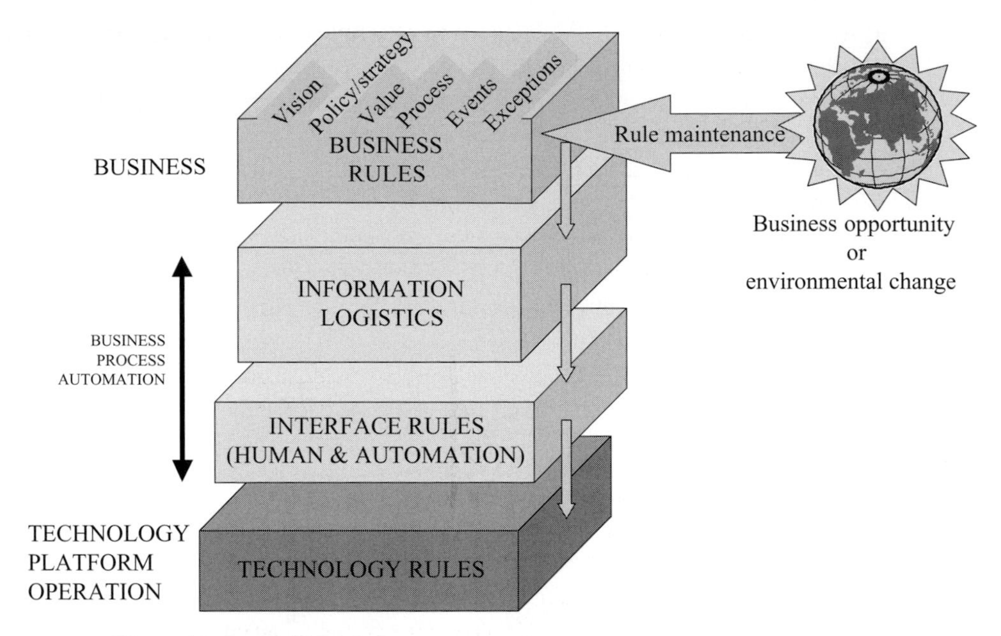

Figure 15 The architecture of knowledge

Change is so rapid today that it is impossible to fully determine all rules and variables of interest in the short time frames available. Scopes and objectives may evolve even as the model is being built. Facts must be added incrementally as they are discovered. Complex black box and node branch models tend to be "chaotic": small differences can lead to dramatically different outcomes when variables become too many and interactions too complex.

This is why it is hard to address change and innovation with those approaches. A different approach is needed. In the new approach, we must start with broad, universally shared rules, and add information in steps as issues evolve and needs become clearer. The new approach will be better suited to tracking moving targets. This approach will refine and evolve our models in incremental steps to absorb change and adapt to new learning. It can only be successful if we recognize the architecture of knowledge (figure 15).

The premises of the architecture of knowledge are:

- Knowledge encompasses a configuration of atomic rules.
- New atomic rules may be configured from older atomic rules by adding information (as new information is discovered).
- Atomic rules of business are different from atomic rules of technology.
- Rules of business are related to rules of information technology through business process automation.
- Each atomic rule may be implemented in information systems by one or *more* information flows. Each information flow must be supported by one or *more*, interfaces. Each interface must be supported by one, or *more*, information technology platform(s)[31] (see box 7).

[31] This multiplicity of choice in how each business rule can be implemented is the basis for building scalable and flexible information systems with reusable components. We will see this later in this book.

- An information system is a *configuration* of atomic rules of business, information flows, interfaces, and technology platforms.
- Rules of business, technology, and process automation are meanings. Meanings are patterns of information, which are *naturally normalized in the real world*, and should be reflected in systems as they are in the real world.[32]

The architecture of knowledge requires that we:

- Recognize atomic rules to configure reusable components of knowledge.
- Distinguish between atomic rules in each of the layers of figure 15 to enable us to configure and reconfigure behavior at each level to support change and new learning across all levels.

(The architecture of knowledge is described in detail on our website, with real life examples.)

Box 7 The architecture of knowledge reuse can help make information systems flexible and scalable (on our website)

Box 7 describes how, in the natural world, a single rule of business may be implemented in different ways with different kinds of automation, and how this gives business enormous opportunity for innovation, and also the opportunity to reuse knowledge resident in *every layer* of figure 15. Box 7 illustrates these principles with business examples.

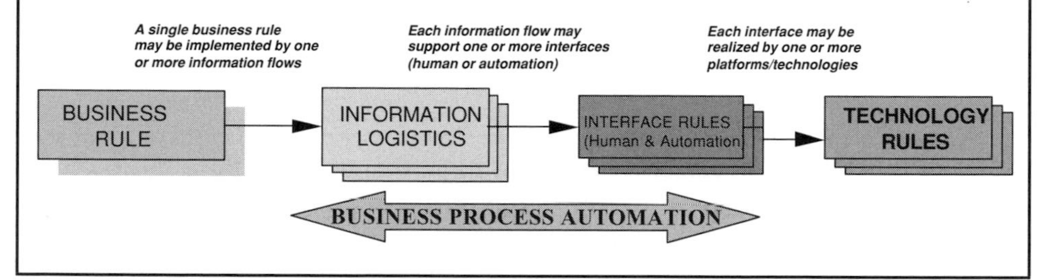

All concepts and objects, tangible or intangible, convey information. Business rules may be implemented by automation. This adds information about how *information* is processed and leads to business process automation. Abstract business rules may also be implemented by other mechanisms. This too adds information to create new irreducible facts from old (see the example in figure 11). The metamodel of knowledge calls them implementation-level business rules. It does not consider them to be rules of business process automation (figure 16).

The focus of this book is on the business rule layer of the metamodel of knowledge, and how it relates to business process automation layers. Supplementary materials under "The architecture of knowledge" on our website discuss business functionality, as well as

[32] Rules may be reflected in systems as they are in the real world if they are normalized and stored in an electronic repository. To optimize computer performance, these rules may be physically replicated. However unless replication is closely controlled, the risk of explosive and chaotic change cannot be managed (see Chapter 1).

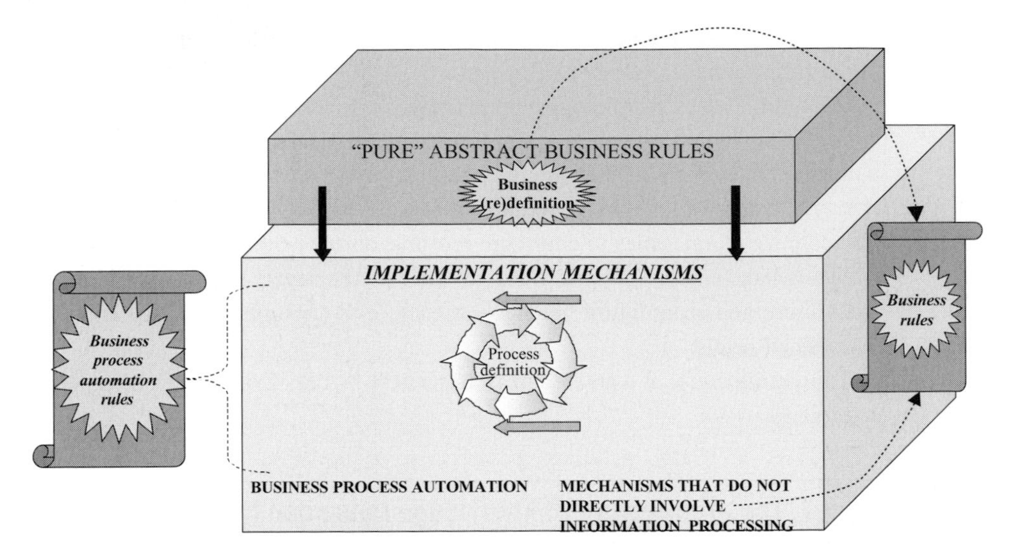

Figure 16 Business process automation is only one of several mechanisms that implement abstract business rules in the physical world

functional and non-functional features in detail. They describe how features of business process automation can shift from business process automation to technology layers in step with technological change.

Box 8 How rules shift between business process automation and technology layers (on our website)

Box 8 describes the evolution of information technology from punch cards to CRT terminals and database management systems to show how features fell from business process automation to technology layers in step with advancing automation.

Combining components of knowledge within a layer in figure 15 leads to increasingly complex and detailed rules in step with new learning in that domain. Combining components across layers leads to increasingly complete automation of components (figure 17).

The biggest opportunity, by far, lies in identifying the key components of shared business knowledge, reused most often, in the topmost layer of figure 17; this layer will drive the design of components in the layers beneath it. Using this approach, one can optimize investment and reuse by anchoring universal meanings in broad, stable, and frequently used patterns of information. This will control chaotic behavior at lower levels. Business processes and information systems may then adapt to change by reconfiguring and reusing stable components, which lead us to the objectives and scope of the metamodel articulated in figure 18.

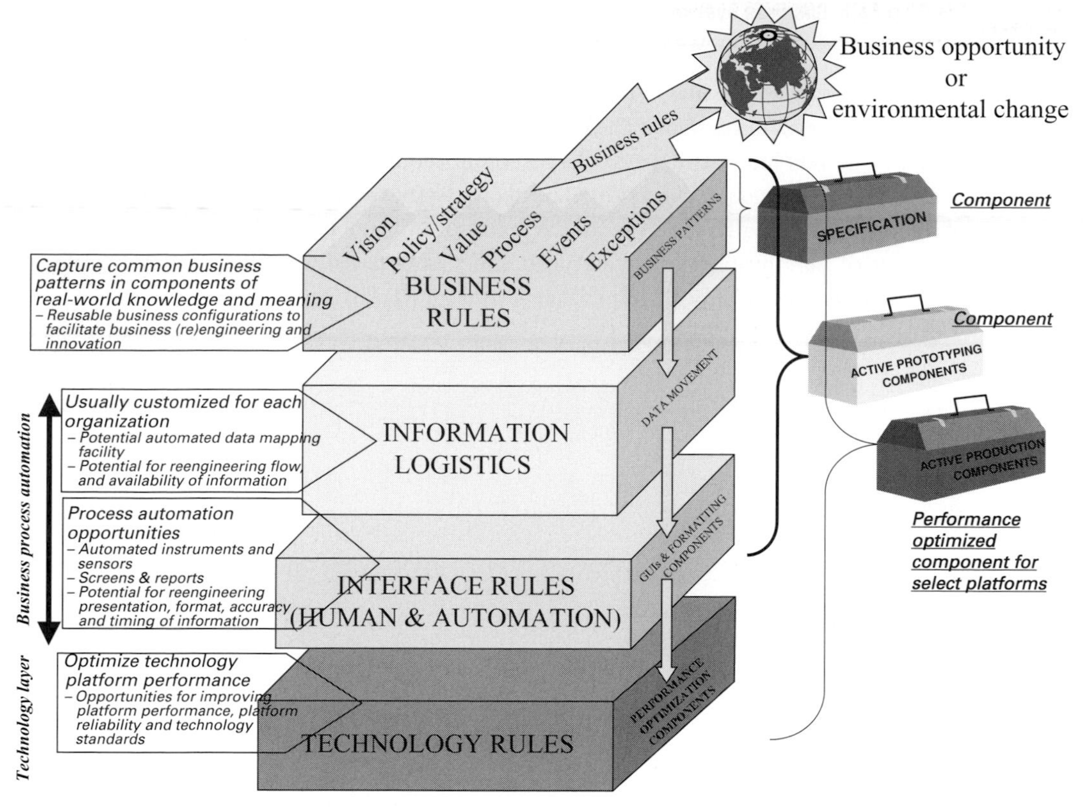

Figure 17 The architecture of reusable knowledge components

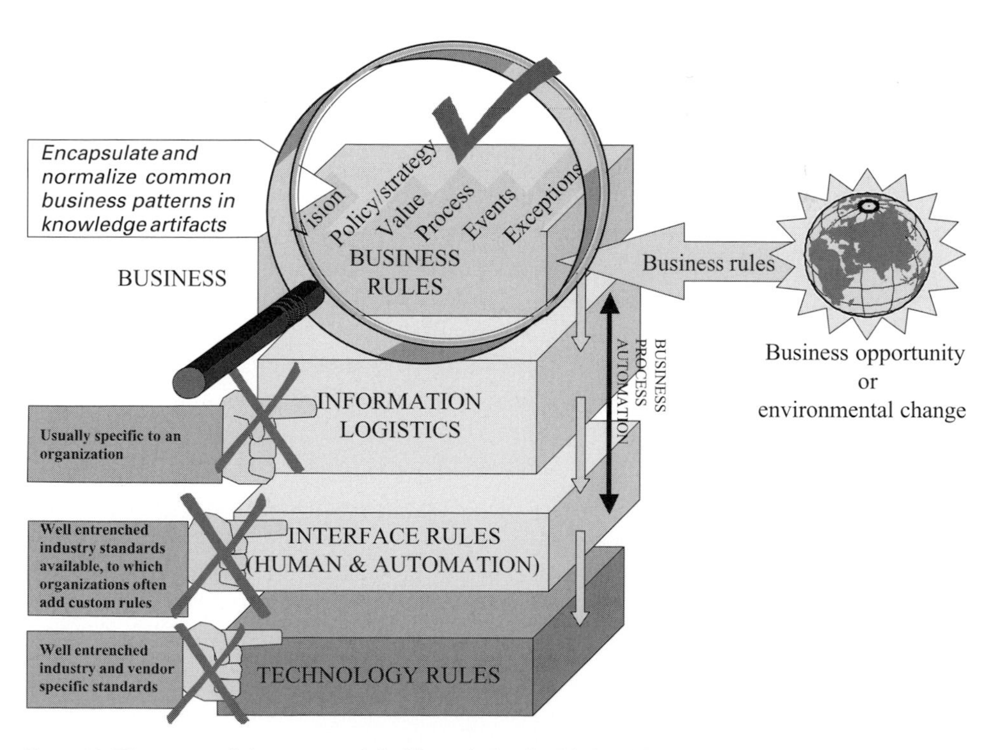

Figure 18 The scope of the metamodel of knowledge in this book is focused on pure business rules

The metamodel of *business behavior* does not care about how business behavior is implemented in information systems. The metamodel is adaptable and behavior can be manifested in many different innovative ways. The rest of this book describes the metaobjects that normalize atomic *business* rules found in the uppermost (business rules) layer of the architecture of knowledge. We will not only learn the behavior of metaobjects that flow *naturally* from the layer of pure business rules, but also understand transforms that turn these rules of business into rules of information exchange and transportation.

2 The object at the root of it all

"The world came out of a single spark; the creator is in the creation and the creation in the creator."

(Kabir Das, a fifteenth century poet-philosopher from India)

This chapter introduces the basic components of knowledge – the object and the state machine. Without recourse to complex mathematics, it demonstrates how meanings are the building blocks of knowledge. It describes how the object paradigm can encapsulate knowledge and automate its propagation through mechanisms such as inheritance. It describes the fluidity of knowledge and how systems adapt to new learning and behavior by reconfiguring components of old knowledge based on new learning. This chapter also introduces the problem of multiple perspectives and adaptation to shifting scopes, and how this may be addressed by reconfiguring the meanings at the heart of knowledge. The chapter uses business and real-world examples liberally to illustrate these complex issues and abstract concepts.

The object is, like the spark, from which all things flow. Objects assume many different roles and, even as they preserve the essential uniqueness and unity of reality, they present it in many superficially different forms. The concept of an object is the core around which meaning is normalized, even as it morphs into different forms and wraps itself around different kinds of meanings to normalize knowledge. The object lurks hidden within all of these forms and formats. It is this core we must understand, the concept of *metaobject*. The metaobject articulates the meaning of *object*. From it, everything else flows.

Box 9 Business definition of an object

An object is a person, place, event, thing, or concept, any behavior of which is within the scope of the model in which it is represented.

This traditional definition is good enough for building object models. However it lacks the mathematical rigor needed to develop the algebra of reusable knowledge components. Box 15 extends this definition with mathematics.

In Chapter 1, we understood that the principal difficulty in identifying reusable components of business knowledge lay in the classification of the behavior into reusable common categories. This was difficult because we had to do this *even when information was incomplete and little detail was available on the complex, large-scale, and close-knit interdependent behavior of real-world objects relevant to a particular process or application.*

Behavior is the key, and real-world objects manifest many behaviors. Real-world objects naturally group real-world behavior; they are the key to reusable common categories of behavior. Our models must therefore mirror *information* about the *behavior* of real-world *objects*. Understanding objects is the first step in understanding behavior. This is where we will start. *Behavior* – hit a glass pane with a hammer. It shatters. That behavior is common to all glass panes. How do we model this information? A very simple model will suffice. The fact is that glass panes can exist in one of two conditions, or *states*: whole or shattered. The *event*[1] – being struck by a hammer – has an *effect*[2] on the glass pane. It changes *whole* glass panes into *shattered* glass panes. In other words, this event changes the state of the glass pane. The scope of our model could be limited to just these two states of glass panes. In our model, hitting a shattered glass pane with a hammer has no effect. (Of course it might have an effect in real life! Remember that the model is *not* reality, *it is only an abstraction of those limited aspects of reality we want to focus on.*)

These then are the atomic rules in our model:

1 There are objects called glass panes.
2 Glass panes are made of glass.
3 There are events called "hammer strikes."
4 Glass panes exist in two states: "whole" or "shattered."
5 The two states are mutually exclusive.
6 The event, "hammer strike," changes the state of a glass pane from "whole" to "shattered." (This is the *effect* of "hammer strike" on the object "glass pane.")
7 The event, "hammer strike," does not change the state of "shattered" glass. It has no *effect.*

These rules together represent our knowledge about the behavior of glass panes. However, the astute reader will note that there is a subtle piece of information still missing. How do we know which individual panes are shattered and which are whole? How do we know which particular *instance* of "hammer strike" actually hit (or did not hit) which individual glass pane?

1 Object class versus object instance

The seven rules we have assembled into our body of knowledge about glass panes describe rules about a *class*, or category, of objects called glass pane and a *class* of events (remember events are also objects) called hammer strike; but to round out our knowledge about glass panes, even at this very limited scope, we must also have information about individual glass panes and individual hammer strikes, or, in other words, *instances* of these *classes* of objects.

This assertion is common sense. It is also a fundamental cornerstone of the entire edifice of knowledge in this book. We will repeat it again: objects can be grouped into categories, or

[1] See [166] for a mathematically rigorous description of *event.*

[2] The endnote on state machines has a mathematically rigorous description of *effect.*

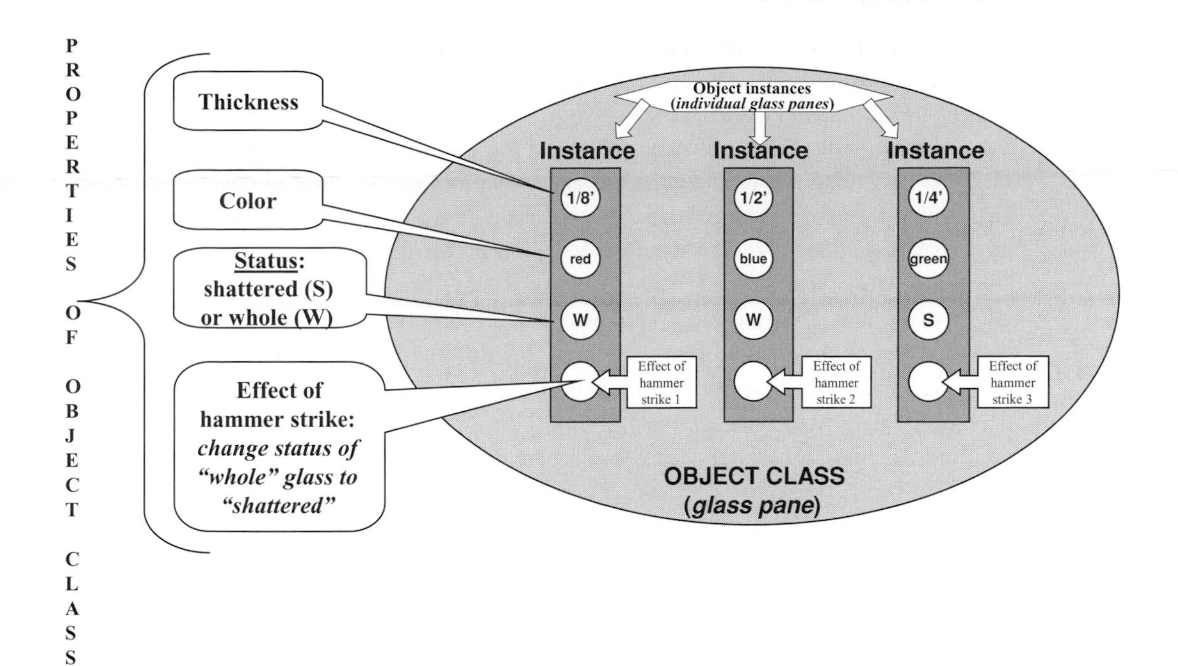

Figure 19 Object classes are collections of object instances with common properties

classes, based on common behavior. An individual member of a class is called an *instance* of the *object class*.

It is also important to consider that all things in the real world must exist for some time interval. They can even exist forever, or for an arbitrarily small instant of time, but they cannot exist for *no* time at all. This is true not only for glass panes, but also for an instance of any object in the real world. This is an irreducible fact that our metamodel of knowledge must recognize. We will consider the full impact of this later. For now it will suffice to keep it in mind as we analyze the behavior of objects.

Some properties of glass panes, such as whether a pane is broken or not, might vary from pane to pane, and also change over time. Other properties, such as the color and thickness of the pane might only vary from pane to pane, i.e. be different for each *instance*, but not change over time for a particular glass pane. Together these properties describe the condition of each pane. Although individual glass panes may have different values of these properties, they are grouped into the same object class because they have the same *kinds* of properties. We might be interested in including these properties in our model as well. The concept holds for any object class. Instances are grouped into classes based on common properties. For glass panes, these could be color, transparency, weight, thickness, strength, hit previously by a hammer, or any other property of interest to the actor or observer[3] (see figure 19).

[3] Actor and observer are synonyms: a person, system, or instrument that accesses or processes (i.e. acts on) information is called an *actor* in object technology. In physics, the same individual or instrument is called an *observer* because it *observes* behavior.

Object *classes* also have properties.[4] They too are objects. The properties of object classes are related to the collection of object instances in them. For example, *an intrinsic property of all object classes is the population (count) of object instances it holds at any given point in time*. Other properties of object classes are related to individual properties of object instances. In our example, the glass pane object class would contain properties such as the average thickness, strength, and transparency of glass panes, total thickness of all glass panes added together, standard deviation of thickness, strength, and transparency. These properties of the object class are derived from the properties of individual object instances in them. They are called *emergent* properties because they emerge naturally from the properties of instances (emergent properties are discussed in Module V, section 2 on our website). For the moment, it will suffice to understand that object classes and collections of objects are also instances of objects, with their own properties, features, and attributes, which are distinct from those of the object instances they contain.

Classes and collections of objects are object instances too. To understand why, let us consider an example – an insurance offering. Assume an insurance firm offers a policy that insures buildings against fire damage. The insurance product being offered is a kind or *class* of policies, and it is for a category or *class* of object instances called buildings. An individual policy, on the other hand, will cover a specific building; this policy may have properties such as the specific building(s) insured, amount of coverage, premium amounts, and the identity of the beneficiary to whom any claims must be paid. The insurance product, or *class* of policies, on the other hand, may have other properties such as the *kinds* of buildings that it will cover, the date the product offering was approved by regulatory agencies, general exclusions (e.g., fire damage from an act of war may be excluded), and *rules* relating premium to various risk factors such as rules about existence of fire alarms in the building and its distance from the nearest fire hydrant. If we stored this common information in each policy, we would be replicating information in each instance of policy, not normalizing it. Therefore, each *class* of policy may be considered an instance of an object that is different and distinct from each *instance* of policy. This instance of a policy *class* is the repository of information common to all instances that belong to the class, and there may be several *classes* of policies, which might store different terms and conditions common to all instances of that *class*. In general, each *class* of policy will have its own unique identity. The class is an *instance* of a kind of policy. In this way, each *collection* of objects is an *instance* of a *set* of objects.

Confused? Think of the collection as though it were a bag full of instances of objects. The bag too is an object instance, quite different from the objects in it, and we might stick a label on each bag to differentiate them from other similar bags. An object class is just one kind of collection of objects. The system described in box 12 is another kind of collection of objects. It too is an object, with its own distinct and unique identity. These objects that are containers of other objects are called aggregate objects. Aggregate objects are discussed in Module V, section 2 on our website.

[4] Properties of classes and other collections of object instances are discussed in detail later in this book. Also see [89].

2 The state of an object

The condition of an object at a point in time is called its *state*. The state of an object is given by the values of all of its individual properties at that time, such as, for glass panes, its thickness, color, weight, and strength.

When we were interested in just two mutually exclusive states of glass panes – shattered and whole – it was easy to define the state of each pane. Now that we have several properties that may vary independently from pane to pane, how can we tell what overall state an individual glass pane is in? To make it simple let us just consider two properties, color and wholeness (i.e. whether the pane is shattered or whole). To make it even simpler, let us assume that glass panes come in only two colors: red and blue. Then there are four possible conditions (states) in which we could find a pane of glass:[5]

1 All whole red panes would be in one state
2 All shattered red panes would be in another state
3 All whole blue panes would be in a third state
4 All shattered blue panes would be in a fourth state

We only extended the scope of our model slightly – just one other property of glass panes, and only two colors at that. Even so, the number of possible states doubled. Had we considered the several other properties and colors, the number of possible states would have exploded. Also, thickness is not restricted to discrete categories. It could vary over a continuum of positive numbers. How can we represent the state of the glass pane when such properties are involved, and circumvent the chaos and complexity that can result from the explosive growth of the number of possible states of behavior?

State charts

David Harel solved the problem in 1988 with the concept of state charts,[6] [80], [81], [82] and higraphs [82]. Figure 20 is an example of a state chart. The big (boundary) rectangle represents the object class. In this example it is "pane." The object class is partitioned with broken lines. Each partition represents a property of the object,[7] and is identified by its label. Mutually exclusive states within each partition are represented by the smaller rectangles inside the partition. The fact that an individual sheet may be in one of two states, *whole* or *shattered*, is represented by the two rectangles in the *wholeness* partition. States separated by broken lines show that they are independent and may exist simultaneously. An unbroken (whole) pane of glass may be red or blue, as might a shattered pane, whereas the shattered and whole states are mutually exclusive. The states within a partition may themselves represent

[5] The set of all possible states of an object is called its *state space*.

[6] State charts were invented in 1984 by David Harel. Harel added the theory of higraphs to enrich the state chart syntax in 1987. [82] has more information [80] compares various techniques for modeling states of objects, their strengths and weaknesses with mathematical rigor.

[7] In mathematical parlance, these partitions are orthogonal, i.e. they present independent perspectives of the object. Orthogonality, or non-overlapping properties imply mathematical independence. Harel's papers in the bibliography describe higraphs, state charts, and their properties with mathematical rigor.

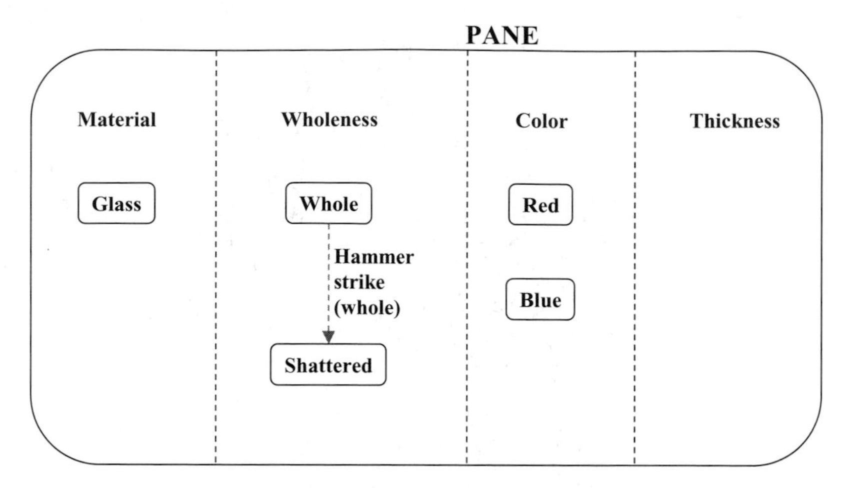

Figure 20 Object classes are collections of object instances with common properties (state chart perspective)

combinations of (sub)states in independent partitions. Broken lines could partition the little rectangles inside a partition and those little partitions within partitions could contain mutually independent and mutually exclusive states too. For example the shattered state might be partitioned by the number of broken pieces, say many or few, and independently by the size of the largest piece, say large or small. These partitions might be partitioned in turn, and the process may be repeated to as many levels as needed to progressively represent all the detail we possess. (We will return to partitions later in this chapter.) In short, partitions help us organize and classify the bewildering numbers of possible states to bring some order into chaos.

The arrow between the "whole" and "*shattered*" states in figure 20 represents the *effect*[8] of the event, *hammer strike*. The arrow tells us that a hammer strike changes the state of glass from "*whole*" to "*shattered*."[9]

(Arrows in most state charts are drawn with solid lines. In this book, we have shown effects with broken lines to show that they represent a "before" and "after" rule involving a sequence in time. In Chapter 6, we will discusses relationships that do not involve time, and are independent of sequence. These will be represented with solid arrows. Our metamodel must distinguish between these two kinds of rules in order to reflect reality.)

The "whole" in parenthesis in the label of the arrow specifies that glass panes must be whole for the state transition to occur. This is called a *guard condition* for the effect.

The guard condition is redundant in this example because the arrow implies that this transition is *only* from the "whole" to the "shattered" state. As such, the guard condition can be safely omitted in this case without any loss of information. However, the concept is useful when more complex behavior is involved. For example, if we added a complex rule that blue glass is shatter proof, and only red glass may be shattered by a hammer strike, the

[8] The endnote on state machines defines *effect* formally.

[9] Effect is a relationship between states, just as process is a special kind of relationship between object classes (Chapter 1). See Module V on our website.

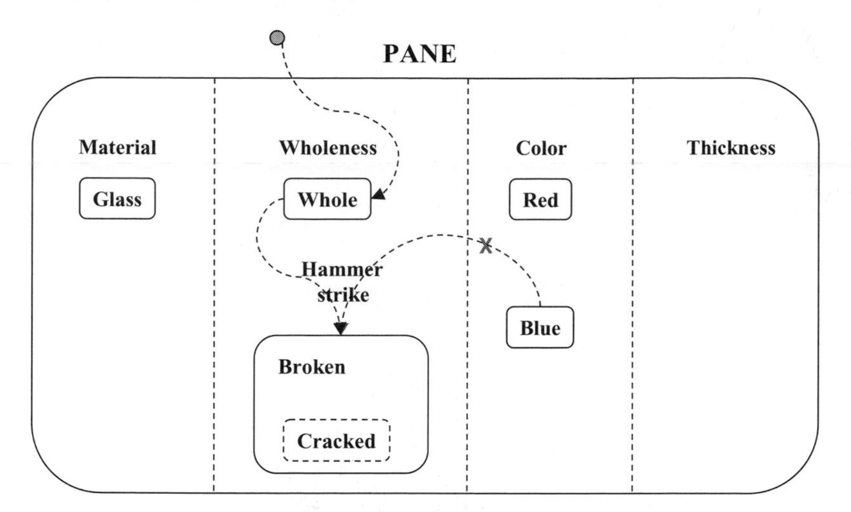

Figure 21 Disallowed effects, default, and "maybe" states

guard condition would read "red," and would not be implied by only connecting the whole and broken states with an arrow.

Harel's state charts provide another way of saying this as well. Consider the state chart in figure 21. We have modified the model of figure 20. In figure 21, the glass pane may be merely cracked, not necessarily shattered. That too is considered broken glass. That is why the cracked state is shown inside the broken state. It is called a *substate* because it is only one of several kinds of broken states. (For example, *shattered* is another kind of broken state. *Broken* is called a superstate of both *cracked* and *shattered* states.) The cracked state is represented by a broken-lined rectangle. (The broken lined rectangle tells us that we are not sure if there are any cracked panes at all.) More important is the crossed-out state transition, or "effect," between the "blue" (color) state and the broken state. This arrow merges into the arrow that represents hammer strike. The X on the link between the blue and broken states of the pane tells us that the effect is not allowed if the glass is blue, i.e. blue glass is not allowed to break (and therefore not allowed to crack either because the cracked state is a kind of broken state).[10]

We also have an arrow with a small round tail pointing at "whole." This arrow tells us that the pane *starts* as an unbroken pane. In other words, it is the *default* state.[11] This might be an irreducible fact in our body of knowledge, or it could be used as an artifice to manage uncertainty: the scope of our metamodel is limited to deterministic systems. This puts us in a bind if we are uncertain about what state we might find glass panes in. If we guess that we are more likely to find unbroken glass sheets before we break them with hammer

[10] Had the rule said that blue glass could crack, but not shatter, the arrow with the X mark would have pointed to a *shattered* substate (not shown in figure 21) inside the *broken* state. This would mean that a hammer strike could shatter *or* crack red glass (because the effect points to the outer rectangle representing the broken state, which contains both possibilities – *shattered* as well as *cracked* substates), but blue glass can only crack (because it cannot shatter, but can break, and the only other substate of broken glass in the model is cracked glass).

[11] The endnote on the state machine defines *initial state* (i.e. default state) formally.

strikes, we could assign a "default," or most likely starting state, to the *information* that we have stored for every *instance* of glass pane. Then the user of our system would change this default state for only those panes that were found broken before the hammer hit them.

Events, effects, and actions

The effect of events must be implemented via a set of procedures that are a collection of actions in information systems or business processes (see "Crossing the chasm" in Module V, section 3 on our website).

Data and attributes: states of behavior

Properties such as color and thickness that represent the *state* of the object at any given moment in time are called "*attributes*" of the object, or "*variables*" of interest to the system. This is what we call *data*. *Effects*, on the other hand, are properties of the object that are *rules* for changing the state of the object.[12]

Must every object have attributes?

Consider the concept of length. It is a ratio scaled domain described in section 3.2 of Chapter 1. What are its attributes? – it has none. Length may be an attribute of objects such as glass panes, rooms, snakes, and roads; these attributes all map to the length domain, but the domain itself is bereft of attributes. Domains are objects too, but the length domain has no attributes! However it does contain information: we know that lengths can be meaningfully added, subtracted, divided, compared, and that they cannot be less than zero. The length domain normalizes this information about length. It normalizes behavior. (We will discuss the properties of domains in greater detail in Chapter 4.)

Behavior of an object is comprised of rules about its properties and valid (lawful) transformations. The effects of events collectively represent one kind of property. Whenever an effect is involved, there must be at least two states (before and after), perhaps more (when guard conditions apply, as in figure 21.) "Before" and "after" states will be mutually exclusive and hence will translate into two different values of a *single state indicator* (i.e. attribute). As such, the object will have at least one attribute if not more when effects of events, i.e. state transitions, are in scope (see box 11).

Can an object have attributes when no state transitions are in scope?[13] Consider a person's gender. There are no effects that change gender in the scope of most models. However, EEOC rules might require firms to report the incidence of female versus male employees. The counts are properties of the *class* of employees, not an *instance* of employee (remember classes are objects too). These counts relate to individuals' genders, i.e. employees who are *members* of the class. Hiring and firing events might cause a net loss or gain of the number of male and female employees (object instances) that belong to the class of employees. This

[12] The endnote on the state machine describes the states of abstract automata mathematically.

[13] All objects must have an instance identifier that captures. The instance identifier is the irreducible fact that the object exists.

may change the *proportion* of male versus female employees. The proportion of employees of a gender is not an attribute of an *instance* of employee. It is an attribute of the employee *class*. Thus, although no gender *changing* effects are in scope for *instances* of employee, events can, and do, impact the gender information for the *class* of employees. The state of the *class* can change. Had we not included gender as an attribute of (an *instance* of) employee, we would not have the information to model state changes of the *class* of employees. Thus, objects can have attributes, even when effects that change values of those attributes are out of scope, because they might be involved in relationships with *other* properties that *do* change. This is why attributes can be in scope even when state transitions for *that* object and *that* attribute are not in scope.

Take another example: a building's address is required for supplies to be delivered. Effects that change the location or the address of the building might be out of scope, but the address object with attributes like street, town, and zip code is required for making deliveries to the building. The delivery system might be out of scope, but it must have this information. Although the address itself will never change state (within the scope of *this* process), it is a valid object class[14] because o*ther* objects such as shipping documents and packing lists may change state by establishing relationships with address in order to *reference* it. Overall, an object could have attributes even when no state transitions are in scope because other objects (or an *actor*[15]) need to reference it.

Consider the example above from a different perspective, one that includes the shipper's systems.[16] The firm might ask a shipper to change the delivery address to a company warehouse instead of the office building for a particular shipment. In other words, instances of objects such as shipping documents and shipments could change their relationship to individual addresses. This means that the conditions (states) of those object instances would have changed because they would now point to a different address. The metamodel of knowledge makes a sharp distinction between metaobjects that normalize rules involving the flow of time, versus those that do not. The concept of state captures the condition of an object at an instant of time, whereas effects involve *change* of state, and hence originate from the flow of time. (Effects are discussed in more detail under "Crossing the chasm" in Module V on our website.) It follows that the fact of an object's participation in a relationship is another kind of state indicator. In short, objects without attributes could exist because they participate in relationships with other objects (or themselves – see box 10).

It is worth noting that a relationship, or more accurately the fact that an object participates in a relationship, is like an attribute because it describes *the condition (i.e. state) of the object at a moment in time*, but effects are not, because effects describe how states *change* from one moment to another. However, the *existence*, i.e. the presence or absence of an effect,

[14] Address could change state in a larger scope. For example, when a building is being built in a new development, it might not have a postal address at first. Its address is drafted, and then formally allocated by the post office. Thus, the state of address changes state in the real world, even if it is frozen in the model.

[15] A person, system, or instrument that accesses or processes information is called an *actor*.

[16] Driven by e-commerce and the World Wide Web, entire supply chains comprising of multiple corporations have started competing for business. As supply chains strive for excellence, efficiency, and competitive advantage, the integration of business operations and supporting systems *across* corporations is becoming as important as integrating systems *within* an enterprise. See [96], [105], [111], [115], [119], [120], [123] in the bibliography.

conveys information about the condition of an object at a point in time and must be counted as a state indicator. For instance, the terms of an agreement may be changed while it is being negotiated, but may not after it is sealed. Accordingly, the "revise terms and conditions" effect will exist for an agreement in the "draft" state, but not in the "sealed" state. The presence or absence of an effect may be shown with a "state indicator"; it will also be an attribute of the object, but an attribute that is tied to the information conveyed by an effect. As such, this state indicator is a computing artifice used programmatically to determine whether to allow or disallow the computer to execute the code for an effect. *It conveys no information independent of the effect.* The mere presence of the effect signals that the change may be triggered by the right business event. The absence of an effect implies that the change is meaningless. The state indicator actually denormalizes information making it redundant. However, it is a useful artifice because it allows us to classify objects and reduce their numbers in our model. For example, with a state indicator, both sealed and draft agreements could be modeled as an agreement object, which may be distinguished by the value of its state indicator.

This is why all three characteristics – effects, attributes, and an object's participation in a relationship – are sometimes called "features" of an object [54], [328]. Attributes, the fact that a given effect exists, and relationships the object has with other objects or itself are state indicators (an object may relate to itself – for example, a person may train himself – therefore the *train* relationship points back to the same object instance, the same *person*). In this book, an object's participation in a relationship will be considered to be on par with the existence of an attribute because both are state indicators. The term attribute will cover both meanings unless an explicit distinction is made.

Box 10 Properties of objects (on our website)

Box 10 discusses the information content and impact of effects and relationships on the state of an object in more detail. It also describes the context of this discussion in terms of the XML standard published by W3C, an industry standards organization.

We will see later, as scopes shift and business processes evolve, not only might relationships sprout new attributes, but existing attributes too may grow into relationships and even object classes. Recognizing this natural property of objects is key to recognizing the flexibility of knowledge and its reusable components. We will return to this concept later in this book.

Box 11 States, attributes, state variables, and type indicators: much ado about nothing (on our website)

Box 11 discusses the impact of scope change on states and state indicators. It discusses how some commonly used data and object modeling precepts are not only redundant, but may also complicate models, making them inflexible in the face of unexpected developments – change in scope, new learning, and new perspectives.

Three values every attribute and relationship may have

There are three real-world values that every attribute[17] can have, regardless of what kind of domain the attribute maps to. These three values may not be obvious if we focus only on software systems and tables of data. However, when we broaden our perspective to consider the real world, they are difficult to miss:

1 Don't know the value: we are sure the attribute has a value, we just don't know what it is (also called the "unknown value" in this book).
2 Null value: we know for sure that the value does not exist, i.e. the attribute has no meaning in a specific context (or a relationship does not exist for a specific instance of an object)
3 Initial (or *default*) value: we know the initial state of the object instance as it is created and this may be any valid value.[18] Every object must exist in some state at every moment of its life (even if that state involves *null* or *don't know* values for some or all attributes). The set of default values of an object's attributes (and relationships), even if some or all of these values are *null* or *unknown*, is the initial state of the object at the moment of its birth.

The following paragraphs discuss the "don't know," "null," and "initial" values in more detail.

Don't know the value

The real world is uncertain. We may know that an attribute has a value, but we may not know what it is. For example, we may know that an individual nicknamed "Sam" exists, and we know for sure that Sam, being a human being is either male or female, but we do not know which. Sam could be either a male named Samuel or a female named Samantha. If we restrict our information system to record only male and female genders, and also insist we record each person's gender because every person *must* have a gender, we will have a problem trying to record Sam's gender. This happened because we did not reflect a fact about the real world in our systems as it is in the real world: *that, in the real world, we may have only partial information and every attribute may assume a special value: "unknown."*

Why is it important to know this? Consider the following example. Some database management systems (especially older versions) do not recognize an "unknown" value. They focus on formats of fields (read attributes) and restrict them to numeric and alphanumeric data formats. Arithmetic operations are permitted on numeric fields, but not alphanumeric fields. These database management systems try to get around the problem of unknown values by assigning a default value of zero to numeric fields and blanks to alphanumeric fields. This is at the root of several kinds of systems problems and is the source of unneeded complexity in application programs, as the following example demonstrates. Assume Sam is using a database of this kind to compute the average seasonal temperature of a town over the last fifty years. Temperature records are missing for several days. The database management system blithely assigns a default zero value to all missing temperatures. Obviously, Sam will get wrong results *because the database management system did not reflect facts*

[17] "Attribute" includes state indicators and participation in relationships.
[18] A valid value is any value in permitted state space. The state space of an object is discussed later in this chapter.

about the real world in its software as it was in the real world. It did not record when the temperature was *unknown.*

Rule: For difference and ratio scaled attributes, the result of any arithmetic operation on unknown values is also *unknown.*

Rule: Similarly, for ordinal, difference, and ratio scaled attributes, the result of any comparisons of unknown values with any other value (for example, an attempt to rank them in order of magnitude) is *unknown.*[19]

Rule: The comparison rule is identical for nominally scaled attributes, only any ranking is meaningless – we can only compare to check to see if the value is the same or different. The answer will always be unknown when unknown values are compared.

Null values

Null values too carry information about meanings inherent in the real world. They are expressions of atomic rules that tell us which properties *cannot* exist.

There is, O monks, a state where exist neither earth nor water, nor heat nor air, neither infinity nor nothingness. It is the Uncreated. (Adapted from the *Gospel of the Buddha* by Paul Carus)

To understand this assertion, remember that *each attribute, relationship and effect is the repository of an irreducible fact: that a specific rule or property of the object class* ***exists****.* Omitting an attribute from the object class implies it does not exist for any instance of the object. Keeping the attribute (or relationship) in the object class, but assigning a special "null" value to it for individual instances of the object implies that the attribute (or relationship) does not exist for those specific object instances. This is very different from asserting that the value of the attribute (or relationship) is unknown. Null values imply that we are *sure* that the relationship, attribute and value does not exist.

To understand why this is important, let us consider an agreement between two parties. The agreement is an instance of an object. The agreement will have attributes such as the date negotiations started, who is bound by the agreement, when it will end, if it is sealed agreement or not, and the effect of the event that will seal the agreement.

When the agreement is sealed, it will have additional attributes such as the date and time it was sealed, who witnessed it, and where it was sealed. Until the agreement is sealed, these attributes will not exist. *It is not that we do not know or are uncertain* of the values of these attributes for agreements being negotiated; instead, we are *absolutely certain* that these values *do not exist.*

Many newer versions of database management systems support null values, but not "unknown" values. Consequently software developers use this null value facility to support both true null values as well as real-world "unknown" values. What penalty do we pay when we do not recognize the difference between "unknown" and "does not exist"? The penalty is replication of information and loss of normalized knowledge. We will illustrate how this happens with two examples.

[19] When we consider the inverse of this situation, it is not necessarily true that the result of an arithmetic operation on known values is always known. Sometimes, even if we know all values involved in an arithmetic operation, the result can be indeterminate, for example dividing 0 by 0 has an indeterminate result.

Sam, the statistician, is writing a fact book about the profile of those who live in Handytown. As a part of her work she needs to count the number of right-handed, left-handed, and ambidextrous people of Handytown. Sam is delighted when she discovers that the town has compiled a database of its residents that shows the handedness of each individual. The mayor of Handytown will be happy to let Sam use their database and she is looking forward to finishing her project early. The record for each individual in the database has a handedness indicator that reads "right," "left," and "ambidextrous."

To her dismay, when Sam browses the survey report, she finds that all residents did not agree to divulge their handedness. The handedness indicator was set to null for such people. However as she thinks it over, she realizes all she has to do is to count the number of left-handed, right-handed, and ambidextrous persons, and include a separate count of persons who would not say. However, the day after Sam publishes her findings in *The Eminent Journal of the Society of Handedness*, the mayor comes rushing into her office. He is looking very upset and threatens to drag Sam to court for misrepresenting facts about the town. Sam is flabbergasted. She denies any fault and reminds the mayor that the data came from his own database. She goes on to explain how she counted the handedness of individuals from the town's database.

The mayor calms down somewhat, but fixes Sam with a quizzical look. "Didn't you know," he says, "we have many veterans in Handytown, and some of them lost their arms in battle. They can be neither left nor right handed, and obviously they can never be ambidextrous! Such individuals were not asked the question and that is why their answers were never entered into the database. This is why their 'handedness' indicator in the database is null. It is not that they would not say – it's just that they had no arms. Sam, your answers were wrong because you did not distinguish between 'do not know handedness' from 'does not apply – has no hands!'."

Here is another example of what can go wrong when "do not know" is not treated distinctly from "does not exist." Let us assume that a global customer of a bank has several accounts. These accounts may be spread over branches and subsidiaries in several countries, and may be denominated in different currencies. The customer checks individual balances of each account, as well as the total on deposit with the bank several times a day in an *ad hoc* manner. All balances must be arranged in account number sequence in the report and expressed in US dollars. Normally this is not a problem. Exchange rates between local currencies and US dollars are available for every market tick. However, there are exceptions. Sometimes the information is delayed, or the information flow is disrupted. When this happens, balances in US dollars for accounts in that country are not known. Therefore the total balance too is unknown. If this occurs, those balances that are known are reported, and those that are unknown are shown as such.

The customer also has accounts that are closed or for a variety of reasons. These accounts too must be shown in sequence in the report, but balances cannot exist for closed accounts, and obviously they will not count in the total balance.[20] Indeed, a null balance (different

[20] The total balance in this example is *defined* as the total balance of all *open* accounts. The total balance of all accounts does not exist – see the rules at the end of this section on null values. Open accounts are a subtype of all accounts (Chapter 2, section 3), an object class.

from a zero balance – some open accounts may have a zero balance or even negative balance if overdrawn) *implies* a closed account.

If account balances are recorded as distinct null values when they do not exist, and "unknown" values when unknown, computing the total balance is simple (if any balance is unknown, the total too is unknown). However, if "unknown" is *not* distinct from "does not exist" (for example, if both "unknown" and "does not exist" are represented by null values as is usual in many database management systems), the process of computing the total becomes more complex. It cannot be calculated from account balances alone. We must set a state indicator for closed accounts to infer which accounts we must exclude from the total balance because they are closed. Had we distinguished "Unknown balance" from "balance does not exist," we could have used the account balance to infer the state of the account. We would not need a separate state indicator. We needed an extra attribute, a state indicator for closed accounts, because we lost information by failing to reflect the real-world situation in our model.

Rule: For difference and ratio scaled attributes, the result of any arithmetic operation on null values is *null* (i.e. "does not exist").[21]

Rule: Similarly, for ordinal, difference, and ratio scaled attributes, the result of any comparisons of null values with any other value (for example an attempt to rank them in order of magnitude) is *null* (i.e. "does not exist").[22]

Rule: The comparison rule is identical for nominally scaled attributes, only any ranking is meaningless – we can only compare to check to see if the value is the same or different. If any value being compared cannot exist, obviously the answer does not exist because it is a meaningless comparison.

Default values

Now let us consider default values again. We had discussed default values briefly in our discussion on the initial state in figure 21. Assigning a default value to an attribute is an atomic rule. It is one of the elements from which our body of knowledge about an application is configured, but we have not yet talked about where it resides in the architecture of knowledge shown earlier in figure 15.

Default values are atomic rules about the initial state of an object instance. Every object (instance) must exist in some valid state at every moment from the time it was created. *This is a fundamental rule in our metamodel of knowledge.* The initial state of an object (instance) is its state at the moment of creation.

Objects of different kinds are found in each layer shown in figure 15. Therefore default values too are distributed between layers. Each object must have one, even if it is null or "Unknown." We may even use them, as discussed in the description of figure 21, to partially compensate for the fact that our metamodel assumes purely deterministic behavior in an uncertain world.

[21] See the endnote on gluing objects together.

[22] It is meaningless to operate on or compare value(s) that do not exist. Under certain conditions, the result of operations on known values might not exist. See the examples in chapter 1 of [308].

Default values are only one of several artifices[23] we use to account for the inherent uncertainty of the real world that we have ignored in our metamodel.[24]

Default values apply to attributes (and relationships) of all instances of a class of objects.[25] (For example, the default state of all glass panes was "Whole" in figure 21.) When d*efault values address uncertainty, they represent our "best guesses" for uncertain values of attributes (and relationships).*

Default states reside in the same layer (figure 15) as the object it is a default for. Screens and other interface objects, data flow and information logistics objects, processes, relationships, and all other objects in every layer of figure 15 must have default states – implicit or explicit. The state of the art of our database management software is such that implicit default values of data are often null. This is a constraint imposed by technology, not the real world.

Why are layers important? Layers are important because they make room for the reuse of common knowledge in a vast diversity of environments. The ability to support default states for different objects in each layer helps us compensate for the different kinds and degrees of risk in different environments. The same object may be attached to different default states in different environments. For example, section 5 of chapter 1 described how similar business processes might be implemented with different default units of measure and yet benefit from shared knowledge. Default values, like any other component of normalized knowledge, can facilitate business agility and innovation. (Box 7 has an example of how shared components can benefit a business.)

The overall state of a system[26]

The concept of the overall state of an object also applies to systems of multiple objects because the system as a whole is also an (aggregate) object. The state of the systcm at any moment in time is the set of states of individual objects in it, and the state of each object, in turn, is the set of values of each attribute. To show this concept, the state chart of the system has been partitioned with broken lines in box 12, one for each constituent object. Each such object, in turn, could be a partitioned state chart – one partition for each component and so on. Ultimately, we would be left with the variables and the same definition of state as at the beginning of section 2. Objects are thus the fundamental building blocks of knowledge.

[23] Recognizing that rules may be violated is another way of providing for uncertainty in an inherently risky world (see exception processes in Crossing the chasm, Module V, section 3 on our website).

[24] Processes that depend on chance are called stochastic processes. Deterministic processes have completely predictable outcomes. This is our scope. Stochastic processes are described in several publications listed in the bibliography: [310] has a basic introduction to probability theory; [312] introduces modeling of stochastic processes; [253] has mathematical techniques for modeling uncertainty; [323] shows how complexity and uncertainty compound each other; and [286] describes techniques for modeling the uncertainty caused by observing or querying complex stochastic systems.

[25] Under subtyping, we will understand how business rules could also assign different default values to different sets of object instances in an object class.

[26] See the endnote on the Bunge–Wand–Weber (BWW) model, or publications under "Knowledge reuse algebras" in the bibliography.

Box 12 The state of a system is the collective state of the objects it involves

This example describes the state of a hypothetical inventory management system. The example has been deliberately simplified to illustrate the principles involved as simply as possible. The following figure illustrates how objects help organize the bewildering numbers of possible states that a system may assume.

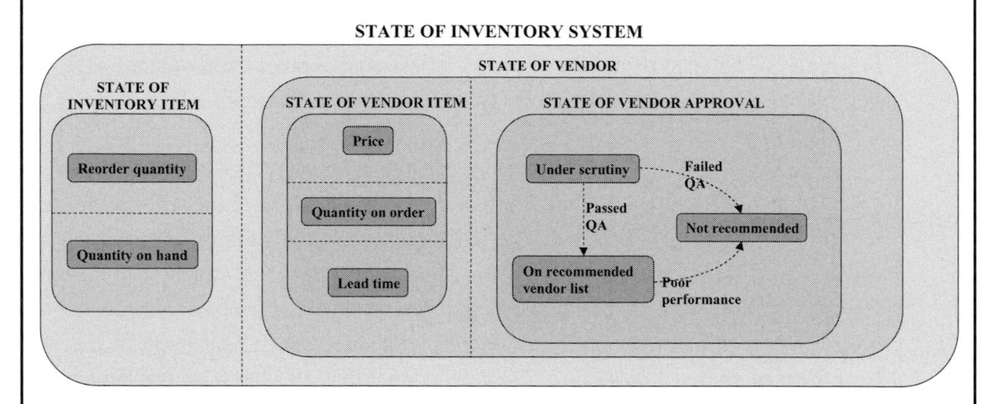

The entire system may be considered to be an object in its own right. The outermost rectangle represents the state of the entire system. It frames the states of the two basic objects the system is assembled from: inventory items and vendors. Inventory items will exist in *some* state at any given moment and so will vendors.

The states of inventory items[27] exist simultaneously with states of the vendors. The combination of the two defines the state of the inventory system at any given point in time. Partitioning the inventory system with a broken line and placing the rectangle representing the state of inventory in one partition and the rectangle representing the state of the vendor in the other implies that each state of the vendor can co-exist with each state of inventory.

Assume, for inventory items, we are only interested in *reorder quantity* and *quantity on hand*. Then the state of an inventory item consists of the values of its two attributes. It is the combination of values of *reorder quantity* and *quantity on hand* that defines the state of inventory item in our model. Both attributes exist simultaneously; if the value of either attribute is changed, the inventory item will change state. One can partition the state of the inventory item with a broken line and place the state of *reorder quantity* in one partition and the *quantity on hand* in the other to show the concept.

[27] The state of inventory is the collection of states of each inventory item. The state chart technique would show this with a box for each item of inventory inside the large inventory box in the figure. Each box inside would be separated from the others by broken lines. If only a few items were involved, this would be easy to draw, but, if many are involved, it becomes difficult. This argument holds for all objects: the utility of state charts is limited in business systems, which usually have large numbers of objects. State charts are useful for real-time engineering systems with fewer variables. The purpose of this book is not to teach state charts; it only uses them to elaborate on state and its ramifications.

The state of a vendor is determined by the state of items supplied, and whether the firm has listed the vendor in its recommended vendors list. Hence the vendor's state is partitioned with a broken line; one partition has the state of the vendor item, while the other has the vendor's state of approval.

The state of a vendor's items is determined by the combination of price, the quantity already on order with that vendor, and the time from order placement to order fulfillment (lead time). Partitioning the vendor item with broken lines and placing each of these constituent states in its proper partition show this state.

In terms of qualifying for the recommended vendors list, the vendor may be in one of three possible states:

1 under scrutiny: the vendor is being examined to see if the requisite qualifications are satisfied;
2 listed in the recommended vendor list after having passed the scrutiny state; or
3 rejected, i.e. is not a recommended vendor

These are mutually exclusive states of vendor approval; hence broken lines do not separate them. The events that may cause these states to change are shown next to the arrows showing the transition from one state to another.

What is an object – really?

Before we forge ahead with our metamodel of knowledge, we need to augment the business definition of "object" in order to establish its properties. To really understand how components of knowledge flow from objects, we must first understand the mathematical concept of a *set* [166], [167], [168], [308].

Box 13 Set membership

A *set* is merely a collection of items or *members*. It is a simple but fundamental concept in mathematics. For example, we can define a set such that all points in the shaded area below are members of one set (set A), whereas points outside the shaded area may be considered members of a different set.

Similarly all persons may be considered to be members of a set, as might the set of glass panes. Object instances are sets of attributes and effects. Object classes[28] are sets of object instances (hence they are sets of sets). Several mathematical operations can be performed on sets to normalize knowledge (see box 19, box 48, [166], [167], [168], [232], [233], [235], and [309]).

Figure 19 shows that an object instance is a *set* of attributes and effects. An object class is a *set* of instances with the same set of attributes and effects. Is this definition complete? Do these assertions carry *all* the information inherent in the *meanings* of object instance and object class? The answer is "no." There is one crucial item of information still missing – that to exist at all, an object (instance) in the real world must exist for a finite period of *time* (see box 14).[29]

Box 14 Object instances must exist for a finite period of time once they are created (often forever after they reach some terminal state)

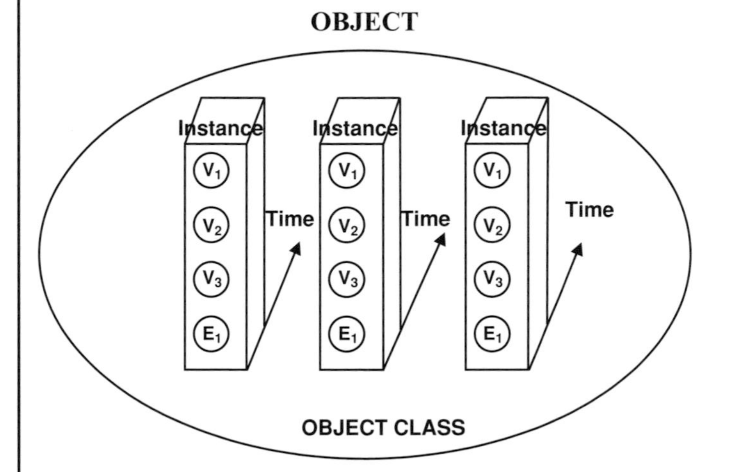

In the figure, items V_1, V_2, and V_3 represent the attributes (and relationships) of the object class, also called variables of the model. Only three are shown. There could be many. Similarly, E_1 represents an effect. Only one is shown, but there could be several in the scope of the model. The arrow labeled "time," perpendicular to the plane of the paper, represents the passage of time.

[28] *Objects are classes*. A mathematical class is a collection of sets that can be unambiguously defined by a property its members share. Classes subsume sets; a set is a class, but all classes are not sets. A category is a class that contains objects, their relationships, and their behavior. See the endnote on the theory of categories or [171], [172], [173], [135], and [185].

[29] The exception. A value in a domain might have existed, and will continue to exist forever (Chapter 4, section 2).

Note that every object instance has a history. This is implied and fundamental to the definition of object. Software designers usually decide how much history should be preserved, for how long, and in what format, files, and tables. There is no separate object called "history," or a "history table," in the metamodel.[30] These tables are technical implementations derived from real-world objects.

We had asserted early on in this book that only discrete change is within the scope of our metamodel. Continuous change was excluded. This implies that we will consider only state changes of the object instance (values of its attributes) in response to discrete events at discrete points in time. This is shown in figure 22. Each slice of an instance represents a different state that holds for a finite time period before another event changes it.

From figure 22, it is clear that we must have some way of identifying each instance of an object so that we can track its history.[31] This is the missing piece of information that Jane will need to satisfy the alien at the Metanesian Zoo at the end of section 3 in Chapter 1. The rule is that *every instance must have a unique identifier that will not change between state transitions – never ever – over its full life history from its birth to death. This identifier cannot have the same value for any other object instance.* It is the **identity** of the object (instance) it describes. Its sole purpose is to distinguish the object (instance) from every other object (instance) as it moves through state space.[32] It records the irreducible fact that a specific instance of an object exists in the real world.[33]

This brings additional requirements into focus – how do we acknowledge the existence of a time slice of an instance of an object? We need a time slice identifier. The design of the time slice identifier is primarily a physical design issue, and we will not dwell on it beyond the basic considerations in the footnote on audit attributes. Some rules of business process automation also flow from the natural structure of a real-world object in box 23. Every state change in figure 22 will naturally be associated with:

1 the time of change;
2 who made the change (the operator as well as process owner); and

[30] Relational databases and logical data models often have distinct "history" entities. A slice perpendicular to the page would be this history table or entity. See the endnote on normalization, or [297] and [304].

[31] All temporal objects must have a history because they must exist in time. However, state changes for some objects may be beyond the scope of a model. This is a design issue; in the metamodel of knowledge these objects *naturally* have histories because the object instance exists in time. However domains are naturally stateless objects – for example, the length domain would never change state. (Even domains like length, time, and mass *have* changed state, and hence have histories in cosmological models. See the books listed in the endnote on the natural zeros for temperature and time.)

[32] Data modelers often look for an attribute that will serve this purpose and call it the "prime key" of a table in a relational database. Prime keys could change as the scope of a model changes. For example, social security numbers could be a prime key if we consider only US citizens and some kinds of foreign workers. However, if the scope of the database expanded to include temporary workers waiting for social security numbers, the prime key must change. These kinds of issues are one reason for the legacy of inflexibility in many information systems. In the object paradigm, the instance identifier identifies the irreducible fact of existence of the object instance.

[33] A null value of the instance identifier implies a non-existent object instance; an "unknown" value implies that the object instance may or may not exist. When an information system has no record of an object instance, it could have either meaning; most current methodologies do not resolve this ambiguity.

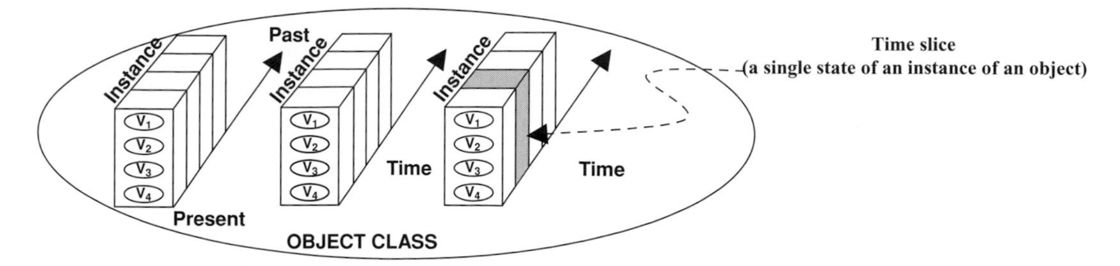

Figure 22 The state of an object changes in response to discrete events

3 the facility (automated or manual system) that was used to make the change (immediate reason, not root cause – the causal chain can always be traced to root causes and sources of change if all these audit attributes are maintained).

We will call these items the "audit attributes" of each object.[34]

Instances of objects may be sets of object instances themselves. (Think of the set as a bag of objects. The bag itself is an instance of the class of bags – see "Object class versus object instance" earlier in this chapter.) For example, consider the set of outstanding claims in an insurance firm. The firm might be interested in the total value of unsettled claims against it. Each unsettled insurance claim is an instance of an object (*unsettled* is a state of an insurance claim). The *unsettled claim* object *class* is a *set* of individual instances of unsettled insurance claims. These instances are not the repository of the total value of outstanding claims, rather they are the object *class* unsettled claim, the *set*, which is the repository of this information. This total value is an attribute, or state of the *collection* of unsettled claims. This collection of unsettled claims is an instance of a package, or aggregation, of individual unsettled claims that must have its own instance identifier. Its state will change as new claims are raised, and some unsettled claims are settled. This kind of object instance occurs frequently in business, and may have complex rules related to the overall financial exposure of the *aggregate*. Each instance of an aggregate object must have its unique identifier and history too. Module V on our website discusses aggregate objects. This definition still does not solve a key problem in forging reusable classes: in the real world, many object instances may share some, but not all, behaviors. Worse, any given object instance might share different kinds of behavior with instances of different kinds of objects.

For example, both people and organizations may be plaintiffs or defendants in a court of law and both may be earning credentials and awards of various kinds. On the other

[34] One way of physically identifying time slices of an instance identifier is by appending the time of state change to the instance identifier. In systems that operate across time zones, one needs to standardize *which* time stamp to append; if the information system is run on physically distributed processors and the system time is used, the relevant processor must also be identified. Then this processor's id would become a part of the time slice identifier. If collaborative business processes are involved in a physically distributed system, different collaborating actors could trigger state changes simultaneously across the network. These technical issues must be considered to ensure consistent, accurate, and timely performance of information systems.

Box 15 Mathematical definition of a business object

- A (business) object class is a collection of object instances with common properties (attributes, relationships, and effects – i.e. behavior) that define a category of persons, places, processes, events, things, or concepts that are of interest to the business.
- A (business) object instance is the set of object states that possess the same identifier.
- A (business) object instance at any given moment in time is a set of attribute values (and relationships), which collectively define the state of the set, of which one attribute is an identifier that stays constant through the history of all state changes of the set.
- The identifier may only assume a null (does not exist) value if a constraint bars its existence (physically assigning a null value to the identifier or not permitting such object in storage is a technical design issue).
- The identifier will always map to a nominal domain because it *only* expresses the irreducible fact that an object exists. (The first of the seven rules in our body of knowledge about glass panes at the beginning of this chapter was an example of a rule like this. It only said that a class of objects called glass panes existed.)

hand, both people and money may be resources for a project. From yet another perspective organizations, people, elevators, and buildings may all hold certificates, and so on *ad-infinitum* and *ad-nauseum*. There may be a bewildering number of ways we could group any given object instance with other instances of objects. How do we group these objects into object classes to maximize reusability? Are we back to square one? Can we really defeat the dark forces of chaos? We can, after we grasp the concept of state, and the abstract fields of meaning that objects traverse in *state space*. We will consider multi-perspectives,[35] [15], [21], [23] in section 4, and the concept of inheritance[36] in section 3 to address this problem.

> Truth was in the beginning
> Truth was before the Aeons
> Truth is here now
> And Truth will be hereafter
> (*Guru Nanak*, the first Guru of Sikhism)

State space

The set of all possible states of an object is called its state space.[37]

[35] The Bunge–Wand–Weber model in the endnotes establishes a mathematically sound framework for describing abstractions and testing the completeness of a set of modeling constructs, a methodology, or language. The endnote on multiperspective modeling addresses multiple classifications for the same object depending on the perspective of the problem.

[36] See the endnotes on polymorphism and inheritance or refer to [239], [90], [91], [328], and [329].

[37] State space is a mathematical *topos*. Topoii may or may not be sets. It is a fine mathematical distinction. Readers interested in a more precise mathematical definition of state space, and why *set* may be an imprecise way of describing state space, are invited to refer to section 6.1 of [178].

Box 16 Domains and measures of distance

Domains extend the traditional concept of distance between objects. The length, breadth, thickness, or distance between physical objects is determined by the distance in space, measured in the length domain. An object's age or the time interval between instantaneous events is another kind of distance; it is measured in the time domain. Similarly, other kinds of domains extend the concept of distance in other ways. In general, closeness between objects refers to how near they are in terms of its attributes mapped to various domains such as temperature, speed, or energy. Even nominal domains measure a kind of distance[38] – whether the gap between two or more objects is zero (they are in the same category in that domain) or not.

When a single attribute is involved, this concept of distance is quite straightforward, but when several attributes are involved, we must consider the object's *state space*.[39]

To understand what state space really means, and why it is important for normalizing irreducible facts, let us take a simple example. Consider only two attributes of the glass sheet: its weight and its thickness. If there were no restrictions on thickness or weight of glass sheets, the state space would be the entire plane bounded by the two axes in figure 23.

An object's location in state space at any given moment is its actual state at that instant.[40] In this example, the location of a 1/8 inch thick glass sheet that weighs 1 1/2 pounds is shown as a point in the state space of figure 23.

We could impose a rule that we will only consider glass sheets between 1/8 and 1/2 inches thick, which weigh between 1 and 2 pounds. The state space would then be limited to only a region of the plane in figure 23. The shaded area in figure 23 would represent this truncated state space. It is sometimes called the object's "lawful" state space.[41]

State spaces are important in helping us to classify objects in an imperfect world and to understand irreducible facts that are constraints of various kinds. These constraints restrict the permitted states of objects. Also, objects trace a trajectory through state space as they change state; state spaces help us to understand how irreducible facts that involve histories can be incorporated into the metamodel of knowledge.

Any constraints on the values of the thickness and the weight of the glass pane would constrain the state space of the glass pane to a region of the two-dimensional plane in figure 23. They do not have to be the kind of closed area bounded by a well-defined perimeter, shown in the figure 23.

For example, in figure 23, the glass sheet's lawful state space would have been an open-ended horizontal plane if only the thickness, and not the weight of the sheet, was

[38] This distance is called a metric. See the endnote on *metric spaces*.

[39] See the endnote on the topic. We will address the generalization of distance in state space in Chapter 4.

[40] The endnotes on metric spaces and items in the section on metric spaces in the bibliography have more rigorous mathematical explanations of state space.

[41] See the endnote on the Bunge–Wand–Weber model: The *conceivable state space* of the object describes its unconstrained state space. The lawful state space of an object is a region of state space that the object has permission to occupy. It is carved out of conceivable state space by *all* constraints on the object acting in concert.

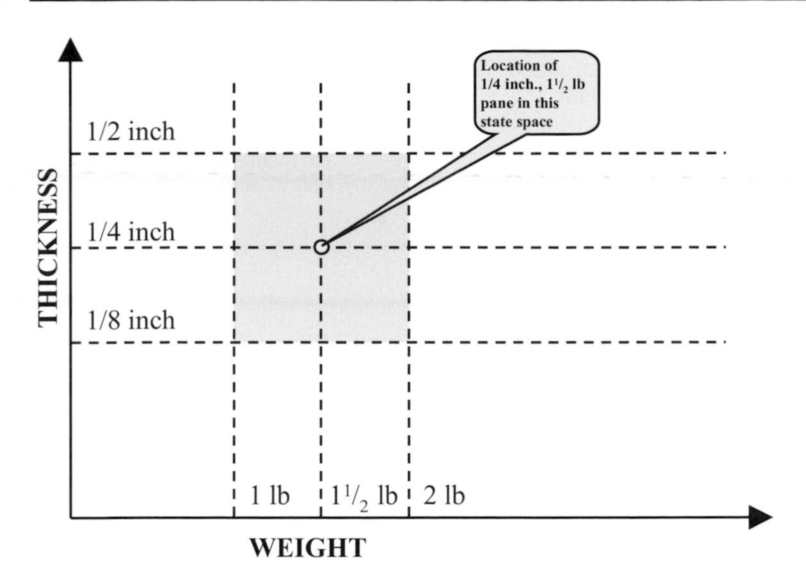

Figure 23 Example of two-dimensional state space

constrained. It might have even been a straight or curved line(s) if there were rule(s) that let us compute the weight of the sheet from its thickness (or vice versa). Figure 25 has an example of this kind of state space.

Indeed, complex rules might be such that the lawful state space could consist of several joint or disjoint shapes and edges taken in any combination. For example, if we constrained the thickness of the glass sheet so that we only considered sheets between 1/8 to 1/2 inch thick, or 1 to 2 inches thick, our lawful state space for glass sheets would consist of two disjoint regions in the plane of figure 23. (Box 17 has examples of disjoint state spaces. Interested readers may like to try this exercise: if we were interested in the number of pieces of each shattered glass pane and the average weight of each piece, what would the shape of the state space be?)

Instead of two attributes, if three attributes were in scope, the state space would have been a three-dimensional volume. For example, if a person's age, height, and weight were in scope, the state space of person (the object instance) would be the three-dimensional volume shown in figure 24.

Just as constraints in the two-dimensional case (figure 23) restricted the state space of the glass pane, constraints on values of the attributes in figure 24 would confine the lawful state space for *person* to bounded or unbounded regions within the volume shown in figure 24. These regions could be disjoint (or not) volumes, surfaces (not necessarily plane surfaces), lines (not necessarily straight lines), or points, or any combination of these shapes.

A *subspace* is a state space that consists of a subset of attributes of a state space. The space from which attributes were selected is called the *super space*. Naturally, a subspace will have fewer dimensions than its super space. Indeed, any cross section of state space will be its subspace. Subspaces of subspaces are also considered as subspaces of the original super space. A plane is a two-dimensional slice, and a subspace of a three-dimensional volume,

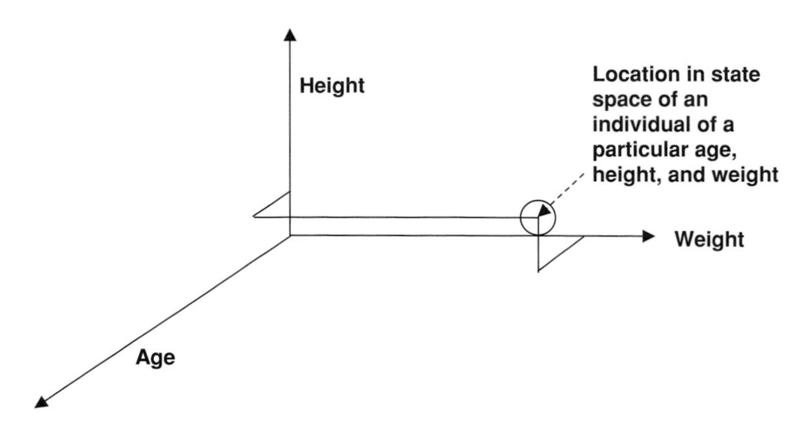

Figure 24 Example of a three dimensional state space

whereas a line is a one-dimensional slice of a plane, and a subspace of both the plane and the volume.

When more than three attributes are involved, it is not possible to draw the state space in our three-dimensional world, or even imagine its shape easily. These would be higher-dimensional spaces (mathematicians routinely deal with such spaces) that we can only understand or try to visualize as analogs of two- and three-dimensional state spaces.[42] Chapter 4, section 1 has an example under *arrays*, of how higher-dimensional state spaces may be visualized and sliced into lower-dimensional subspaces.

Knowledge in state space

Knowledge is manifested in state space. State space is also a component that can be reused to create new configurations of knowledge.

Each object is a collection – a bag so to speak – of attributes, relationships, and effects naturally packaged together. Thus, each object can be a reusable component. The states and behavior of these natural objects may potentially be used in several business processes.

These attributes and behaviors are attached to the instance identifier. (Think of the instance identifier as the label of the "bag.") Thus, each instance is a structure (called a "tuple" by mathematicians) in which each attribute, relationship, and effect of event is attached to its instance identifier. This structure will help us partly, but not fully, to normalize knowledge. It helps us to normalize knowledge when several business processes reuse the natural behavior of natural objects. Then they can share the information about an object stored in a repository and do not each have to replicate the same data and effects. On the other hand, several similar objects may share some kinds of common behavior (see "What is an object – really" Chapter 00, section 1). Sharing the same object will not normalize this kind of shared behavior. Corresponding attributes and behaviors will then have to be packaged into reusable

[42] [255], [260], and [268] describe higher-dimensional spaces. [273] and [275] describe dimensionality of patterns in space.

Box 17 State spaces with qualitative attributes

In our examples we have shown only state spaces for attributes that map to quantitative domains (ratio or difference scaled domains). What would happen to the state space if some (or all) of the attributes involved were ordinally or nominally scaled? The following figure shows how this can happen.

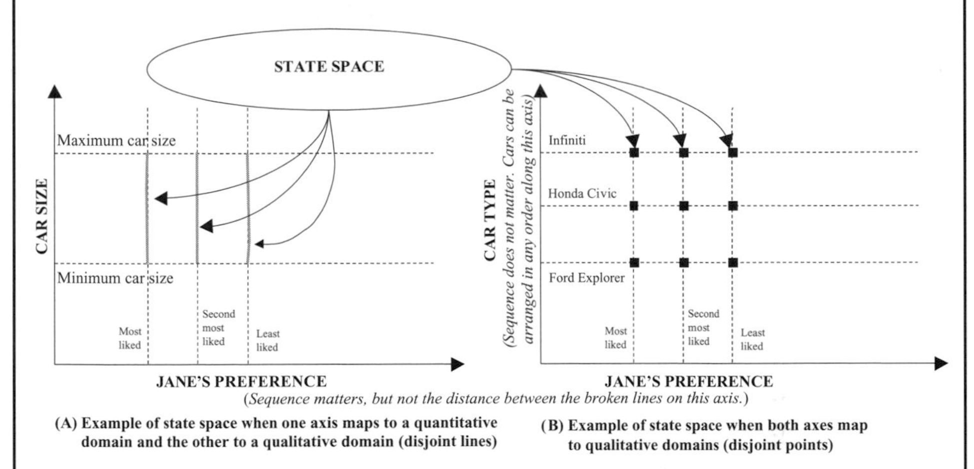

(A) Example of state space when one axis maps to a quantitative domain and the other to a qualitative domain (disjoint lines)

(B) Example of state space when both axes map to qualitative domains (disjoint points)

Examples of state spaces that map to qualitative domains

If the attribute were ordinally scaled, we could still show its value on an axis in state space, but the state space would be restricted to a set of discrete points on the axis. The points would be arranged in increasing order (for example, ranking of Jane's color preference for cars in section 3 of Chapter 1), but the degree of physical separation between points would not carry any meaning.

If the attribute were nominally scaled, the state space would be restricted to discrete points on the corresponding axis, one for each category. However, neither the sequence in which points are arranged on the axis, nor the degree of physical separation between points would carry any meaning. The only material consideration would be whether two or more instances of the object (instance) are in the same category or not.

sets of their own. In terms of state space, it implies one or more axes may be chosen to create reusable state spaces, which are called facets or subspaces of the original state space.[43] Let us therefore understand how knowledge is manifested in state space and its many facets.

Each axis, i.e., dimension of an object's state space, is an irreducible fact – that the attribute (or relationship) *exists*. The state space describes a *configuration* of irreducible facts: the collection of irreducible facts that each attribute (and relationship) exists. There are two ways we can build on this collection of facts:

1 add new attributes (or relationships) to the collection;
2 constrain permitted values of attributes (or relationships).

[43] See the endnote on multiperspective modeling, [21], and [23].

Adding new attributes: take figure 24. Let us assume that only an individual's height and weight were in the scope of the model to begin with. The state space of a person would then be the plane of the paper, bounded by the height and weight axes. The movement of individuals on this plane in response to events will represent their behavior. Now if we wish to create a new model – one that includes the individual's age, height, and weight – we would attach the age axis to the plane as shown in figure 24 to create the new state space for an individual. (This is similar to the loose collection of facts described for figure 10 in the example of section 5, Chapter 1.) Their movement in this volume will then represent the behavior of individuals in the new model.

Now, if we need a second model that involves gender, in place of age, we could reuse the height–weight plane (subspace) of the three-dimensional state space of figure 24, and attach a gender axis to it (instead of age) to create our new state space. Therefore, state spaces (and their subspaces) can be reusable components that represent *reusable configurations*, or subassemblies, of knowledge.[44]

Constraining permitted values: irreducible facts about permitted values will constrain an object's lawful state space, as described in figures 22 and box 17, to one or more regions as discussed earlier. Constraints can be complex and involve interactions between several attributes (and relationships). For example, we may have two different models, which involve only individuals' heights and weights. The models differ, in that they attach different constraints to permitted heights and weights. We could use the plane in figure 24, bounded by height and weight in both models, but we would need to carve it up into different regions and to attach different constraints to each. Thus, in this situation too, state space can serve as a reusable component that represents a *configuration*, or reusable subassembly, of knowledge. (This is similar to the structure in figure 11 of the example in Chapter 1, section 5, where the weight domain was constrained.)

Mixing both kinds of constraints: indeed, we could reuse these structures to assemble even more complex mixed structures. For example, we might say that we will attach the age axis (and special age dependent behavior) of figure 24 only for children, i.e. individuals less than 18 years old. Now we have the third dimension attached only to a region of the height–weight plane. The height–weight state space (or facet) may be reused in several ways. The example, under *states of behavior*, of how some attributes of agreement may be null under certain conditions is another example of how state spaces may add or lose dimensions for some, but not all, instances of a class of objects. (Null values imply that the attribute does not exist and hence corresponding axes of state space do not exist for those instances of the object.) We will revisit the overall implication under the discussion on inheritance and subtyping.

[44] One way of *physically* representing this common facet might be to define an object that has all attributes (height, weight, age, and gender in our example), but to force those attributes that are not common to all scopes (age and gender in our example) to have null values. After all, a null value implies "does not exist"; hence, it is one way, albeit a rather clumsy way, of saying that the attribute does not exist in the common state space. A better approach might be to let the object have all attributes that any scope needs, so that those who need it may use it, whereas those who do not are not really affected. The attribute's value will be "unknown" for them. This would provide for scopes shifts and new behaviors. Providing for unknown attributes and effects will facilitate even more flexibility. Although these designs might be flexible, they may adversely impact computer performance.

The current discussion serves to emphasize that modeling real world behavior can be exceedingly complex. Indeed, it is this complexity that leads to tangled inheritance and chaotically replicated knowledge scattered through our processes. This has been the main stumbling block to nimble information systems and a major limitation of conventional object technology. To transcend this barrier, we will have to solve the problem of multiple perspectives *and* the problem of automatic recognition and inheritance of common behavior. In order to do so, we must first understand not only how knowledge is represented in state space, but also how knowledge is represented in the *movement* of objects through state space.

Moving through state space

Each adjacent slice of an individual object instance in figure 22 occupies a different point in the object's state space. If we joined these discrete points, it would give us the trajectory of the object (instance) through its state space. (If we had not restricted our scope to discrete change, the path would be a line instead of a set of discrete points.)

The set of *all possible paths* through state space will be the collection of effects of events in the scope of the model. That too is a configuration of knowledge. The *specific* path an object instance actually takes represents its history. Why is this important? It is important because behavior often depends on history in the real world. Box 18 describes how this can happen (the trajectory described in figure 25 is an example of this). Module V, section 3 (on our website) revisits the significance of history.

To illustrate these concepts, let us take a simplified example of the behavior of an imaginary firm, Shenanigan's Services Inc. Assume that we are interested in only Shenanigan's income and the total amount of money the firm has borrowed. (*Shenanigan's Services* is an instance of an object, whereas income and borrowing are its attributes.) We discover that income rises in step with borrowed money at first because the *Shenanigan's Services* uses its borrowings to scale up its operations.

However, beyond a point, interest payments start eating into its income, which falls steeply. When the firm reduces its borrowing, income starts rising again. The company then increases its borrowing to finance growth. (The left most loop in figure 25(a) represents this behavior.) This time, its income keeps rising in step with borrowing.

However, after a small but significant spurt in income, interest starts eating into income again. The firm repeats the strategy it had employed earlier successfully, and starts growing again (shown by the upper loop in figure 25(a)).

Figure 25 is a graph of the firm's income versus its borrowing. It is an example of the firm's trajectory through state space. The firm moves along this curved and looping line with the flow of time.

If we had discovered a formula (an irreducible fact) that related the firm's income to its borrowing, then, at every point in time, the firm's income would have been determined by its level of borrowing alone (and vice versa). The state space of the firm would then become a line that traces this relationship between income and borrowing. The firm's trajectory through state space would trace this line as its income and borrowing changed over time.

The concept of movement and speed along its trajectory in state space can be replaced with only the concept of position if we included time as the third axis. For example, in

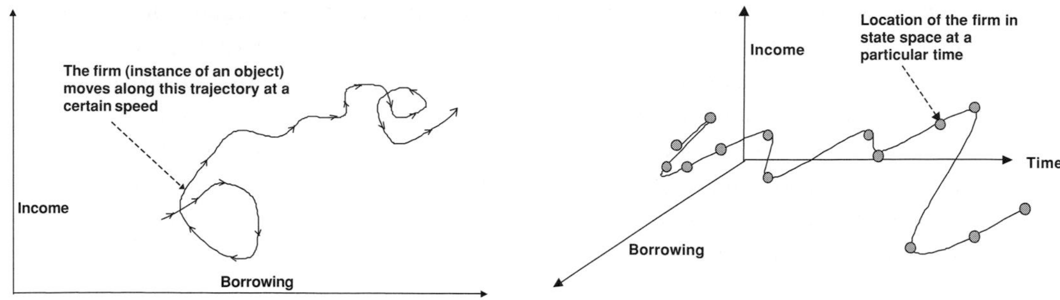

(a) An object will move along a trajectory in state space as its state changes with the passage of time

(b) The object's trajectory can be reinterpreted as a region in state space (a line in this case) when the time axis is added to its state space

(When only discrete changes are considered, the region consists of a sequence of discrete points (●) on the trajectory.)

Figure 25 Example of trajectory through state space

figure 25(b), as we move forward along the time axis, each point represents the firm's state at that moment in time (the amount of borrowing and the firm's income at that point in time). Effects at each point in time have changed the state of the firm to the next point in the figure. Business rules will dictate whether subsequent effects and guard conditions will depend on the shape of this trajectory (see box 18).

Box 18 Chaotic behavior

Points where the state space crosses itself in figure 25(a), when the line loops back on itself, are unstable (chaotic) points. The path can branch, and both mutually exclusive branches are legal (allowed) paths. The object (instance) may follow any one of available paths as it moves through state space. The object's actual trajectory as it crosses these points of intersection (which branch it actually takes at the intersection, since the state space permits either) would depend on pure chance and tiny differences in the object's history and velocity as it travels through state space.[45] This is called chaotic behavior. The theory of chaos analyses this kind of behavior and readers may refer to [292] and [323] in the bibliography.

Spontaneous state change

At the beginning of this chapter we discussed how events [166] have effects that change states objects. Events are also objects (Chapter 1). We have seen how objects themselves can be collections of other objects. It follows that some objects in the collection might be events. The overall state of the aggregate is determined by the states of its components. These internal events may change the state of the aggregate object (or objects inside it, which amounts to the same thing). These internal events might not be in the scope of the model,

[45] Chaotic behavior occurs because even infinitesimally small differences in state and history may change the trajectory of an object in state space. [220], [222], and [231] show that these differences may be infinitesimally small – small enough to be considered zero – and still display chaotic behavior.

or even be visible to an observer, who will therefore see the aggregate object changing state spontaneously. For example, in figure 14, v_4 and v_6 might be attributes of an object, which would continually and spontaneously keep changing state (Chapter 1, section 6). There are many examples of spontaneous state changes in the physical world. For example, the filament of a light bulb might spontaneously burn out and render the bulb inoperative. Several species of fish change gender spontaneously (see the endnote on gender change.) Thus, when we think of moving through state space, we must remember that this movement may not always be in response to external effects. It could be spontaneous movement.

Lost in space: the curse of change

The state of a system at any point in time is the set of values of every attribute (and relationship) of every object in it (see box 12). Large and complex business systems can have large numbers of objects with many attributes and relationships that traverse complex paths through complicated state spaces – sometimes in response to events, and sometimes spontaneously. The number of possible states and trajectories can be mind boggling, and, unless we have robust criteria for grouping behavior, we can easily lose our way.

When the scope of a large and complex system changes, we must involve new behavior, objects, and rules. State spaces, initial conditions, and trajectories can all change simultaneously. Seasoned analysts and systems developers understand the challenge through bitter experience. In Chapter 1, we saw how easy it is to get lost in state space without objects to guide us. Without objects to classify behavior, the system will have too many states, and developers will have to keep track of more rules, interactions, and unintended side effects than is humanly possible.

If rules are not normalized, changing a rule in one place will not guarantee that the new behavior will be reflected everywhere; if we do not know every impact, the system will deviate from its intended trajectory and often take a totally different and strange trajectory that will be hard to diagnose. Thus, the impact of a single change, forgotten because we could not keep track of all places where a rule might be replicated, might cascade through the system as its path in state space deviates chaotically from the intended trajectory. The larger the system and the more complex its rules, the greater will be the risk. We must normalize knowledge to manage this risk.

Section 5 of Chapter 1 had an example of how knowledge was automatically normalized through inheritance. This property of objects, behavior inherited from other objects, is the key to order in the chaotic world of complex systems. Inheritance is the magic bullet that organizes knowledge into reusable components based on common behavior. Therefore let us understand how inheritance happens.[46]

3 Inheriting behavior – subtypes, supertypes, and partitioning of objects

We have seen how objects help to normalize real-world behavior and can be reused in many business processes. For example, employees may be resources for a project, and also individuals who earn paychecks and benefits. "Employee," the object class, may therefore

[46] The endnote on inheritance describes several kinds of inheritance. See [239] and [328].

be (re)used for both purposes, and serves to normalize information about the states and behavior common to both processes such as cost, productivity, length of employment, and job function. Similarly, customers are individuals who buy products, and also return defective products. Thus, "customer," the object class, may be (re)used for both purposes and serves to normalize information about the states and behavior common to both processes.

We also understood that object classes by themselves are necessary, but not adequate to normalize common behaviors of similar objects. For example, both customers and employees are individuals who age, have a gender, have credit ratings, are known by names and social security numbers etc. The "employee" and "customer" object classes will not normalize this behavior. Instead it will be replicated in both. As such, if an individual's name or credit behavior needed correction, we would need to remember to change it for each object class separately.

In order to normalize behavior common to similar objects, we must create more generalized object classes that could serve as the repository of their common meaning. For example, aging, gender, credit rating, names, and social security numbers are common to all persons in the US. Persons may be both employees and customers. "Person," the object class, serves to normalize behavior common to people, such as aging and gender regardless of whether the person is an employee, customer, or both. If we acknowledge that the set of customers and the set of employees are subsets of the set of persons, we know that all behavior common to persons will also be the behavior of employees and customers (see definition of subsets in box 19. Box 20 deals with inheritance in a more rigorous manner.) For this reason, *employee* and *customer* object classes are said to *inherit* the behavior of *person*. *Person* is the repository of their common behavior.[47] This is why we need inheritance to normalize the common behavior of similar objects. (Employees and customers may have other behaviors as well, that are specific to each. For example, employees may be promoted, which does not apply to customers, and customers may be offered product warranties, which does not apply to employees.)

Objects may inherit their behavior from more than one object class. This happens when they are subsets of several sets simultaneously. For example, in box 19, sets A and B overlap. This overlapping set is their intersection. Assume set A is the set of products in inventory. Inventory items will be characterized by attributes such as quantity on hand, quantity on order, and quantity issued from stocks. Assume set B is the set of toxins. It will have attributes such as toxicity, precautions, instructions for emergencies, and OSHA reporting obligations. The intersection of the two sets will be the set of toxic inventory and will inherit the behavior of both inventory items (set A) and toxins (set B). Toxic inventory items may have specific behaviors in addition to inherited behavior. For example, the firm might insist that toxic substances must be isolated in a special warehouse. When an object class inherits behavior from two or more object classes, it is called multiple inheritance (see box 20 for a more rigorous discussion of multiple inheritance.)

[47] The object class *person* is a *superset* of both the set of employees and the set of customers: The set of employees are those persons who have an employment relationship with another object class: *employer*. The set of customers are those persons who have a purchasing relationship with the *product* object class. Thus both *employee* and *customer* object classes are subsets of person.

Box 19 Set operations (on our website)

Box 19 describes set intersection, set difference, set union, subsets, proper subsets, set negation, set equality the Cartesian product of sets, the concept of a tuple, a list, the empty (or null) set, and the power set.

Supertypes and subtypes

Inheritance does not involve any logical *flow* or physical *transmission* of data. Inheritance just *is*. It is the *nature* of the object, and springs from the logic of set operations and the *meaning* of real-world objects.[48] The class (or classes) that an object inherits behavior from is called its *parent class*(es) or *supertypes*. The object classes that inherit behavior are called *child classes* or *subtypes*.

The supertypes in our examples so far have all been physical objects such as people, inventory items, and toxic materials. However, common behavior not only springs from the laws of nature, but also from the laws of man – laws and common practices in business, such as ownership, trade, and legal processes. State spaces created by these practices may be reusable components (see "Knowledge in state space"), but are sometimes difficult to identify as such, because they are more abstract and less tangible than physical objects.[49]

For example, people and organizations are two very different kinds of concepts that share common behavior because the law considers some kinds of organizations to be surrogate persons. Both persons and firms may buy and sell products and services, both may be parties in legal processes, make agreements of various kinds, have bank accounts, own assets, employ or be employed by other people or organizations.

An abstract object, person/organization, captures this common behavior. Persons and organizations both inherit their shared behavior and relationships from person/organization. Each adds its own specific behavior to behavior inherited from person/organization. For example, persons get married and divorced, whereas organizations may be bought, sold, and be restructured. The state space of person/organization is a common component from which state spaces of both people and organizations can be developed by adding behavior special

[48] Section 2.25 of [328] describes inheritance in detail. Section 2.25.5 of [328] has different kinds of inheritance. Inheritance and polymorphism (which we will discuss later) emerge from the mathematics of partial order and λ-calculus (lambda calculus). [217], [239], and [217] describe partial order succinctly, and [239] describes both untyped and typed λ-calculus and how they lead to inheritance and polymorphism.

[49] Identifying abstract supersets has been a major problem in building robust enterprise or cross-enterprise object models. These generalizations are usually subjective and unstable. They tend to change as scopes, perspectives, and priorities shift, or as new business rules are considered. The universal perspective addresses this problem. Also see the endnote on multiperspective modeling and bibliography items on facet and multiperspective modeling.

Box 20 Inheritance, state space, and polymorphism

Remember that state space is just another name for the set of attributes and relationships of an object, and facets, or its subspaces, are subsets of this set.[50]

In terms of the set operations in box 19, the state space is a set derived from the Cartesian product of all attributes (and relationships) of the object class it belongs to. As such, if there are "n" attributes (including relationships), each point in state space is an n-tuple. Each element of the n-tuple is a possible value of an attribute (or relationship).

The path of an individual object instance through state space is its history of state changes, and the set of all possible paths is another name for the set of all possible histories, which is the same as the set of all possible effects.

Strictly speaking, inheritance involves shared *facets* or *subspaces* of state space (shared attributes, relationships, and constraints) and shared sets of permitted paths (shared effects) in these facets and subspaces, not subsets of object *classes*. Multiple inheritance emerges from a facet or subspace of the state space of the child object, created by the union of attributes and effects of each parent object.

When an object class is a true subset of the parent object class, such as the set of small firms is a subset of all firms, the subset will naturally reside in a region of the state space of the parent class (as discussed under state space) and must share facets (i.e. attributes) of its state space, as well as of some or all permitted trajectories through this shared facet. Thus, it will always inherit these attributes and effects from its parent class. This kind of inheritance, in which the state space of the subtype is constrained to a region *inside* the lawful state space of the supertype, is called *restriction inheritance* (see the endnotes on kinds of inheritance). It is a special kind of extension inheritance (described below) in which the subtype has no additional attributes, and its state space is not augmented with additional dimensions. The subtype's state space is merely constrained to a region inside that of the supertype. For example, the annual revenue of a firm is an attribute of an object class called *firm*, and that of a small firm may, by definition, have a ceiling.

Even when the state space of a subset is extended into additional dimensions by including behavior specific to the subtype (described earlier for the set of employees, a subset of the set of all persons), the state space of the parent object must be a subspace of the extended state space of the child object (i.e. the set of attributes of the parent object will be a subset of the set of attributes of the child object), and the child object may share some or all possible paths through this subspace with its parent. Thus, it will inherit these shared behaviors from its parent. This is called *extension inheritance*.[51]

[50] The number of possible sets of common attributes and effects may be unmanageably large, hence the number of possible ways an object can be generalized may be many. For example, both people and publications age; both apples and cars have colors. Does this mean we need common parents for these very different object classes? No! Similarly there may be many combinations of common attributes and hence an unmanageable number of ways in which subtypes can be defined from supertypes. The solution to this problem will be discussed in section 4. Mathematically, this involves the cardinality of the power set of the set of common attributes. The power set of any given set is the set of all possible subsets of the set. The cardinality of a set is a measure of its size, and may be infinite. See [170], [202], [203], and [206].

[51] See the endnote on kinds of inheritance or [328]. Extension inheritance flows from supertypes that are the *intersection* common properties of their subtypes.

Subtypes are object classes, as are supertypes. The subtype object class is the repository of *unshared*, or special, irreducible facts about similar object instances. Supertypes, on the other hand, normalize shared irreducible facts. Instances of a subtype are said to *inherit* shared facts from their supertype and add special facts germane to the subtype via the set union operation. Their properties are the *union* of the set of shared properties of the supertype and the set of special properties of the subtype. An instance of a subtype must naturally be an instance of the subtype (but not necessarily vice versa.) Subtypes are object classes that are related to their supertypes (also object classes) with a special *subtyping relationship*.[52] What irreducible fact does this relationship normalize – after all shared facts are normalized by supertypes and special facts by subtypes, so what is left? The subtyping relationship asserts that a subtype *exists*, and must therefore inherit from the supertype it connects with.

Relationships are objects too, and the algebra of objects will apply to the subtyping relationship equally. This is important to the metamodel of knowledge – see Module V on our website.

to each.[53] This is called extension inheritance because the subtype inherits all behavior from the supertype and then extends the repertoire of behavior by augmenting it with the subtype's special behavior (as described under "knowledge in state space" and box 20).

This is how person/organization normalizes knowledge – with extension inheritance. Person/organization is an abstraction that emerges from shared facets of state space. Indeed, it is not a traditional concrete object, rather it is a container of meaning shared in state space.

Object partitions and role modeling

In real life, objects may play many different roles simultaneously. For example, you, the reader of this book, might *simultaneously* be a spouse, a process engineer, a parent, and a business person. Recognizing multiple simultaneous roles of an object is key to understanding and representing real-world behavior in real world business processes. Object partitions are a powerful tool towards this end.

In the example above, the set of customers are those persons who have a purchasing relationship with another object class: *product* (the set of products). Therefore, *customer* is a subtype, or *role*, of *person*. All subsets must necessarily be based on some *criteria* for partitioning the parent set (in this example the criterion is the existence of a purchasing relationship with *product*). These criteria reside in the *partition*. A set may be partitioned in as many ways as needed to define business rules. Figure 26 describes how objects can be partitioned to show several simultaneous roles.

[52] See box 22.

[53] The person/organization object class is not the union of the sets corresponding to person and organization object classes (members of which are mathematical *tuples*). Rather, person/organization's state space contains the common *properties* of person and organization, and hence is the intersection of corresponding sets of *properties*, not tuples. Thus extension inheritance flows from supertypes that are the *intersection* common properties of their subtypes.

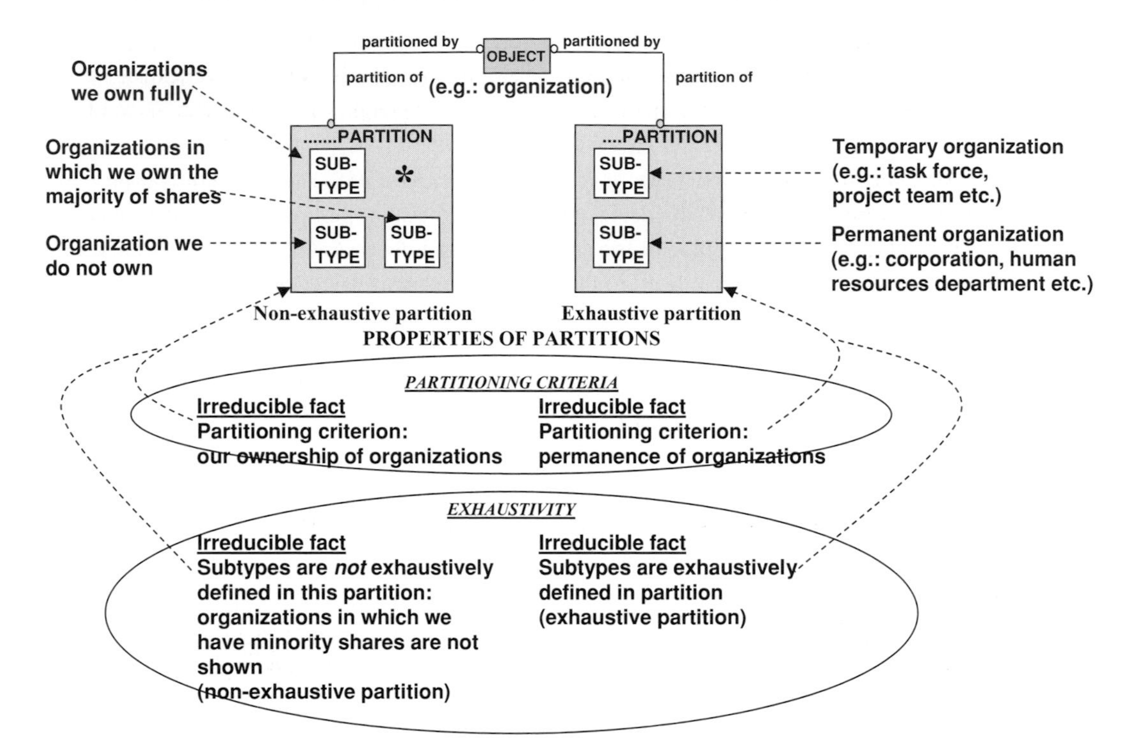

Figure 26 Example of object partitions

A specific object instance may concurrently exist in several subtypes located in different partitions. For example, in figure 26, an object class, organization, has been partitioned in two ways: the partition on the left is based on the ownership of the organization, whereas the partition on the right is based on the expected life span of the organization.

A specific instance of an organization may be a permanent organization, as well as a firm we do not own. It belongs to the class of organizations we do not own, a subtype in the partition on the left side of figure 26; and it simultaneously belongs to the class of permanent organizations, a subtype of organization in the partition on the right. Thus it plays both roles at the same time.

Organizations in both partitions will share common behavior like structure, reporting hierarchies, reorganization, and charters. However, organizations that are not owned may have special behavior like being candidates for acquisition. Similarly, temporary organizations may have dates or events when they will be dissolved, whereas permanent organizations may not. An instance of an organization that plays several simultaneous roles inherits the behavior of each role it plays.[54]

[54] If we used relational database technology to implement partitions, each object might be a table and each partition could be an independent state indicator. The subtypes within a partition would correspond to valid values of corresponding state indicators. Partitioning rules, state space constraints, and specific attributes of subtypes would all translate to complex validation criteria based on values of state indicators. Effects could be implemented as program code or as database stored procedures.

Properties of partitions

Subtypes *within* the same partition are always mutually exclusive subsets of the object class being partitioned. *Partitioning criteria* divide the subtypes into mutually exclusive object classes. For example, an organization we own fully obviously cannot be an organization we do not own. On the other hand, we have just seen how subsets in *different* partitions permit an object instance to play multiple roles simultaneously.

All subtypes within a partition may or may not cover all possible instances of the parent object class being partitioned. This property of partitions is called *exhaustivity* and it happens because some kinds of behavior may not be in the scope of the model. For example, in figure 26, organizations in which we have minority shares have been omitted in the partition on the left. Thus, object classes in this partition do not include all possible object instances in the object class being partitioned. It does not include organizations we do not own.[55] Such partitions are called *non-exhaustive* partitions.

Subtypes in other partitions may include all possible object instances in the set being partitioned.[56] Such partitions are called *exhaustive* partitions. The partition on the right of figure 26 is an example of an exhaustive partition.

There is one other subtle, and somewhat arcane, property inherent in partitions – the property of *exclusivity* (versus inclusivity – the absence of exclusivity). The partitions we have discussed so far assert that subtypes exist. They are *inclusive* partitions. We could, by the same token, specify what *cannot exist*, and put mutually exclusive subtypes that *cannot exist* in an exclusion partition.

Yes, an exclusion partition is an arcane concept, but consider its utility in normalizing knowledge. Sometimes we define an object class by a process of elimination; we define it by what it is *not*. For example, the Department of Motor Vehicles in New Jersey defines class A, B, and C vehicles based on various properties, and then rules that class D vehicles are those motor vehicles that are not A, B, or C. Thus class D vehicle would be a subtype in an *exclusion partition* of an object that covers class A, B, and C motor vehicles. The subtyping criteria for such a partition would say what *cannot be*: that subtypes in this partition *do not* have the properties that make a vehicle class A, B, or C.

Exclusion is the foundation of *variation inheritance* and "*unaffecting*," i.e. excluding specific properties of an object from its subtypes (see box 21 and the endnote on kinds of inheritance). Thus, exclusion partitions are the foundation of variation inheritance.

Exclusive partitions can also be used to assert what parts of an object's state space are *not permitted*. For example, take the case of non-negotiable agreements. They are a subclass of agreements that cannot be negotiated, and hence can never be in a state called "under negotiation." Non-negotiable agreements may thus be a subtype in an exclusion partition – one that excludes an effect called "negotiate."

The discussion on derived attributes and constraints in Chapter 3, section 1 elaborates on exclusion partitions and their relationship with "unknown" values. In the remainder of this book, unless it is specifically mentioned, *partition* will mean *inclusive partition*.

[55] In non-exhaustive partitions, the union of all subtypes defined in the partition does not exhaustively cover all possible object instances in the object class being partitioned.

[56] In exhaustive partitions, the union of all subtypes defined in the partition exhaustively covers all possible object instances in the object class being partitioned.

Box 21 Exclusion partitions, variation inheritance, and polymorphism (on our website)

Box 21 discusses how partitions affect inheritance. It discusses different variations of the subtyping relationship and how each may be used with examples. It describes how variation inheritance flows from subtypes based on excluding information in the parent object, versus extension inheritance, which flows from supertypes anchored in the intersection of common properties of all its subtypes. Box 21 discusses why variation and extension inheritance should not be used together, and how such usage could potentially lead to inflexible systems under the pressure of scope creep. Box 21 uses the following figure to discuss different kinds of subtyping mechanisms:

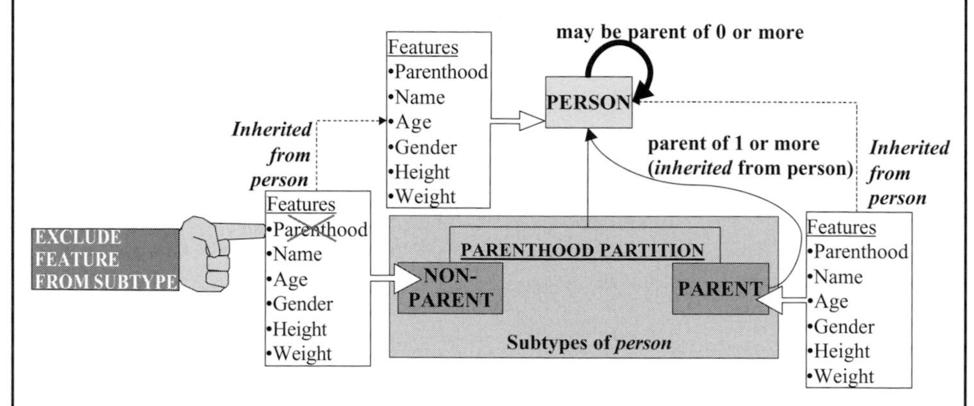

(a) Partitioning criterion: *exclusion* of parenthood relationship ("*parent of*" is inherited by one subtype but not the other)

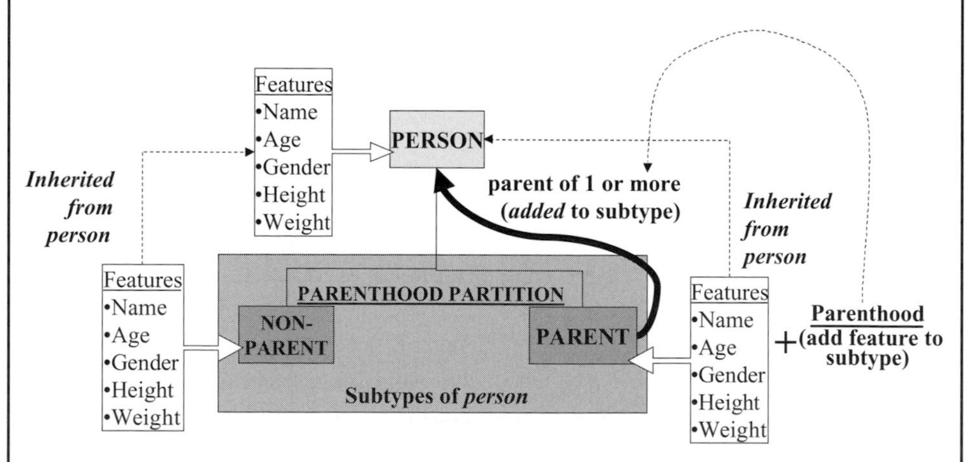

(b) Partitioning criterion: *addition* of parenthood relationship (all features are inherited by all subtypes)

Box 21 also describes why subtyping by adding information is normally the preferred option. It discusses inclusion polymorphism and the relationships between different kinds of inheritance. Inclusion polymorphism is illustrated by the following example: engineering firms manufacture engineering products. A relationship called *manufacture* relates *engineering firm* to *engineering product*. An *auto parts manufacturer* is a subtype of *engineering firm*. *Automobile part* is a subtype of *engineering part*. The two subtypes inherit a restricted form of the *manufacture* relationship between the corresponding parent objects, in which only automobile parts are made. Since the general manufacture relationship includes manufacturing automobile parts, this kind of subtyping is called an *inclusion polymorphism* – the relationship may be inferred from the subtypes it connects. These subtypes may then be considered to be parameters of the parent relationship. Changing these parameters would change the kind of manufactured item, i.e. the manifested polymorphism of the parent relationship. It is a frequently used form of inference supported by the metamodel.

Partition is a metaobject, which, by its very nature, has three properties:

1 partitioning criteria (the *discriminator* in box 22);
2 exhaustivity; and
3 exclusivity.

Partitioning criteria are important for subtyping and inheritance, whereas *exhaustivity* tells us that scope changes might bring hitherto unrecognized subtypes into a partition. Exhaustivity also tells us about validity of completeness checks, i.e., whether it makes sense to match object instances in the parent class against those in subtypes within a partition. Of course, in the real world, the object instances in an exhaustive partition *must* always be the same as those in the parent object class, but no system is flawless and information *could* get lost in software systems or manual procedures. Exhaustivity and exclusivity provide the basis for testing the integrity of business systems through checks and balances.

Atomic rules that span more than one partition

At any given moment in time, a single instance of an object may play several roles (belong to several subclasses) simultaneously, as long as these roles (subclasses) are in different partitions. However, additional business rules might constrain this behavior. These are best represented by relationships between subtypes. (Module V discusses relationships in detail.) Whereas an object instance may belong to only a single subclass within the same partition, two kinds of rules can constrain the existence of specific object instances that could simultaneously belong to two or more subclasses in different partitions as elaborated below:

1 All objects that belong to a given subclass *must* simultaneously belong to one or more (specific) subclasses in other partitions (not necessarily true the other way round – see figure 27).

This rule implies that the first subclass is a subset of the second subclass, even if they are in different partitions. If a similar rule is simultaneously true in the reverse direction, then the subsets will be equal to each other, even if they are in different partitions.

A business rule that asserts that all military recruits (a subclass of person based on employment status) must be 18 years old (a subclass of persons based on age) is a

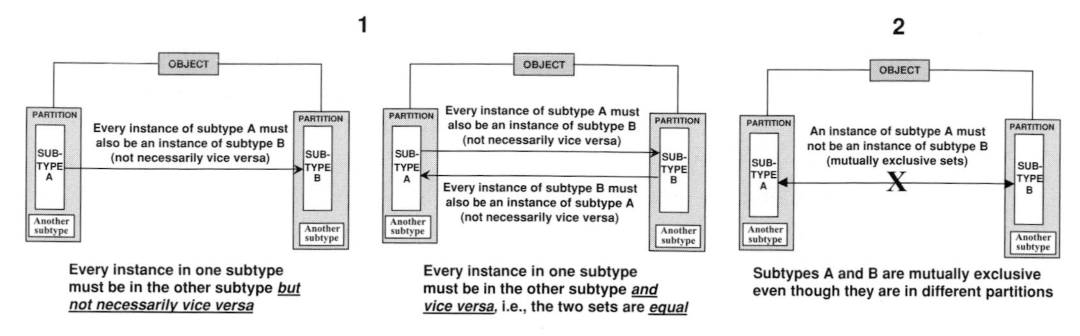

Figure 27 Constraints across partitions

rule of this kind. It is an irreducible fact and a separate component of knowledge in its own right. If we imposed the same constraint in the reverse direction, that all 18 year old individuals must be recruited in the military, it is a separate irreducible fact (and independent component of knowledge) that would ensure that there is no difference between the set of new recruits and the set of 18 year old individuals. One implies the other.

Software developers do not always recognize that this kind of set equality flows from two or more irreducible facts acting in concert. They assume that any one of the two sets implies the other, hence it does not matter which they choose to recognize for processing purposes. For example, there may be a requirement to test reflexes of military recruits. It would not be correct to attach this rule to the (sub)class of 18 year old persons. If we did this, our system might select the right people to test reflexes today, but knowledge would not be normalized and it would lead to inflexible systems: if the drafting age changed, the system for checking reflexes would have to be changed.

The correct approach would be to attach the rule to the subclass of military recruits – i.e., to represent knowledge in systems *as it is in the real world*. If we did this and requirements changed so that one of the two rules were removed, and the two sets were not bound to equal each other any more, it would not impact the reflex checking procedure. The reflex checking procedure would not be impacted because knowledge was normalized.

Rules could get more complex if we added more irreducible facts. For example, new recruits might have to be male, less than 21 years old, etc. Then more partitions (e.g., gender based in this case) and different subsets in the same partition (18–21 years old, not 18 year old persons) will be involved.

2 All objects that belong to a given subclass *must not* simultaneously belong to one or more (specific) subclasses in other partitions (the pairs of subsets are mutually exclusive even if they belong to different partitions – see figure 27).

For example, a business rule might assert that no one over 55 years old may serve in the military. This rule states that two subtypes in different partitions (the set of persons older than 55 in the age based partition of the person object class, and the set of persons in military service, a subtype of the person object class, based on occupation) are *mutually* exclusive. The configuration of components on the extreme right-hand side of figure 27 represents rules like this.

The meaning and syntax of behavior-based partitions

The kind of partition we are now discussing is more general than the partitions in figure 20, and the partitioning of state spaces we discussed earlier. In figure 20, object instances were partitioned in terms of single attributes. Then we discussed partitioning state space in terms of constraints on values of attributes and relationships. Now we are discussing partitions that might involve several properties[57] – one or more attributes and relationships (partitions

Box 22 UML syntax for partitions and subtypes

The unified modeling language (UML) has been adopted as a standard by the object management group (OMG), a cross-industry forum for standardizing object technology and syntax. UML has a robust and rich set of constructs and a broadly accepted syntax for expressing concepts related to modeling objects, subtypes, and partitions. In the following figure, objects are labeled as rectangles that contain attributes and effects (operations in UML parlance), the arrows are subtyping symbols (the direction shows the direction of *generalization* – from subclass to parent class), each partition is a separate comb that hangs from an arrow.

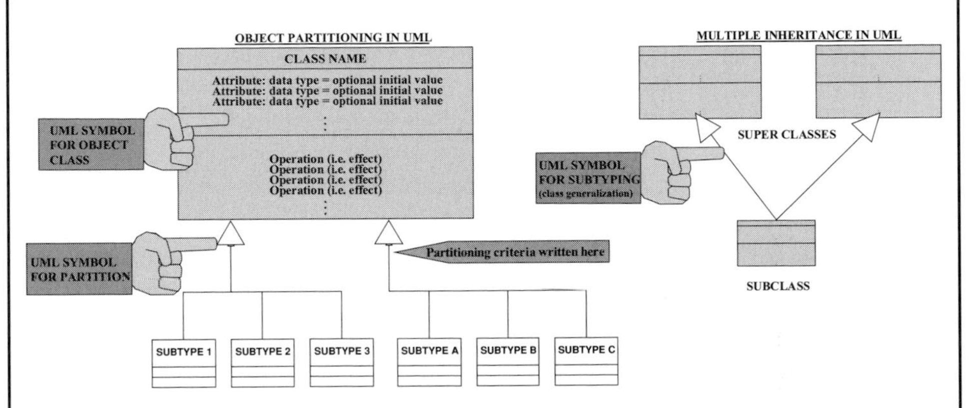

Partitions are called *discriminators* in UML because they contain the subtyping criterion. Readers interested in UML may refer to several papers under that heading in the bibliography, or to [329]. The symbols and syntaxes of other methodologies might be different, but may express the same concepts. For example, in NIAM [297], object classes are oval shapes and subtypes are nested within the supertypes. An asterisk in a supertype that contains subtypes is the symbol for non-exhaustive partitions. When there is no asterisk, it is considered an exhaustive partition by default. The reader should be aware that many methodologies are not complete, i.e., they do not have *all* the constructs needed to normalize knowledge.[58]

[57] The partitions discussed earlier were special cases of this generalized partition.

[58] The Bunge–Wand–Weber model tests methodologies for completeness and redundancy of symbols and concepts [14] includes a report on the health of UML subject to BWW testing. [18] has a summary one-paragraph report on the completeness of UML. See also the endnote and items under "Knowledge reuse algebras and test beds for techniques" in the bibliography.

of state spaces), as well as one or more effects and trajectories through state space (i.e. the object's history and the set of possible state changes). These partitions can be the basis for creating subtypes that will support complex, very specialized behavior and business rules. The subtyped object instance gets its special behavior from the subtype, and adds the more general behavior shared with other less special objects by inheriting them from its parent class. This is how partitions and subtypes support knowledge normalization, even as they facilitate modeling of special behavior and business rules tailored to specific conditions.

Default states, subtypes, and variation inheritance

Should subtypes inherit the default (i.e. initial) state of their supertypes? To understand the importance of this question, remember that our metamodel assumes that the real world is deterministic. We used default values for two reasons:

1 to express an atomic fact about an object's initial state; and
2 as an artifice to compensate for ignoring the inherent uncertainty of the real world. We did this by making assumptions about initial states to minimize risk (see the discussion on default values in section 2).

Let us consider how initial states can impact subtyping with examples.We know that fish must live in water – or more accurately, there is an overwhelming probability that they must. Still, there are six species of fish that can survive without water. These fish that survive on land are called lungfish because they have lungs to breathe air.[59] If we assert that a class of animals called *fish* must live in water by default, all instances of fish, regardless of subtype or species, would inherit this atomic rule when they are first recorded in the information system. We would be almost, but not always, right: we would be wrong when the fish is lungfish.

We could correct it in three ways:

1 Exclude that default state (that fish must live in water) from the class of all fish, since it is not always true (even though there are overwhelming odds in its favor). This might be theoretically correct in terms of encapsulating common behavior in components to make them reusable, but practically it will impose a heavy burden on users. They will have to enter the fact that the fish needs/does not need to live in water to stay alive for every instance of fish.
2 Since the odds overwhelmingly favor purely aquatic fish, include this as a default state for all fish, but override the default (manually or automatically) for a subtype of fish that can live outside water, when and if they occur. This is a workable solution, even though it is not theoretically correct from a purist's perspective. This technique is *variation inheritance*.[60] In variation inheritance, new subtypes are defined in terms of *differences* from supertypes. Variation inheritance can make it harder to configure knowledge by *assembling* atomic rules (i.e. assembling rules with *set union* operations) – we would have to

[59] The endnote on lungfish has more information on these strange amphibious fish.
[60] Variation inheritance is described in more detail in the endnote on kinds of inheritance.

consider *set differences*, i.e. conditionally removing some structures, from subassemblies of knowledge. This can become complex.

3 Exclude the default state from the class of all fish *and create two subclasses of fish: those that must live in water and those that can survive on land*. This is the theoretically correct approach that will not burden users. However, this possibility may not have been foreseen when the model was first built and systems built with traditional technology are hard to change after code is written: risk is high, there may be several unforeseen side effects that ricochet through the system, hence changes must be thoroughly tested and proven for every contingency before it is deployed. Foreseeing every contingency is not humanly possible in our imperfect world.

Let us understand how we would make the change in a system built with *knowledge artifacts* that normalize knowledge. The third approach is not only theoretically correct and user friendly, but will also be the simpler approach when we configure systems from components of knowledge.

- We would create two partitions:
  1 A partition based on species. In this partition, we would create a subclass called fish that would contain behavior common to all fish, such as: all fish swim with fins and all of them use gills to breathe in water. Each species of fish would be a subtype of the subclass of fish, and we would attach behavior of the species to the species subtype.
  2 A new partition based on whether animals need water to survive or not. This would be created by partitioning the set of *animals* (instead of only fish), thereby allowing reuse of the subtypes for other kinds of animals as well. In this partition we would create new subtypes:
     (i) Animals that do not need water to survive.
     (ii) Animals that do need water to survive. We would detach the assertion that the animal must live in water from the class of all fish and attach it to the class of purely aquatic animals *as a property of the subclass* instead of a default state. In the case of purely aquatic animals, it will not be a default state, rather it will be the defining criterion for the subtype, i.e. the *role* of the atomic rule will shift from default value to defining criterion.
- Then we would attach each species of lungfish to the class of animals that do not need water to survive. Each species of lungfish would then automatically inherit the behavior common to the subtype of animals that do not need water to survive. For example, the ability to extract oxygen from air, and if bugs are out of scope, the fact that they will use lungs to breathe air.

  Each species of lungfish would also automatically inherit all behavior common to all fish from the subclass of fish in the other partition. Each species of lungfish would thus inherit both kinds of behavior from each subtype it is linked to in different partitions:

  using lungs to breathe air etc. from the class of animals that do not need to breathe water to survive;

  using fins to swim and gills to breathe water etc. from the class of fish.

  We had earlier attached the special behavior of each species to species of fish. Inherited behavior will be added to this species behavior.

Each instance of lungfish would reflect the general behavior of fish, the general behavior of animals that do not need water to survive, and the specific behavior of the particular species of lungfish.

- Similarly each instance of aquatic fish would automatically inherit the common properties of all fish from the *class of fish*, the assertion that they cannot survive without water from the *subclass of aquatic animals*, and species-specific behavior from the *species* of fish.

In this third approach to changing the scope *after* building the system, reconfiguring old components, not rewriting program code, created new knowledge. Knowledge components were re-organized into new configurations to keep knowledge normalized as new knowledge was added. Knowledge was not only normalized but also *stayed* normalized. Done right, it can also be faster, cheaper, and more responsive to changing needs of business – today and tomorrow. This technique is called *refactoring*.[61]

In the example above, the default state was detached (removed) from the superclass, and a partition was attached (added) instead. In some situations, partitions may be based on the likelihood of different initial states occurring in different subtypes. For example, the likelihood that cars exiting a gas station have full tanks is high, whereas cars entering the station are likely to be low on fuel. Thus, exiting cars may have a default status "gas tank full," whereas entering cars could default to "low on gas." However, it *is* possible that a car that was low on gas exited the gas station without taking gas for some reason, and a car with a full tank stopped at the station to take air, oil, or for some other purpose. The default may have to be overridden for some instances of each kind of car. In this case, two subtypes of the class *car*, namely *exiting car* and *entering car* will have different default values. The default in this case must be attached to each subtype, and not the parent class in order to normalize knowledge.

These are examples of how default states may be different for instances of objects in the same object class (cars or fish) and how subtyping can facilitate normalizing knowledge when this happens. Object classes may be partitioned by default state (i.e. likely initial state) in order to normalize real-world knowledge because the *chance* of finding objects in different initial (i.e. default) states in the real world is high.

Subtyping criteria – dividing to conquer

Subtypes specialize the behavior of their parent objects by adding to them, or constraining them in special ways.[62] They add components and constraints to the state space of the parent object, changing its shape and dimensions. The parent is the repository of common behavior, and the subtype of special behavior. Effects of events change individual object instances, moving them in or out of these sets and classes. Box 23 shows different kinds of information that subtypes may contain to distinguish special behavior of a subset of objects from behavior common to the entire (parent) class.

Subtypes may be based on the *existence* (or not) of relationships as well as attributes. This is no different from any other state indicator, and both are mutually equivalent and mathematically indistinguishable in the metamodel of real-world knowledge.

[61] See the endnote on refactoring.

[62] Items in the bibliography, under theory of categories, discuss mathematics of subtyping.

Box 23 Subtyping criteria

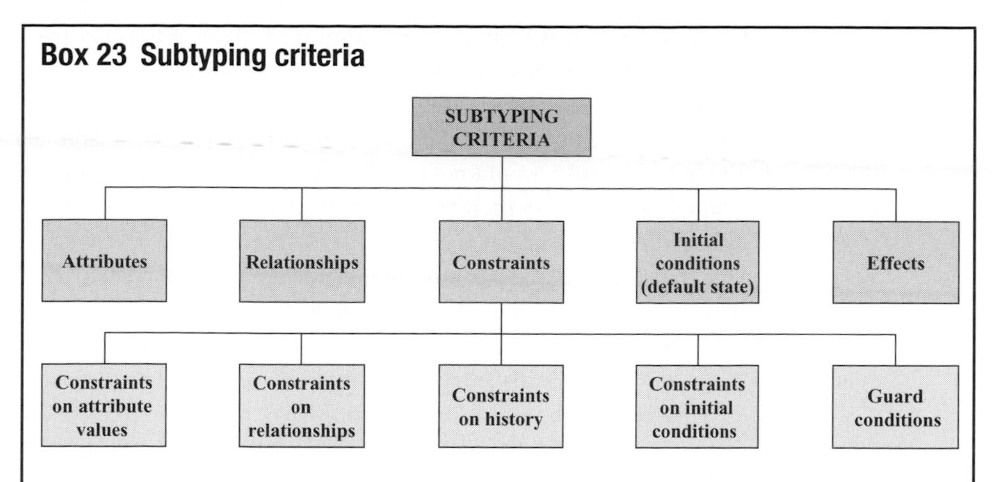

Effects of events on subtypes

Object instances may respond to events by changing their state. A change of state might make the instance a member of a subclass, or remove it from a subclass. For example, an entering car could become an exiting car in the example above, or an unemployed *person* who is hired becomes an *employee*. *Employee* is a subset of *person* (see the discussion under *inheriting behavior*). Similarly, an *employee* who is fired skips out of the employee subset. Thus, individual objects skip in and out of subclasses in response to events, i.e. their *roles* change (see the discussion on subtypes and roles in box 11). In addition to changing roles, effects of events on subtypes may also be subtypes of corresponding effects on the parent object. Just as relationships and features of subtypes may be inclusion polymorphisms of corresponding features of the parent objects (see box 21), so too might effects on subtypes be polymorphisms of corresponding effects on the parent. For instance, a generic *move* effect, for a *vehicle*, may become a more specific movement like *fly*, if the vehicle is an *aircraft*, or *sail* if the vehicle is a *ship* because *aircraft* and *ship* are both subtypes of *vehicle*.

Subtypes may also be based on *values* of attributes or *specific* relationships with *specific* object instances – the two are equivalent and indistinguishable in the metamodel of knowledge. For example, an investor may not want stocks of a specific company in his or her investment portfolio for personal reasons. The investment portfolio is an object instance. So is the stock. The rule implies that no relationship can exist between the two objects, i.e. the relationship must be null. Partitioning an object class on criteria of this kind, based on relationships between specific object instances, is no different from partitioning the object class based on specific values of specific attributes. It all boils down to limiting its *lawful state space*.[63]

Subtypes inherit the common behavior of their parent objects and add special behavior of their own, namely specific states, effects, constraints, and other properties not shared with their supertype(s) (see box 23; also described in box 10). Only some of these properties (features in XML terms) are attributes or relationships. Others are effects and constraints. Subtypes can thus result from partitions based on pure behavior.

[63] Lawful state space: see the endnote on the Bunge–Wand–Weber model.

For example, in figure 21, we might have partitioned glass panes into two mutually exclusive subtypes based on behavior – breakable and not. Had we done this, the guard condition would not be needed.[64] The effect of the hammer strike would not be generic to glass pane (the supertype); instead it would be a property of only breakable glass. Thus the *effect* of the hammer strike would not be inherited by, and *would not exist at all, for unbreakable glass*. If we had to assert, like in figure 21, that all blue glass is unbreakable, we would do so by a relationship like that on the left side of figure 27. The relationship would be between blue glass in a color-based partition and unbreakable glass a breakability partition. This is why *guard condition* is not among the metaobjects in figure 32. Guard conditions are covered by the *other* metaobjects in the figure. It is redundant, and including a *guard condition* in the metamodel of knowledge would replicate, not normalize, behavior.[65]

Contrast subtyping with process decomposition and the node branch methods of Chapter 1. Process decomposition started by creating hierarchies of poorly defined concepts that actually encouraged replication of atomic rules, which it blithely ignored. Subtyping starts by seeking concrete common behavior and separating differences in behavior based on well-defined irreducible facts. Thus, subtyping is the key to reusable components of knowledge.

The node–branch method lost its way in the complexity and scale of industrial strength business systems. It needed too much detail before it could represent behavior of business systems accurately. It had no means of extracting and normalizing common knowledge before plunging into detail. It could not recognize and normalize common irreducible facts. Subtyping makes no assumptions about obtaining all detail up front. The focus of subtyping is on modeling *shared* behavior first. Detail and differences may be added in steps, and components of knowledge reconfigured to keep business knowledge normalized (as described in the example on lung fish in the discussion on default states and variation inheritance). This prevents the chaotic and uncontrolled impact of change on business processes and systems. Otherwise consequences of change will ricochet and ripple through unintentionally replicated and unmanaged knowledge hidden in business processes and the systems that support them. Thus not only can systems assembled from objects and subtypes be more responsive to business needs, but the process of testing and quality assurance can be simpler too!

4 The problem of perspective

> Truth fails not; but her outward forms that bear
> The longest date melt like frosty rime
> (William Wordsworth in *Mutability*)

Can real-world objects really anchor reusable components of business knowledge even as scopes and rules shift? Are subtyping and inheritance the crack team that will truly defeat

[64] The guard condition is a kind of *variation inheritance* – that the effect exists for all glass panes *except* the unbreakable kind.

[65] See construct overload and redundancy in the endnote on the Bunge–Wand–Weber model.

Figure 28 Perspective is a point of view

the dark forces of chaos? Our world is complex, driven by learning, change and opportunity, threats and competition. To defeat chaos we must normalize knowledge in the *right* objects and subtypes. Otherwise objects will not inherit the behavior they must, and might inherit behavior they should not. We will call this inheritance by mistake; see the example in box 25. How can we identify the right objects in a shifting world? This is the problem of perspective.

As many practitioners know from bitter experience, objects alone cannot normalize and encapsulate reusable knowledge, nor can subtyping and inheritance by themselves defeat the forces of complexity and chaos because both come up against the problem of perspective. The world is a chimera, and so is our perception of it.[66]

What is perspective?

We understand the world around us by experiencing its behavior. We seek its meaning by forming *concepts* of what behavior is shared by what objects, and what is special to each. These concepts are based on our individual experiences and perceptions. Not only are our experiences and perceptions different, but also two people will never think exactly alike. Therefore our concepts, i.e. generalizations of what is shared and what is special, are naturally different. For example, a physicist might say that a ball thrown by a child and a shell fired from a field gun are similar because both follow trajectories with similar shapes, under the influence of the same forces – gravitation and air resistance. On the other hand, a general might say that the shell and field gun are similar, but not the ball, because both the gun and shell are complementary weapons of war that must be issued from the inventory, whereas the ball is a child's toy. Figure 28 is another example of the problem of perspective, in which the same object, a wavy line in this case, may have multiple interpretations depending on similarities in shape, use, and association.

[66] See the endnote on multiperspective and facet modeling.

What do differences in perspective mean to objects that try to anchor knowledge? Perspective is a point of view. Our concepts of shared behavior are generalizations. They are object classes, sometimes abstract, like person–organization in section 3. These object classes and subtypes anchor the shared behavior of objects we perceive. Since each one of us generalizes and specializes our perceptions differently, many of our concepts may not match those of others (figure 28). Objects and relationships are sets of properties based on classification of common behavior of things we experience (box 15); hence we may not agree on object classes and relationships themselves – a fundamental problem on which many software projects have foundered.[67]

Box 24 Perspective is an object

Do you see two people in a private conversation or a chalice in the figure above? What you see depends on how you *classify* the white and black spaces in it – which color is empty and which is solid? ***Perspective is a point of view. It is also a model.*** It is the entire *structure* of interconnected objects that anchor knowledge – classes, aggregations, relationships, constraints, state spaces, domains, effects, and all the other metaobjects we have described, glued to each other in a structure we call knowledge, or, more modestly, our perspective of knowledge. It is also an object in its own right – an aggregate object with a structure. Each individual's perspective is an instance of a model. If the model changes in response to new information or an insight, it has *changed its state*.

[67] Some analysts have proposed that we do not try to classify objects intuitively. Instead, they suggest that we mathematically analyze similarities between objects in terms of their properties to group them into object classes and subtypes [283]. While this approach may be useful, it will not guarantee stable object classes. If the scope of the process changes so that some properties under consideration change, so might the classification scheme. Inclusion or exclusion of behavior may change affinities between object instances, which in turn can change the taxonomy of objects and relationships. This is because we did not address the root problem – we only mechanized it. Facet modeling, described in the endnote on multiperspective modeling, is another approach in which *aspects* of an object might be reused. For more information, see [15], [53], [13], [21], and [23].

Often the changes are minor – a new attribute, a new relationship, an additional effect or a new subtype in an old or new partition. However, sometimes the change can be fundamental. The classification scheme – the objects and relationships themselves, change.

Consider why the same underlying reality can appear very different from different perspectives. Object instances are things or concepts that have properties. Some properties are shared with one set of things, and other properties with other sets of things. It follows that the same thing might belong to different object classes when perceived from different perspectives. Changing perspectives can change entire classification schemes, which can have a very profound effect on the model.

Consider what happens when classification schemes change. Object classes are classification schemes based on *similar* properties of object instances – *all* properties of instances in a class do not always match. Matching properties are shared, and the others are not. We manage shared properties with the concept of *superclass* and unshared properties with the concept of *subclass*. A subclass is meaningless without a superclass; if a superclass disappears, so must all its subclasses. If entire objects vanish, they take with them all their relationships, constraints, and subclasses; and *their* relationships and constraints as well.

When classification schemes (i.e. taxonomies) change, they can have a domino effect on the entire model, sweeping away entire subclasses and myriads of relationships, constraints, partitions, and all other structures that relate objects and subclasses into a consistent and cogent configuration of knowledge; new structures might have to take their place. New and old structures are objects too. The appearance of the new and dissolution of the old may impact other structures, which in turn have other impacts. Change can ripple through the entire structure of knowledge till it settles into a new configuration (and the possessor of the change perspective thinks "Aha – *now* I understand!") – it is called a paradigm shift – a different model of the world, or a perspective that has changed its *state* quite radically.

This is why we need the *universal perspective* with its universal object classes and relationships to pin down widely shared ideas about business and reality. The secret of these universal objects[68] that anchor knowledge firmly from every possible perspective is not hidden in some arcane and abstract detail; rather it is explicit in the sweeping generalizations that can withstand the incessant pounding of continual change and the immense diversity of creative thought and innovation. The *universal perspective* consists of objects and structures that masquerade as *apparently* different objects in different perspectives, but are actually different states, roles, and compositions of universal objects. By gaining an understanding of universal objects and the universal perspective, one can acquire a better feel for the essence of universal reality and the unity of all perspectives. All perspectives are states of the universal perspective. Paradoxically, the universal perspective is changeless because it underpins change. It is this universal perspective we seek in order to solve the problem of perspective.

[68] Remember, relationships are objects too.

Indeed, as scopes shift, and new behavior is recognized and old constraints are retired, the *same individual* may change the way he or she classifies common behavior. Did that just happen to you in box 24? Object classes themselves become chimerical, and the object model a chimera; this is yet another cause for chaos instead of a firm anchor of reusable knowledge. Almost all data and object modelers have experienced this problem.

We have replaced the domino effect of change ricocheting and rippling through unintentionally replicated and unmanaged knowledge in the system with *two* other problems (see the example in box 25):

1 Inheritance by mistake – wrong behavior was inherited because our object taxonomy was incorrect, or *became* incorrect when scopes and perspectives changed.
2 Inheritance deficiency – behavior that should have been inherited by an object was not, because the object taxonomy was defective, or became deficient in a new scope and a different perspective.

The example in box 25 is too simple to be real. It involves only two analysts, you and Jim; three objects, *bill*, *payment*, and *document*; and one subtype, *bill*, with two parents, *payment* and *document* – but even in this simple example there is ample room for both kinds of mistakes: *inheritance deficiency* and *inheritance by mistake*. In the real world, many object instances may share some common behavior, but not all behavior. Worse, any given object instance might share different kinds of behavior with instances of different kinds of objects. Teams of analysts may be large and each analyst will have his or her own unique perspective. How much greater would be the risk in the real world: larger teams, more perspectives, more objects, and more multiple inheritances from larger numbers of supertypes!

These problems stem from the problem of categorizing behavior coherently, the legacy we discussed in Chapter 1, section 6. It is a problem left unresolved for 40 years from the time of the first formal business models. We are still stuck with it.

Does a universal perspective exist?

What is the solution? To group behavior cogently, we must have cogent objects; to get cogent objects, we must solve the problem of perspective; to solve the problem of perspective, we must seek common ground, and we can seek common ground because we know individuals perceive the world partly from their own unique point of view and partly from widely shared ideas generic to the world of business, or imposed by the physical world. Without these shared ideas, each one of us would be forever condemned to our own private universe. We would not understand each other, nor would we be able to work as a team. We know our perspectives can converge quite rapidly when we model simple situations because of these shared ideas. It is much harder when our models are broad in scope and complex in detail. To handle the industrial strength models of today that span complex corporations and even cross-corporate boundaries, we need a more robust anchor. We need a standard universal perspective – one that subsumes all individual perspectives.

Without stability or change, Eternal, it has no origin and no end. (Adapted from *The Gospel of the Buddha* by Paul Carus)

Does the universal perspective exist? Is it possible to define universal classes or must we be forever chained to the chimera of perspective? Widely shared ideas about business and reality underpin our perceptions, and, because of these widely shared ideas, it is possible to define universal classes of objects that encapsulate shared knowledge.

Box 25 An example of the problem of perspective

Consider a bill in an accounts receivable system. You might be justified in considering it to be a request for payment, and hence a state of an object called *payment*. You have just generalized two key business concepts: *bill* and *payment*. The bill is now a subtype of payment and inherits various properties of payment, such as currency of payment, the payee and payer, due date, goods and services being paid for. You are quite satisfied that you have normalized and reused the behavior of payment and are certain that other applications will be able to reuse this intelligence. You store it as an artifact in an electronic repository of business components.

In the meantime, your employer has expanded the global operations and has key customers in non-English speaking countries. Speedy international cash flows are critical to growth. Raising electronic bills in the language of the customer is the key to strengthening the relationship with international customers and getting paid on time. The plan is to send bills to customers, in the language of their choice, by email. An electronic copy of emailed bills will be retained in your employer's database.

The billing system must be enhanced so that each customer's bills can be formatted in the language of choice. Your repository of knowledge artifacts has an object called *document* with special translation behavior attached to it. Jim, a billing analyst, finds your knowledge artifact, called *bill* in the repository classified as a kind of payment. He is puzzled. "A bill," he thinks to himself, "is not a payment – it is a document we send to customers! I know documents already have an automatic translation facility attached to them; if I make bill a subtype of document *instead of a kind of payment*, my translation problem will be solved." He proceeds to do just that. To his dismay, bills can now be translated into the customers' languages, but have lost all payment information – amounts, currency, due dates – because bill is not a subtype of payment any more.

Jim brings the problem to you. You realize at once that it is an *inheritance deficiency* caused by a deficient taxonomy of objects. You look Jim in the eye and sagaciously suggest, "Why not make bill a subtype of both document and payment? That way we will inherit all the behavior we need." Jim is impressed, thanks you, and does just that. He is quite happy until Joan, the billing manager, approaches him. Joan tells Jim that her staff have a problem trying to understand foreign language bills. She wants two copies in the firm's database – one in English and the other in the customer's language. The *document* object is a supertype of *bill* and has copying behavior attached to it. Jim thinks Joan's request is as good as done. The bill would have automatically inherited this copying behavior from *document*.

All hell breaks lose when copies of a bill are made in English. From the perspective of bill being a request for payment, duplicate requests for payments are being logged against customers in the firm's accounts receivable system. Customers are understandably upset, and the firm's global position has become vulnerable to competition. This is an example of *inheritance by mistake*, and just one instance of how the problem of incompatible perspectives can cause chaos.[69]

[69] See patterns of buying and selling in the universal perspective for the answer.

These universal object classes anchor knowledge firmly and coherently from every possible business perspective. They constitute a pattern – a standard perspective that other perspectives can add to, but one they will not have to change to satisfy their requirements. This pattern, summarized in the universal perspective on our website, is a component on its own. It is described in detail in a companion book by the same authors, *Agile Systems with Reusable Patterns of Business Knowledge – a component-based approach* [337]. Like the chassis of a car, it is a component that can connect standard and custom parts to make the whole work.

In this new architecture, the standard parts are the universal object classes in it. These objects normalize shared ideas. Custom components will inherit this shared wisdom, and will add the special behavior and creative ideas that innovative businesses formulate to prosper and excel, even as the universal pattern of shared ideas in our "chassis" automatically and naturally integrates special behavior with other processes within, and even beyond, the firm.

This "chassis" of shared perspective can provide a firm anchor for *virtually every possible* perspective. The standard perspective springs from the metamodel. The metamodel will provide definitions of objects, properties, states, inheritance, subtypes, partitions, and the other paraphernalia of inheritance along with properties of other metaobjects we will describe later in this book. We will need them to group common behavior and to keep knowledge normalized. Thus, to solve the problem of perspective, we will need to understand both the universal perspective and the metamodel of knowledge that is its fountainhead.

The metamodel and the universal perspective together can potentially defeat the forces of chaos, but there is another practical, and equally important problem that we must overcome to make them one team. It is the tyranny of words.

The tyranny of words

> As shadows wait upon the sun
> Vain the ambition of kings . . .
> To leave a living name behind
> And weave but nets to catch the wind
> (John Webster in *Vanitas Vanitatum*)

What is in a name? Everything! Names are labels for our concepts, and our means of communicating them to others. Every data administrator knows the tyranny of words, and those that work for large corporations know how intractable it is. Different groups and organizations often need very similar concepts, but call them by different names, or, worse, different groups have the *same name* for very different concepts. It is a recipe for confusion when organizations merge, or integrate business processes and systems. It is also a culture – attempts to standardize names for concepts across organizational borders usually generate more heat than light – a seemingly trivial problem of syntax can take up disproportionate amounts of organizational time and resources.[70]

[70] Recently there has been an interest in standardizing vocabularies across value chains (for information on value chains, see "supply and demand chains" in Module V, section 3). VCML, an acronym for the value chain markup language from Vitria Technology, Inc, is one such initiative. VCML defines a value chain as "a network of all of the business partners and transactions in a supply and demand chain from raw materials and subassemblies to the consumer. A value chain spans vertical and horizontal relationships within and across

Box 26 Synonyms and homonyms

Different names for the same concept are called *synonyms*, and a single name for different concepts is a *homonym*.[71]

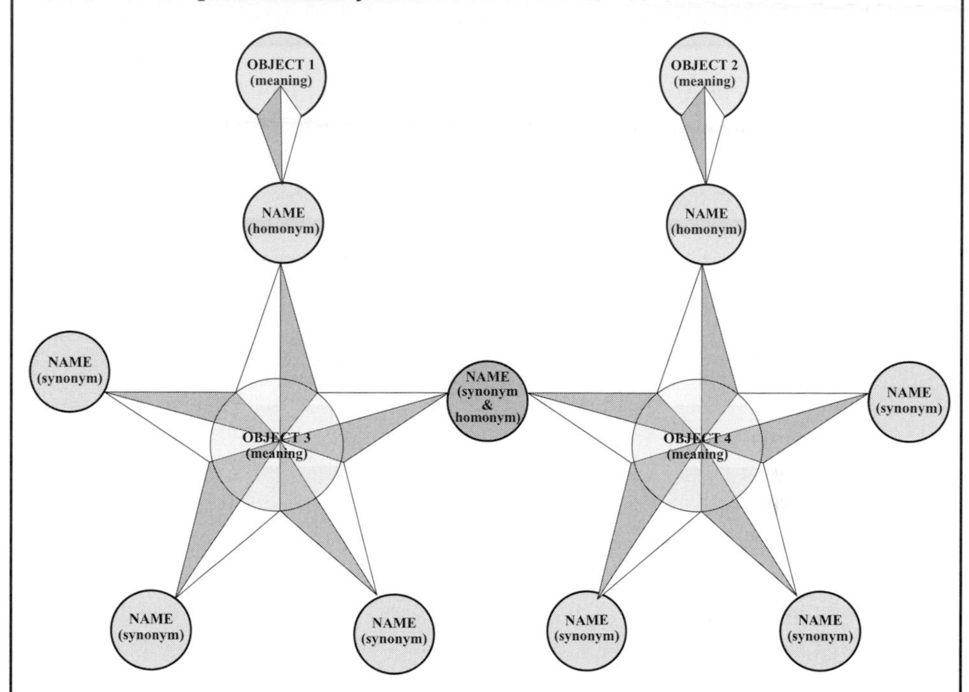

B*ackbone* is a network of broadband connections between switches for the telecom industry, and the spinal bone in our backs for the rest of us. It is a homonym. Similarly, *SDLC* is an abbreviation for *systems development life cycle* for the information systems professional, and *synchronous data link control*, a kind of data transmission protocol, for a network professional. Thus SDLC too is a homonym.

A word can be a synonym and a homonym at the same time. For instance, most of us know that an *account* for a sales person is a synonym for customer, but to an accountant it means a category of expense or revenue. Thus *account* is both a synonym and a homonym.

Synonyms are common even within the same industry, and even in the same firm. For example, in a major telecommunications company, the operations departments identified

industries. It addresses relationships with all parties participating in designing, manufacturing, financing, marketing, delivering, and supporting a product or service." VCML standardizes vocabularies to facilitate B2B collaboration. VCML models have been published for aerospace, automotive, banking and finance, education, energy, government, healthcare, insurance, petrochemical, retail, telecommunications, and transportation industries. See [65].

[71] Synonyms and homonyms are *states* of names. When a single name has a naming relationship with more than one concept, it is a homonym. When a single concept has a naming relationship with more than one name, each name is a homonym. See figure 29.

central offices[72] for telephone switching facilities with a *CLLI code* (pronounced "silly" code), whereas commercial departments called the identifier *Sensor Id*. Senior managers and professionals, including many who had been with the firm for several decades, had no idea that the other half of the firm used a different word for the same concept – a concept that was central to their business.

The tyranny of words emerges from the metamodel of knowledge – that the concept is different from its label and the same object may have many names, *i.e. name itself is a class of objects that consists of individual instances of name*. Different (instances of) names may be preferred in different contexts. The context is the perspective.[73] These concepts provide the anchor for the rules in figure 29; read it as you did figure 8.

In many situations it might be best *not* to standardize names. The problem can be quite intractable when deeply entrenched, long-standing interest groups clash over choices of names – and there are so many concepts to name in any real business! Consider too that battling over names may not only be an exercise in futility – it can bring the entire exercise into question – but also that unfamiliar names might actually sow confusion and become a barrier to creativity. Remember our intent is to make change easier – not harder!

Instead of standardizing names, the group that administers the universal perspective might have its own label or name for each concept (object) that can be the hub around which all its synonyms revolve. See figure 30; it shows three perspectives of two objects with several names each. The primary names (and objects) at the core are shared, but hidden from all three perspectives. Perspective 1 has one name for object 2, a name that means the same thing to perspective 2 as well, and six synonyms for object one, one of which is a name for object 2 in perspective 2. Thus it is a homonym. Perspective 2 has six synonyms for object 2, one of which is a name perspective 1 uses for object 1. Perspective 3 has a name for object 1 that it shares with perspective 1, and another for object 2 that it shares with perspective 2.

The primary name, or *concept id*, can pin down the concept and anchor all its other names and synonyms. In this way, not only can default names be different in different perspectives and still map to the same concept, but also users can, at any moment, see and understand how other groups have named their concepts in the repository of knowledge. *These synonyms and other names of an object will be aliases for the object*. For clarity, the concept id should be crisp, and never a homonym. The *concept* is the key to meaning and it is *meaning* we must focus on.

If this structure of knowledge is stored in an electronic repository, each stake holder need only be aware of names in his or her own perspective; there may be synonyms in a single perspective too! The homonym between the two objects has a different meaning in each context (perspective), but those who hold one perspective can be aware that it is a homonym, and will be free (to use the repository) to look up its meaning to others.

[72] A *central office* (CO) in the telecommunications industry is a switching center in which telephone trunks and loops are terminated and switched. Some synonyms for central office in the telecommunications industry are *telephone exchange*, *switching center*, *switching exchange*, and even *switch*.

[73] In facet modeling, all properties of an object, not just its name, are said to belong to an *aspect* of the object. Instead of linking a property of an object directly to the object, the aspect is linked to the object. For more information, see the endnote on facet modeling or [13], [15], [21], [23], and [53].

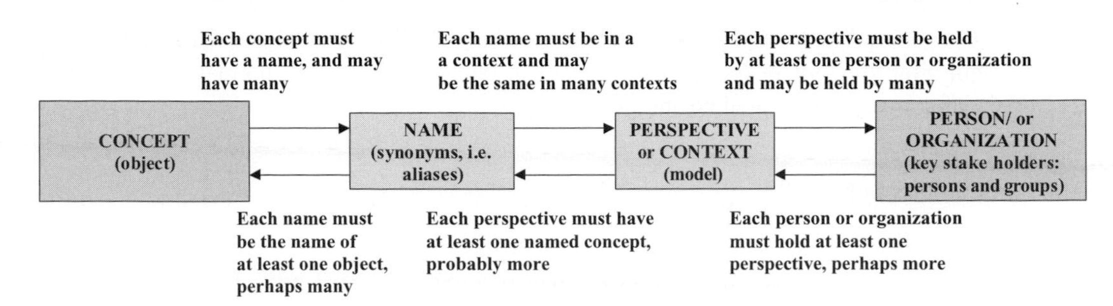

Figure 29 Name is an object class linked to perspective

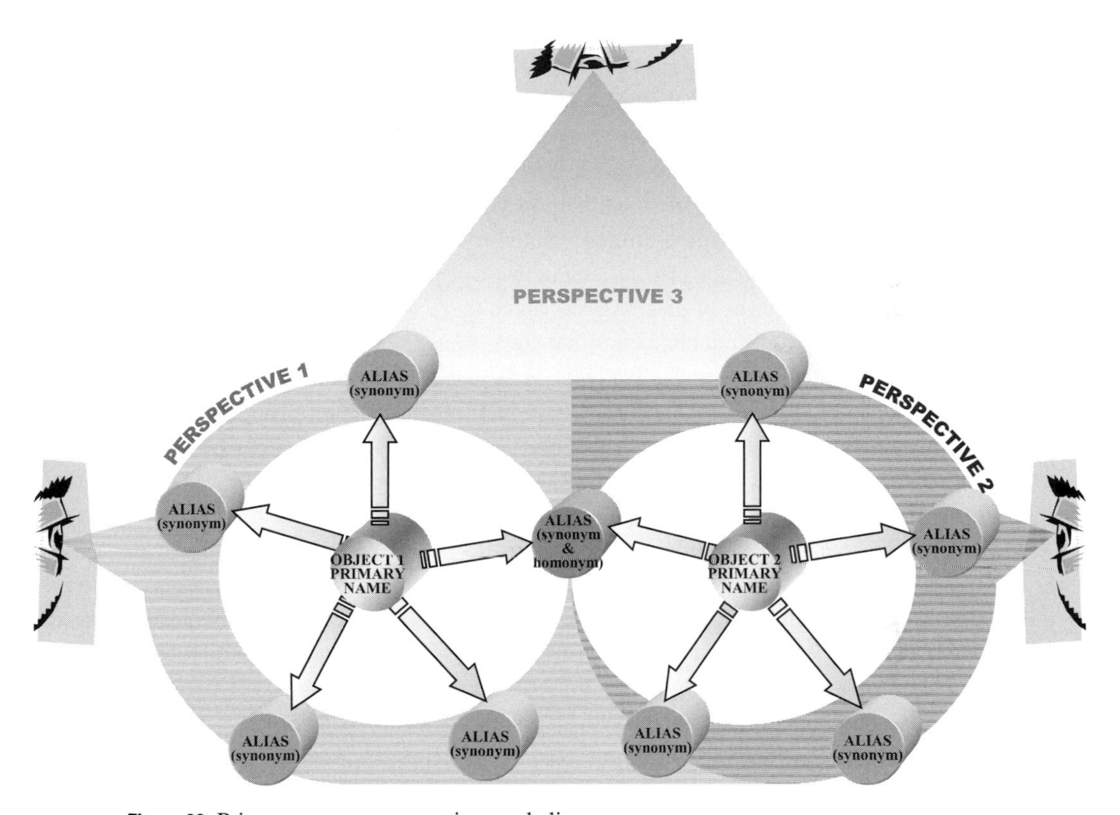

Figure 30 Primary names, perspectives and aliases

5 Repositories of meaning

Each object we have discussed so far has been the repository of a meaning. Metaobjects are objects too (indeed, they *are* our model of *object* and its properties). Each metaobject inherits the universally shared properties of objects. For example, the metaobject called "state" (figure 32) has subtypes called "substates." We saw how this happened in figure 21. Remember the common properties each metaobject inherits when we discuss it in this book. All properties and operations we have discussed for objects will apply to each. We will not repeat shared properties each time we describe an object in the metamodel of knowledge (like the multitudes in figure 32). Thus knowledge will be normalized. Different entities will wrap themselves around meanings of different kinds as they normalize knowledge, adding nuances to shared meanings.

The metamodel of object

Real-world objects are known by their behavior. Behavior carries within it the seeds of information. It is this information we perceive, integrate, and classify when we crystallize our intuitive understanding of the world around us, and the objects in it. This information may be our experience of tangible objects, or our understanding of intangible concepts; it does not matter which – both are objects. Objects are patterns of information. Each pattern is an instance of an object. These patterns are classified based on common behavior. Thus, object classes normalize behavior instantiated by object instances. Objects are patterns, and the metamodel of pattern is the metamodel of object.

Objects are at the root of knowledge; they are its basic building blocks. The real world is complex, and so are the patterns that help us perceive and understand it. Therefore objects can be complex. The infinitely diverse behavior of objects reflects the richness, diversity, and complexity of the real world, as well as those of patterns – the internal structures hidden within objects – patterns of meaning that make them what they are.

These patterns are irreducible facts that engage each other, like the gears of a machine, to produce new and related irreducible facts; facts that can be causes and effects, and facts that form patterns of knowledge. These patterns constitute the internal structures of objects manifested as meanings and behaviors of different kinds, conveyed by a huge diversity of chimerical objects – meanings that melt, merge, twist, and change in uncountable ways in uncounted dimensions. The metamodel of pattern *is* complex and so is the metamodel of object.

Chapter 4 has the metamodel of pattern. However, there are allied concepts we must cover before we can understand it. Figure 32 shows the hierarchy of object types we have discussed so far. In this chapter we will add many more. Each will be a stepping stone towards reusable components of knowledge – the knowledge that will help us build reusable *knowledge artifacts*. However, the journey must start with object – the basic concept. Figure 31 is the basic metamodel of object, and the hierarchy of figure 32 inventories its basic subtypes that inherit these properties.

Figure 31 articulates some of the most fundamental rules about objects. We have discussed these in Chapter 1 and earlier in this chapter. Read figure 31 as you did figure 8. Relationships read backwards include rules too. They are the inverse of rules read forward along the arrows, and are enclosed in square brackets [like this], near names of relationships. The inverse of a relationship is a relationship (a rule) that maps the object at the arrowhead back to the object at the tail of the arrow.

The backbone of figure 31 is the relationship between object class and object instance. Object classes are object instances too (section 1). Figure 31 articulates this.

Figure 31 divides patterns of objects into two classes. One is a *set*, and the other a *list*.[74] Sets are aggregate objects that do not count multiples of the same object among their

[74] *Pattern* is a broader concept that subsumes *set* and *list*. See the endnote on the theory of categories, [171], [172], [173], [183], [184], [185], and [186].

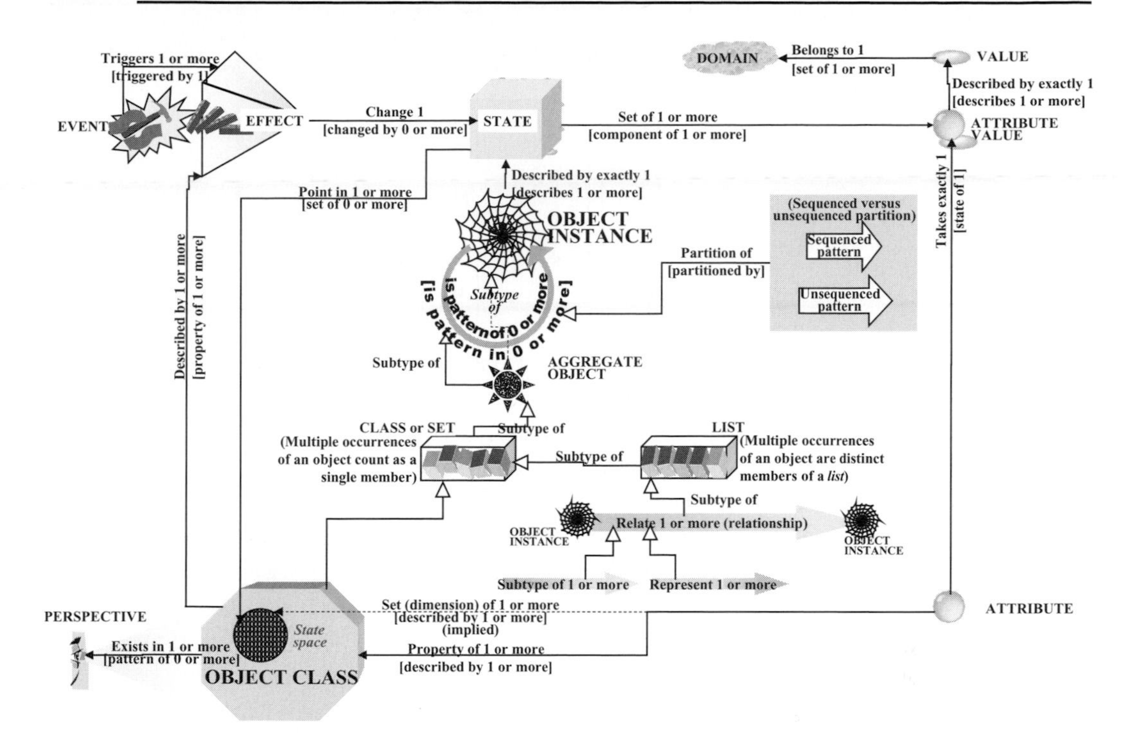

Figure 31 Fundamental metamodel of object

members. Lists, on the other hand, do. As such, a list with two of the same object instances among its members will be considered different from a list that is identical in every other way, but counts only one instance of the object among its members. A list adds information on distinguishing numbers of identical members of a set. It is therefore a subtype. Box 30 and the subsection on patterns in Chapter 4 will expand on differences between sets and lists. Box 65 on our website tells us why relationships are special kinds of lists that normalize *interactions* between listed objects. An example of how lists of domains are naturally manifested in the metamodel of knowledge may be found under "attaching value constraints to format", in box 38.

Furthermore, patterns may be partitioned on whether they are sequenced or not. In unsequenced patterns, only membership counts, whereas, in sequenced patterns, it is not just the fact of membership that distinguishes one pattern from another, but also the sequence of members (see Chapter 4, section 1 on patterns, and Module V, sections 1 and 2 on our website).

Object classes are obviously aggregate objects that are unsequenced sets[75] of object instances. However, they are special sets. They are sets based on common attributes. The objects that are members of the set have been granted membership based on shared attributes and effects. As we have seen, this makes the object class a very special kind of aggregate object.

We have discussed how events trigger state changes through their effects on objects. An event may have different effects on different objects, and, indeed, different events will affect

[75] Objects are mathematical classes. The mathematical concept of *class* is broader than *set*; sets are a kind of class. The distinction is subtle. We have not made this distinction in order to simplify the discussion.

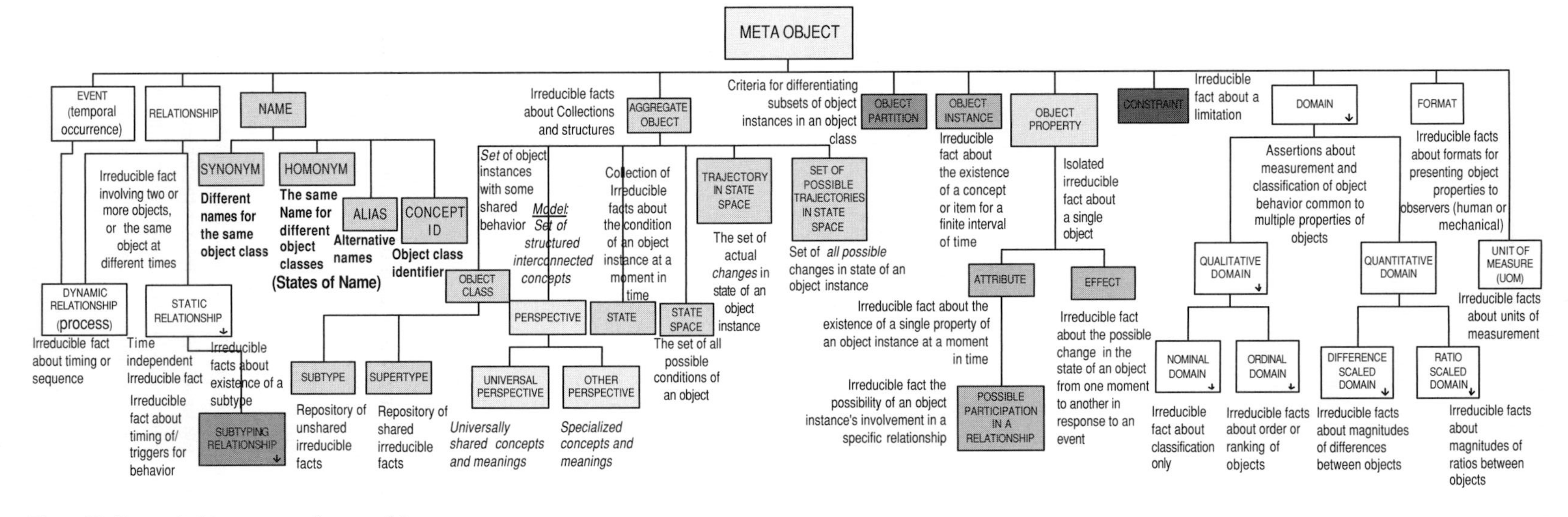

Figure 32 Expanded inventory of meta-objects

objects differently. The relationships between events, effects, and states of objects describe these rules.

The state of an object instance is an unsequenced set of attribute values. We have discussed this in section 2. The attribute carries the meaning of the value in the context of the object class (we will discuss this in depth in Chapter 3). The value, by itself has a meaning based on its domain. We have touched on this in Chapter 1, and will expand on it in Chapters 4 and 5. Thus, attribute value is an aggregate object. It is the conjunction of *attribute* and *value*. The state of an object instance is a set of attribute values. Figure 31 articulates these concepts as well.

An object is a pattern of information – information in a pattern of objects. A collection is a kind of pattern. An aggregate object is a collection of objects. This pattern of object instances, the aggregate object, may repeat the same object instance more than once, but an object class may not. However, this *pattern* has an identity of its own that identifies it as a unique collection of object instances. No other instance of aggregate object may contain exactly the same object instances, and convey exactly the same information on their sequence (if indeed sequence matters) and still retain its distinct identity. If two patterns are identical in every way, they are considered to be the same pattern – the same object instance. The *aggregation* is also an object instance. The very assertion that an aggregation (and an aggregate object) may consist of duplicates of an instance of the same object is based on this fact. On the other hand, since the aggregate object is a collection, it could also be the empty set we discussed in box 19. This is why figure 31 tells us that the aggregate object may be a pattern with *no* object instances in it, and it may also be a pattern with multiple instances of the same object.

A pattern is an instance of an object. An object instance is a pattern and so is an aggregate object. The subtyping relationship between the pattern and the object instance implies and includes the subtyping relationship between the object instance and the aggregate object in figure 31. They are not independent irreducible facts, and treating them as such will only denormalize the information. The subtyping relationship has only been duplicated in the interests of clarity, and this is why one is drawn as a broken line. It is actually implied by the other, and is redundant.

A relationship is a special kind of aggregate object. It conveys the meaning of an association between objects – perhaps even the meaning of an interaction an object instance has with itself. For example, a person may represent himself (or herself) in a court of law and thus interact with himself (or herself). Thus, a relationship may list an object multiple times. This is why it is a kind of list, not set. At a minimum, a relationship must relate at least one object, and it could relate more. This is why a relationship cannot be the empty set, and figure 31 tells us that it is not.

The subtyping relationship is a special kind of relationship – a relationship in which two object classes share the instance identifier (section 3).

The basis of all systems, indeed of all symbolic thought and human language is the *represent* relationship between objects. It is also a special relationship, where one object is a token for another. This relationship will be discussed in box 36, and its many manifestations will be discussed in Chapter 4.

An object class is set apart from other aggregate objects by the fact that it is a collection of those attributes, the values of which define the state of object instances it classifies.

Figure 31 shows these rules. Naturally, instances of relationships and aggregate objects will also belong to a class.

Object class is a token that tells us that instances have been classified, based on their common properties. The state space of the object fills out the detail. It tells us what these properties are, and how they are scaled. The existence of a state space is implied by the existence of the object class. State space is the collection of attributes that defines the class. Figure 31 shows this. We could have shown this relationship equally effectively with a relationship between *object class* and *state space*. However, the relationship would be redundant, given the relationship between *object class* and *attribute*.

Indeed, all objects, be they classes or instances, exist only in a perspective. The perspective is a model. It determines how items are grouped and related. For instance, each item in figure 31 is an object. The entire pattern in figure 31 is also an object and a perspective (see section 4).

The remainder of this book examines the internal structure of each subtype of the universal metaobject in figure 32. It is these patterns that forge the differences. Figure 31 articulates rules common to all objects. The objects in figure 32 and the others in this book are all merely different states of the metaobject.

The hierarchy in figure 32 does not necessarily show mutually exclusive states. Some like static and dynamic relationships are mutually exclusive, whereas others like synonym and homonym can coexist at the same time.[76] How much our basic inventory in figure 9 has expanded! The metamodel of knowledge is configured from these metaobjects, a structure that emerges from their natural relationships, states, and properties (see Module VII).

Windows into objects

How can we look inside this structure, at contents, instances, individual states, and relationships? For this, we need a *view* – a window into the object. This window is an *interface*. It does not belong to the business rule layer of the architecture of knowledge (figure 15). It is a *mechanism* for *accessing* and *presenting* information to an actor, and hence the mechanism belongs to the business process automation layers. The window into an object is an aggregate object with a structure that consists of components across two layers – a connection in the information logistics layer to the object being viewed, and presentation mechanisms in the interface rules layer.

The presentation mechanism will consist of a rule about the access sequence and another about its presentation format. The presentation layer will also have screens or other mechanisms for the physical display of the contents of the object (for example, instead of being displayed on a screen or printed report, the information may be presented in spoken words). We will call this the display method. Display methods may be screens or other methods that can *present* data to actors. In figure 33, it is labeled "Display." Every view must have at least one display method. Without it information cannot be sensed, and hence the view cannot exist.

[76] Guard conditions are subtyping criteria that signal the absence of an effect. *Subtype* and *guard condition* are therefore redundant in the metamodel. This is why *guard condition* is not shown separately.

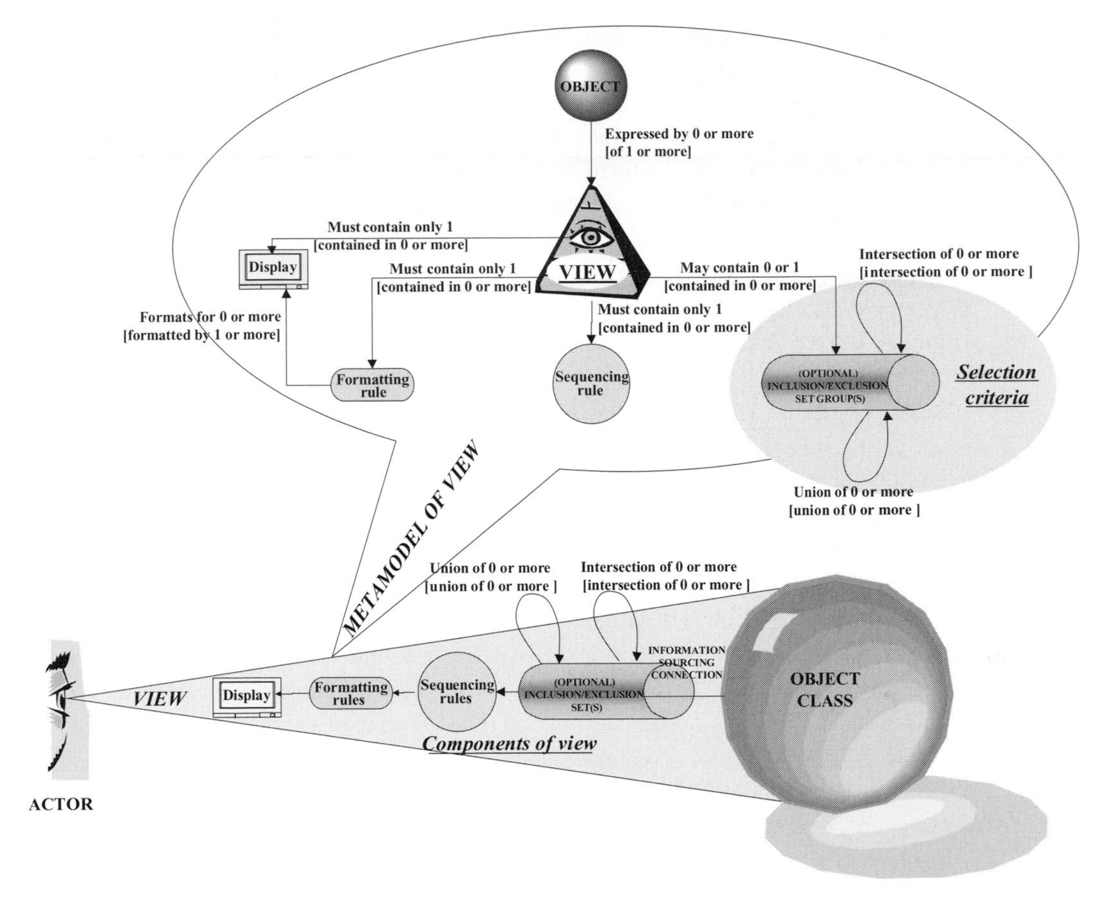

Figure 33 Views are aggregate objects

Sequencing rules are not necessarily rules about time sequences.[77] They could be rules about sort sequences that depend on values of the instance identifier and values of attributes of object(s) involved.[78] Sort sequences are mapping rules that map the item to a position, or rank in one dimension. In general, the mapping rules could map the contents of objects to two- or three-dimensional displays or multidimensional arrays. Display formats may be simple, like reports and lists, or sophisticated like diagrams, maps, and charts. However, the focus of this book is on the uppermost business, not process automation layer of figure 15, and we will not dwell on this discussion of format any more.

Inclusion and exclusion rules may also be optionally attached to these views, either in the form of selection criteria or as access permissions. Selection criteria and access permissions may be either for entire object classes or sets of attributes. Sets may contain only one attribute, in which case it becomes an attribute level permission/access criterion. These

[77] UML recognizes sequences in views. It applies "{ordered}" to the label of a relationship to show sequenced access to an object (UML: see box 22).

[78] Sort sequences could be a form of coercive polymorphism. See coercive polymorphism in the endnote on the theory of categories.

components are identical to the inclusion and exclusion sets that will be discussed in Chapter 3, section 2.[79] The only difference is its role and usage. Here it is an intermediary between the data mapping mechanism in the interface rules layer and the data flow mechanism in the information logistics layer, rather than a constraint on permitted values of attributes (as in Chapter 3). Thus, as for many other structures, constraints are *polymorphic*, i.e. they can masquerade as different objects when they are combined, or attached to different structures. We will expand on polymorphism in Chapter 3, section 2 and Chapter 4. For now, it will suffice to understand that the same object can play different roles as components in different structures, and the role could disguise its substance; we must look beyond a role within a structure or perspective, and understand the substance of the object – the pattern of information that makes it what it is – to forge reusable components.

Selection criteria in views may be complex. Objects may have several attributes and relationships. Therefore selection criteria for views and permissions may involve intersections and unions of inclusion/exclusion sets attached to individual attributes. Selection criteria may be components of the view that are derived from other selection criteria in it.

Selection criteria and views may be dynamic. Processes may not only keep changing parameters, like bounds and values inside inclusion and exclusion sets (see Chapter 3, section 2), but also value sets within the view, and the views themselves. Indeed, actors might set these parameters, or change display formats and sort sequences within a view dynamically (i.e. change the state of a view) depending on a host of factors such as the actors involved, formatting rules (see box 38), and the state of the object being viewed.

The view is an aggregate object that depends on key components for its very existence. If the view loses its display method, formatting rules, or the sequencing rules, it ceases to exist – all three are needed to make the object and its contents visible to an actor. It is worth noting that the sequencing rule may be a random access rule (although this is unusual, the metamodel permits it) and that a view may be used with distinct information flow rules if the object is fragmented or replicated in multiple files. The link to an object must be physically implemented by rules of information flow (see the information logistics layer).

Although the view is an aggregate object that depends on key components for its existence, the components, in contrast, do not owe their existence to any view. They are independent subassemblies of knowledge and do not need a view to give them meaning or existence. They are reusable components and may be reused in several views as well as in other subassemblies of knowledge we will discuss later in this book.

Note also that:

[79] Inclusion and exclusion sets in views may be more general than the "value constraint" of Chapter 3, section 2:

- Value sets need not be proper subsets of the domain (proper subset: see box 19)
- Value sets might be generalized to include formatting and unit of measure rules. Unit of measure rules will only apply to quantitative attributes, i.e. in addition to values, the set (and hence selection criteria) may contain attribute expressions (attribute expression: see box 35. See also the endnote on lambda calculus and functional programming)
- The rule expression in figure 48 would also involve formatting and unit of measure rules
- An optional *permission* component might mediate between the generalized value set above and the *constrain* relationship of box 28.

The attributes of the permission metaobject would be: Permit/disallow visibility; Permit/disallow update.

1 An object may not have any views attached to it. However, if this is the case, the object is invisible to process automation and is only an abstract concept.
2 There may be several inclusion and exclusion sets attached to the view. The *group* of inclusion and exclusion sets involved is an aggregate object that is a component of the view. Members of this group cannot be mutually contradictory. They must be consistent with the rules for combining inclusion and exclusion sets described in Chapter 3, section 2 to co-exist in harmony (in some cases may even merge and blend into each other)
3 Sequencing criteria inside views can be complex when we consider views that drill into aggregate objects (see Module V, section 2 on our website).
4 In addition to the display methods, sequencing, formatting, and inclusion/exclusion rules shown in figure 33, a view may have information update permission attached to it. That is a state of the *view*, the aggregate object.

We will not dwell further on this aspect of business process automation. It will suffice to understand that several views may be glued to objects, and that these views are aggregate objects themselves. Components inside views may be reused in other views, and, indeed, this provides us a basis for subtyping and partitioning views of objects, which are objects in their own right, albeit in the business process automation layers of figure 15. Our focus is on normalizing *meaning*, not process automation. Indeed, as we understood in Chapter 1, unless we normalize meaning, rules of process automation cannot be normalized (also see box 7).

The universal metaobject

Remember, in the metamodel of knowledge, the universal metaobject of figure 32 rules every other metaobject (and object). Indeed, it *is* every object – it underpins the essential unity and meaning of objects. Objects look different only because the metaobject masquerades as different objects in different perspectives and states. It is these states of the metaobject that normalize different perspectives of knowledge.

Remember also that the state of an object at any moment is the *set*, or *collection*, of the values of its attributes and relationships at that moment. The key to an object's behavior is its state, and attributes are the basis for recording state. Attributes provide the most common basis for partitioning object classes to differentiate the behavior of objects in business processes. State indicators distinguish the behavior of subtypes, even when objects are partitioned purely on their behavior or history. State indicators are attributes too.

Therefore, we must understand the nature of attributes to understand the nature of reality. Attributes are the key to the door that leads to domains, relationships, processes, effects, constraints, and all the other entities of figure 32. Understanding the nature of attributes will be our next step on the road to the metamodel of knowledge.

3 The nature of attributes

"Look to the rose that blows about us – 'Lo, laughing,' she says, 'into the world I blow: At once the silken tassel of my purse Tear, and its treasure on the garden throw.'"

(From the Rubaiyat of Omar Khayyam)

This chapter elaborates on properties of objects and constraints. It describes the metamodel of *State*, its components, configurations, patterns and constraints.

An attribute is a special kind of object. It is the *repository of an irreducible fact that a specific rule or property of an object class* ***exists***.[1]

Every object must have at least one attribute – the instance identifier that asserts that *the object itself exists*.[2] This instance identifier must always be a nominally scaled attribute because it asserts a nominal irreducible fact – the existence of an object instance – whereas other properties of object may map to nominal, ordinal, difference, or ratio scaled domains.

1 The structure of attributes and states

Every attribute has three items of information associated with it – the *object* it describes, the *kind* of property it represents, and its *value*.[3] The value is the actual property the object instance possesses. For example, *length* is a kind of property and the length of a room is an attribute of an object class called *room*. The fact that a room is 20 feet long asserts the (value of) *length* of a specific room, i.e. an instance of *room*. Formally, this model of attribute is

[1] See Chapter 4, section 3 and the endnote on how attributes emerge from relationships between temporal objects and domains.

[2] See Chapter 2: "*What is an object – really*".

[3] Our metamodel is a purely deterministic model. We associate three items of information with each attribute – its existence, domain, and value. We do not recognize the inherently uncertain nature of the real world, although we try to compensate with default initial states and unknown values of attributes. Metamodels that support uncertainty would have to associate additional items of information with each attribute. For ordinal and nominally scaled attributes we would assign the probability of the attribute taking that value. For difference or ratio scaled attributes, we would substitute value with *tolerance*. *Tolerance* is a range of values that the attribute may take. For each such range, we would assign a probability or degree of certitude that the value will fall somewhere in this range. For more information, see Chapter 1, section 5, Chapter 7 of [309] or "Representing uncertain facts" in Chapter 4 of [298] under "Representing knowledge".

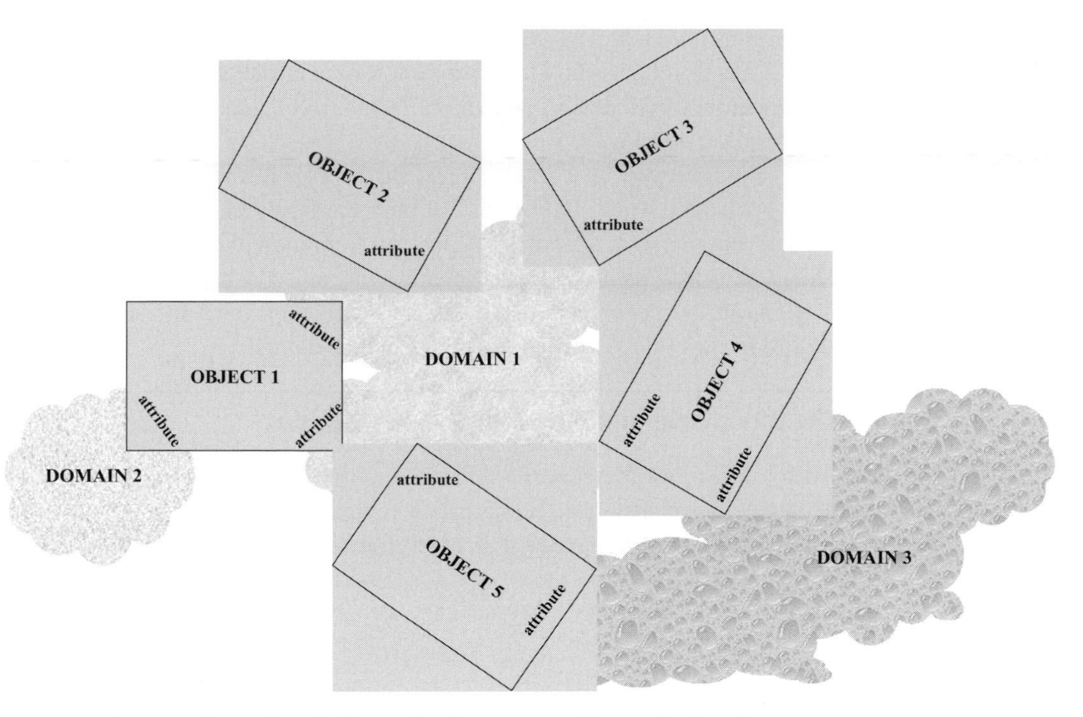

Figure 34 An attribute is the overlap between an object and a domain

called the object–attribute–value model, or OAV model for short. The object, its attribute, and value are together called the OAV triplet.[4]

Every attribute must map to a domain. The domain determines the *kind* of property the attribute describes. The domain also tells us how the attribute must be measured and what properties of the attribute are shared with others of its kind.

For example, *car* and *paper* are objects. The color of a car is a property of a car that maps to the domain of colors, as does the color of a sheet of paper. Thus, two properties of two very dissimilar objects can map to the same domain.

Indeed, an attribute might be thought of as an overlap (intersection) between the domain with shared properties and an object with specific properties. The attribute inherits all properties of the domain and adds properties specific to the object class it describes. An attribute is a piece – a subtype – of the domain embedded in the object class.

For example, *person* and *room* are two distinct object classes. A person's height is an attribute of *person* that maps to the *length domain*, as does the width of a room. The length domain is a ratio scaled domain with units of measure and conversion rules that will apply to both the height of a man and the width of a room. On the other hand, some constraints may apply to a person's height, but not to the width of a room. For example, a person's height may be constrained to be less than 10 feet but the width of the room might be unconstrained. The constraint on height is a specific rule added by the object class *person* to the *subtype* of

[4] For more information, see object–attribute–value triplets in Chapter 4, of [298] under "Representing knowledge".

the length domain embedded in it. This is similar to the two different intersections domain 1 has with objects 2 and 3 in figure 34. One intersection is an attribute of object 2 and the other of object 3, but they both map to domain 1. Each object could add its own rules to those it inherits from the domain 1.

Sometimes several attributes of a single object might map to the same domain. For example, a room has height, width, and length. All three map to the same (length) domain. Special rules could apply to each attribute, independent from those that might apply to the other two. This is similar to the two distinct intersections object 1 has with domain 1 in figure 34. Each is an attribute of object 1 and both map to domain 1. Each intersection could independently add its own rules to those it inherits from the domain.

Box 27 The OAV model and the structure of attributes

Figure (a) below describes the concepts in figure 34 with greater precision. It is a fragment of the metamodel that shows how attributes emerge from relationships between metaobjects. Read it as you did figure 8. It is clear from the figure that the attribute is a

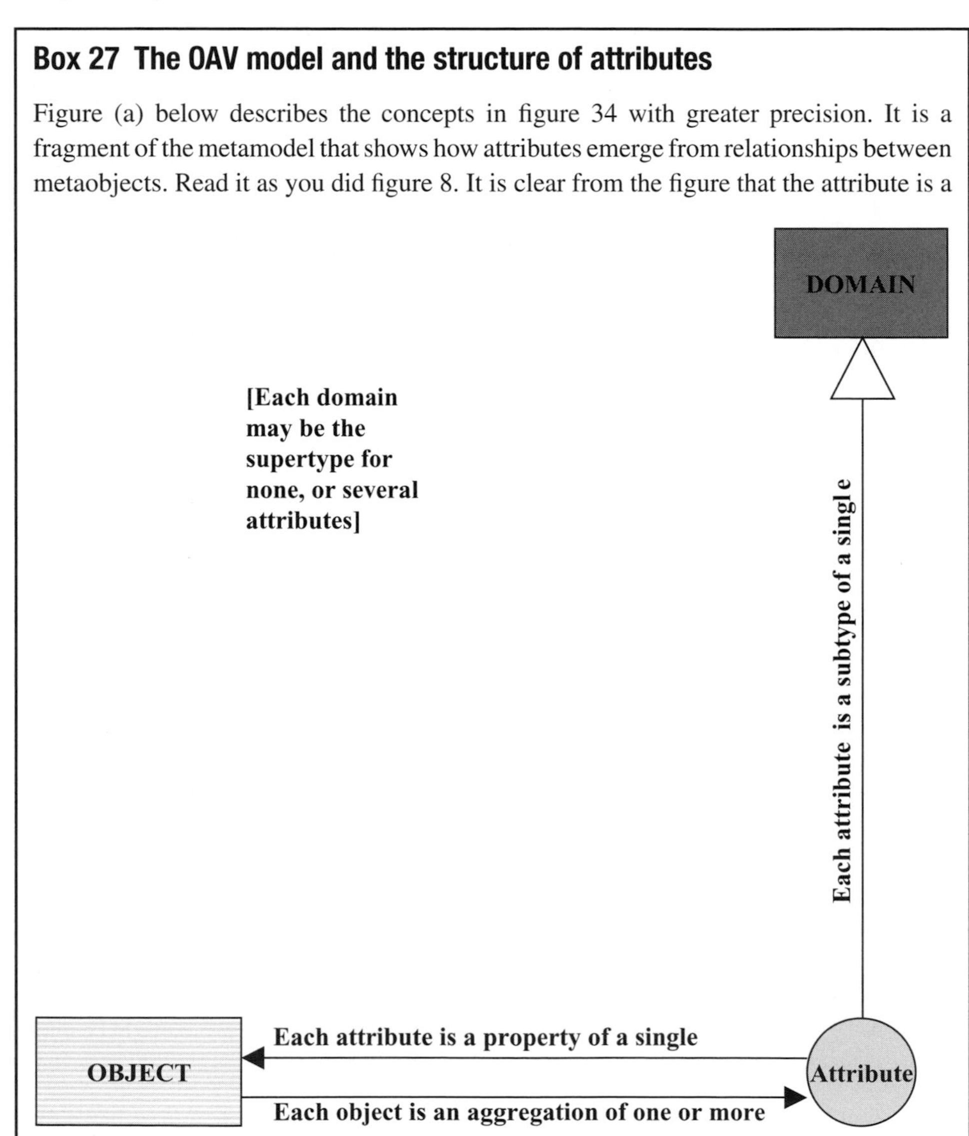

(a) An attribute is the intersection of an object with a domain

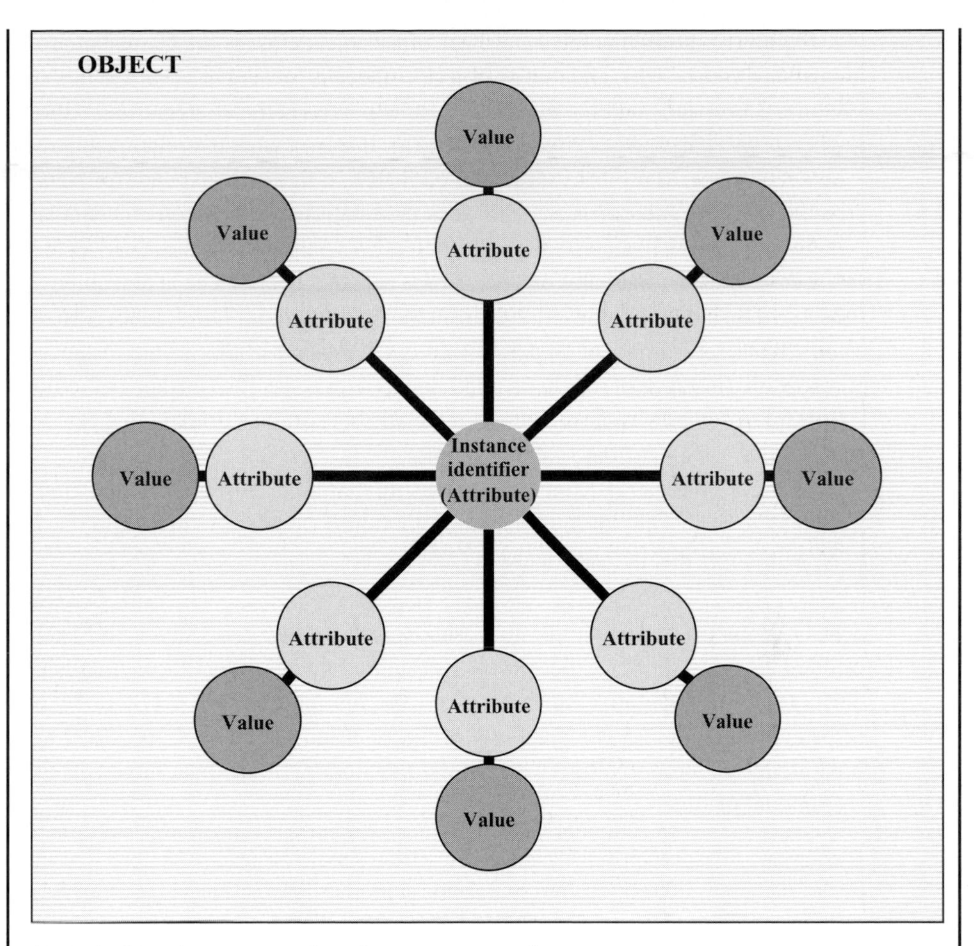

(b) Each object is an aggregation of one or more attributes

role of domain that emerges from its relationship with object class.[5] It is also clear that an object class is a collection of attributes. Figure (b) shows this graphically. In figure (b), a special attribute, the instance identifier, signals the existence of each object instance and all the other attributes represent its properties. Each attribute also has a value. Values and attributes collectively describe the state of the object. The instance identifier is the hub of this pattern. Without the hub, the entire structure falls apart and ceases to exist. This is what happens when perspectives change. Old objects dissolve and new objects crystallize out of the same attributes.[6] Thus, attributes are reused between perspectives and continue to map to the same domains, as they *must*, because they *are* fragments of domains.

[5] Chapter 2, sections 2 and 3 describe subtyping and aggregation briefly; Module V, sections 1 and 2 on our website have more detail. Chapter 4, section 3 discusses relationships between objects and domains in detail.

[6] See Chapter 2, section 4.

Figure (b) shows that the state space of an object class is the Cartesian product of its attributes (see box 20) and that an object instance is located at a point in this state space. Figure 35 expands on this. Figure (b) also expands on the aggregation relationship between *object class* and *attribute* in figure (a). It shows what the relationship means to an object instance. Many an astute reader might ask, if figure (b) expands on the relationship between *object* and *attribute*, what does the relationship between *attribute* and *domain* in figure (a) mean to an instance of the object? An example will best answer the question.

Let us consider the color of a car. *Car* is an object class. The actual color ("value" of color) of an individual car is drawn from the *color domain*. The domain is an object class too – it is a set – the set of all possible colors. Members of this set are instances of *color*. A specific (instance of) car has a relationship with a specific (instance of) color. It is this (instance of) the relationship between the attribute and the domain that is manifested as the "*value*" of an attribute called car color.

Thus, *values* emerge from the relationship that maps attributes to domains. Domains normalize values and things we can do to values (as we saw in Chapter 1, section 3). We will cover this in more detail in Chapter 4.

The OAV model is too simple to normalize all atomic rules about attributes. We have omitted the finer structures that relate to formats as well as those that distinguish quantitative attributes from qualitative attributes from the figures in box 27. Figure 35 includes this information. Figure 35 is a synthesis of figure 8 and figure A of box 27. It integrates the metamodel of domain in figure 8 with the metamodel of attribute in box 27 to give us the metamodel of *state*.

(An inverse relationship shows how a relationship should be read in the opposite direction from that of the arrow. In figure 35, inverse relationships are shown in square brackets like this: [inverse relationship]. Module V discusses inverse relationships. Also see *inverse of a function* in the endnote on the theory of categories.)

Figure 35 differentiates the special structure of qualitative from quantitative attributes. Box 27 described properties shared by both, whereas figure 35 describes how qualitative and quantitative attributes are different. The subtyping relationship between *attribute* and *domain* is identical to the relationship between *attribute* and *domain* in figure 27. Figure 35 adds key subtypes of this relationship.

In figure 35, it is clear that the state of an object is a collection of attribute values. A relationship identifier is an attribute too (see *participation in relationships*); hence relationships too are determinants of *state*. It is also clear that qualitative attributes emerge from the subtype that maps to *qualitative domain*. This relationship too has subtypes, one of which maps to *ordinal domain*, from which ordinal attributes emerge, while the other maps to *nominal domain*, from which nominal attributes emerge. Quantitative attributes, and its two subtypes, difference and ratio scaled attributes, similarly emerge from subtypes of corresponding relationships between attributes and domains.

Quantitative attributes are a subtype of quantitative domains, which are expressed in units of measure and formats. Quantitative attributes inherit these relationships and must therefore be expressed in units of measure and formats to become tangible in the real world.

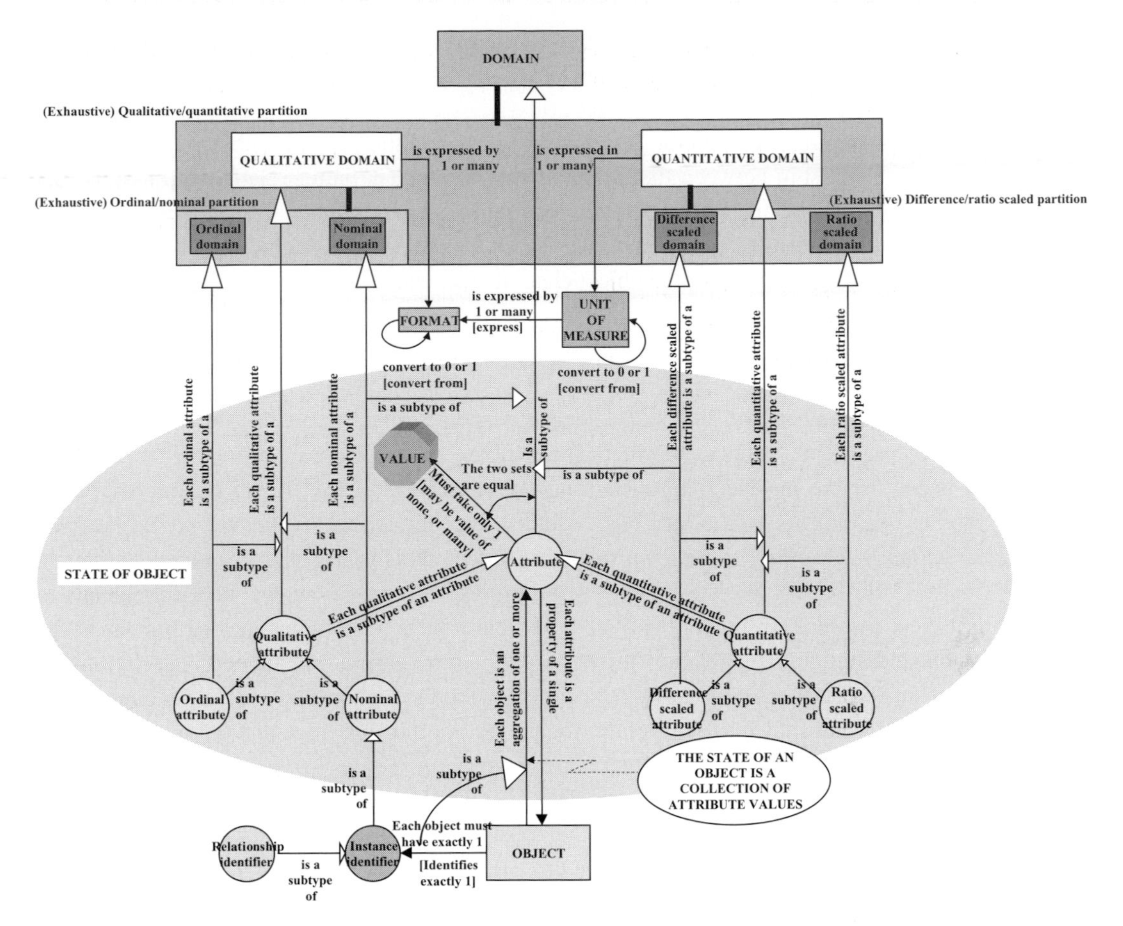

Figure 35 The structure of attributes and metamodel of state

On the other hand, qualitative attributes are subtypes of qualitative domains, which are expressed only with formats. Therefore qualitative attributes only need formats to become tangible in the real world and units of measure have no meaning for them.

Instance identifier

Ordinal and nominal attributes are subtypes of qualitative attributes. A special kind of nominal attribute, the instance identifier, has a special relationship with *object*: Each object instance must have *only* one of each, *no more, no less, because it is the identity of the object instance*. The instance identifier labels the "bag" of attributes and their values. It is the hub of the pattern in Figure (b) of box 27. For this reason, not only must every object have one, but also no other object instance may share it, i.e. it cannot have the same value for any other object instance.[7] (Note that an object instance may simultaneously be a member of a

[7] See Chapter 2, section. 2, "*What is an object – really*".

subtype in one or more partitions. Subtypes are object classes too. Hence identical values of the instance identifier could be found in more than one object class, but *never* in more than one object instance.)

This relationship between the *instance identifier* and the *object* is subsumed in the more general relationship between the *attribute* and *object*. The latter relationship asserts that each object must be an aggregation of *one* or more attributes. The instance identifier is that *one* attribute that must be present. Once it exists, the *option* of including other qualitative and quantitative attributes in the object class also exists.

Participation in relationships

The relationship between an instance identifier and an object asserts that an object instance can have only one instance identifier. The attribute cannot share its instance identifier *role* by becoming the instance identifier of another object. However, it can be an *attribute* of another object. This kind of attribute asserts an irreducible fact – *that the object instance participates in a relationship with the other object*. This is why the *instance identifier* in figure 35 has a subtype, the *relationship identifier*. Figure 36 shows what this means to an object instance. Instance identifier 6 of figure 36 is a relationship identifier. It is the instance identifier of the relationship between the three objects it connects – a relationship is also an object. Remember, as with the other attributes, each instance identifier has a value, although it is not shown to avoid cluttering the figure. It is this value that uniquely identifies the instance of the object class under consideration.

Some relationships can be complex. They can involve multiple instances and objects. At this stage, it will suffice to understand that attributes can represent an object's participation in relationships, and hence its state (see box 10).

Attribute value

Now we are ready to address the question of *value*. Value is at the heart of the attribute, and indeed the reason it exists. *Value* is also an object class in the metamodel of knowledge. We know that attributes normalize irreducible facts about the state of an object and values of an object's attributes determine its state. We also know that attributes are subtypes of domains and domains are sets of values.[8] As we saw in box 27, *value* is buried in the relationship between *attribute* and *domain* and a value must always be associated with an attribute. Figure 35 makes this explicit. The relationship between *attribute* and *value* shows that an attribute must always have only one value.[9]

Asserting this is important because it is not implied by the relationship between attribute and domain. Domains are sets of values and also object classes. Subtypes, and hence attributes too, are object classes. Object classes are sets (collections) of object instances.

[8] Domains consist of values *and* valid operations on those values. To focus on *attribute value*, we have ignored these operations. See Chapter 4, section 3.

[9] The inverse relationship asserts that (1) there may be values that no attribute has assumed at the time and (2) several attributes may assume the same value. For example, the length and breadth of a room are two different attributes of an object called *room*. They have the same value when the room is square.

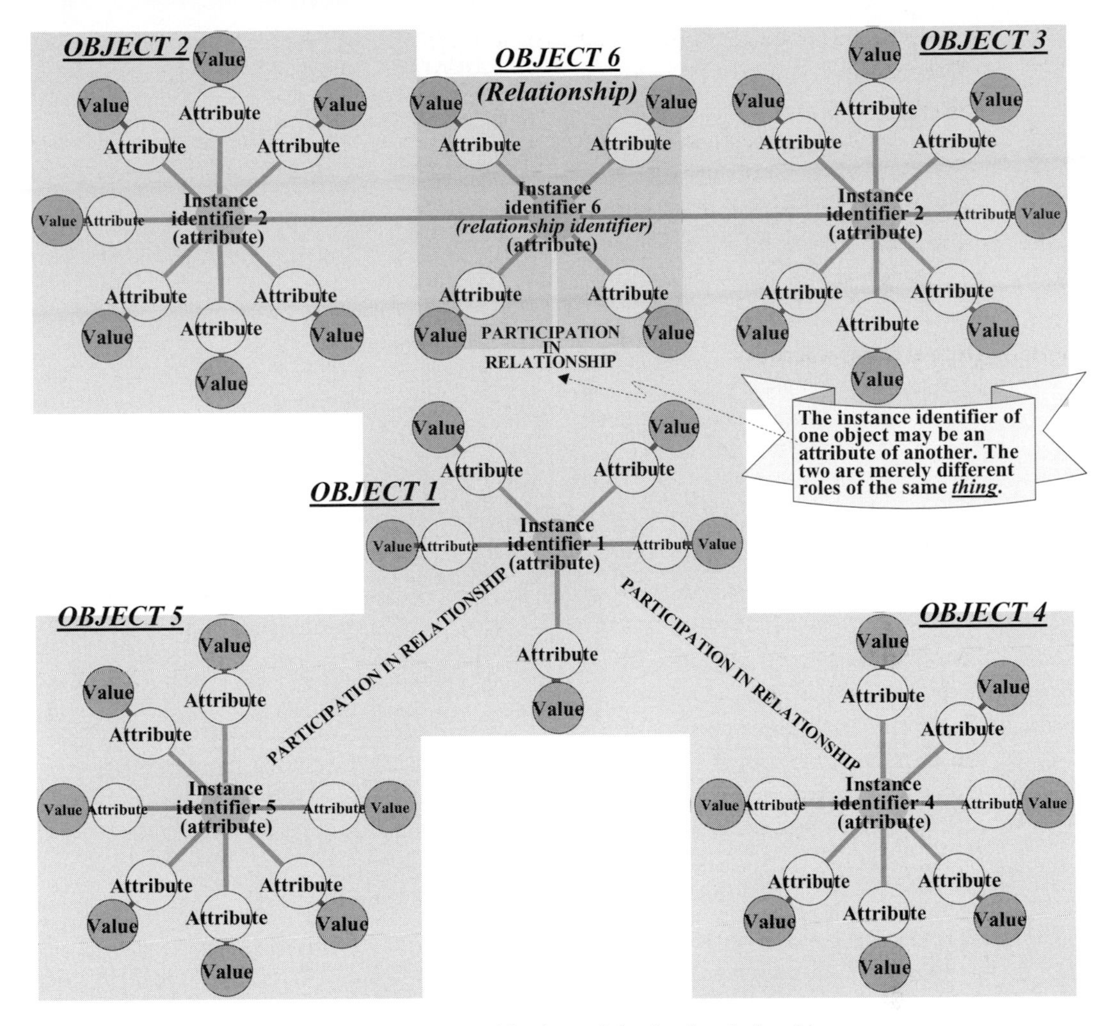

Figure 36 Attributes can represent an object's participation in relationships

Therefore, merely asserting that attributes are subtypes of domains does not exclude the possibility of an attribute being a collection of several values. This is patently false; an attribute has only one value. The relationship between *attribute* and *value* makes this clear.[10]

The subtypes of the relationship between *attribute* and *domain* also imply the existence of *qualitative value* and *quantitative value* (not shown in figure to avoid clutter), which will be subtypes of *value*.

Quantitative and qualitative values have subtypes too. Just as *quantitative* and *qualitative value* emerge from *value*, *nominal*, and *ordinal value* (not shown in figure 35) emerge from subtypes of *qualitative value*, whereas *difference* and *ratio scaled value* (also not shown) emerge from subtypes of *quantitative value*.

[10] In a non-deterministic metamodel, the relationship between *attribute* and *value* could have other attributes like the probability of an attribute assuming that value.

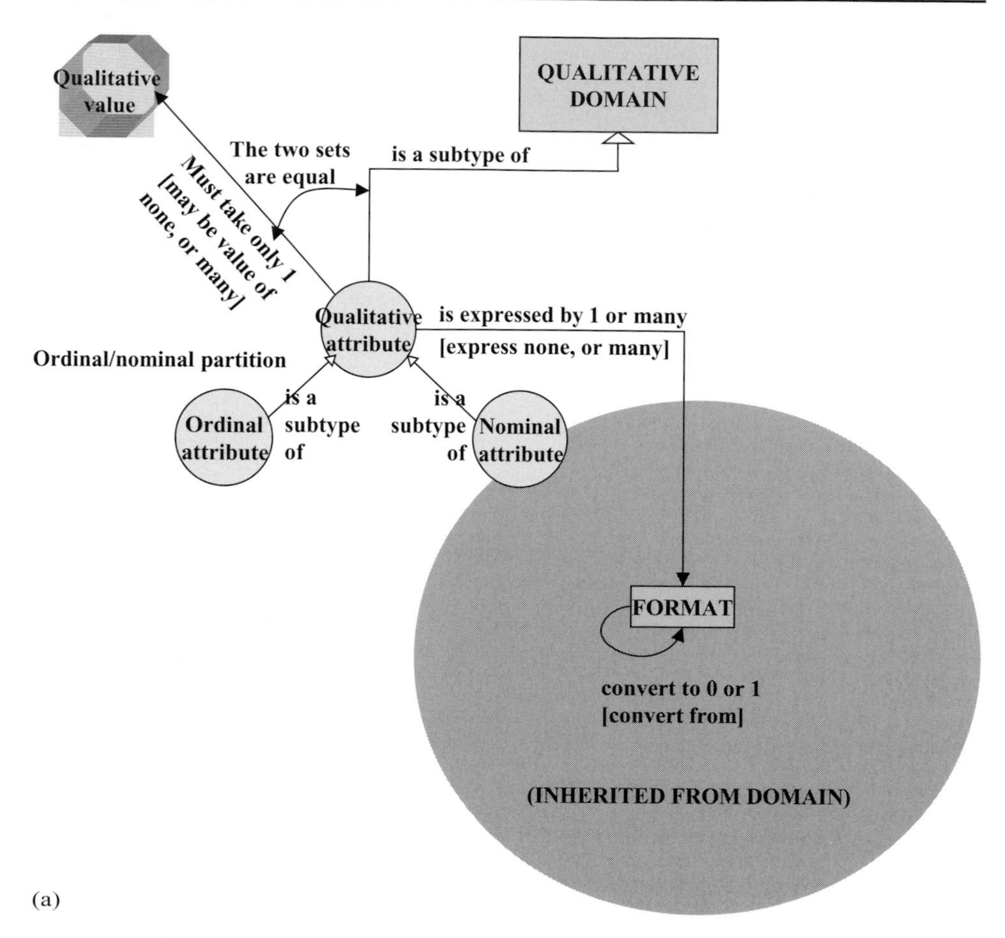

Figure 37 Attributes and values inherit properties of domains
(a) Qualitative attributes need only formats
(b) Set equality implies qualitative values need only formats
(c) Quantitative attributes need UOMs and formats
(d) Set equality implies quantitative values need UOMs and formats

Whenever an attribute maps to a domain, which it always must, this association between *attribute* and *value* will exist. From a slightly different viewpoint, we could assert this mandatory co-existence in the reverse order: whenever an object has a value, which it always must, the attribute must be a subtype of domain from which the value is selected. Each relationship implies the other and the two are inseparable. The double-headed arrow connecting these two relationships implies that neither relationship can exist independently of the other. In other words, since the two relationships are objects, and therefore sets, we can say that the two sets are equal (see box 19). Figure 35 asserts this.

The implication of this set equality is that quantitative values must be expressed in units of measure and formats, just as quantitative attributes are; whereas qualitative values, like qualitative attributes, need formats only. This is how structures in the meta-model of knowledge manifest themselves in the behavior of the real-world objects we

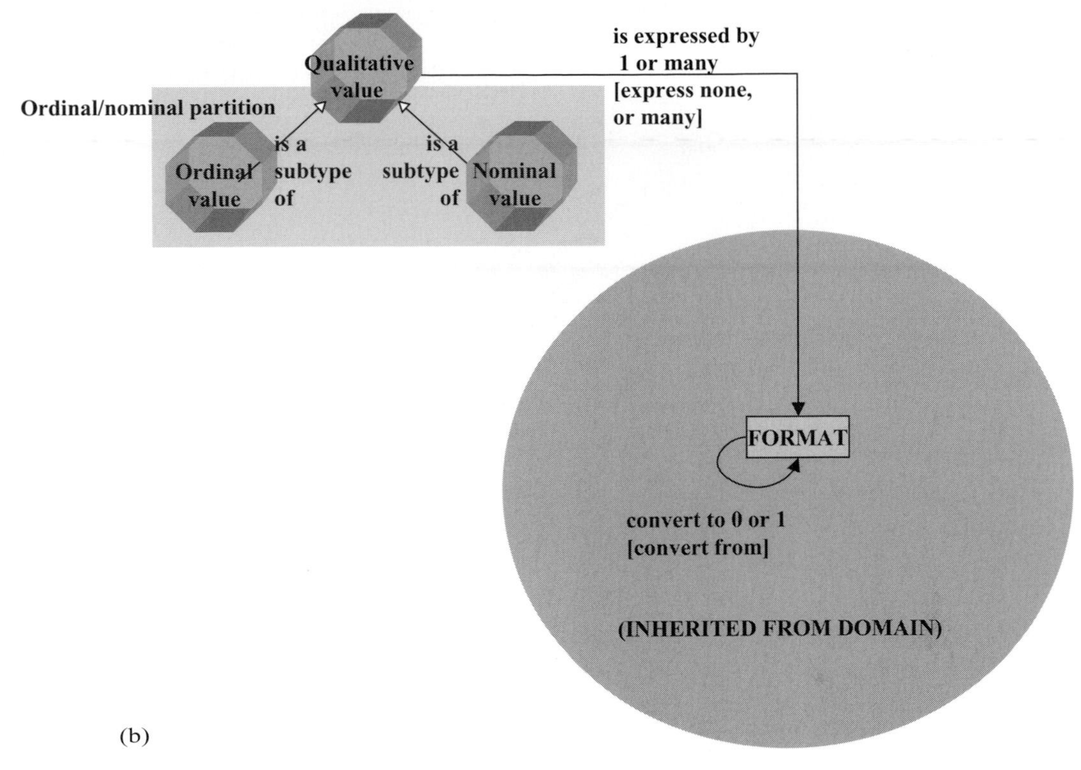

(b)

Figure 37 (*cont.*)

discussed in the parable of Metanesia (section 3 of Chapter 1). Figure 37 highlights this issue.

The metamodel of attribute and the metamodel of state

The metamodel of attribute is closely related to the metamodel of state because the state of an object is a collection – a set – of its attribute values. Both are represented by figure 35. The shaded area of figure 35 represents the metamodel of state, which is a part of the metamodel of attribute. The instance identifier too is an attribute of object, a special attribute that is its very identity. The state (value) of the instance identifier distinguishes an object instance from every other object instance.

However, the metamodel of attribute cannot be complete without considering constraints on states of objects. States are constrained by *collections* of one or more attribute value constraint. It is from this collection, a set of rules, that information about an object's lawful state space flows to it. Figure 38 shows how value constraints flow to objects through their attributes. Figure 38 is an attachment to the metamodel of attribute and an important subassembly in the metamodel of knowledge. To understand limitations on an object's state space, we must understand the structure of value constraint, and more specifically attribute value constraints. This will be our next step towards the integrated metamodel of knowledge – the pattern from which all meaning flows.

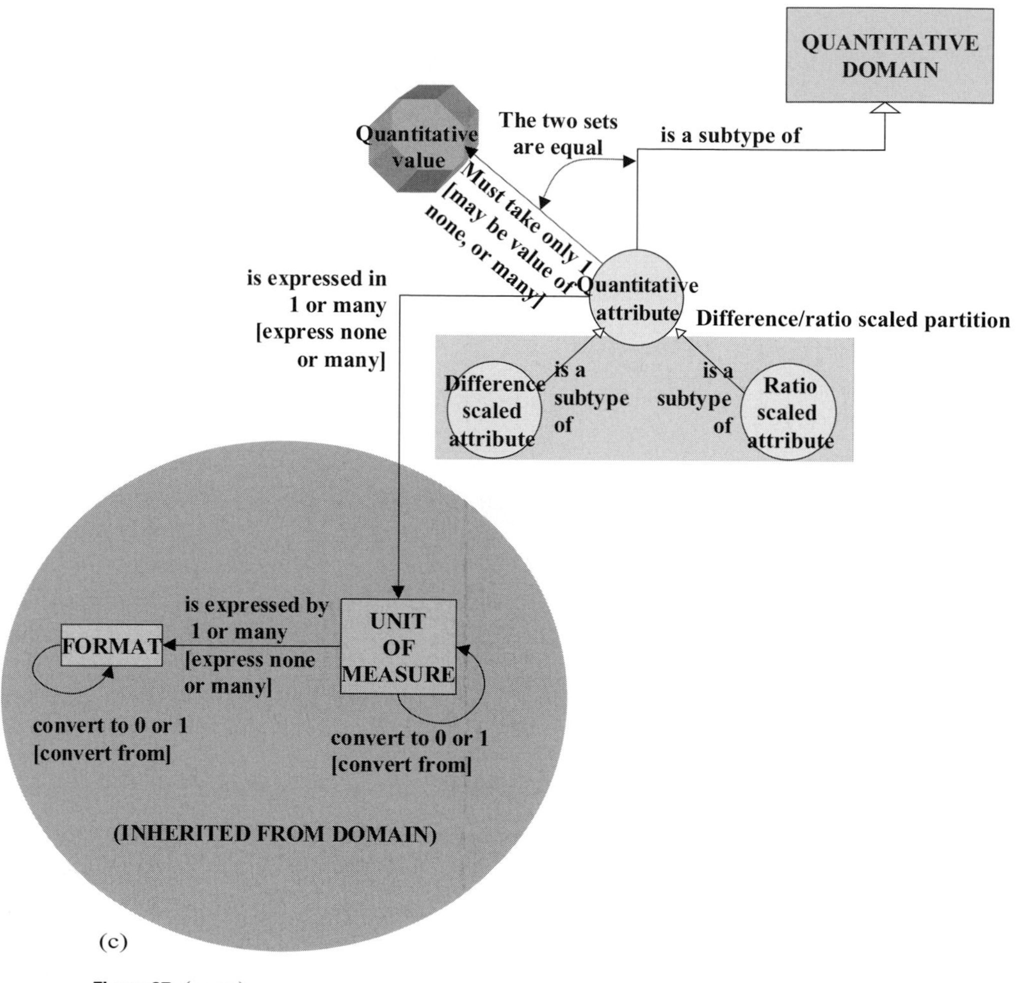

Figure 37 (*cont.*)

2 Attribute constraints

Here the bones of birth have cried –
'Though gods they were, as men they died.'
Here are sands, ignoble things,
Dropt from ruin'd sides of kings;
Here's a world of pomp and state,
Buried in dust, once dead by fate.
(Francis Beaumont, 16th century English poet,
"On the Tombs of Westminster Abbey")

Constraints certainly limit object states. They stop objects from being all they can be and doing all they can do. They stunt capability, destroy opportunity, and deny potential.

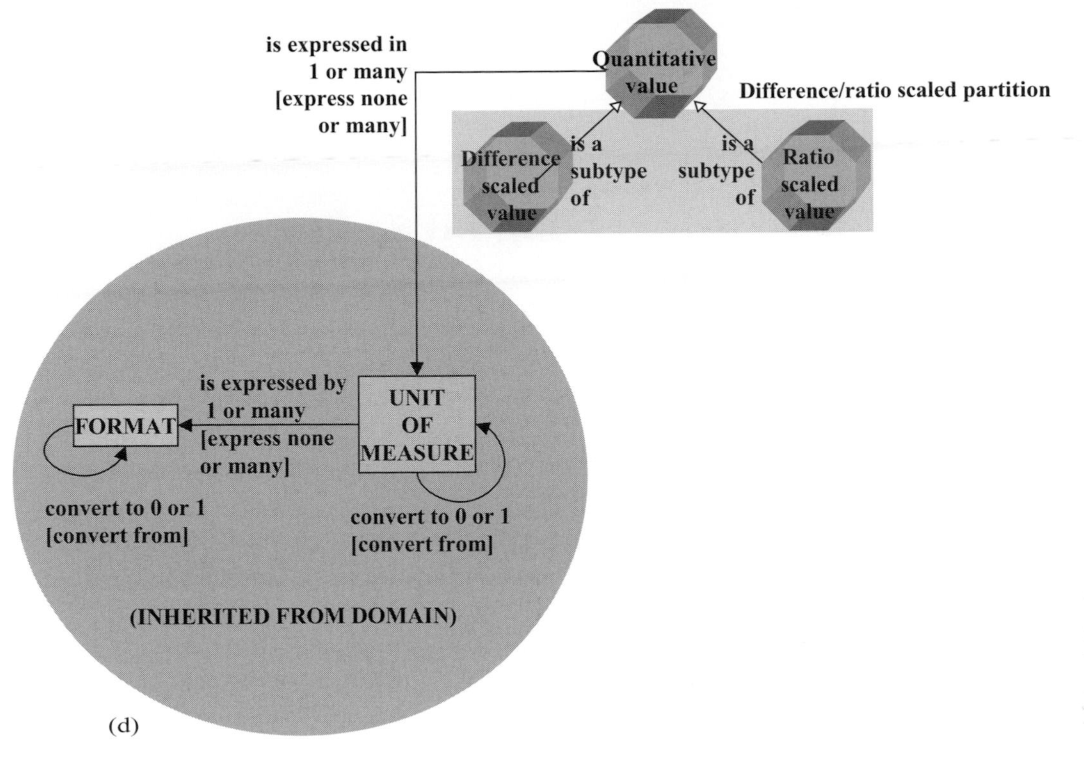

Figure 37 (*cont.*)

However, they do not lead to the kind of ruin Beaumont's verse suggests. Instead, constraints lead to business rules. Constraints frame reality, like death frames life, and pave the road to business procedures and information systems. Value constraints are special because they hem in an object's lawful state space, and state space, as we saw in box 12, is the door to information systems. Figure 37 tells us how attributes are expressed, and how each must have a value, but says nothing about how they constrain state spaces of objects. Physical reality and business rules sometimes dictate that some values of attributes are not permitted. This too must emerge from the metamodel of knowledge.

For example, physical reality dictates that the length of physical objects cannot be negative. Indeed, it cannot even be zero, but it can be too small to measure, hence we could *call* it zero, but it can *never ever* be negative. We must make room for laws like these in our metamodel. Constraints like these are attached to the domain. They constrain values of all attributes that map to a domain, and are inherited by each attribute from the domain, just as units of measure and formats were.

Other constraints may be specific to attributes of objects. For example, a standard shipping container might come in only two sizes, 40 feet or 20 feet. This constraint would apply only to *shipping container length*, an attribute of the object class *shipping container*. Constraints that apply only to specific attributes are not attached to the domain. Instead they are attached to attributes they constrain. Figure 38 (a) shows the fragment of the metamodel that attaches generic and specific constraints to attribute values (attaching the

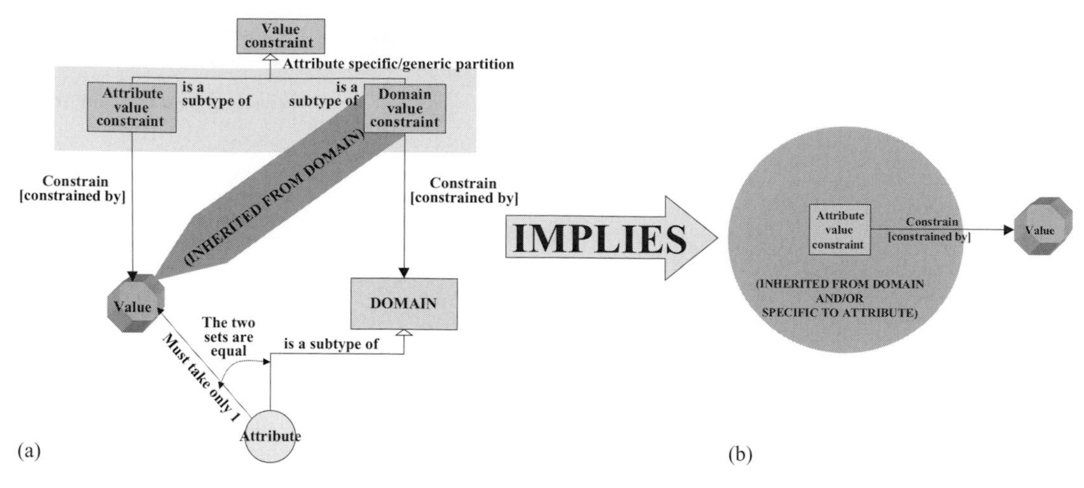

Figure 38 Attributes values may be constrained

constraint to *value*, instead of *attribute* means the same because of the set equality shown in figure 38(a)). Figure 38 (b) is a less formal, simplified diagram that highlights the meaning of figure 38 (a).

Constraints may depend on other attributes too – even attributes of other objects. For example, the height of a room in a building is constrained by the height of the building. However, we must understand simple constraints before we go on to those that are complex. Our intent in this section is to discover those structures that will normalize knowledge about constraints on attribute values. Let us start with the simplest of constraints – those that do not involve any other attributes or objects – and the simplest of attributes – nominal attributes.

Constraints on nominal attributes

Nominal attributes have only discrete values – values that carry no information on magnitude, absolute or relative. Therefore, constraints on nominal attributes too must be discrete values. For example, the gender of an earthworm can *only* be "hermaphrodite."[11] The *constraining relationship* may be of only two kinds (two mutually exclusive subtypes of the "constrain" relationship in figures 37 – also object B in box 28): values that are permitted (*only* values in the set are permitted) or values that are excluded (only those in the set are not permitted). We will call this set of permitted values an *inclusion set*, and the set of impermissible values an *exclusion set* (see box 28). For example, the inclusion set for the gender of a person has only two values – "male" and "female" (provided we ignore "hermaphrodite"). If inclusion sets sound like partitions, it is because they *are* partitions. Inclusion and exclusion sets are a *type* of partition based on the *value* of an attribute, i.e. certain *states* of the object. Inclusion sets are *inclusive partitions* and exclusion sets are *exclusive partitions*[12] (see "Properties of partitions" in Chapter 2, section 3).

[11] See the endnote on gender.

[12] Partitions play a different *role* in the metamodel when they constrain an object's lawful state space than when they discriminate between mutually exclusive subtypes. When a partition is an inclusion or exclusion set, the partitioning criterion (*discriminator* in UML terms) discriminates between *distinctly different values* of a *single attribute* of an object.

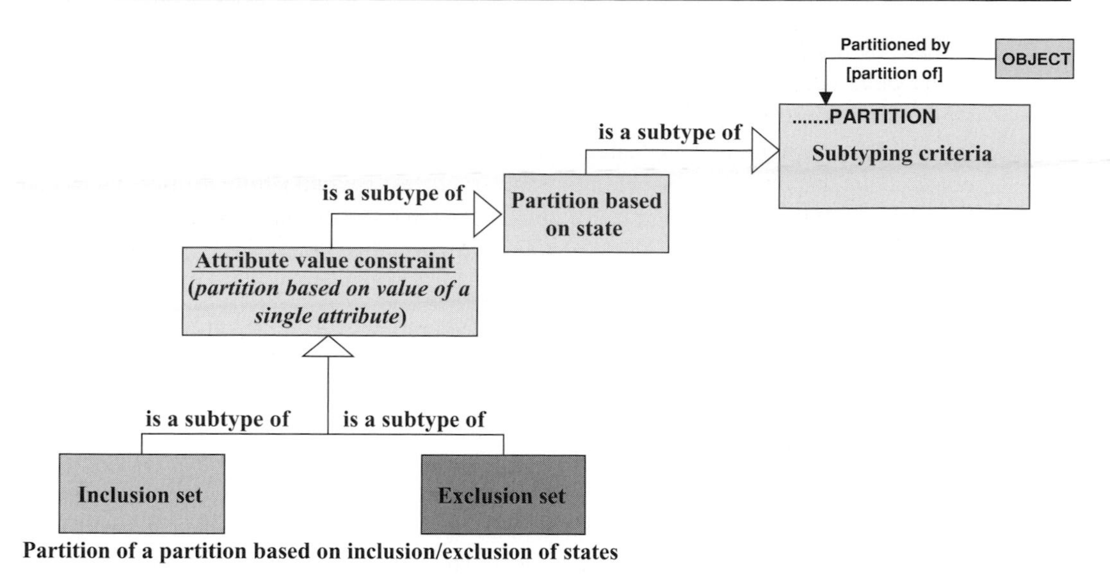

Figure 39 Inclusion and exclusion sets are mutually exclusive partitions

Inclusion sets

In Chapter 2, section 2, we saw how constraints shape and reshape the *lawful state space* of an object. Each value in an inclusion set is an atomic rule that asserts that the attribute could assume that value. For example, the assertion that an earthworm must be a hermaphrodite is an atomic rule. The rule cannot be divided into parts without losing information. Similarly, the fact that a person may be a male person is an atomic rule, as is the irreducible fact that a person could be a female person. However, perceptive readers might have noticed that there is a vital item of information missing from these assertions about gender. We have asserted that a person may be male or female, but where have we asserted that people can be *only* male or female and *nothing else*, or that the earthworm can *only* be a hermaphrodite? This assertion belongs to the *inclusion set object* (object C in box 28). The inclusion set is an aggregate object. It is also a set. The members of the set are permitted values of the attribute in question (the attribute the set is attached to). The *set* (aggregate object) itself asserts an atomic rule, that the attribute may take *one* of the values listed in it, and *no other*, i.e. it is an exhaustive partition.

What if we have incomplete information – we know that other values are permitted, but do not know what they are? There are two possibilities:

1 we know *how many* other values exist, but not what those values are; or
2 we know that the set (i.e. partition) is not exhaustive, but have no information on the number of unknown values (the number is *don't know* – see Chapter 2, section 2).

Both cases are theoretically possible. They do represent a kind of business knowledge, so we will discuss them, but they are of limited practical value – we cannot use inclusion sets like these to validate attribute values in an information system, or to define lawful state spaces of the object in question – so this discussion will be brief (see box 28 for more detail).

Take the first possibility. Each value, including those unknown, is a distinct value object in the value set (object A of box 28). Each value object that we are unsure of assumes the "unknown" value[13] inside this value set. As such, the inclusion set is an exhaustive partition, but one or more values in it are "unknown." This could be the state of an inclusion set under construction, but it has little value in a system or process beyond letting its users know that some states of the attribute (and hence object) exist, but their values are not yet known.

Now consider the second possibility. It is no different from a non-exhaustive partition.[14] However, the partition is not being used to classify special behavior, as it was when it was used to develop subtypes. Its only utility in its role as an inclusion set is to assert that it is an inclusion set in an incomplete state – perhaps one under construction.

In both cases, the attribute may assume an "unknown" value. Such attributes are sometimes called optional attributes. *However, the optional attribute by itself cannot distinguish between possibilities 1 and 2, which are two distinct pieces of knowledge about a business system, and therefore optionality of an attribute cannot, by itself, normalize knowledge.*[15]

There is another subtle difference between the optional attributes described here and those found in many information systems prevalent today: the latter do not distinguish "unknown" values from "does not exist."[16] The optional attributes in this section *must* exist, but their *values* may or may not be known. On the other hand, if an attribute has a null value, it does not exist and actors[17] need not even be *aware* of the attribute.

The structure in box 28makes the meaning of optional attribute more precise than most systems do today. If, for an inclusion set, the null value were among values in the value set (object A of box 28), it would imply that the *existence* of the attribute is optional.[18] This precision of meaning was not required when systems and processes were not assembled from components of knowledge, but hand crafted individually for each system. However, when an object and its state space represent a component of business knowledge stored in an electronic repository, and either automated agents (see box 36) or human analysts assemble systems from such objects, the designer of the object should be in a position to assert what aspects of the object's state space are optional, and what are mandatory. The ability to specify that the *existence* of an attribute is optional, independently from the ability to assert that the *value* of an attribute may be unknown, provides the requisite precision and capability.

Exclusion sets

Exclusion sets are exclusive partitions that are similar to inclusion sets. The same considerations apply, but there are two key differences in behavior we must recognize. Exclusion sets can validate attribute values even when the sets are (i) not exhaustive and (ii) we know

[13] See "Data and attributes" in Chapter 2, section 2.

[14] See "Object partitions and role modeling" in Chapter 2, section 3.

[15] See the endnote on the Bunge–Wand–Weber model: if we merely recognize the optionality of attributes, but not the structure in box 28, the methodology has a *construct deficit*. Attribute optionality cannot, by itself, tell us whether we do or do not know *how many* distinct attribute values exist – values that are unknown.

[16] See "Three values every attribute and relationship may have" in Chapter 2, section 2.

[17] Actor: a person, system, or instrument that accesses or processes information.

[18] If in an inclusion set, the null value is the only value inside the value set (object A) of box 28, it will imply non-existence of the attribute. Business models will never have this attribute. On the other hand, if the null value were among the values inside Object A of box 28 in an exclusion set, it would imply mandatory existence of the attribute in the model.

certain values are excluded, and we know how many, but we do not know what they are. In both cases, we know for sure that the attribute cannot assume a *known* value in the exclusion set. We can trap *some* violations on this basis, if they occur, but, even if the attribute value passes muster, we cannot be *certain* that the attribute (and therefore the object) has not violated its lawful state space because we do not know *all* its impermissible values. The known values in an exclusion set of this kind can only reduce, not eliminate, risk.

There is another subtle point we must not overlook – the case of a constraint that asserts that an attribute cannot assume the "unknown" value.[19] Asserting that an attribute's value must always be known is different from saying we do not know all values that are illegal for the attribute. To understand how structures in our metamodel distinguish between these two very different assertions, we must recognize that inclusion and exclusion sets are aggregate objects, and each member of the set consists of two components – a *subset of a domain* attached to a *relationship class*.

Consider the value set (object A) in box 28. If attribute 3 assumes the "Unknown" value, we will not know *all* values that might constrain the value of the attribute, *even if the "unknown" value is not inside object A*. On the other hand, if attribute 3 of object A is known (i.e., it has not taken the "unknown" value), we know for sure that the exclusion set is exhaustive. Then, if a constraining value *inside* object A is "unknown," it implies that the attribute is barred from ever assuming the "unknown" value, i.e. the attribute value must always be known (these are sometimes called *mandatory attributes*).

Mandatory attribute is a misnomer in a mathematically precise sense. What we really mean is that it is mandatory that the attribute have a *value* – not that the *existence* of the attribute is mandatory. As discussed under inclusion sets, this distinction is useful when systems and processes are electronically assembled from *knowledge artifacts*.

Constraints on ordinal attributes

Ordinal attributes, like nominal attributes, have only discrete values, but, unlike nominal attributes, they do contain information on relative magnitudes. Jane's car color preference in Chapter 1, section 3, was an example of an ordinal attribute. Therefore, like nominal attributes, it is meaningful to speak of restricting values of ordinal attributes to a set of discrete values, but, unlike nominal attributes, it is also meaningful to speak of restricting values of ordinal attributes to a *range* of values.

Take Jane's car color preference. She could meaningfully say that she adores blue cars, likes red cars a lot, green cars a little, is neutral about white cars, and detests black cars. While it is possible to rank Jane's color preference in order of magnitude, it is impossible to say *how much* she prefers one car color above another.

If Jim, the sales manager of a car showroom knows Jane's car color preferences, he could ask Charles, the car sales man, to show Jane cars of colors she is *at least* neutral about. Thus *neutrality* in preference for car colors is the *lower bound* of the *range* of car colors he will show Jane. In other words, the attribute is *person's car color preference*. The range is *neutral or greater car color preference*, and the *lower bound* of the range is *neutral car color preference*.

[19] See the subsection on unknown values in Chapter 2, section 2.

Box 28 Components of inclusion and exclusion sets

Exclusion and inclusion sets consist of two components that normalize different irreducible facts:

1 Set of values (object A in the figure). It is an aggregate object that contains each permissible or impermissible value. We will call it a ***value set***.[20]

2 Set of constraining relationships between these values and the constrained attribute (object B in the figure). It is an aggregate object that contains the set of relationships that assert whether the value in question is permitted or not.

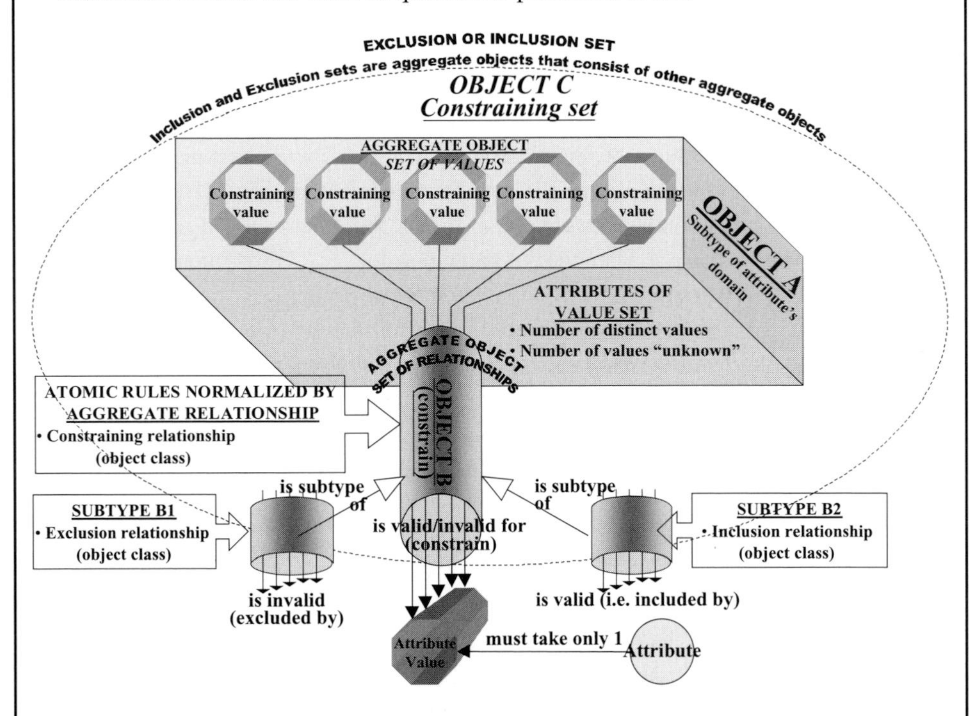

An attribute value constraint is an aggregate object

Every aggregate object must have at least one attribute, viz. the number of objects in it. We also understood that the set of constraining values (permitted or debarred values) might contain items that we know exist, but do not know what values they have.[21] Object A in the figure will therefore have at least three attributes (object A is an aggregate object):

1 The number of distinct constraining values that we know for sure exist, but we do not know what the actual values are.

2 The number of distinct constraining values that we know exist, and we know what these distinct values are

3 The total number of distinct constraining values in object A
(The sum of 1 and 2 above yields 3.)

[20] Inclusion and exclusion sets correspond to Ron Ross' "any" constraint. See [294], Chapter 7.

[21] See the subsection on "inclusion sets".

When the attribute 3 of object A takes the "unknown" value, the partition (and constraining exclusion or inclusion set) is non-exhaustive. Otherwise it is exhaustive.

The set of constraining relationships in the figure (object B) comes in two different flavors. Each is a subtype and they are mutually exclusive. One subtype (B1 in the figure) contains relationships that exclude the constraining value it is attached to, whereas the other subtype (B2) contains only relationships that permit the corresponding constraining value. B1 makes the constraint an exclusion set, whereas B2 makes it an inclusion set.

Note that the values in set A are called constraining values only because of the (instances of) *constraining relationships* in set B, i.e. these values are playing this *role* via those relationships. They could just as easily play the role of being the value of a different attribute at the same time, or even that of an instance identifier. Then the constraint would be more complex. The value of the attribute in the figure would be constrained by the value of another attribute (not shown in the figure). Thus, *to normalize knowledge, it is appropriate to drop the qualifier "constraining" for the values in set A*. They are just values. The qualifier is implied by the relationships in set B (We have retained "constraining" in the names of values in object A only because, in this section, we are focusing on the role they play in limiting the state spaces of objects.)

Domains too are sets of values; therefore object A, the *value object*, is nothing but a subset (subtype) of a domain. Thus, in the metamodel of knowledge, an attribute value constraint is a structure in which an object class – constraining relationship – is glued to a subtype of the domain that the constrained attribute maps to.

The inclusion or exclusion set is an aggregate object that consists of the set of values and the set of constraining relationships. The inclusion (exclusion) set is the metaobject that normalizes the atomic rule that asserts that all valid (invalid) values are within the inclusion (exclusion) set. This is object C in the figure. Object C is thus a supertype of exclusion and inclusion sets and is called the constraining set. It is also called the "*constrain*" relationship.

The values in a value set can be values of other attributes too, and may change in step with the state of the system (state of a system: see box 12). When this happens, consider that the value set – the aggregate object – has changed its state. It is not a new and different instance of the value set. Instead it is the same (instance of the) value set with different values in it. Remember how history is implicit in every object (figure 22). Thus complex rules that might constrain how value sets can change may be based on relationships between time slices of the value set (see Module V, section 3 on our website).

Under inclusion and exclusion sets, we understood that constraining sets can normalize some kinds of knowledge about incomplete information – value sets could be exhaustive or not, and they could count "unknown" values among their members. *Not knowing whether the constraint is an inclusion or exclusion set, but knowing a constraint exists, is also knowledge about another kind of uncertainty that a constraining set (Object C) normalizes.*

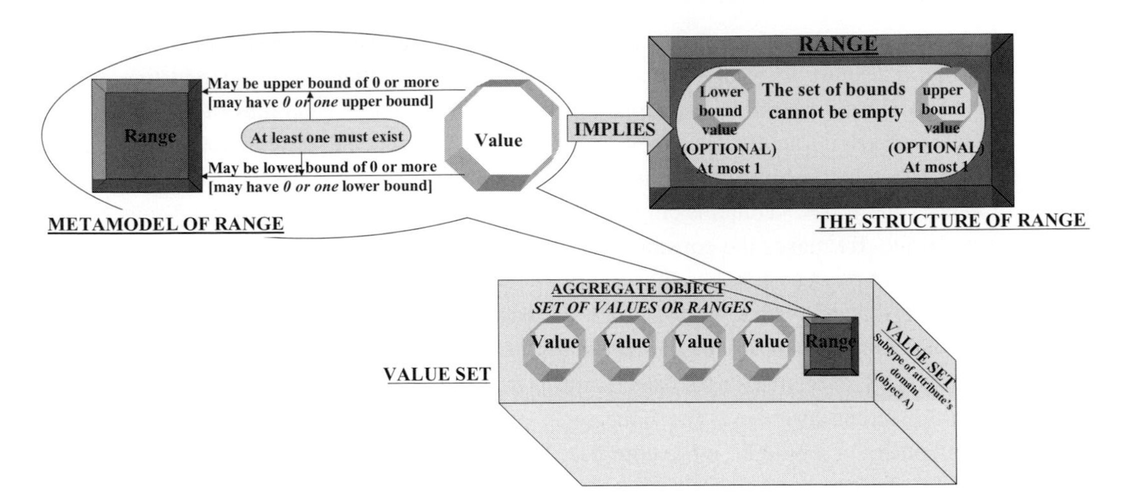

Figure 40 Subtypes of ordinal domains may contain both discrete values and ranges

Charles, unfortunately, loathes Jane secretly because she had once turned him down for a date. Charles is a petty man who does not want to show Jane cars with colors she will like. On the other hand, he is afraid of disobeying Jim. He strikes a secret compromise with himself. He decides he will show Jane cars with colors which, at worst, she will dislike only a little and, at best, she will like only a little. Charles sets the *upper bound* of the car color preference *range* at *likes a little* for cars he intends to show Jane. *Thus a range may have an upper bound, a lower bound, or both.*

This range constraint was an *inclusion set*. The cars Charles intended to show Jane *had to take a value from in this range* of car color preference. Similarly, Ranges can also be *exclusion sets*. For example, Jim could have asked Charles *not* to show Jane cars of colors she neither likes nor dislikes, or worse (neutral car color preference or car colors she dislikes even a little). This would then be an *exclusion set* for car color preference – Jim would have *excluded* this range of Jane's car color preference from cars he would show her, i.e. values in the range would not be permitted.

When we consider ordinal attributes, we need only add ranges to the repertoire of objects that could reside in the value set (aggregate object A in box 28). As such, the *value set* in box 28 may contain one or more values *and* one or more ranges. Figure 40 illustrates this.[22]

Figure 40 also contains a fragment of the metamodel for *Range*. Consider the two arrows from *value* to *range* in this diagram. The arrow on top asserts that a value may be the upper bound of a range. Actually, it is the inverse relationship that asserts this in a stilted, but mathematically precise, way. It asserts that there may be no (*zero*) upper bound, but, if the range does have an upper bound, there can be only one value that is its upper bound (*or one*

[22] Inverse relationships in the figure are shown in square brackets like this: [inverse relationship]. An inverse relationship shows how a relationship should read in the opposite direction from that of the arrow. See "Inverse of a function" in the endnote on the theory of categories.

upper bound). Similarly, in the direction of the arrow, it asserts that a specific value may not be an upper bound of any range, but, if it is, there is no injunction against the value being the upper bound of several (*many*) ranges.

In the same way, the lower arrow asserts that the lower bound of a range is optional, but, if the range does have a lower bound, it can have only one. A given *value* on the other hand may not be a lower bound, but, if it is, nothing stops it from being the lower bound of several ranges.

Some ranges may not have lower bounds and others may not have upper bounds. But, if a range has neither, then it stops being a range.[23] Where is the rule that asserts that both upper and lower bounds cannot be absent in a range? Remember, relationships are objects too, and, like any object, they may be related. The oval between the arrows in the figure is this kind of relationship. Its purpose is to normalize the assertion that both upper and lower bounds of a range cannot be simultaneously absent, although each may or may not be missing individually. It asserts that *at least one* of the two relationships between *value* and *range* must exist (and of course nothing stops *both* from existing).

These structures together normalize atomic rules about *range*, and assert that each range may have optional lower and upper bounds (values), and at least one bound, if not both, must exist for every range.

Constraints on quantitative attributes

Constraints on quantitative attributes are just like constraints on ordinal attributes, except for one subtle difference in the kinds of upper and lower bounds that might be imposed on ranges. Consider the following examples in which we impose an upper bound constraint on the length of a string.[24] There are two ways we can do this. We can say that:

1 The string must be two feet long *or less*; or
2 The string must be *less than* two feet long

The two constraints are almost, but not *quite* the same – in (1), the string *can* be two feet long, whereas in (2), it *cannot* be 2 feet long. In the second example, the string could be *close* to being two feet long – as close as we like – but can never actually *be* two feet long. It must always remain just *short* of two feet.

(1) is an example of a *closed* upper bound, whereas (2) is an example of an *open* upper bound. Similarly there may be open and closed lower bounds, or, in general, *open and closed bounds* of a range of values. The value of a closed bound lies inside the range, whereas the value of an open bound lies *just* outside the range.

Open above means either the upper bound is open, or there is no upper bound, and *open below* means the lower bound is either open or non-existent. Similarly, *closed above* means that the upper bound is closed, and *closed below* means the lower bound is closed.

[23] If both bounds are missing, the range becomes a domain (in the absence of bounds, all values of the domain will also belong to the range). An unconstrained range in an inclusion set is purposeless, and in an exclusion set it implies a nonexistent attribute.

[24] *String length* is a quantitative attribute of *string*, the object.

Ranges with both upper and lower bounds are called *intervals*. When both bounds are closed, it is called a *closed interval*. Otherwise it is called an *open interval*. Like ranges, intervals too may be open or closed from above or below, but, unlike ranges, they cannot be *unbounded* above or below.

One may ask whether the property of being an open or closed bound applies equally to ordinal values as well. After all, just as we restricted the string length in example (2) to being *less than two feet*, in the example on Jane's car color preference, we could have said that Charles decided to show Jane car colors she would like *less than a lot* instead of *equal to or less than a little*. However, had we done this, the *range* would not have changed. It would have only been a different way of specifying its upper bound. This demonstrates that ordinal ranges, like quantitative ranges, *can* have open and closed bounds, but, unlike quantitative ranges, the type of bound does not necessarily affect the actual ordinal range. Open and closed ordinal bounds are merely different ways of specifying the same ordinal range.

Ordinal domains always contain discrete values.[25] By restating the value of an open bound (either by reducing an open upper bound to the adjacent lower value, or by increasing an open lower bound to the next higher value), we can always define an ordinal range in terms of closed bounds and leave the range unchanged. The property of being open or closed *exists*, but has little value for ordinal ranges and bounds. On the other hand, a quantitative range may contain a continuum of values, and we cannot always close an open range, or open a closed range, by merely restating its bound(s). Therefore, the distinction between open and closed bounds is more useful for quantitative ranges.

Combining inclusion and exclusion sets

As businesses continually flex and reconfigure under the combined pressures of markets, technology, and regulation, constraints too shift and flex in step with business. This kind of change often involves changing, merging, adding, or eliminating constraints like exclusion and inclusion sets. Let us therefore understand how inclusion and exclusion sets can merge, clash, and flex as they shift in step with business.

Value sets are the core around which inclusion and exclusion constraints are built. Value sets are sets of values and value sets can be merged with other value sets. Mathematically speaking, when this happens, the resultant value set will be the union of all merging value sets (set union: see box 19). Ranges are sets of values too. Therefore, they too are a kind

[25] Quantitative domains are polymorphisms obtained by adding information to ordinal domains because it is mathematically possible to envision domains with an infinite number of ordinal values in any arbitrarily small interval between values. Unlike quantitative domains, mathematical operations like addition, subtraction, multiplication, and division may not exist in these domains (see items in the bibliography under set theory and metric spaces). The only requirement is that these domains be metric spaces (see the endnote on generalized distances). When arithmetic operations are added to these domains, they become quantitative domains (see Chapter 5, section 3). Thus, being subtypes of ordinal values, quantitative values inherit all roles of ordinal values, including the potential to be the bound(s) of a range. The metamodel of knowledge would infer this behavior, normalized by ordinal domains, from the subtyping relationships between domains discussed in Chapter 4.

Box 29 The structure of bounds and ranges (on our website)

Box 29 describes the metamodels of bound and range shown in figures (a) and (b). It discusses how the metamodel of knowledge formulates upper, lower, open, and closed bounds.

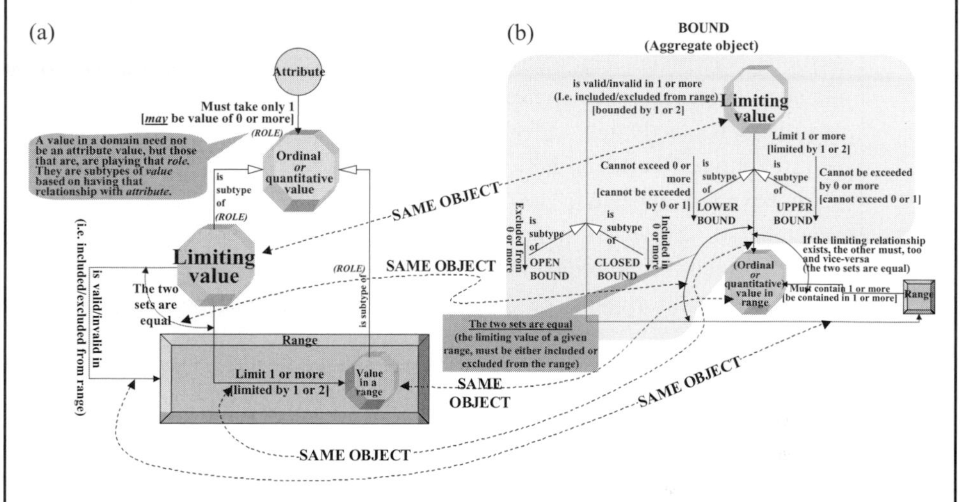

(a) Establishing the limits of a range is only one of several roles an ordinal or quantitative values may play. (b) A bound is an aggregate object

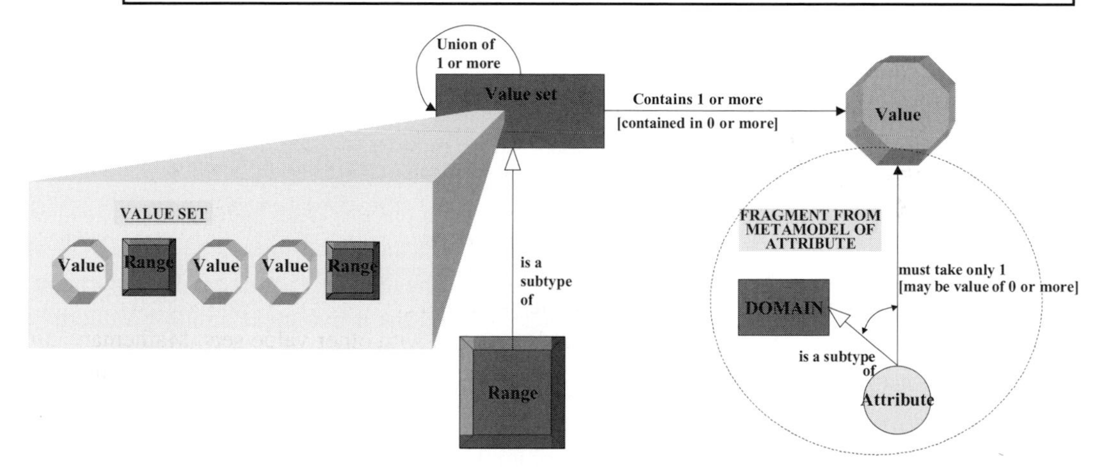

Figure 41 Metamodel of value set

of value set. Like other value sets, ranges too may merge with other ranges and value sets. When ranges merge with ranges, the merged value set is a range. When value sets and ranges merge, the resulting value set will contain both discrete values and ranges. Thus, value sets may contain discrete values, ranges, or both (figure 41).

When inclusion sets are combined with other inclusion sets, their value sets merge, i.e. the result of the combination is also an inclusion set, and the value set in the combined inclusion set is the *union of component value sets*. The process of merging exclusion sets is very similar; only the merged constraint is then an exclusion set. Bear this in mind as you examine the metamodels of value set (figure 41)[26] and value constraint (figure 42).

Box 30 Set versus list

Remember that a set is not a *list* of items – it is a *collection* of items. The difference is subtle but crucial. The members of a set *define* the set. The concept of "duplicate members" of a set is meaningless – an item is either a member, or it is not. A list on the other hand is exactly that – a list of items, and items may be repeated in a list. Lists may be differentiated based on not only the items they list, but also how many times each might have been repeated. The result of a set union is a set, not a list. Even if an item is a member of several sets in the union, it will appear only once in the unified set. Based on the principle of subtyping by adding information in box 43, a list is a subtype of a set; a list adds information on distinct occurrences and numbers of identical members in a set.

What if sets overlap or a single value set in a merger is a subset of another? Subsets and set intersections are all subsumed in the merged set. They lose their individual identities in the union. When overlapping ranges merge, only the highest upper bound and the lowest lower bound remain in the merged range.

If the ranges had not overlapped, they would have retained their identities in the merged value set. An inclusion set that contains a merged value set with disjoint ranges would permit disjoint ranges of values and bar those in the gaps between ranges. (If discrete values exist separately in those gaps, then only those values would be specifically permitted.) For exclusion sets, the opposite would happen: values in ranges would be barred and values in gaps permitted; specific values in gaps would also be barred.

If we had left the bounds of overlapping ranges in place (via set aggregation instead of union – see box 32), they would have been inactive in any case, superseded by the highest and lowest limiting values in the merged value set. The inactive bounds would lie dormant, and perhaps forgotten, until another change reduced the active upper limiting value to a value below a dormant upper bound. Then we could get a nasty surprise as long forgotten, dormant, and unneeded limiting values suddenly spring to life. (This would apply equally to lower bounds as well.) Set union will not permit this.

Does this kind of merger (via set union) imply that we have lost information on bounds and limiting values in the old value set? No, we have not. The metamodel of knowledge has a special place for it. In Chapter 2, section 4 we established that a perspective is an object – a model of reality. The unmerged value sets exist in a different perspective (or a previous state of the current perspective).[27]

[26] To reduce clutter, several structures in the metamodel of range (box 29 and figure 44) have been omitted from figure 41. Remember that these links exist – they are merely hidden in the diagram.

[27] Access to older perspectives is key to best practices in configuration management.

Box 31 Dormant ranges and containers for value sets

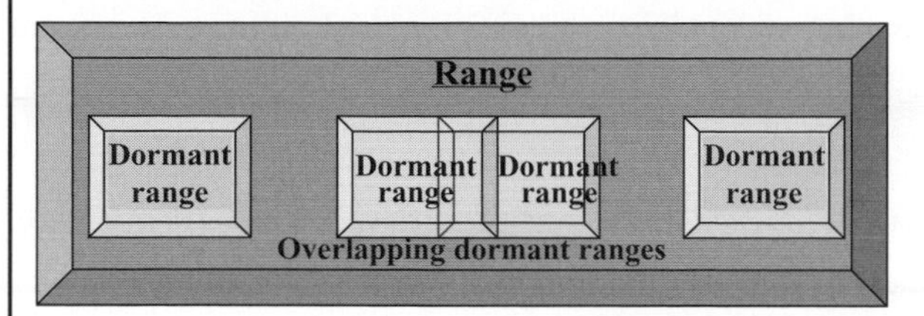

What if, for some reason, we *did* want to keep layers of dormant limiting values in our merged constraint? Set aggregation would help us do this. Remember, an aggregate object is like a bag (Chapter 2, section 1). Putting bags into bags (set aggregation) is different from emptying the contents of individual bags into a new bag, i.e. set union. In the former case, the bags retain their identity *and* contents. In the latter case, the contents of each bag remain, but the individual bags themselves are lost (except, of course, the new bag – the set that results from the union).

Thus, we can have an aggregate object that is not the union of the two value sets, but a set (object) in its own right. The members of the set (instances of this object) will be individual value sets (aggregate objects themselves). Dormant value sets could be extracted and made active when they are required in different contexts in systems and processes of a large and complex organization.

For example, an insurance company might insure 16–70 year old drivers in one country, and 21–79 year old drivers in another. There might also be a corporate policy that the firm will only insure 15–80 years old drivers. Different subsidiaries might be responsible for operations in different geographical footprints. Subsidiaries could carve out their ranges for ages of insurable drivers from the corporate range. If each range is linked to countries that subscribe to it, different ranges can become active in different countries; each country's range would be extracted from the aggregate range. The aggregate range could serve as a corporate asset that is reused as needed.

This kind of aggregate set is only a convenient physical storage mechanism for ranges and value sets. It is not a part of the metamodel of knowledge and has no role in normalizing knowledge for the reasons described in the subsection on clashing constraints.

When non-exhaustive inclusion (or exclusion) sets merge, the merged set may be either exhaustive or non-exhaustive – we cannot know which on the basis of the exhaustivity of the merging sets alone. On the other hand, if either set is exhaustive, it suggests that we know *all* values that are included (or excluded). Therefore, from a purely logical perspective, it might seem that such a merger will result either in a union that is identical to the exhaustive constraint, or in a conflict between contradicting constraints.

The key to understanding mergers between exhaustive constraints is to understand whether or not the constraint is exhaustive in *universal perspective* or not (perspective: see Chapter 2, section 4). If either merging set is exhaustive in the *universal perspective*, then the merger will be either redundant (and hence pointless) or in conflict (and hence impossible – at least one value set is erroneous – see *clashing constraints*.) On the other hand, the merging constraints, although exhaustive in a narrower perspective, might not be exhaustive in the broader, merged perspective, in which case we can treat them as such.

Merging open ranges

Subject to the exhaustivity issues we just discussed, there is one other issue that we must be aware of when we merge ranges of quantitative values. In box 30, we understood why ranges with open bounds cannot be merged – if the limiting value is excluded from the range, it will always divide a range immediately above it from that below it, because the value itself cannot be included in the range. Therefore, ranges cannot be merged across open bounds via set union. (They could, of course, be included in a container of convenience like that in box 32, which will maintain the separate identity of each range.)

Open *overlapping* ranges cannot merge because inclusion and exclusion sets cannot merge (see the subsection on clashing constraints), and in an open inclusion set, at least one limiting value is excluded, while in an open exclusion set, at least one limiting value is included (box 30). The metamodel of value constraint in figure 42(a) has this injunction against merger of inclusion and exclusion sets, which also implies that open overlapping ranges cannot merge.

However, set union *can* merge open and closed non-exhaustive ranges with the same limiting value. The merger is not *across* bounds in this case; it stops *at* the bound.[28] If a limiting value of an open range is the same as that of a closed range with which it is being merged, the closed bound will supersede the open bound in the resulting range. The open bound will flow from outside to inside the range via the set union that combines value sets of merging ranges as follows:

> If open and closed upper bounds coincide, the union will result in a closed upper bound. Similarly, if open and closed lower bounds coincide, the union will have a closed lower bound. On the other hand, if an open upper bound of one merging range coincides with the closed lower bound of the other, the bounds will vanish, and the limiting value will become a member of the merged value set. Similarly, if a closed upper bound of one merging range coincides with the open lower bound of another, both bounds will vanish and the overlapping limiting value will become a member of the merged value set in the resulting constraint.

Clashing constraints

When we merge inclusion sets with inclusion sets, or exclusion sets with exclusion sets, we merge not only corresponding value sets, but also the entire constraining set (object C of box 28). The union of value sets is only one aspect of this merger of constraints. Merger of constraining relationships (B1 or B2 in box 28) is the other. When inclusion sets merge with inclusion sets, corresponding inclusion relationships (object B2 in box 28) also merge

[28] From a set-theoretic perspective, the intersection of non-exhaustive open and closed ranges with overlapping limiting value(s) is empty, i.e. the sets are disjoint because a value at one end of the range is excluded in one of the sets being merged. Therefore, their merger is similar to merging disjoint ranges, only, in this case, the ends of the ranges get pinned together after the merger.

through set union. When exclusion sets merge, it is the exclusion relationship (object B1 in box 28) that merges.

These constraining relationships (inclusion or exclusion) of the merging sets can merge via set union, and still remain instances of the merged object *class*, because the relationships are all of the same *kind* – all inclusion constraints, or all exclusion constraints. If the merging sets are inclusion sets, the merged set will also be an inclusion set, and, if the merging sets are exclusion sets, the merged set will be an exclusion set.

However, if we tried merging exclusion sets with inclusion sets, they would not mix.

Mixing completely certain constraints

Consider merging inclusion and exclusion sets that are exhaustive partitions with no "unknown values" first: The inclusion and exclusion sets in such a merger will retain their individual identities in an aggregate object like that in box 32 – and the object would not normalize knowledge – indeed it could replicate it as explained in the following paragraphs.

An exhaustive inclusion set asserts that values *not* in it are excluded – the attribute can only assume a value *inside* the inclusion (value) set. Therefore, if an exhaustive inclusion set and an exclusion set were to simultaneously constrain the value of an attribute, the exclusion set would merely repeat the exclusion rule for a few attribute values or ranges; this would be the rule already established by the inclusion set.

Similarly, an exclusion set asserts that values *not* constrained by it are permitted. Therefore, if an exhaustive exclusion set and an inclusion set simultaneously constrained an attribute, the inclusion set would only repeat rules already established by the exclusion set. Thus, this kind of aggregate set of merged constraints has no place in the metamodel of knowledge. It would replicate, not normalize, knowledge.

An organization might consider storing both inclusion sets and exclusion sets in a container object of the kind described in box 32, and associate different constraints with different contexts. However, bear in mind that *if the inclusion/exclusion sets are exhaustive, and have no "Unknown" values, multiple inclusion/exclusion sets can never be attached to the same attribute in the same context* because an exhaustive constraint says it all – every value in it is included and all else is excluded (exhaustive inclusion sets), or every value in the constraint is excluded and all else is included (exhaustive exclusion set). Container objects that contain exhaustive inclusion or exclusion sets are merely expedient containers of components.

Mixing incomplete or uncertain constraints

Now consider merging inclusion and exclusion sets that are either non-exhaustive partitions or have "unknown values" in their value sets.[29]

Constraints, in expedient containers like the one above, may be linked to attributes in different perspectives. Unlike the case above, when constraints were exhaustive and certain

[29] If the constraining relationship does not specify the kind of constraint (inclusion/exclusion), then:

1. If the value set is exhaustive, but has "unknown" values in it, it implies we know *how many* distinct constraining values exist, but not *what* all of them are.
2. On the other hand, if the value set is non-exhaustive, it means we do not even know how many distinct constraining values exist.

(with no "Unknown" values), we *could* consider attaching several constraints to a single attribute in the same perspective, *provided the constraints are not contradictory* – i.e., they do not share values in their value sets. In this way, although we know we have incomplete information, we can represent what we *do* know, and also what we *do not know*, or know incompletely.

Consider what would happen if value sets of different constraints being combined *did* have values in common. If inclusion sets had values in common, the two would merely be merged via a set union, and there would be no problem in representing the combined knowledge in a normalized form. The merger of exclusion sets would be similar. On the other hand, if we imposed both inclusion and exclusion sets simultaneously on an attribute, and the sets had one or more values in common (in their value sets), then we would have a conflict. The attribute cannot be *both* permitted and barred from assuming the shared value. This contradiction would happen, not because of any inherent contradiction in our metamodel of knowledge, but because of the way we have *stored* our components in our container. The container with mixed constraining sets has no place in our metamodel of normalized knowledge. It can only be a storage mechanism – a physical "bucket" for storing and locating various unrelated items of information.

Even though we cannot meaningfully aggregate conflicting inclusion and exclusion sets and attach the aggregate to an attribute value in our metamodel,[30] we *can generalize* the two kinds of constraining relationships to assert that they both *constrain* attribute values, but not *what* this constraint is (see box 28). Unlike the aggregate object of box 31, this generalization would have a role in the metamodel of knowledge. The generalized relationship (object B in box 28) asserts that *we know that attribute values are constrained by values in the merged set value set, but we do not know how – i.e. which are included, and which excluded.* When we break out the "*constrain*" relationship in box 28 (or figure 42(b)) into its exclusion and inclusion subtypes, we will have this information.

Unlike conflicting inclusion and exclusion sets, we *can* use set union to merge generalized constraining sets. Corresponding value sets of the merging constraints would be merged via set unions to yield the value set of the merged constraint. Similarly, the "constrain" relationships (object B in box 28) of the merging constraints would be merged via set unions to yield the merged constraint.

If the merged constraint is exhaustive, we know that the corresponding inclusion and exclusion sets are *collectively* exhaustive (the merged constraint is the natural home of information of this information.) On the other hand, the inclusion set might be exhaustive, but not the exclusion set, and vice-versa. However, if they are *both* exhaustive, then the generalized constraint *must* also be exhaustive. Note that this does not necessarily apply in the opposite direction. We could have an exhaustive generalized constraint, but might not

[30] Inclusion and exclusion sets are mutually exclusive, disjoint sets. Their intersection (overlap) is the *empty set* ∅ of box 19, i.e., if constraints are in conflict, the perspective will not exist (another way of saying this is to say that conflicting constraints cannot exist in any perspective) – a mathematically perfect assertion, but flawed in the real world, where it is more likely that a mistake was made. If automated agents (see box 36) assemble systems and business processes from components of knowledge, inclusion (exclusion) sets should not be merged solely by the automated agent that attaches them to a perspective. Further, inclusion and exclusion sets must also be validated against each other to ensure consistency.

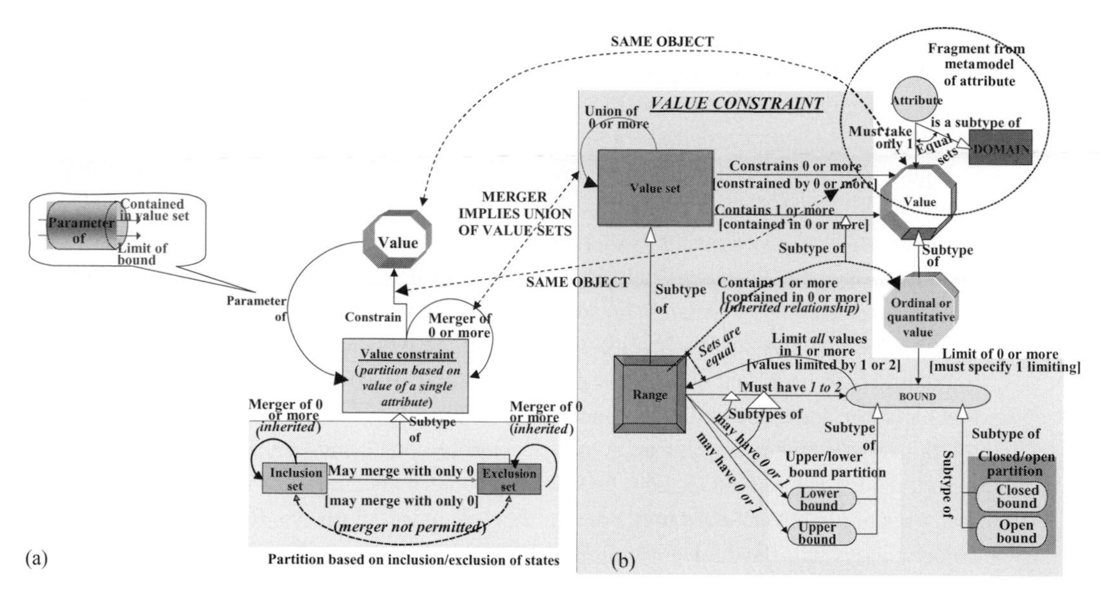

Figure 42 Metamodel of value constraint

always know which individual relationships of the constraint (the fibers inside the "tube" of set B in box 28) are inclusion constraints and which are exclusion constraints. Corresponding inclusion and exclusion sets would be non-exhaustive even though the generalized constraint is exhaustive.

The irreducible fact, that the value set of a merged constraint is the union of value sets of merging constraints, is represented by the recursive "union" relationship on *value set* in figure 42(b). This relationship will be inherited by the two subtypes of *value constraint* – exclusion and inclusion sets. Value sets in each subtype will be merged separately, resulting in merged inclusion constraints that are distinct and separate from merged exclusion constraints as illustrated in figure 42(a). The injunction against merging inclusion and exclusion sets ensures that this kind of merger will not happen in this partition or any other partition of the value constraint. Every instance of value constraint will inherit this rule.

The merger of meaning and the metamodel of value constraint

Figure 42 is the model of value constraint. It is a model of what a value constraint means and how this meaning is assembled from components of knowledge. Of course, the value may be the value of an attribute as the fragment of the metamodel of attribute in figure 42(b) illustrates. Figure 42(a) represents the basic concept, and figure 42(b) elaborates on it.

Figure 42(a) expands on the basic model in figure 38(b). It gives us a window into the object (box) labeled "attribute value constraint" of figure 38. That object is identical to the partition in figure 39 with the same name. "*Attribute value constraint*" has been abbreviated to "*Value constraint*" in figure 42(a). Both refer to the same metaobject. As we will see later in this book, *value constraint* may constrain not just attribute values, but values of domains and other objects as well.

Figure 42(a) recognizes that value constraints may have parameters. Parameters are values that have a role in describing the constraint. This is represented by a relationship between *value* and *value constraint*. There are two basic parameters of value constraints: members of value sets and the limiting values of ranges. Thus, the "*parameter of*" relationship is an aggregate object that contains two relationships: "*contained in value set*" and "*limit of bound*".

Figure 42(a) also recognizes that constraints may merge. The recursive relationship in figure 42(a) is identical to that in figure 42(b). Earlier, we understood that the merger of attribute (value) constraints imply merger of corresponding value sets. Figure 42(a) also has an injunction against merging inclusion sets with exclusion sets. It is a relationship between *inclusion set* and *exclusion set* that reads "*inclusion set* may merge with *zero exclusion set*" – a strange but mathematically precise way of saying that the two kinds of sets cannot merge.

It could have been said in several ways. For example, it would have sufficed to simply say that inclusion and exclusion sets cannot merge. We used the syntax we did to emphasize that relationships between objects are actually a kind of constraint – an *occurrence* constraint. Relationships not only state how two or more object instances are related, but also how many instances of one class may relate to how many of another.[31] Relationships between object classes will have lower (and optional upper) bounds on the number of instances of the relationship that may exist at a moment in time. (This property, the range, is called the *cardinality*[32] of a relationship.) If the upper bound is zero, then the relationship is banned. (The lowest possible lower bound is also zero. If the upper bound is not zero, but the lower bound is, it implies that the relationship is optional, i.e. object instances in the class at the source of the relationship may or may not be related to an instance of the relationship's target class.)

It is usually not necessary to show a relationship that cannot exist. However, in this case, its subtypes (the inclusion and exclusion set in figure 42(a)) will inherit the recursive *merger* relationship on *value constraint*, the supertype. We know this is okay – inclusion sets can merge, and so can exclusion sets. We must not replicate this inherited relationship on the subtypes. Replicating inherited information denormalizes knowledge – the rule has already been stated once by the supertype and is automatically valid for subtypes. However, the inherited relationship also leaves open the possibility that inclusion sets may merge with exclusion sets, unless we specifically ban this. We know that inclusion sets cannot merge with exclusion sets; hence *we must have this relationship to bar the possibility of such mergers*. Therefore, this injunction adds, not duplicates, key information. *The relationship that carries the injunction will normalize, not replicate, knowledge.*

Figure 42(b) is a window to the detail inside the *value constraint* object in figure 42(a). Inclusion and exclusion subtypes of "*constrain*," the relationship between *value set* and *value*, have been omitted to avoid cluttering the diagram. Remember that they exist, but are only hidden in the diagram. The structure inside the shaded area, consisting of bounds, value sets and ranges, is the value constraint. It is an object with an internal structure.

Figure 42(b) is a metamodel that formally represents the rules we have discussed in this section. Constrained values connect to the constraint via the *constrain* relationship. This

[31] Relationships may also relate object instances in the same class. These relationships are called recursive (or *homomorphic*) relationships. See Module V, section 1 on our website, isomorphism and homomorphism in [240], [237], and the endnote on the theory of categories.

[32] [202] and [206] discuss of cardinality, countability, and the theory of cardinal numbers.

relationship has two subtypes: *exclude* and *include* (not shown in figure 42 to avoid clutter). At first glance, it might seem strange that a value set might constrain no values. To understand this, consider that value sets may be required for other reasons as well. Participating in a constraint is only one role of value set. It is just a set of values and might be used in several ways. It does not necessarily have to be bound to a constraint for its existence.

A range is a kind of value set. A range may have two bounds at most, and must have at least one to be considered a range. *Bound* is an aggregate object with a structure too, and this relationship between *range* and *bound* is identical to that in figure (b) of box 29. Bounds may be partitioned into upper and lower bounds, and also open and closed bounds. The relationship is inherited by each subtype of bound. These inherited relationships are subtypes of the parent relationship as shown in figure 42(b). Only the inherited relationships to upper and lower bounds have been shown. Similar inherited relationships to open and closed bounds have been omitted to minimize clutter. Although they are hidden, remember they exist. However, the cardinalities (i.e. the range(s) of *instance level occurrences of the object at the end of the arrowhead*) of the supertype and subtype relationships are different. This is okay. The subtype relationship need not inherit the cardinality of its supertype, but cannot *violate* it.[33] The cardinality of the subtype relationship must lie *inside* the range of the supertype, or, at best, may equal it. The subtype relationships in figure 42 do not violate the cardinality of their supertype.

Box 32 Cardinality, cardinality ratio, and object counts

When two objects are related, an object instance at one end of the relationship may be tied to a single object instance at the other end, or to several of them. The number of target object instances that a single object instance at the root of the relationship (the beginning of the arrow in our diagrams) is tied to, is called the *cardinality ratio* of the relationship. It is often abbreviated to cardinality. The population of object instances is also called cardinality. The two are obviously different but related concepts that measure the size of a class of objects. For example, at a given moment, there may be ten instances of a relationship class with a cardinality ratio of 2. The cardinality of the relationship *object* will be 10 at that moment in time, and the cardinality *ratio* of that class of relationships will be fixed at 2 at all times. To avoid confusion between the two meanings of cardinality, we will distinguish *cardinality ratio* from *cardinality*. We will call the population of an object class its population or cardinality, and the cardinality ratio by its full name, *cardinality ratio*, or abbreviate it to CR.

The "*limit*" relationship between *bound* and *range* is identical to those in figures (a) and (b) of box 29. The *limiting value* of box 29 is a role of the object called *ordinal or quantitative value* of figure 42(b). You could think of the *limiting value* of box 29 as being buried inside the

[33] Subtypes inherit constraints from their supertypes. Therefore, constraints on subtypes may be more restrictive, but cannot violate constraints inherited from supertypes. This is why any constraints on occurrence imposed by the subtype's cardinality ratio must lie within constraints imposed by the supertype's cardinality ratio. If the subtype's cardinality ratio(s) is not the same as the supertype's, the ratio must fall *inside* the range established by the supertype's cardinality ratio(s).

relationship from the *ordinal or quantitative value* object and *bound* in figure 42(b). Note that the inverse relationship asserts that every bound must have at least one limiting value.

In figure (b) of box 30, the **relationship** between *limiting value* and *value in range* and the containment **relationship** between *value in range* and *range* are related by an equality constraint, i.e. neither relationship can exist without the other. This is also true in figure 42(b); only, the containment relationship is inherited from that between *value set* and *value*. This happens because *range* is a subtype of *value set* and *ordinal* or *quantitative value* is a subtype of *value*. (Note that we could have replaced the object named "*ordinal or quantitative value*" in figure 41 with "*ordinal value*" without impacting the behavior or meaning of "*value constraint*." We may do this because, as demonstrated in Chapter 4, quantitative values are subtypes of ordinal values, and inherit their properties.)

The metamodel in figure 42 is a key fragment of the overall metamodel of knowledge. Attributes, domains, and value sets tap into the knowledge in constraints through the *constrain* relationship (via the value object), and the value constraint taps into the knowledge in other components of knowledge via the other relationships that link values to the objects inside the shaded area of figure 42(b). The structure in the shaded area manifests itself as constraints on real world values.

The "*constrain*" relationship is the key to value constraint. So far, we have only understood its role in permitting or barring specific values. We have not discussed how it is the object from which derived values and mathematical formulae also flow. To do so we must look deeper into *constrain*, the relationship. Only then will we see that *constrain* possesses an internal structure that normalizes meaning. This is what we will do next.

Derived attributes and relationships between attributes

The constructs we have discussed so far can normalize rules about many kinds of constraints – some simple and others more complicated. However, the metamodel of attribute (value) constraints is still not complete – there are important gaps left unfilled – gaps in which many commonly found business constraints reside. Our metamodel has no place yet for *joint constraints*, nor does it have room for constraints that involve ranking, arithmetic, and other mathematical operations. In this section, we will fill these gaps and complete our metamodel of attribute (value) constraint.

Joint constraints

A joint constraint is a rule that constrains attribute values based on the *interaction* or conjunction of two or more attributes. For example, take a check – a check issued by a business. It might require two signatures – both the CFO and the CEO might need to sign the check before it becomes payable. The state of the check – whether it is payable or not – depends jointly on two nominal attributes of the check – the CFO's signature and the CEO's signature.

This too is a constraint, but it is a joint constraint between three items of information, i.e. attributes of the check (an object) – the CEO's signature, the CFO's signature, and the

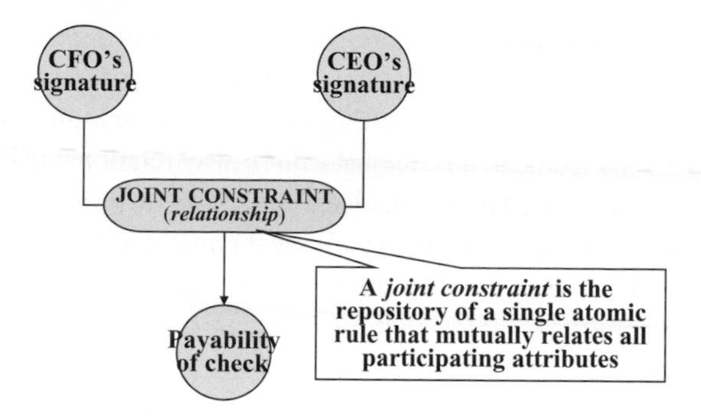

Figure 43 The state of the check is constrained jointly by two other attributes

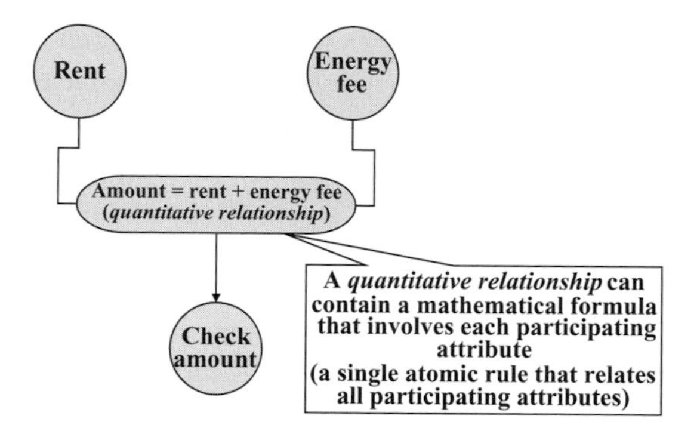

Figure 44 A quantitative relationship relates check amount, monthly rental and energy charges

payability (i.e. validity) of the check. It is a three-way *relationship* between three attributes.[34] We could also call it an *interaction* between three attributes. Unlike the value sets we have discussed earlier, the value sets involved in joint constraints may contain values from several domains [169]. Inclusion and exclusion sets will not, by themselves, normalize these kinds of atomic rules (see figure (c) in box 33).

Magnitude constraints

A magnitude constraint is a limitation on the magnitude of an ordinal or quantitative attribute. It is a special kind of joint constraint. A magnitude constraint conveys more information than a nominal constraint of the kind in figure 43 does; it conveys information on the interaction of magnitudes as well as the interaction of occurrences. Based on the principle of subtyping by adding information (box 43), a magnitude constraint is a subtype of a joint constraint; it implies the joint constraint *and more*. Note that a magnitude constraint does not *have* to involve three or more attributes. It could also relate two attributes (as the value constraints in our early discussion in Chapter 3, section 2 did). Overall, a magnitude constraint may even be an information-rich subtype of a binary relationship between attributes.

[34] Module V discusses normalizing knowledge with multiway relationships.

Figure 44 shows an example of a magnitude constraint between three attributes. Assume that the check is payment for two separate items, a flat fee for monthly rental of office space and variable charges for energy consumption. We know that the check amount must equal the sum of the month's rental and energy charges. It is not just a three-way relationship between *check amount*, *monthly rental*, and *energy charge*, but also a constraint on the magnitude of *check amount*. The relationship dictates that *check amount* must equal *monthly rental* **plus** *energy charge*. Inclusion and exclusion sets alone cannot express magnitude constraints like this that involve mathematical operations – not even if they are constraints between merely two attributes.

Naturally, the constraints on magnitudes of ordinal attributes will only be in terms of ranks of values, whereas constraints on magnitudes of quantitative attributes may be in terms of both rank and arithmetic. Further, other mathematical operations (all four kinds of attributes, nominal, ordinal, difference, and ratio scaled), may be constrained in terms of *existence* of specific values or ranges of values. We have discussed these kinds of constraints earlier. We will call the *rule* that expresses what values may be excluded or included as a *rule expression*.

The inclusion set permits the constrained value to equal the result of a (evaluated) rule expression. The exclusion relationship bars it. For example, in the example above, the rule expression was "*Monthly Rental + Energy Charge.*" This expression was glued to an inclusion relationship to create the joint constraining relationship from *monthly rental* and *energy charge* to *check amount*.

Indeed, equations like *check amount = monthly rental + energy charge* serve as a kind of attribute value constraint (see box 33). This is how the metamodel of knowledge provides room for equations and inequalities too. Inequalities are rules about what may not equal what. Like equations, an inequality would involve a rule expression, but, unlike equations, the rule expression in an inequality would be attached to an exclusion relationship instead of an inclusion relationship.

Rule expressions

Unlike the *constrain* relationship in box 28, the constrain relationships in figure 43 and figure 44 consist of a rule expression *and* an inclusion/exclusion set. Such rule expressions relate values to each other with formulae that may include arithmetic and higher mathematical operators, ranking rules as well as the Boolean operators "*or*" and "*and*." We will call these kinds of *constrain* relationships *rule constrain* relationships. For example, the rule expression in figure 43 involves the *and* operator, whereas that in figure 44 involves addition, an arithmetic operation. Both constraints are rule constraints.

The logical operator "*not*" is provided by inclusion/exclusion sets. We could include it in the rule expressions attached to inclusion or exclusion sets to make it convenient to express a rule. Further, we can express *every possible rule* even if we restrict the expressions like those in figures 43 and 44 to only *or*, *and*, ranking, and arithmetic operators, and judiciously combine them with inclusion and exclusion sets. Indeed, when the rule expression contains only *or* operators, the *constrain* relationship is identical to the kind in box 28. The simple constraints are just special cases, or subtypes, of the more general constraint of the kind in figure 45.

The rule expression lets us *plug* one (as in box 28) or several (as in figure 44) constraining values to the constrained value to describe complex interactions. Both *jointly constrain* and

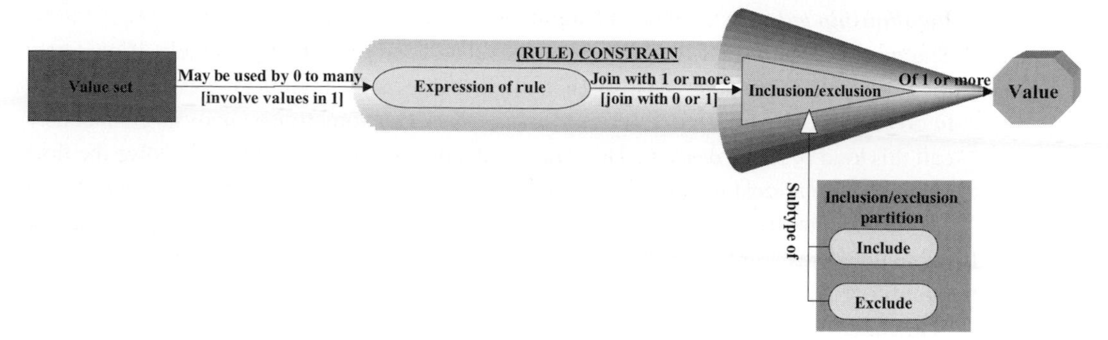

Figure 45 A rule constraint is an aggregate object

constrain magnitude must involve rule expressions, and hence are subtypes of *rule constrain*. Note also that these subtypes are not mutually exclusive; they belong to different partitions of *rule constrain*: one is based on the *number of attributes* (and values) involved in the rule, and the other is based on the operators in the rule. A constraining relationship that involves a rule expression could be both a joint constraint and a magnitude constraint, like the relationship in figure 44.

Statics versus dynamics of derived attributes

The state of the check in figure 43 was derived from the fact that the check had (or did not have) both requisite signatures. Like the state of the check, adding the monthly rent and the energy charge derived the amount of the check in figure 44. The state and amount of the check were *derived attributes* of object class *check*. It is perhaps less obvious that there is not one, but two separate aspects of derived attributes that we must consider in order to normalize information about them – rules about the sequence of events that leads to assigning a value to the derived attribute, and rules about constraints between attributes that do not involve any time sequence:

1 ***Process rules***

Processes normalize rules about sequences in time; "before-and-after" rules, for example, rules like derived attribute(s) must precede the attributes they are derived from. Rules like these that involve the flow of time are processes.

Many analysts would show these rules in black boxes (see Chapter 1, section 6). The output of the black box would be the derived attribute, and its inputs would be the attributes it is derived from. For example, in the case of the check that requires two signatures to become payable, the inputs would be the CEO's and the CFO's signature and the output would be the state of the check.

The rule is a process if it asserts that two individuals must sign the check *before* it is paid. This sequence of steps in time is easier to visualize than a timeless rule, that the *presence*, not the *precedence*, of the two signatures is sufficient to make the check payable. Stated thus, the rule is a time independent relationship between three attributes. In this section, we will focus on these non-temporal relationships between attributes. Module V on our website discusses processes.

2 ***Relationship rules***

The mere presence of two signatures makes the check payable. Conversely, if we know that the check is payable, we also know that it bears both signatures. This rule has little to do with the flow of time. It is a rule about *existence*. Sometimes, analysts erroneously call this kind of rule a *process*. This rule is not a process. Processes must involve the flow of time.[35] Relationships between attributes are repositories of atomic rules about how attributes mutually constrain each other (see box 33).

Static rules represented by joint constraints and relationships between attributes are quite common in business. For example:

- *Percentages must add up to 100* is one such rule, a multi-way relationship between individual percentages that many of us have frequently encountered.
- *The duration of a process cannot be less than the duration of the longest subprocess in it* is another joint constraint between the duration of a process and that of its subprocesses. In this case, the joint constraint sets the lower bound of the permitted range of the duration of the process.
- *The area of a rectangle is the product of the length of its sides* is a three-way constraint between the length of each side and the area of a rectangle. (In box 33 we will understand how equations are a kind of constraint.)
- *The height of a room cannot exceed the height of the building that contains it* is a relationship between the height of buildings and heights of rooms in it. More precisely, it is a two-way relationship between the height of the building and an upper bound, i.e., the upper limiting value of the range of values for heights of rooms in it.

Box 33 Relationships between attributes, meanings, and expressions (on our website)

Box 33 elaborates on the behavior and the structure of relationships between attributes and the possibility of time dependent, non-stationary relationships, and constraints. It discusses multiway, conjoined interactions, recursive interactions, rules, and the difference between a meaning and its expressions. It discusses, with examples, the mathematical relationship between a meaning and its possible multiplicity of expressions.

Box 33 describes the impact of models articulated by the following figures:

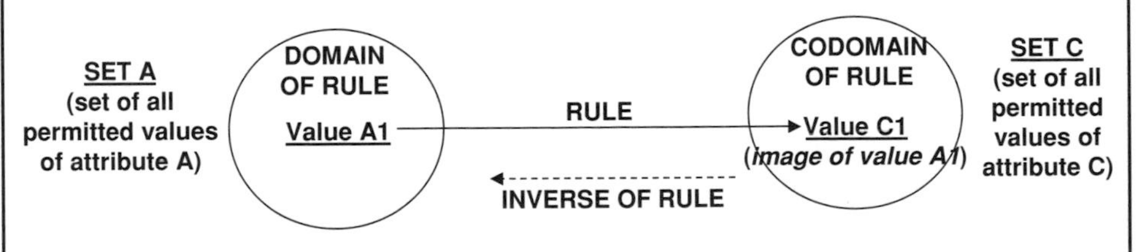

(a) A relationship between attributes is a mapping rule

[35] The *workflow* that results in the check being signed is a process because it *does* involve the flow of time – a sequence of activities in time derived from a non-temporal relationship by adding components to it. See Module V, section 3 on our website.

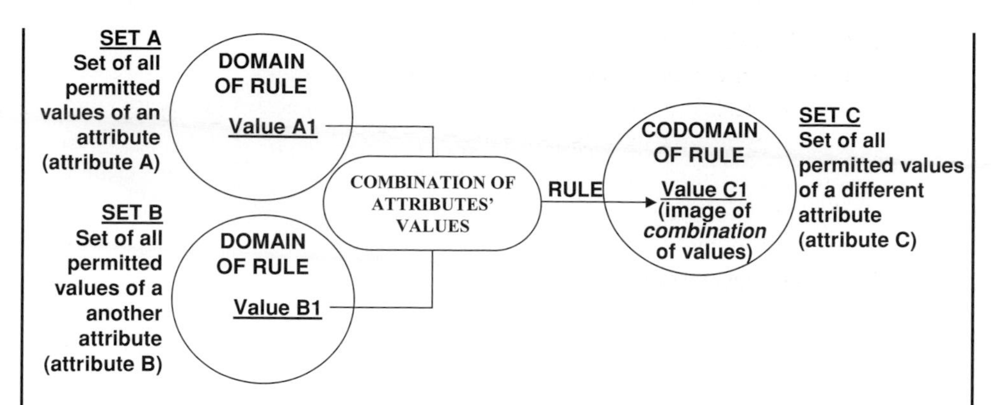

(b) Multiway relationships between attributes are joint constraints that involve three or more interacting attributes

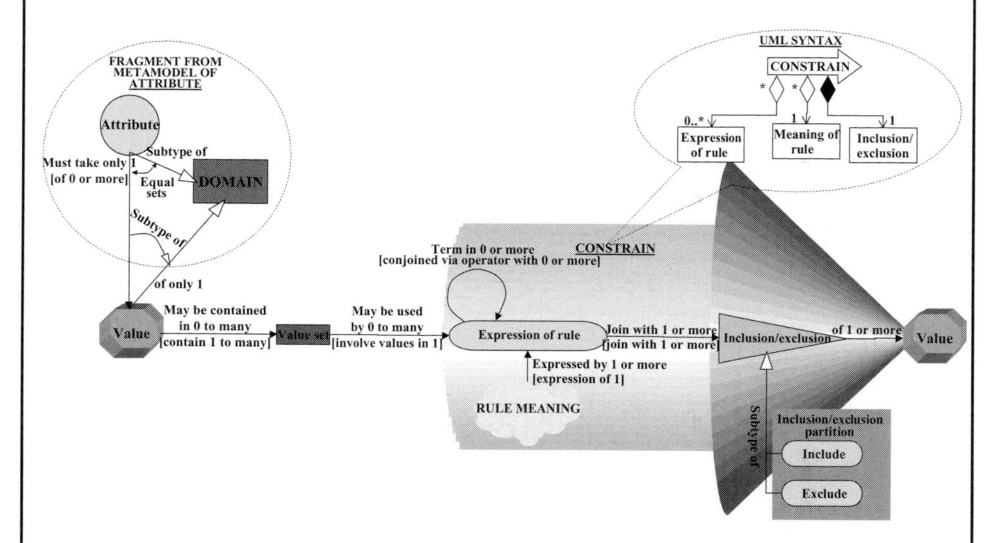

(c) A "rule constrain" relationship is an aggregate object with a structure

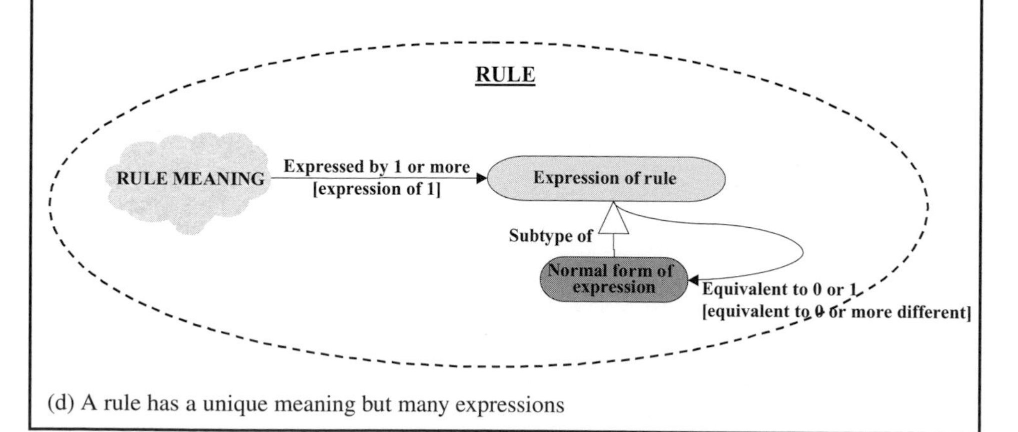

(d) A rule has a unique meaning but many expressions

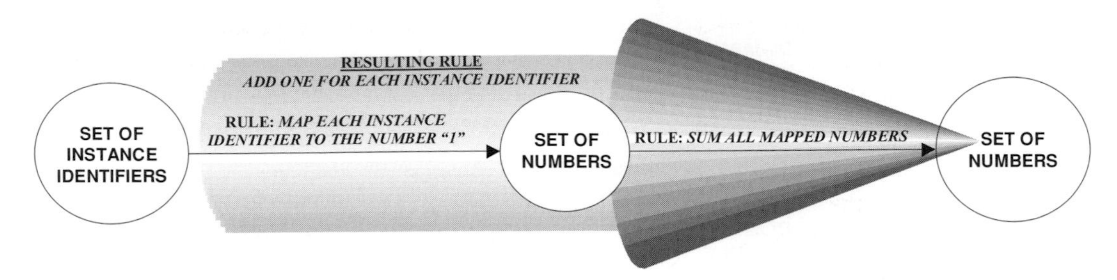

Figure 46 Intermediate rules for mapping object instances to object population

Implicit and intermediate rules

Relationships between attributes may be assembled into daisy chains. These daisy chains become relationships in their own right.[36] Indeed, these daisy chains may consist of values linked together with value constraints. These values and value constraints would be subsumed (i.e. become "intermediate results") in the resulting relationship, and be irrelevant to an external observer interested in the "whole" rather than its parts. (Those interested in more information may refer to the endnote on lambda calculus.) The following examples show how relationships can sweep through, and subsume other relationships.[37]

Enumeration

Consider the population of an object class again. From box 33, we understood how enumeration is a quantitative relationship between a nominal attribute of an object instance, the *instance identifier*, and a quantitative attribute of the object class, the *object population*. We can actually show this as two connected relationships.

Each instance identifier maps to the number 1, which in turn is summed. The mapping and summation are two different atomic rules that have been glued to each other to result in the enumeration rule.[38] Each intermediate relationship is an irreducible fact, a component. The enumeration relationship is a composition, a conjunction of objects (Module V, section 2 on our website), and an atomic rule. It consists of a sequence of relationships; this is not necessarily a time sequence, even though it might be implemented thus in a computer system. A non-temporal sequence is inherent in this aggregate relationship, a pattern.[39] Enumeration

[36] The glue in this case is a connective/associative operator like the "∗" and "✣" operators in the endnote on gluing objects together. If the overall relationship cannot exist (is "null") when any component object in it is missing (i.e. is null), the connective is like the "∗" operator. It is like "✣" when a missing component results in the resultant being equal to the remaining component(s). The behavior of the overall relationship depends only on the sequence in which its components are strung together. It does not matter how we subdivide, or group subsequences of objects within this sequence. Therefore, the connectives *must* be associative operators (associative operations: see the endnote on theory of categories).

[37] Module V on our website discusses compositions of relationships in detail.

[38] The endnote on gluing objects together describes the kinds of operations that join objects.

[39] See Chapter 4 on patterns and the endnote on gluing objects together. Rules may be rules of interactions between values (attributes), or interactions between rules. Then the scope of the value set in figure (c) of box 33 will increase: the value set would become a *rule set*, a generalized object that can contain both values and rule expressions (a value set is a subtype, a special case of *rule set* that contains only values). Rules as well

might be implemented by a counting procedure that involves a sequence of activities. That would be a process. The aggregate rule in figure 46 does not care what instance identifier is counted when, or in what sequence. It is an immutable and timeless rule used frequently in the physical world – a static rule and a reusable component of knowledge.

Implied relationships and rule normalization

The existence of the two intermediate relationships in figure 46 implied the existence of the enumeration relationship.[40] If we simultaneously show all three, we will be replicating, not normalizing, knowledge. Indeed, the existence of *any* two of the three relationships in figure 46 (mapping to number 1, summing all mapped numbers and enumeration) implies the existence of the third. Relationships like these are called transitive relationships. When daisy chains of relationships are involved, the overall relationship between attributes at the beginning and the end of the daisy chain may be considered a relationship in its own right. We might or might not be interested in drilling down into the relationship to understand its intermediate links. However, if we do, we must understand that, although the overall relationship stands on its own, it is *implied by the sum total of individual links in the daisy chain* and is a derived relationship between attributes. Asserting both overall and component relationships *independently* will replicate, not normalize, knowledge. (Indeed, this is true for any part of the daisy chain, as well as for the whole chain.)

Just as derived attributes exist in their own right, and relationships like those under *magnitude constraints* normalize rules of derivation, derived relationships too stand on their own; intermediate relationships, like the individual terms of rule expressions, express the overall relationship. Module V (on our website) discusses transitivity and derived relationships in detail.

Sometimes mathematical formulae and algorithms are assembled from intermediate terms, which may in turn be reused in other formulae and algorithms (see the examples in rule the expressions of box 33). The property that lets relationships consist of other relationships[41] is the metamodel for this kind of reuse. In general, relationships between objects can be aggregated simultaneously in different and distinct conjunctions (see Module V and Module VI, section 2 on our website).

Polymorphism

Polymorphism is adaptation of behavior to its context. (Module V, section 4 has a more precise definition of polymorphism, and Module VI, section 2 describes its generalized form; both on our website.) This means that generic behavior may be subtyped, depending on states

as values could then be parameters of the rule expression in figure (c) of box 33. The rule expression would return either a value or another rule expression. Lambda (λ)-calculus (see endnote) supports this. Both values and rules are generalized parameters of rule expressions, which may be parameters of other rule expressions. See the endnote on λ-calculus or items in the bibliography on λ-calculus. *Functional programming*, based on λ-calculus, is emerging in support of these concepts. See the endnote on functional programming or [242], [254], and [306].

[40] See transitivity of relationships in Module V, section 1 on our website and the endnote on "morphism" and "Infix ring" in the "Theory of categories".

[41] See "Infix ring" in the theory of categories and λ-calculus in the endnotes.

of other objects. These states are its context[42] and the subtypes are special variations of common behavior. The supertype normalizes common behavior, and the subtypes normalize context-specific behavior. Together, they normalize variants of behavior. For example, *movement* is a generic concept. Moving up, down, sideways, hopping, walking, crawling, and other kinds of movement are subtypes of *movement*. Thus *movement* is polymorphic. It depends on context.

Figure (c) of box 33 inherently supports polymorphism. The values in the *value set* of figure (c) could be values of attributes. Several attributes may be attributes of the same object. *Value set* may be a set of object states. The constraint on the *value* on the right-hand side of figure (c) of box 33, the constrained value, might depend on states of other objects. The constraint may be polymorphic. It depends on states of other objects, which are the parameters of a polymorphic constraint.

Consider how the metamodel of enumeration in figure 46 can be polymorphic. If we replace the first rule in the sequence so that we map each instance identifier to the number 2, instead of the number 1, the conjunction would become the rule for counting by twos, instead of the enumeration rule. This is a variant of the enumeration rule. We could generalize the rule to map each instance identifier to a number "*n*" (that is identical for each instance). Then specific subtypes of the rule would be rules for counting by ones, twos, threes, and so on. Each subtype is based on a subtyping criterion, or parameter – the value of the number "*n*". This is an example of parametric polymorphism.

A parametrically polymorphic relationship has several variants, i.e. subtypes and is therefore "manifested in several forms" (see box 21). It applies to all values of the domain, not just those assumed by attributes of specific objects, and will therefore be inherited by all attributes that map to the domain (see the endnote on polymorphism in the theory of categories).

Just as the enumeration rule can have variants, so too can other relationships between attributes. Like any other object, relationships between values or attributes can be generalized. These generalized components can be links of the daisy chain that comprise the overall relationship. Since these generalized relationships may have several subtypes, so too may the daisy chains that contain them.[43] The subtypes of these generalized components will be parameters in the daisy chain that reflect variations in behavior between subtypes. Common components in the daisy chain will normalize behavior common to all variants. (Those interested in more information on, and examples of, parametric polymorphism may refer to parametric polymorphism in the endnote on theory of categories.)

Consider the enumeration relationship again. We could have made the mapping rule even more general than we did. For instance, the mapping rule for instance identifiers need not necessarily have to map to the same number. There could be complex rule expressions that determined which instance identifiers mapped to what numbers. Eventually any relationship can be generalized to merely assert that we know "*some* relationship" exists, but *what* it is will be specified by parameters.[44] So where do we stop generalizing? Indeed, this question

[42] In programming terms, these objects are "parameters" passed to the relationship.

[43] These generalized, conjoined relationships are λ-expressions. See the endnote on λ-calculus.

[44] λ-calculus and functional programming generalize relationships and functions. See the endnotes.

will apply not just to the rule for enumeration, but also to every relationship between attributes or objects. Module V and the universal perspective, summarized on our website, answers this question. Another book by the same authors, *Agile Systems with Reusable Patterns of Business Knowledge – A Component Based Approach* [337], elaborates on this theme.

Some relationships between values will apply to every value in related domains, and will therefore be inherited by all attributes that map to those domains (see figure 35, figure 37, and box 27). Variants of these relationships too will be inherited by every attribute that maps to the domain. These are the variants (subtypes) that will be manifested as parametric polymorphism, i.e. polymorphism that depends only on parameters, not object classes (see *parametric polymorphism* in the endnote on the theory of categories for an example). In contrast, other constraints and relationships will apply only to attributes and states of a specific object class. Inclusion polymorphisms are subtyped relationships that apply to corresponding subtyped objects. That object could also be an attribute, which we know is a special kind of subtype of a domain (see figure 35). Box 33 shows that relationships between objects are special cases of relationships between attributes; it does not matter whether the relationship applies to the instance identifier or some other feature – both are instances of inclusion polymorphism. From box 21, we can appreciate how this could happen (also see "Polymorphism" in Chapter 4, section 3; Module V, section 4 and Module VI, section 1 on our website).

Recursion

Consider the enumeration rule again. The statement of the rule could read:

enumeration of 1 item = 1 and

enumeration of N items = 1 + Enumeration of ($N-1$) items

We have stated the rule in terms of itself, i.e. the rule is a term in its own statement. Like the components in figure 46, the rule is also a sequence of terms (not necessarily a time sequence – see the end note on gluing objects together) and the domain (figure (a) box 33) of each term has a relationship with the domain of the preceding term. Rules like this are called recursive rules (box 33). Recursive rules are subtypes – a special case of an object composition – a conjunction of objects (see Module V, sections 1 and 2 on our website) like figure 46. The aggregate is a recursive rule if it consists of a sequence of the following kind:

1 the relationship must represent the conjunction of a sequence of like terms; and
2 each term in the series may be expressed in terms of the term *before* it.

Incomplete rules

As we saw under *mixing incomplete or uncertain constraints*, we might not always have complete information on constraints that we know exist. The constructs in box 33 let us represent not only what we know, but also what we do not, with a great deal of precision and fine granularity.[45] For example:

[45] λ-calculus can normalize granular information with precision. See [307].

- We might know two or more attributes *are* related, but we do not know *how* they are related. The rule expression would then be "unknown" but values in the value set would be values of attributes that we know are related.
- Knowing or not knowing which attribute values participate in a rule is different from not knowing a value of one of those attributes at any given time. If it turned out that we did not know the value of an attribute while executing a process, it could drive its value in value sets to "unknown."
- Sometimes we might have even less information. We might know the meaning of a relationship, i.e. *what* must be calculated, but not how to do it, or even all the variables involved. In other words, even if the *rule meaning* is known, both attributes and the *rule expression* involved may be unknown.

  There are numerous examples of this in the real world. For example, we might know that exposure to advertisements will increase sales, but not how to compute the increase, or even all objects and attributes that might be involved in predicting consumer behavior.

  We can then accurately normalize our ignorance if we include the attributes we know matter in a non-exhaustive value set and attach the latter to an "unknown" rule expression (see Module V, section 2 on our website; the value of the rule expression is "unknown.")
- If the co-domain of the relationship is a quantitative attribute, distinctions between what we know and what we do not may be granular indeed.

  For example, we might know that a mathematical formula relates the domain(s) of the rule to its co-domain and we might even know that the relationship contains a quantitative formula, and even know some, but not all the terms and operators in it.

  We can then assign the "unknown" value only to items we do not know. Chapter 2, section 2, under "don't know" values describes why this will always result in an overall "unknown" for the relationship.

The structure in figure (c) of box 33 would normalize, and accurately reflect, the state of our knowledge, or lack of it, in the real world.[46] The relationships between the attributes allow us to represent degrees of incomplete knowledge with consistency and resilience, at the appropriate level of granularity.

Rules about rules

Value constraints may also be constrained by other constraints. Constraints that govern other constraints may change with the flow of time, events, and processes.

The order of a constraint

There are no injunctions against various values in inclusion and exclusion sets, like limiting values and members of value sets, from being constrained by inclusion and exclusion sets. These are called second-order constraints because they are restrictions that constrain other

[46] Real-world chaotic phenomena almost always fall into this category of laws, in which we know the meaning of an effect and often all involved attributes, but not how to calculate the outcome. See the examples in [323].

constraints.[47] Indeed, there is no injunction against parameters (members of value sets and limiting values) of second-order constraints, in turn being constrained by other inclusion and exclusion sets, resulting in third-order constraints and so on. The metamodel of knowledge thus makes room for "*n*th" order constraints, where "*n*" is a number that describes the order of the constraint: 1 for simple inclusion and exclusion sets, 2 for second-order constraints, 3 for third-order constraints, and so on. Higher-order constraints are rules about rules. Although these higher-order constraints are complex and infrequently used in business models, the metamodel of knowledge can cater to them if and when they are needed – as it must.

Stationarity of constraints

Effects of events might change constraints. Constraints could also change or evolve spontaneously (see *spontaneous state change* in Chapter 2, section 2.) Constraints that change or evolve in time are called *non-stationary constraints*. The processes that change them are rules about rules, and are higher-order constraints that govern other constraints. When spontaneous change is involved, the rule expression of non-stationary constraints will involve time dependent terms. Constraints that do not depend on time or the tide of events are called *stationary constraints* (see box 33).

Constraining state space

Each constraint we have discussed is the repository of an atomic rule. As we saw in Chapter 2, section 2, a set of constraints can limit the lawful state space of an object. This set is an aggregate object – a container or bag of atomic rules. The bag contains and constrains the state space of the object(s) involved via constrained attributes. Each atomic rule in the bag shapes an edge or boundary of an object's lawful state space. Indeed, if one or more constraints in the bag are non-stationary, the state space of the object will evolve and shift with the time and tide of events.

When these bags merge, so do constraints in them. Merged constraints must conform to the rules and injunctions described in the section on merging inclusion and exclusion sets.

Subtypes of objects inherit the bag of constraints from their supertypes.[48] Constraints on subtypes may be more, but not less restrictive, than the constraints on supertypes because a subtype may not violate the constraints on its parent. Constraints restrict the lawful state space of an object, and if constraints on the subtype are more restrictive than the subtype, the lawful state space of the suptype will fall *inside* the boundaries of the lawful state space of its parent. Thus, subtypes must always stay *inside*, or at the boundary of the facets of lawful state space they inherit from their supertype(s). Therefore, if a subtype has more restrictive constraints than its parent, constraint(s) inherited from the parent can become redundant for the subtype in much the same way that a range inside a range makes the

[47] The endnotes on *n*-morphisms and *n*-categories under the "Theory of categories" generalize governance of rules by rules. [174] has more information.

[48] Subtypes inherit all constraints on the supertype if we exclude variation inheritance. Factoring of components is usually easier without variation inheritance. See variation inheritance and refactoring in the endnotes.

outer range redundant. For example, we know that the height of a cubicle cannot exceed the height of the building it is in. Now if we divide the building into rooms, we will have an additional constraint that the length of a cubicle cannot exceed the length of the room it is in. We also know that the length of the room cannot exceed the length of the building. So, the earlier limitation on length – that the length of the cubicle cannot exceed that of the building – becomes redundant.

When constraints on subtypes involve inclusion or exclusion sets, bounds and ranges, we can eliminate redundancy as described in the subsection on combining inclusion and exclusion sets. When constraints are more complex, lambda calculus (see endnote) can help to simplify rule expressions and to eliminate redundancy.

The integrated metamodel of value constraint

Consider how constraint is not only an aggregate object, with a structure (object composition), but also a recursive relationship. In figure (c) of box 33, a sequence of objects and relationships related the value (object) on the left with the value (object) on the right. The two value objects represented the same object class. They had been shown separately only in the interests of diagramming clarity. The sequence loops back on value as illustrated in figure 47. This represents a recursive relationship. This relationship expresses the irreducible natural law that values may constrain values in the real world via rules that might involve one or more (attribute) values. This relationship is the backbone of value constraint, shorn of bounds and partitions, simplified to show only its core.

The recursive relationship in figure 47 shows the daisy chain of objects that it contains.[49] Like the enumeration relationship in figure 46, the aggregate is the overall result of interactions between components glued end-on-end. The "constrain" relationship of figure 42 (and figure (c) of box 33) is identical to the part of the recursive relationship in figure 47 that loops forward from the broken line between *value set* and *expression of rule*. The containment relationship in figure 42 (and figure (c) of box 33) is identical to the containment relationship in figure 47.

The glue that binds components of value constraint into the loop in figure 47 is a connective like the "∗" operator in the endnote on gluing objects together. Like the expression in the endnote, the overall recursive constraining relationship cannot exist (is "null") even if one of the objects in it is missing, i.e. is null. The connectives are also associative operators, as they must be when relationships are linked in chains to produce aggregate objects that are also relationships. (See the endnote on gluing objects together for the meaning of connectives, and the endnote on the theory of categories for the meaning of associative operations. The subsection on implicit and intermediate rules describes how relationships may be aggregated to result in new and different relationships.)

Figure 48 expands on the loop in figure 47 to show additional detail. It adds the detail in figure (c) in box 33, and also shows how the co-domain of the relationship in figure 47 interacts with rule expressions via inherited components.

Figure 48 explicitly shows that, like the inherited "manufacture" relationships in box 21, inherited relationships between rule expressions and values are (inclusion) polymorphic.

[49] This is the same loop as that from *value* to *value set* to *value* via the containment and constraining relationships of figure 42(b).

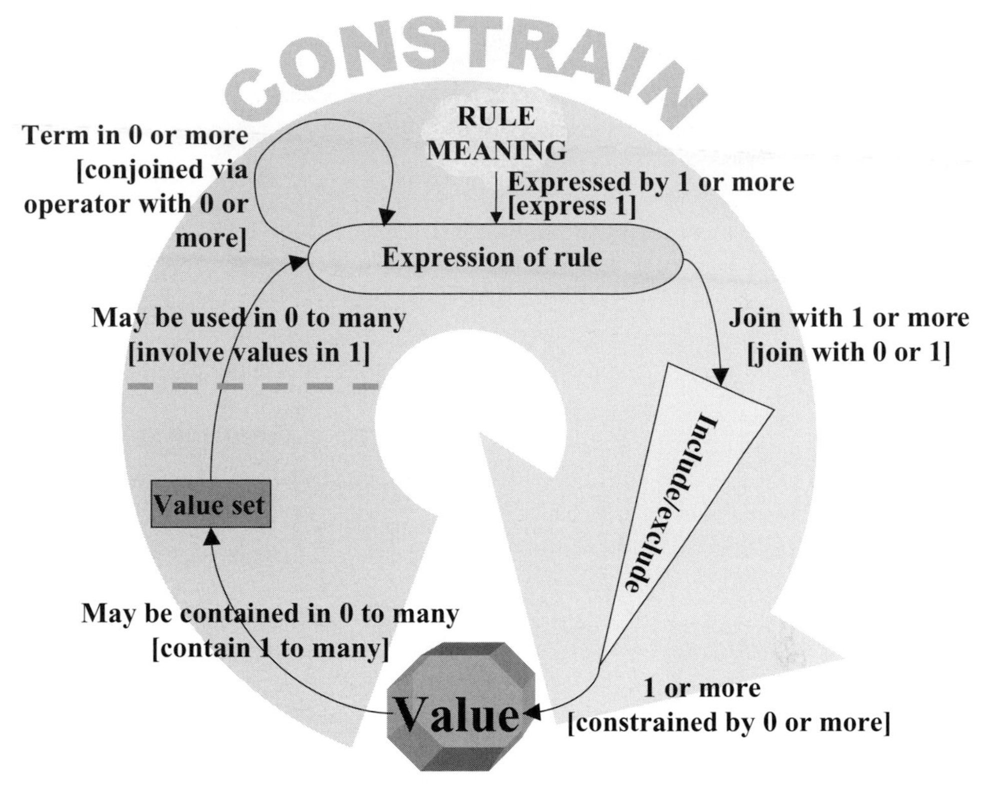

Figure 47 Constrain is a recursive relationship

Although rule expressions, in general, link to constrained values via include/exclude relationships, the figure makes it explicit that quantitative values may be constrained by Boolean, ordinal, and the full range of arithmetic rules, whereas ordinal values may only be constrained by ordinal or nominal rules. Restrictions on nominal values are even more severe. They may only be constrained by nominal rule expressions.

Note the use of exclusion partitions in subtyping rule expressions in figure 48. The general rule expression is the set of all possible rule expressions – those with the full range of terms, arithmetic, logical, and ranking operators, or any combination of these.[50] The subtype (subset) immediately under it excludes all arithmetic and higher mathematical operators as *connectives* between terms (note that these operators may exist *inside* individual terms of the rule expression).[51] That implies that these rule expressions will have only ranking (sequencing) operators and Boolean operators ("*and*," "*or*," and "*not*"). Only rule expressions with ranking or Boolean operators (or both) may connect to ordinal values. It is

[50] Variation inheritance and inclusion polymorphism occur because the rule expression object class was defined as the union of all rule expressions, not as the generalized intersection of common properties of rule expressions. In variation inheritance, subtyping is based on exceptions, i.e. set differences (see box 21 and the endnote on "Kinds of inheritance").

[51] For example, $(A + B = 5)$ or $(A - B = 5)$ is a Boolean rule. It must have Boolean connectives, but its individual terms, $A + B = 5$ and $A - B = 5$, have arithmetic operators *inside* them. Note that each term is also an equation – a kind of value constraint (see box 33).

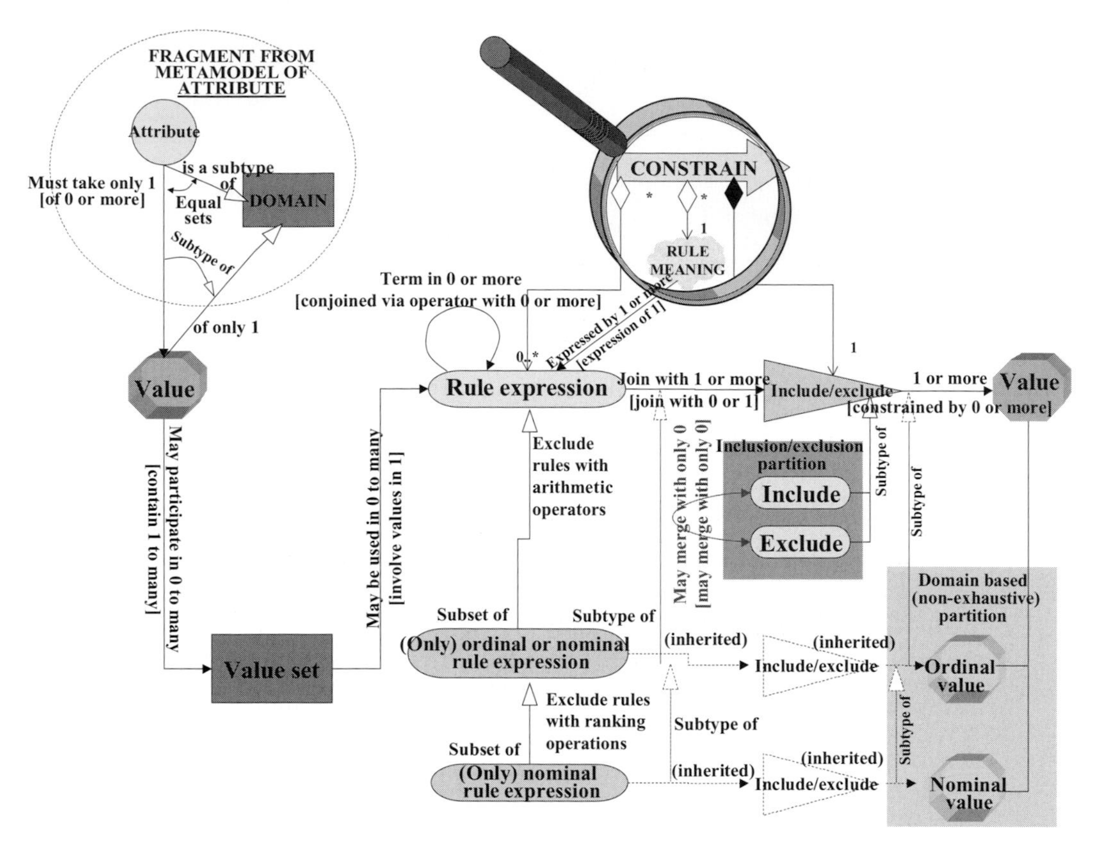

Figure 48 Metamodel of "rule constrain"

a subset of the more general set of rule expressions obtained by *taking away* the set of rule expressions with arithmetic (and other higher mathematical) operators from the (general) set of all rule expressions.[52]

Similarly nominal rule expressions are a subset of rule expressions that contain only terms with Boolean operators (*and*, *or*, *not*).[53]

This is the internal structure of "*constrain*," the relationship between *value set* and *value* in figure 42. It is also the relationship going forward from the broken line in figure 47. Figure 49 juxtaposes and integrates this information.

Just as constrain is an aggregate object with an internal structure, *bound* in figure 42 has an internal structure. It is the structure in box 29. Together, figure 42, figure 48 and the metamodel of bound in box 29, constitute the integrated metamodel of value constraint. Figure 49 integrates and juxtaposes all three metamodels. It represents the integrated metamodel of value constraint. This metamodel is a key fragment of the integrated metamodel of Module VII on our website.

[52] Since the subtype was obtained by *taking away* some rule expressions, from the superset of (all) rule expressions, the rule expression subtype does not inherit all terms from its supertype (see set difference operations in box 19). It excludes the terms that were taken away (see box 21).

[53] Subtypes like these were discussed in box 21.

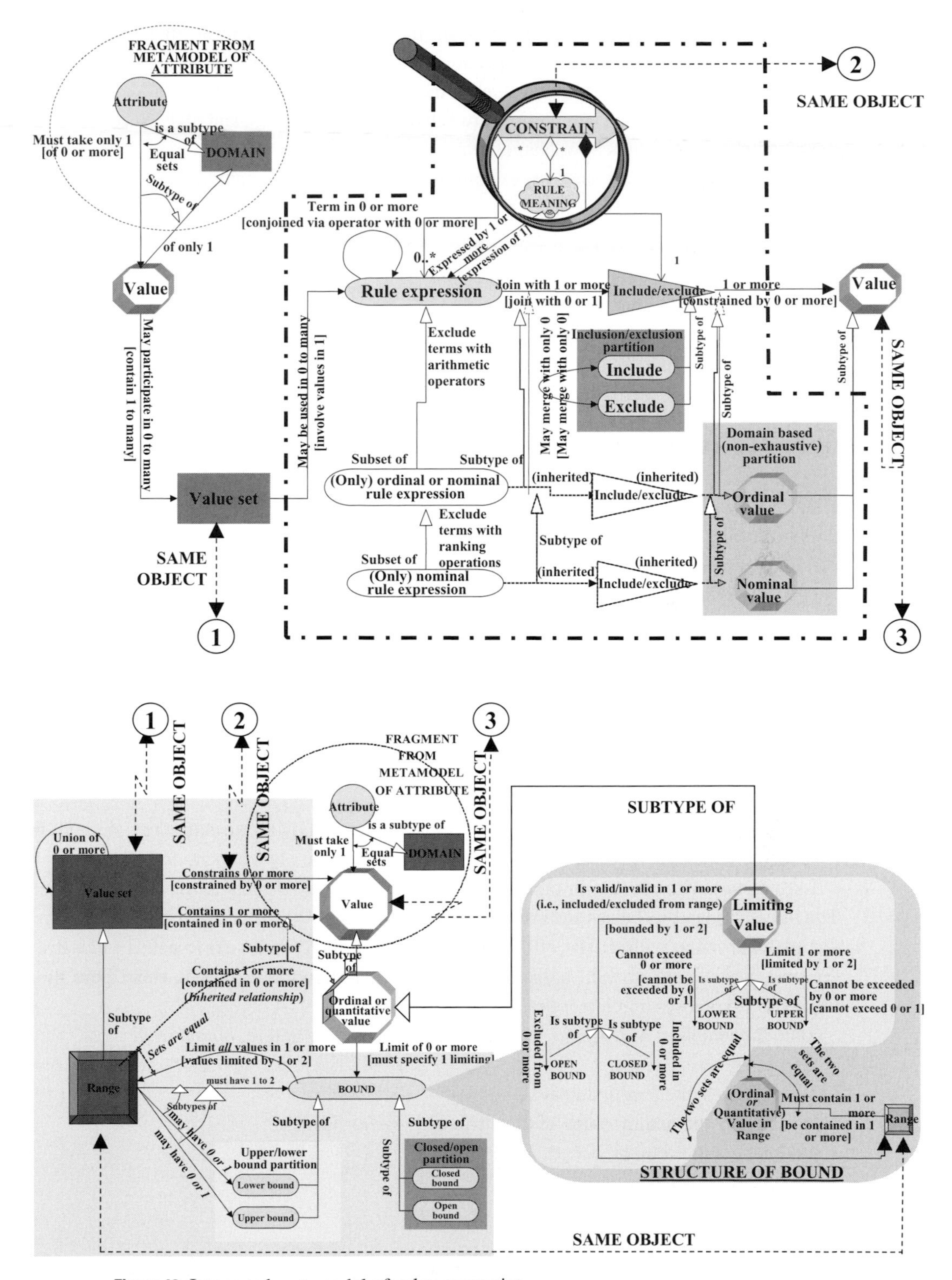

Figure 49 Integrated metamodel of value constraint

The integrated metamodel of value constraint applies not only to attribute values, but also to domains. Domains are those shadowy containers of meaning that govern values, attributes, and measurement as they anchor the existence of objects. In the next section of Chapter 3, we focus on the interaction between domains, attributes, and objects from which the meanings of attributes flow. Names express meanings. The next section will cover how attribute names, meanings, and expressions all converge in the metamodel of knowledge, a fountainhead of meaning.

> For in and out, above, about, below,
> 'tis nothing but a magic shadow show,
> Play'd in a box whose candle is the sun,
> Round which we phantom figures come and go.
> (From the *Rubaiyat of Omar Khayyam*)

3 Naming and expressing attributes

Naming conventions for attributes have often been a topic of heated debate among practitioners. It might therefore be worth noting that attribute names emerge *naturally* from the metamodel of knowledge. This happens because natural names reflect the *meaning* of the attribute and the metamodel of knowledge is the repository meaning. It follows that natural names will flow from its structure.

Let us start with the OAV model. True, it is an oversimplification that omits vital detail, but it is only our first step. It will make it easier to comprehend how names flow from the detail in figure 35 if we add detail a step at a time.

Naming a meaning

The meaning of an attribute is manifested in the metamodel by the structure in figure 34. It reflects the fact that every attribute is an overlap between an object and a domain.

Let us start with a very simple case – when a domain intersects an object only once. The intersecting object class and domain can then uniquely identify the attribute. Therefore the attribute name will consist of the object class and the domain name pair. In fact it is the object class and domain name *sequence*.[54] The following examples show how this happens.

Take the length of a string:

- the object class is *string*
- the domain is the *length* domain

Consequently the natural name of the attribute is *string length*.

Take a person's weight:

- the object class is *person*
- the domain is the *weight* domain

[54] The object name may also be in possessive form; the attribute names in the three examples could read *string's length*, *person's weight*, and *car's color* respectively. This is also consistent with the structure of *attribute*.

Consequently the natural name of the attribute is *person weight* (if we had used the footnote variant of the naming rule, the attribute would have been called *person's weight*).

All nominal domains are subtypes of a domain that may be called the *type*, *class*, or *category* domain. *Type*, *class*, and *category* are synonyms. Sometimes subtypes of the *type* domain have specific names, like the *color, gender*, or *language* domains. However, they are often unnamed. If the nominal domain has no specific name, we may name it *type*, *class*, or *category*. It does not matter which. Logical names will flow from all three.

For example, a manufacturer of cars might categorize the products into sedans, hatchbacks, and SUVs. Then:

- the object class is *car*
- the domain is the *type* domain

Consequently the natural name of the attribute is *car type*.

Not all ordinal domains are named. When an ordinal domain has no specific name, we can call it the *rank* domain, because, like all nominal domains are subtypes of the *type* domain, all ordinal domains are subtypes of this *rank* domain (in contrast the color preference domain in Chapter 1, section 3 was a named domain). For example, take titles in a hierarchical medieval plutocracy. We know that some titles were more exalted than others: King was greater than Earl, but level with Caliph. What do we call this hierarchical attribute of *title*? The naming rule would assert:

- the object class is *title*
- the domain is the *rank* domain

Consequently the natural name of the attribute is *title rank*.

Now let us consider an intangible object – a concept – insurance coverage. Insurance policies provide coverage against various perils and contingencies. Thus *insurance coverage* is an object. It protects us (financially) from various perils. Individual insurance policies consist of insurance coverages we have elected to include, and others that the insurance company might be obliged to offer. Many kinds of coverage that insurance firms offer their customers have been standardized by ISO, an insurance organization,[55] for rating (pricing) purposes. These standard classes of coverage are called ISO classes by the insurance industry. Thus, one attribute of *coverage* is its ISO class code. What domain does this attribute map to?

ISO Class is a *classification scheme* for coverage. No magnitudes or ranking are involved.[56] Therefore the attribute must map to a nominal domain. The naming rule for nominal domains would suggest that calling this attribute *insurance coverage class*, *insurance coverage category*, or *insurance coverage type*. Some readers might find these names logical but intuitively ambiguous. You are right, dear reader – even if logically precise, these names *are* ambiguous (and verbose for another reason we will discuss soon) because the behavior of all but the simplest business systems is complex. Most objects may be categorized and subtyped in several different ways to represent complex behavior. Indeed, there *are* other

[55] Insurance Services Office, Inc. (ISO) is the leading supplier of statistical, actuarial, and underwriting information for the property/casualty insurance industry in the US.

[56] ISO classes will help rate the coverage. *Rate* maps to a ratio scaled domain, not the ISO class. The ISO class only categorizes the coverage.

Box 34 Identifying domains

Identifying the domain an attribute maps to usually follows from common sense. However, the following flow chart establishes a formal process that will identify the kind of domain the attribute emerged from. Remember domains of derived attributes may be different from domains of attributes they were derived from. Therefore, derived attributes too must go through this process independently, as must those they were derived from.

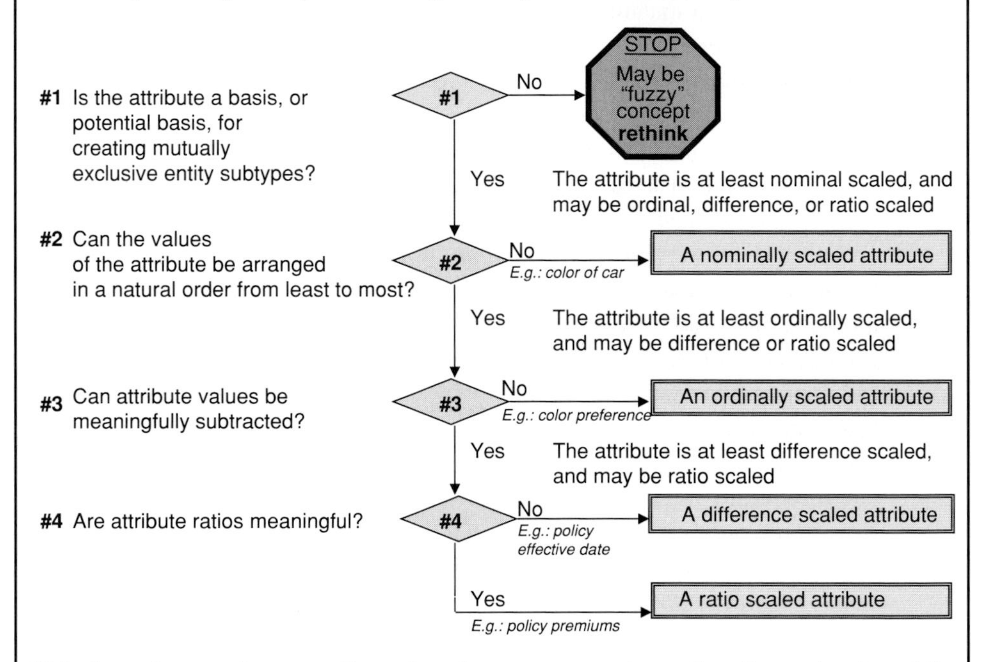

This flow chart analyzes *attribute* domains, not *expressions* of attributes (see box 35). When we consider *expressions* of attributes, differences in units of measure and formats might suffice as the basis for subtyping in step 1. This can pose a problem because formats and units of measure are not domains. In legacy systems, attributes are typically not separated from their expressions. Therefore, before we can use the flow chart, attributes must be stripped of units of measure and formats (and coding schemes) to extract the *essence* of their meaning. *Expressions* of attributes are objects in the business process automation layers of figure 15, or in the implementation layer of figure 16. Stripping them of formats and units of measure is one way we "generalize" attributes as we normalize knowledge or *refactor* software (see the endnote on refactoring).

classification schemes for coverage. ISO class is only one of several ways we can classify insurance coverage. *insurance coverage class* and its synonyms are therefore ambiguous. They do not say *which* classification scheme we mean.

This example demonstrated how, in all but the simplest systems, it is very likely that the *type domain* will intersect the object not once, but several times – one intersection for each category or partition. When this happens, the object name – domain name pair – alone cannot identify the intersection of the domain and object we have picked. We must qualify *which* intersection we mean when we pick an attribute. In this case the qualifier is *ISO*. The attribute is therefore *insurance coverage ISO class*. Clumsy and verbose? – Yes, but logically precise.

Why ISO *class*? Why not ISO *type* or ISO *category*? The answer is simple – ISO class is correct because that is the convention. When we restrict our context to insurance products, insurance coverage is redundant because it is implied. We can drop the clause *insurance coverage*. This too is convention. Convention *usually* does not clash with logic – it only limits the "correct" choice of name from several possible logically correct synonyms and constructs. However, this is not consistently true – sometimes this alignment between convention and logic may only be partial, and, occasionally, convention and logic may even clash. Let us see how this happens:

Take the color of a car. Logic and convention go hand-in-hand when we consider only the body of the car:

- The object class is *car*
- The domain is the c*olor* domain

Consequently the natural name of the attribute is *car color*.

Let us now consider the case when a domain intersects an object not once, but several times. When this happens, recognizing only the intersecting pair – the object class and domain – cannot uniquely identify the attribute. We must now qualify *which* intersection we mean. We need a *qualifier*. The qualifier is the aspect of the object we are interested in. For example, we might be interested in the color of the dashboard of the car as well as its body. Then one attribute of car will be the color of its dashboard and the other would be the color of its body. Take the color of the dashboard first:

- The object class is *car*
- The qualifier is *dashboard*
- The domain is the *color* domain

Consequently the natural name of the attribute is *car dashboard color*.

Now take the color of its body:

- The object class is *car*
- The qualifier is *body*
- The domain is the *color* domain

Consequently the natural name of the attribute is *car body color*.

Of course, in English, we can leave the word "*body*" out – it is implied. *Car color* would do just as well. Every language has its own rules, conventions, and peculiarities that are not always logical. Many things can be implied or understood by convention, and conventions differ between languages, but *car body color* is the logically correct and mathematically precise name of the attribute. It expresses exactly what the attribute means and does not depend on any quirks of English or any other language.

Although in this case *car color* would do, and *car body color* might sound a trifle verbose, it is not bizarre. However, sometimes convention clashes with logic with more drastic results. Blindly using the object name, qualifier, domain name sequence to name an attribute (or, if only one intersection is involved, the object name, domain name sequence) can result in logically correct, but bizarre, and even amusing names.

For example, take a person's height. The object class is *person*, and the domain, *length*. Consequently the logical name of the attribute is *person length*. English demands that we call it "height" instead of "length," even if the individual lies down, but there are other languages that demand we call it "length,"[57] even if the individual stands up! If you did not speak English, you might have felt *person length* was better syntactically than *person height*. To you, the English name would have sounded bizarre.

Quirks and inconsistencies of language can be even stranger. Take the length, breadth, and height of a room. They are all attributes of an object class called *Room*. *Room width length*, *room length length*, and *room height length* are three examples of *logically* sound but *syntactically* strange attribute names. Convention in English dictates that we know that attributes with the words "width," "height," and "length" must always map to the length domain, and we must not be explicit about the fact. Adding the domain name, *length*, is not only redundant, but also incorrect in the English language.

In the examples discussed so far, both logical and conventional English names have been clear. The conventional English name has been either identical to the logical name, or a syntactic improvement on that name (an arguably subjective view point of proponents of English). However, there can be occasions when ambiguities of languages can be even stranger and only logic can lead to clarity.

Take the length and width of a square. Which side is its length and which its width? There is no convention to guide us. Two different individuals might talk about doing different things to the different sides of a square shaped playing field, each may call his or her side the "length," and not know that the other person is interpreting it as the side(s) he calls "width." Processes and automation sometimes demand tremendous precision and speed. For this, they need complete clarity. There may be little scope for human intervention or corrections after the fact. Ambiguity can be intolerable in such situations and a logically precise name will definitely help.

In the high noon of the knowledge economy, in a world of global commerce and seamless integration of automation of unprecedented complexity and scale, the need for multilingual support, translation, and cross-cultural collaboration can only increase. We stand at the dawn of this era. The demand for diversity of names will grow, not shrink. The structures in figures 29 and 30 can link homonyms and synonyms of departments, divisions, and corporations to a hub in order to facilitate collaboration between them; they can also link syntactically sound names of attributes in different languages to their logically sound universal name to facilitate large-scale, global, and cross-cultural collaborations.

Meaning versus expression

September 17, 2001 – MSNBC news item – "Standards Body pushes accessibility – proposed guidelines would help disabled use the web:

Advancing its initiative to make the Web more accessible to people with disabilities, a major standards body [The World Wide Web Consortium (W3C)] has issued guidelines for designing browsers,

[57] *Hindi*, the predominant language in Northern India, is one such language.

multimedia players and other Web-based user interfaces . . . making the Web accessible becomes an increasingly urgent task. Federal Web sites must conform to Section 508 . . . requires technology procured by the federal government to be accessible to people with disabilities . . . improve accessibility for color-blind people . . . devices such as screen readers.

Industrial strength global systems, which flex with the global economy in support of a vast and diverse global society driven by rapid advances in technology, must manage meaning in harmony with meaning's diverse expressions. Indeed, the need to recognize multiple names of an attribute can become imperative when we step beyond the OAV model. This imperative flows not from the peculiarities of any language; rather it flows from reasons of pure logic – that several distinct and different *expressions* can manifest the same meaning in the real world. Therefore, to normalize meaning, we must divorce it from its *expression*. We have done this in box 35. There we have stepped beyond the bounds of the simple OAV model and recognized that attributes not only have values, but may have several formats and units of measure as well. These considerations must impact expression of attributes because they too are parts of its structure.

Quantitative attributes

Quantitative attributes not only have formats, but also have units of measure. To understand the impact of units of measure, let us return to the example of a person's weight. A person's weight is an attribute of *person* that maps to the *weight* domain. The weight domain is a ratio scaled domain. Therefore a person's weight may be expressed in several units of measure. *Person weight* was the name derived from the simple OAV structure in figure (a) of box 27. Box 35 elaborates on it to show how *person weight* can involve several units of measure and formats. We know a person's weight could be simultaneously expressed in kilograms and pounds. Expressed in kilograms, we would call (the expression of) the attribute *person weight in kilograms*; and, expressed in pounds, the expression would read *person weight in pounds*.

The *object name, qualifier (if more than one attribute maps to the domain), domain name, unit of measure* **sequence** results in the logical name of the **expression** of a quantitative attribute, provided we insert the word *"in"* between the *domain name* and *unit of measure*. The object name and domain name must always be singular, and the unit of measure must be plural.

The following examples demonstrate this rule.

A person's weight in kilograms:

- The object class is *person*
- The domain is the *weight* domain
- The unit of measure is *kilograms*

Consequently the natural expression of (the attribute) *person weight* is *person weight in kilograms*.

A person's weight in pounds:

- The object class is *person*
- The domain is the *weight* domain
- The unit of measure is *pounds*

Consequently the natural expression of (the attribute) *person weight is person weight in pounds*.

Similarly, the length of a string would be expressed as follows:

Length of a string in feet:

- The object class is *string*
- The domain is the *length* domain
- The unit of measure is *feet*

Consequently the natural expression of (the attribute) *string length is string length in feet*.

Length of a string in meters:

- The object class is *string*
- The domain is the *length* domain
- The unit of measure is *meters*

Consequently the natural expression of (the attribute) *string length is string length in meters*.

The rule applies even when convention clashes with logical precision. For example, as long as we defer to the English convention of calling the length of a person his or her height, the rule yields syntactically correct English expressions for names for a person's height in any unit of measure. The expression of the attribute would be *person height in meters*, *person height in feet* and *person height in inches* when the units of measure are, respectively, meters, feet, and inches. (Exercise: How do mixed units in expressions like "5 feet 6 inches high," fit this framework? Hint: See the discussion under *Patterns of symbols, patterns of objects* in Chapter 4.)

Box 35 The many faces of meaning

Is *person weight* a single attribute or several? Does each expression of *person weight*, like *person weight in kilograms* and *person weight in pounds* qualify as a separate attribute? These questions arise when we recognize the inherent multiplicity of domains and formats in which the innately unique *meaning* of an attribute can be expressed. Is the unique *meaning* the attribute (i.e., the object–attribute pair), or is each *expression* of the object–attribute pair an attribute? Most state-of-the-art software (programming languages, database management systems, CASE tools, and process design tools) does not address this question, because it does not recognize that a single meaning can have multiple expressions. Indeed, this is one of several reasons why unmanaged and chaotically replicated denormalized knowledge exists in our processes and systems.

In this book, the meaning will be called *attribute* and its expression will be called *expression* of the attribute. *Person weight* is an attribute and *person weight in pounds* is its expression. However, readers should keep in mind that most software platforms, such as modeling tools and database management systems, will call each expression of meaning an attribute and may replicate the "true attribute" in each distinct expression. As new systems design techniques evolve, technologies will also evolve in step. We hope that emerging implementation technology will contain the "hooks" to better incorporate the ideas that we have proposed in this book.

Although the current state-of-the art may not support it, the metamodel described in this book is the blue print we have envisioned for the near future. This blue print will help leverage information and knowledge assets effectively as systems are upgraded over time. We can normalize knowledge if we follow the metamodel, and we can build knowledge artifacts that will normalize components of business knowledge using that metamodel. After all, our intent in the new paradigm is to store normalized knowledge in an electronic repository and to configure business processes and software from them.

The multiplicity of ways in which a quantitative attribute may be expressed is compounded further when we consider that each attribute's expression in a unit of measure can, in turn, be expressed in several formats. For example, the expression *person weight in pounds* must be elaborated further before it can be tangible in the real world. In the real world, *person weight in pounds* may be expressed in digits, printed words, spoken words of several languages and scripts, images of various kinds such as bar graphs and dials. These are all formats.[58] Formats too are objects and can have subtypes, and each subtype would add a qualifier to the name of the format. For instance, if *person weight* is expressed in written words (as opposed to numerals or spoken words), it might be in English or Japanese. Similarly, if it is expressed in numerals, it could be in Arabic, Roman, or other numerals. Thus, if *person weight in pounds* is expressed in written words, the expression of the attribute would become *person weight expressed in written words*. *Written words* is the format in which the attribute is expressed. If we further subtype the format to specify that it is Japanese words we are using, the name of the expression would become *person weight expressed in written Japanese words*; the qualifier "Japanese" must be added to the name of the parent format. In the same way, if *person weight* were expressed in Roman numerals, the full expression of *person weight in pounds* would read *person weight in pounds, expressed in Roman numerals*. The clause "*expressed in*" is inserted between the unit of measure and the format to describe the complete expression. Through its complete expression, the attribute is manifest and tangible in the real world. For this reason, we call this complete expression of the attribute, its ***tangible expression***.

The rule is that the *object name, qualifier (if more than one attribute maps to the domain), domain name, unit of measure, format sequence*, results in the logical name of the *tangible expression* of the attribute, provided we insert the word *"in"* between the *domain name* and *unit of measure* and the clause *"expressed in"* between the *unit of measure* and *format name*. The object name and domain name must always be singular, and the unit of measure must be plural.

The following examples demonstrate this rule:

Take the length of a string in feet. A business process might demand that this length be stated verbally in English:

[58] Domains and units of measure structure quantitative values in the real world. Formats map reality to information systems. Formats therefore belong to the interface rules layer of figure 15. See the discussion on differences between *domain* and *format* in Chapter 4.

- The object class is *string*
- The domain is the *length* domain

Consequently the attribute name is *string length*

- The unit of measure is *feet*

Consequently the expression of *string length* in feet is *string length in feet.*

- The format is English *speech*

Consequently the (natural name of the) tangible expression of *string length* is *string length in feet, expressed in English speech.*

Instead of spoken words, if we had required the string length expressed in numbers, its *tangible expression* would have been different, but not the *meaning* of the expression:

- The object class is *string*
- The domain is the *length* domain

Consequently the attribute name is *string length*

- The unit of measure is feet

Consequently the expression of *string length* in feet is *string length in feet.*

- The format is *numeric digits*

Assume that the *default state* of the domain of numeric digits is western script. However, it is worth bearing in mind that there could be several other subtypes of numeric formats, based on non-western scripts (such as the Devnagri script of India, which uses a different script, but the decimal system to express numbers) or even non-decimal number systems like the still widely used system of Roman numerals.

Consequently the (natural name of the) tangible expression of *string length* is *string length in feet, expressed in numeric digits.*

As described earlier, English convention can sometimes clash with logic in the naming of attributes. If this happens, these syntactic clashes will be carried forward to the names of *expressions* of these attributes as well. For example, take the total distance traveled by a car. Assume that an automated system must state this distance in miles and express it in English speech. Then:

- The object class is *car-travel*[59]
- The qualifier is *total*
- The domain is the *length* domain
- Consequently the attribute name is *car-travel total length* (instead of total car-travel distance)

Although most of us are so used to English that we do not need to consider English grammar and convention consciously when we express ourselves, it is not difficult to realize that this syntactically awkward attribute name is logically sound.

- The unit of measure is *miles*

Consequently the expression of the attribute name is *car-travel total length in miles* (instead of total car-travel distance in miles)

- The format is English speech

[59] Some readers might ask why car-travel, not car alone, or travel by itself is the object in question. Module V, section 1 has the answer.

Consequently, the (natural name of the) *tangible expression* of car-travel total length is *car-travel total length in miles, expressed in English speech* (instead of total car-travel distance in miles, expressed in English speech).

Usually the choice of *format* is specified or constrained by business process automation (see "The architecture of knowledge" on our website), whereas the unit of measure depends solely on business convention and process design. Note also that the format is not necessarily restricted to formats of real-world expression. Formats can mean formats of data internal to technology platforms, such as jpeg or bitmap formats for graphics, ASCII or binary formats for characters and numbers. However, this book focuses on business processes and requirements. Formats in the technology layer of figure 15 are beyond the scope of this book.

There may be many variants and a huge diversity of users of large-scale business systems in a global environment. Components may be reused in widely diverse contexts for diverse processes. Expressions and attribute names will be used not only by groups that administer the repository of knowledge artifacts and those that configure systems or business processes with them, but also by users of systems in diverse contexts, cultures, and languages, each with its own conventions. They may need help files, explanation, training, and other descriptive information. This information must therefore be rendered in equally diverse ways to be clear, succinct, and precise. Clear, succinct, and precise rendition is important because it is the key to rapid, effective, and reliable utilization of new systems and processes. Rapid ramp up is vital in an environment of continual change under intense competitive pressure.

These diverse renditions can be hard to manage unless we underpin them with a logical hub of the kind in figures 29 and 30. It will not matter if this hub, hidden from users, is verbose or clumsy in some language or context, as long as it is unique, precise, logical, and *complete*. If it is unique, precise, and complete, it will facilitate accurate, timely, and syntactically elegant renditions of information in diverse environments. Thus, all renditions can be tied to a unique, precise, and complete expression, which in turn can, with the metamodel in figure 35, be tied to unique, precise, and complete *meanings* of these expressions.

Qualitative attributes

Qualitative attributes have no units of measure, but they do have formats. Consequently, naming tangible expressions of qualitative attributes is similar to, but simpler than, naming tangible expressions of quantitative attributes. To express qualitative attributes, all we must do is drop the clause for unit of measure, because units of measure are meaningless to qualitative attributes. The rule for naming tangible expressions of qualitative attributes is:

> The rule is that *object name, qualifier (if more than one attribute maps to the domain), domain name, format sequence*, result in the logical name of the *tangible expression* of the qualitative attribute, provided we insert the clause *"expressed in"* between the *domain name* and *format name.*

The following examples demonstrate this rule. Take the color of a car again. Like the example on the length of a string, the color may be expressed by automation in spoken or written words, or just a colored shape, say a bar. Take spoken words first:

- The object class is *car*
- The domain is the *color* domain

Consequently the natural name of the attribute is *car color*

- The format is *English speech*

Consequently the natural name of this expression of car color is *car color expressed in English speech.*

Take written words next:

- The object class is *car*
- The domain is the *color* domain

Consequently the natural name of the attribute is car color

- The format is English writing

Consequently the natural name of this expression of car color is *car color expressed in English writing.*

Take the colored bar next:

- The object class is *car*
- The domain is the *color* domain

Consequently the natural name of the attribute is *car color.*

- The format is *colored bar*

Consequently the natural name of this expression of car color is *car color expressed in colored bar.*

Take the gender of a person. It may be coded in numbers, say 1 for male and 2 for female, in letters, say M for male and F for female, or expressed in words, "Male" and "Female." These codes are not gender, but merely *represent* gender in information systems (manual or automated – it does not matter which). Therefore, they are not only codes, but also *formats* of gender.

Let us call the first format *digits* (we could just as well have named it the *1 / 2 code*, or anything else that captures its meaning), the second format *letters* (we could just as well have named it the *M/F code*) and the last format *written English word* (to distinguish it from English speech format). Then the expressions of gender would be derived as follows:

- The object class is *person*
- The domain is the *gender* domain

Consequently the natural name of the attribute is *person gender* (or person's gender if we use the variant in the footnote on the naming rule for attributes).

- When the format is:

  digits, the expression would be called *person gender in digits*;

  letters, the expression would be called *person gender in letters*;

  written English word, the expression would be called *person gender in written English word.*

Formats are objects too, and may be partitioned, subtyped, and aggregated. This can give us a rich repertoire for expressing meaning in multiple media, in different languages and character styles such as fonts, colors, codes. However, we will defer that discussion to Chapter 4.

Attributes emerge from domains and are expressed by formats. Domains anchor meaning, measurement, existence, and value. Therefore, our next step towards the integrated metamodel of knowledge is to understand the nature of *domain* and its expression in the real world through meaning and measurement of properties of objects.

4 Domains and their expression

Silent and void,
It stands alone and alters not,
It moves but does not tire.
. . . I know not its name
. . . I call it the way
(Chinese philosopher Lao Tzu, 6th century BC)

This chapter discusses the concepts of pattern and measurability. More importantly, these concepts are delineated in intuitive manner, without requiring mathematical sophistication. The chapter addresses the spectrum of meanings, from those that precisely quantify and measure numerically, to those that are purely qualitative. It describes components, configurations, and patterns of information that derive these meanings and eventually lead to the very concept of existence and meaning itself. The concepts are illustrated with examples from diverse areas.

Attributes have meanings. So have their values. Domains are the wellspring of meaning. Domains are sets of values. These values are measures of a meaning.

What does being a measure of meaning mean? Consider a room. Assume it is 30 feet long, 20 feet wide, and 10 feet high. These numbers describe the dimensions of the room. However, by themselves the numbers are only numbers. They *mean* nothing. Numbers only measure magnitudes, not meaning. Moreover, different numbers might describe the same dimensions. For example, measured in inches, the same room would be 360 units long, 240 units broad, and 120 units high. The larger numbers do not mean that the room has some how expanded; they only mean that the *values* of length, breadth, and height, i.e., the *meaning* of these values, have been expressed by *mapping* them to different sets of numbers. We distinguish between the two different maps between meanings of values and the set of numbers by calling them different measures (feet and inches). Units of measure are merely names of maps (like figure (a) in box 33) from the set of meanings to the set of numbers that express relative magnitudes of the (*meaning* of) values.

Different meanings carry different amounts of information. This was obvious and intuitive in the parable of Metanesia in section 3 of Chapter 1. To recapitulate and summarize:

- Nominal domains carry only information on existence and distinctness of (meanings of) values.[1] They carry no information on relative or absolute ranking of magnitudes of these meanings (values).
- Ordinal domains carry not only information on the existence and distinctness of meanings (values) but also information on relative ranking of magnitudes of value meanings.[2] However, they carry no information on meanings of differences or ratios between these meanings (of values).
- Difference scaled domains carry information on existence and distinctness of meanings (of values), as nominal values do; information on relative ranks (of magnitudes of value meanings), as ordinal values do; as well as information on the meaning of the *magnitude of **differences*** between meanings.[3] However, they carry no information on meanings of ratios between these values (of meanings).
- Ratio scaled domains carry information on existence and distinctness of meanings (of values), information on relative ranks (of magnitudes of meaningful values) as well as information on meanings of differences and ratios of meaningful values.[4]

We will call domains that lend pure, abstract meanings to real-world properties like the length and volume of a room, "*domains of meaning*." Abstract meanings must be expressed in symbols to give them tangible form. We will call sets of symbols that represent abstract meaning *formatting domains* in this book. Values in difference and ratio scaled domains convey not only the meaning of the domain, but also information on quantifiable magnitudes that may be expressed with numbers. We must distinguish between the *meaning* of a number and its *expression* with a symbol. For example, "X" in Roman numerals and "10" in Arabic numerals are different symbols that express the same number. We will call sets of numbers that are divorced from symbols that might express them as the *domains of numbers*.[5] Thus, there are three kinds of domains that lend meaning and expression to information in the tangible world of business. They are:

1 formatting domains;
2 domains of numbers; and
3 domains of meanings.

When we do not explicitly qualify a domain into one of the three kinds of domains above in this book, it will mean that the domain is a domain of meaning. In this chapter, we will describe the role each plays in the metamodel of knowledge. Of the three, formats are the most tangible and arguably the easiest to grasp. Therefore we will start with the meaning

[1] Since nominal domains carry only information on existence and distinctness of (meanings of) values, only Boolean operators are valid in these domains (see Chapter 3, section 2 or the theory of mathematical groups and rings in the endnotes).

[2] Since ordinal domains carry information on existence, distinctness, and ranking of (meanings of) values, Boolean and sequencing operations are valid in ordinal domains. See the theory of mathematical groups and rings in the endnotes and ordinal value theory in [171].

[3] Boolean, ranking, and arithmetic subtraction are valid operations between values in difference scaled domains.

[4] All Boolean, arithmetic, and ranking operations are valid in ratio scaled domains.

[5] Our everyday concept of "number" will suffice to understand this section. T*he domain of numbers* is a pattern that is more than just the symbols and digits used to express a number (see the section on set and number theory in the bibliography). The number domains of relevance to this book are those that constitute a one-dimensional continuum in an *ordered field* [216]. See [206], [204], [219], [220], [221], [222], [224], [225], [230], and [231].

of formats and formatting domains before we proceed to abstract domains of meaning, but bear in mind that domains of meaning are the key to the metamodel of knowledge.

Meaning flows from domains of meaning into the metamodel of knowledge, shaping objects and seeping through numbers, symbols, and relationships into the world of tangible things. Meaning orchestrates the tangible world of business from the shadow world of concepts. Formats and numbers are merely spans of that bridge, between abstract knowledge and its realization in concrete processes – automated or manual.

1 The meaning and architecture of format

Pure meanings, by themselves, are abstractions. To give substance and communicability to a meaning, we must express it in some form. This means each meaning must map to a symbol that can be sensed by one of our five senses. These maps are similar to those in figure (a) in box 33. The *rules for mapping meanings to symbols are formatting rules*, and *the symbol is the format*. Thus, format is the image (see box 33) of the meaning in the domain of expressed and perceived symbols.[6] In this book, we will call these *formatting domains*. (They are actually the *co-domain* of the formatting rule.) This is how formats emerge from the metamodel of knowledge.

Formatting rules, like the map in figure (a) of box 33, are maps between a set of meanings and a set of symbols that are subject to the following constraints:

1 Each meaning in a domain must map to only one symbol in the co-domain. If it maps to several, each will be a synonym (see Chapter 2, section 4, "The tyranny of words").
2 Each symbol in the co-domain must map back to only one meaning. If it maps to several, each will be a homonym (see box 36 and Chapter 2, section 4, "The tyranny of words")

If both conditions are violated, the representation is ambiguous. In practice, these conditions are satisfied only in a limited context, and can be the cause for confusion between different perspectives. However, sometimes symbols like audible alarms may be deliberately ambiguous because they may be needed only to draw the user's attention to another representation (e.g., a visual display). Boxes 36 and 38 discuss this aspect in more detail. Indeed, there are an infinite number of possible formats (symbols) for each meaning, but only five kinds of symbols are fundamentally aligned with our *perception* of meaning.

Five fundamental formats

Formatting rules are *relationships* between domains of meaning and formatting domains of symbols; they take us from abstract business knowledge to its expressions in business systems. For example, when a meaning is expressed through speech, the spoken word is the symbol, or format, that the meaning maps to, and *speech* is the map, or formatting rule for expressing the meaning. Formatting domains bridge abstract meanings in the business rules layer of figure 15 with concrete representation(s) in the interface layer – see the examples of *tangible expressions* in Chapter 3, section 2.

[6] The science of mapping meanings to symbols is called *semiotics*. [325] discusses semiotics lucidly, with humor, without mathematics.

Box 36 When one object represents another

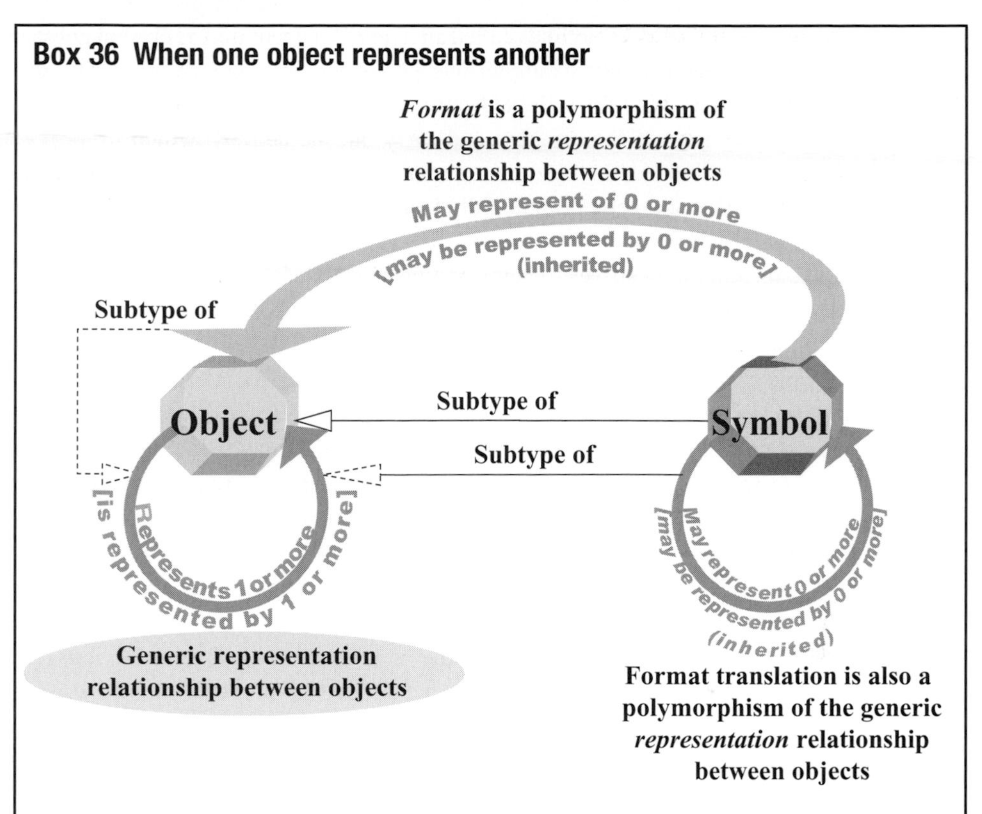

Other objects may represent an object. Strictly speaking, all objects represent at least themselves. That is why the generic represent relationship has a lower bound of 1. It connects, at a minimum, an object instance to itself. However, self-representation conveys no information, and should therefore be discounted. (Relationships like this are called reflexive relationships – see Module V on our website.)

Symbols are objects that are perceived with our senses and exist in space and time. They may *represent* other objects, like *value*, that convey only meanings. When they do this, they are formats of the abstract objects they represent. This *represent* relationship was inherited from the generic object, and is a (inclusion) subtype of the more generic recursive *represent* relationship on *object*. Thus *format* is a polymorphism. Of course, a symbol might not represent *another* object, and hence *represent* becomes optional in this role (see the discussion on inheritance of cardinality ratios between supertype and subtype relationships under "The merger of meaning and the metamodel of value constraint" in Chapter 3, section 2). Similarly, when one format maps to another, it is a format conversion, i.e. *translation* relationship. We will elaborate on the internal structure of this relationship in box 38 and on its generic representation in Chapter 4, section 3. In this section, our focus is on its formatting polymorphism.

Homonyms and synonyms also flow from this *Represent* relationship. When two or more objects represent an object, each is a synonym. When an object represents two or more objects, the object is a homonym. When an object is both a homonym and a synonym, its meaning is ambiguous.

A representative is sometimes called an *agent*. An agent may represent another object in a limited context. For example, one person or organization may give the power of attorney to another for a limited purpose. All software programs are agents. They represent the user of the program and carry out the actions, which by proxy are the user's actions. Indeed, the *represent* relationship is also the basis of *agent* technology, a computing discipline emerging rapidly from the shadows of academia into the glare of business applications.

Agent technology strives to automate the process of representation. An agent adopts the goals of the object it represents and sets its own goals and strategies to achieve them. Autonomous agents also strive to make their goals and processes context sensitive. Multiple dimensions such as multiple variables, complex interactions, and properties of objects represented (like user preferences, demographics, interests, past behavior, and other factors) provide the context in question. Autonomous intelligent agents strive to represent intelligently, i.e. they try to adapt and reason. In this chapter, we describe some of the laws that weave components of knowledge into components of reason.

Agents may be autonomous in that, if they are given a goal, they will find the necessary resources and forge ahead with minimal intervention from the objects they represent, especially when those objects are people or organizations. Agents obviously communicate with the objects they represent, but it must be borne in mind that agents may also communicate and collaborate with other agents, some of which might represent other objects. Agents could also be gatekeepers for the objects they represent, and their objectives might clash with those of other agents. Agents may negotiate on behalf of the objects they represent, matching intentions, objectives, and resources with other objects and agents. Agents might even create new generations of agents, spawning variations or "cloning" themselves in a massively parallel or distributed effort to meet their goals. Agents may not only forge blindly ahead on set goals following the dictum of process maps cast in stone, but may also adapt their goals and processes to their environment and its constraints by encapsulating and incorporating governance processes within themselves. (Process maps and governance are described in Module V, section 3.) Lastly, an agent is only a representative because the object it represents has delegated a part or all of its behavior to the agent that is now a mediator for it. An agent has variables, procedures, a *state machine* (see the endnote on "state machines" on our website) and may contain other agents. [1], [2], and [3].

Formats are symbols we must perceive with one of our five senses. Therefore, our fivc senses naturally divide formats into the following broad classes:[7]

[7] The partitioning of formats based on our five senses is natural for humans. Entities with other kinds of senses would find other kinds of partitions more "natural." Since exobiology and science fiction are not in the scope for this book, we will stick to the basic 5 to normalize knowledge about *perception* and *expression* of meanings (distinct from the *meanings* themselves).

- visible (visual) formats
  - script
  - graphics
- audible (audio) formats
- tactile formats
- olfactory formats
- taste formats

The domain of visible formats naturally normalizes behavior common to visual perception, such as three-dimensional, movement, and rotation in space, viewpoints from different locations, color, size, contrast, and brightness. Written symbols such as alphabets, numerals, and words belong to the class of visual formats, as do "graphics," such as diagrams and pictures. Each is a subclass of the class of visual formats.

Just as visible formats normalize rules of visual perception, the domain of audible formats normalizes behavior common to audible perception, such as loudness (volume) and pitch. In the same way, tactile formats will normalize behavior about touch, such as the feeling of pressure, roughness, or smoothness, heat or cold, hardness and softness, sharpness or bluntness, and friction.[8] The other formatting domains similarly normalize behaviors natural to senses of smell and taste.

Of the five basic formatting domains, current technology is most adept at managing visible and audible formats. Expressing meanings in tactile and olfactory formats is an area of current research. Expression in the taste domain is still far in the future.[9]

Individually, or in combination, the above set of formatting domains can support powerful and sophisticated multimedia, biometric, and virtual reality capabilities of today, as well as the more sophisticated business systems of the future.

Patterns of symbols – the architecture of pattern

> A star at dawn,
> A bubble in a stream, a flash of lightning . . .
> A summer cloud, a flickering lamp,
> A phantom, a dream
> (Diamond Sutra 32 of Buddhism)

The fundamental metaobject is a pattern of information. Objects may be abstract concepts or tangible things we can see, touch, smell, hear, and taste. Meanings are concepts, which are abstract patterns of information, whereas symbols used to format information are patterns we can see, feel, hear, smell, and touch. Both meanings and symbols are polymorphisms of the fundamental metaobject. In this section, we will understand this metaobject. It is an abstraction and an inchoate pattern of information. The architecture of *pattern* captures the common meanings from which both abstract concepts and tangible symbols flow. *Pattern*

[8] The science of *haptics* addresses tactile sensations. The application of haptics to automation is still in its infancy, but is of growing importance in robotics.

[9] The "display" object in figure 33 (metamodel of object view) is not necessarily a visual display. It could be in any of the five types of displays (formats) corresponding to our five senses.

captures the common essence of all meaning, and the metamodel of *pattern* is the common root of every concept in this book, including that of *knowledge* itself. The metamodel of *pattern* is also the metamodel of *object*.

Symbols are easier to visualize than abstract information, and the physical space we live in is easier to understand than the abstract information space in which meanings are manifested. For this reason, we will start the discussion with patterns of symbols in the physical world.

Information is formatted with patterns of symbols. Symbols are physical objects, which are also patterns, perceived in space and time. The five fundamental formatting domains each have characteristics that normalize behavior of perception, symbols, and patterns. These characteristics are attributes and effects that describe the behavior of patterns and symbols in that domain. As for object, each fundamental domain may also be subtyped on the basis of special characteristics. For example, visual domains may be partitioned into one-, two- and three-dimensional spaces.

Symbols (and patterns) in visual formatting domains will have attributes such as color; brightness; relative location in one, two, and three dimensional space; mutual distance (that maps to the length domain); shape; orientation in space; length; area (in two or more dimensional spaces); volume (in three-dimensional space), and relative angular separation (in two or higher dimensional spaces). These domains also normalize effects such as movement in space (changing locations), rotation (for spaces of at least two dimensions or more), and changing intensity (blinking patterns are a special case of this kind of effect). When we add the time dimension to space, we can add characteristics such as movement, speed, and acceleration. From these examples, it is obvious that some attributes of formatting domains will be nominally scaled, such as shape, while others may be ordinally or difference scaled (like location), or even ratio scaled, like area, volume, brightness, or speed.[10]

Similarly, the other formatting domains will each normalize different kinds of perceptual information – both attributes and effects – and these attributes could be nominal, ordinal, difference, or ratio scaled. For example, the audio domain will have attributes such as tone, cadence, and loudness (volume).

In order to normalize information about patterns, it is important to distinguish the *meaning* of these attributes of formatting domains from their *expression*. For example, a musical note is a *concept*. It is also a tone. The tone may be expressed as a sound. It may also be written down as a musical symbol. It can even be expressed as a set of numbers and stored digitally. These are all expressions or *formats* of the tone. The *meaning* of the note stays the same in each format, only the symbols that express its meaning change (remember the monster of Metanesia in Chapter 1!).

What is a pattern?

Symbols may consist of arrangements of other symbols. For example, each alphabet is a symbol. So is a string of alphabets. The string is a symbol that is also a sequence of symbols

[10] Formatting domains have at least one ratio scaled attribute, *intensity*, in common. Intensity is a polymorphic attribute. In the visual domain, it will be brightness. In the audio domain, it will be loudness. We will discuss this in Chapter 4, section 3.

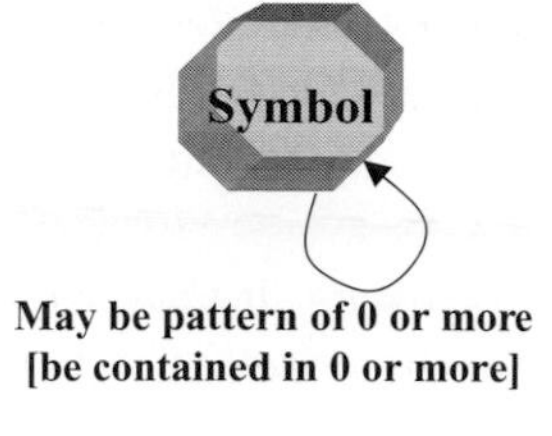

Figure 50 Patterns are symbols that consist of symbols

arranged in one dimension. Similarly, written sentences are sequences of written words, which in turn are sequences of alphabets. Symbols may be aggregates and conjunctions of other symbols, which may consist of yet other symbols that are themselves aggregates and conjunctions of symbols, and so on. These symbols within symbols may be patterns that can be reused across more than one set of symbols. The recursive relationship on symbol in figure 50 represents this fact.

Symbols that consist of other symbols may or may not be patterns. Collections of symbols are patterns only if symbols in the pattern conform to a law (the law must not increase the total information conveyed by the collection of symbols). The law determines the identity and meaning of the pattern. For example, the spelling of a word determines its identity. The spelling is the law. Patterns need not always be one-dimensional strings of characters like words and sentences. They could be multidimensional visual, auditory, and other patterns in any of the five fundamental formatting domains, *or their combinations*. Symbols within a pattern need not all belong to the same kind of formatting domain. Patterns may consist of symbols and symbol sequences (which too are symbols) synchronized across one or *more* of the five kinds of formatting domains. Audio-visual presentations are examples of symbols that consist of mixed patterns of visual and auditory symbols.

Patterns in physical space and time

Meanings are manifested in state space. Formatting symbols project abstract meanings into the physical world framed by space, time, and perception. Therefore, all formatting symbols and patterns must be located in space and time. These symbols may be located in a three-dimensional space, at points in time (or a span of time), in a one-, two-, or three-dimensional document, in cyberspace – or in a place[11] made of combinations of these locations. For example, a musical note is an audible symbol that must sound at a particular point or a volume in space, at a given moment (or span) of time.

Because formatting symbols (including patterns) are located in, and occupy regions of space and time, they may have boundaries. These boundaries *delimit* symbols and are symbols themselves. Indeed, a delimiter is only another role for a formatting symbol.

Circles (circumferences) delimit disks (the shape, not the storage device!) and blank spaces delimit words. Thus, circles and blank spaces are symbols that also play the role of being delimiters in some contexts.

[11] *Place* and *document* have been described in the universal perspective.

Delimiters may be varied, and rules of use complex – governed by convention, context, and the patterns of symbols. For example, sentences are delimited by periods, as are words at the end, but not middle, of a sentence. This is an example of a complex pattern of delimiters rooted in convention.

The *pattern delimiter* object in figure 53 represents this feature. It shows whether the symbol is a delimiter or not, and, if it is, what kind of delimiter – a delimiter that indicates the beginning, end, or mere existence of the boundary of a pattern.

In some patterns, the order of their constituent symbols may count, whereas, in others, only their existence might suffice. It all depends on how we define the pattern. Similarly, there may be patterns in which distances between symbols count,[12] not their relative order in space or time. There are several fundamentally different ways of partitioning the rule for locating symbols in a pattern.[13] The rule for locating symbols is called the *law of location* and its partitions are all indicators of state. The fundamental states of the law of location are described in the following paragraphs.

(A) Patterns of sequenced versus unsequenced association

A fundamental aspect of *pattern* is the pattern of association between its constituents – which symbols (or objects in general) are associated with which. Even an abstraction like space is subject to this kind of law. Adjacent points in space (points that are infinitesimally separated) are mutually connected, whereas others are not – we can only get from one point in space to another by traversing points in between. The metamodel we are developing in this book is another – it is a pattern of associations between objects of various kinds. There are several other patterns that are patterns of association. In this subsection, we will understand association to be a key feature and state of *pattern*.

Consider a simple pattern that consists of a particular tone that always sounds with a given graphic icon, or a more complex pattern that consists of a set of tones, in which each tone is sounded with a corresponding graphic icon. It does not matter what comes first, the tone or the icon. They might even occur together. All that matters is that, if the tone sounds, the graphic will appear and vice versa. The tone is an audio symbol. The graphic is a visual symbol. The tone and graphic together are also a symbol. This audio-visual symbol is a pattern, in which an audio icon (tone) and a visual icon (shape) go hand in hand. The pattern has no information on how its constituents must be sequenced. It is a pattern of existence, not order. It is also a pattern in which gaps between symbols do *not* count. Only association counts. It is a pattern of association – unsequenced association.

On the other hand, if the law for locating symbols decreed that the tone and the visual icon had to appear within 30 seconds of each other, but did not decree which must occur

[12] The law of location must locate symbols in a pattern so that the distance between symbols is a *metric*, governed by the rules in the endnote on generalized distances.

[13] To locate symbols in physical space and time, we must consider two items of information – sequence in space or time, and the magnitude of separation between symbols. By recognizing "negative" separation, we *could* combine both items into one mathematical concept but we would still be implicitly recognizing two items of information – magnitude (of separation) and sign (direction), instead of sequence and magnitude. The content, not expression, of information is the key to normalizing meanings and concepts. We use order (sequence) and magnitude (separation) to describe rules of location.

first, the pattern would remain an unsequenced pattern, but one in which separation between symbols mattered. We will discuss these kinds of patterns of cohesion in the next subsection.

Unlike the unsequenced patterns above, written words and sentences are sequenced patterns. The sequence of letters in a word, and the sequence of words in a sentence are integral to their identity. Words (and sentences) are patterns of order, not mere existence of symbols. Words are delimited by blank spaces. Therefore words are sequenced patterns where distances (gaps) between symbols *do* count.

Consider another example of sequenced pattern – a different kind of sequenced pattern – a pattern in which gaps between symbols do *not* count. Imagine a set of tones that always sounds in the same sequence. The tones are located in time by a law that tells us which tone will follow which, but not how long after. In this pattern, the *sequence* of symbols (tone), not their *degree of separation* in time, is the key. It, too, is a pattern of sequence, but not one of separation.

Indeed, the relationships between objects we discussed in this book were also patterns of sequence. The direction of the relationships mattered. However, the aggregation of attributes in figure B of box 27 (where the object was merely the label of the "bag" of unordered attributes) was an example of an unsequenced pattern. There was no "direction" of association involved. Module V on our website revisits these issues.

This aspect of pattern is the container of information of "connectedness" between its components – which components are associated with which, and whether direction of association matters to the pattern.[14] It is also an indicator of the state of a pattern – partition A in figure 53.

Incomplete order: Since multiple dimensions frame the physical world (three spatial dimensions that map to the length domain, and one time dimension), the law for locating symbols in a pattern might force order in only some, but not all dimensions.[15] Consider a pattern that does not distinguish between mirror images. In such patterns, the relative distances and angular separation of points count, and even their order of placement counts in all dimensions but one – the spatial dimension that distinguishes left from right. In this dimension, separation still counts, but not the left to right (or right to left) order of symbols. It is an example of a law of location that demands only incomplete order in the pattern; it is also a law of location in which relative separation between symbols matters.

(B) Patterns of separation: distance and distinction

Take the example of audio patterns above a step further. An audio pattern might consist of a set of tones that follow each other at fixed intervals. In this pattern, not only does the order of succession of tones matter, but also their separation in time. It is a pattern of sequence and *separation*.

[14] Two or more shapes are said to be topologically equivalent when we care only about connections between points (not angles), separation or dimensionality. Topologically equivalent shapes can be deformed into each other without "tearing" or "breaking" them. This property of connectedness in the metamodel of knowledge is its connection to topology, a major branch of mathematics. ([262], [264] and chapter 11 of [314] introduce topology).

[15] [211], [212], [213], [214], [215], [216], and [217] describe the mathematics of order.

Contrast this pattern with the pattern in the example where the visual symbol and tone had to occur within 30 seconds of each other (in subsection A above). That too was a pattern in which relative separation of symbols counted, but the *sequence* of symbols did not matter.

Relative separation, or cohesion, of components in a pattern is therefore another indicator of the state of a pattern. It is independent of the sequences in which these components may be arranged.

So far, we have discussed three kinds of patterns of separation in physical space and time. We will expand our repertoire to cover state space later in this section. The three kinds of patterns are as follows:

1 Patterns of quantified separation

These are patterns in which magnitudes of separation count. The audio patterns we just discussed were examples of patterns like this. They were patterns of quantified separation in time. The distance between two points is an example of quantified separation in space.

2 Patterns of ordinal separation

The *magnitude* of separation between symbols is irrelevant to patterns of ordinal separation – the separation of components is merely ranked.

Consider an audio pattern in which the interval between notes with a higher pitch is always less than notes with a lower pitch. We do not know (or care, in the context of this pattern) how much time elapses between the end of one note and the start of another. We only know that the higher the pitch of the note, the more time will elapse before the next note sounds. Only the ranking of time intervals between notes matters to this pattern. It is a pattern of separation that merely ranks the separation between its constituent symbols (notes), and does not care about the actual quantum of separation. It is a pattern of ordinal separation.

3 Patterns of distinction

These are patterns that only consider the collocation (or not) of symbols.

Imagine an eternal train of waves in an infinite ocean. We know that every crest must be adjacent to a trough. It is a pattern of troughs and crests. We do not care how far the crest of the wave is from troughs on either side, nor do we care about which distances between troughs and crests are larger or smaller. It is a pattern that only cares that the trough of the wave is different from its crest. It distinguishes between the two. It is a pattern of distinction.

Interaction between partitions: Consider the audio pattern in which the visual symbol and tone had to occur within 30 seconds of each other again. Had the law of location in this example decreed that the visual icon and tone must be simultaneous, it would still be a pattern in which the extent of separation mattered (it *must* be zero), but is it a sequenced or unsequenced pattern? The answer is ambiguous because the sequence of symbols becomes meaningless when their separation is zero (asserted by a subtype in partition B). Therefore, sequenced patterns exist only when symbols are not collocated.[16] Figure 53 illustrates this.

[16] Situation 2 of figure 27 shows how the existence of a subtype in one partition might bar the existence of a subtype in another partition. That sequenced patterns can only exist when locations are distinct and separate.

When shape, not size, matters: In the same way as there are patterns of incomplete order, there are patterns of incomplete size. Consider a visual pattern in which the shape, but not the size of the icon matters. In such patterns, the law of location decrees that angular separation between symbols matters (angular separation is illustrated in figure (c) of box 37), not their mutual distances in the pattern – i.e. angles must be preserved, regardless of distances between symbols in the pattern.[17] This is a special case of laws of location that preserves some, but not all information about size. In this case, it considers separation (size) in terms of angles between constituents (see figure (c) in box 37), but not separation in terms of linear distances between constituents of the pattern.

(C) Patterns of inclusion versus patterns of exclusion

Many patterns are patterns of co-existence. They are rules about what objects or symbols must go together, or *coexist* in a pattern. The examples above were all patterns such as this. Such patterns are patterns of *inclusion* because they mandate items that must be *included* with other items to create a pattern.

In contrast, consider a rule that asserts that Roman and Arabic numerals *cannot* be mixed in a written number. It is an example of a pattern of existence that says what symbols *cannot* co-exist in a formatting pattern. It is a pattern of exclusion.

These examples demonstrate how the property of inclusion and exclusion is a fundamental feature, as well as a basic partition, of *pattern*.

(D) Patterns of shapes: dimensions of freedom

Shapes occupy, enclose, and delimit regions of space. Dimensions are integral to space,[18] and hence to shapes in space. The number of dimensions of space determines which shapes can exist in that space. To understand the metamodel of patterns of shapes, we must consider the meaning of dimension – its abstract information content. Every point in physical three-dimensional space may be located by three independent values (see box 37) called its coordinates. Similarly points in two-dimensional spaces need only two coordinates, whereas one-dimensional space needs only one coordinate. In one-dimensional space, the only coordinate of a point in space will be its distance from a point of reference (called the origin – see box 37). In general, the number of dimensions needed to describe a shape is an attribute of pattern.

Consider written English words. They are symbols. They consist of strings of alphabets located only by the position of the alphabet in the word. This position is a single number, a sequence number relative to the first alphabet of the word. Therefore, a word is a pattern in one dimension. On the other hand, geographical maps and engineering blue prints are patterns in *two dimensions*. They too are symbols. They are symbols that represent meanings in a format that we can perceive with our sense of sight. Therefore they are also visual

[17] Angles will apply only to two- or higher-dimensional patterns. Only distance, not angles, count in a single dimension. In two dimensions, symbols may be located by their mutual distance *and* angular separation; in three dimensions, symbols must be located by distance and *two* angles (see box 37). In higher-dimensional spaces, the number of angles required to locate an object will increase commensurately.

[18] [273] and [275] define the dimensionality of space.

formats. Similarly three-dimensional geographical models and mathematical graphs are *three-dimensional* visual formats.

We could just as easily represent these maps, models, graphs, and shapes as strings of numbers or as patterns of coded colors. These patterns would merely be alternative (harder to visualize) formats for the *same information* about the same objects. The same abstract meanings can be represented with different patterns of symbols located in space and time. These patterns need not all have the same dimensionality, but can convey the same information. The dimensions of patterns are attributes of *formats* – they are not intrinsic to the *meanings* conveyed by those formats.

Symbols and patterns of symbols by themselves *mean* nothing. They may be only shapes, sounds, and odors. However, symbols (and patterns of symbols) can *connect* the world of meaning to the physical world of perception. Imagine that they are pipes that convey meanings from the abstract meta-universe to the concrete physical world around us. Like all pipes, their capacity to convey information is finite. A symbol cannot express a meaning that contains more information than it can pipe, as we will understand from the following discussion.

Information carrying capacity of formats: In past systems, customers seldom used or interfaced directly with a corporation's information systems. Standard operating environments could be mandated or controlled. The information carrying capacity of a format was not as important, because formats were relatively simple. They were merely printed or displayed numbers, characters, or simple graphics. Neither was normalizing and assembling meanings into behavior important. Adequate systems could be built, maintained, and debugged without worrying about normalizing meanings. The integration of business process and information systems was less important. Business was smaller and simpler, competition less intense, the leeway for error larger, and the cost of error much less.

Future systems will not have these luxuries. The customer will interface directly with the corporation's information systems and will be king or queen. Business processes and systems will have to follow. Few businesses will have the power to dictate customers' standard operating environments. Customers will be diverse, and users varied. Operating environments will often be as diverse and uncontrolled as the customer base is large. Complex systems will be the very fabric of business process, and will need to flex at the speed of thought to support innovation within corporations and across large, complex and globally integrated supply chains. Information will be expressed in a complex mix of multimedia formats in multilingual, multidimensional, and technologically diverse environments. These future systems will need to explicitly deal with the information carrying capacity of formats to support the complex formats of tomorrow that will flow from integrated industrial strength global systems. For this reason, we must understand the information carrying capacity of formats. To understand the capacity of a format to convey information, we must understand the concept of variability of a pattern – how much leeway does a pattern have to change its state and still retain its identity. The law that defines the pattern determines this leeway. To understand this aspect of play within a pattern, we must understand how the definition and the dimensionality of a pattern are interrelated.

Consider the curved *surface* of a three-dimensional graph. The surface is a pattern. It is two dimensional, but may curve and twist in three dimensions. Is it a pattern of two or three dimensions? For example, the *surface* of the sphere in figure (c) of box 37 is a symbol, a delimiter for the sphere, and a pattern in its own right – a pattern that is different from the *solid* three-dimensional sphere it encloses. The solid sphere and its surface are different patterns because the pattern of points that make the curved surface of the sphere is not the same pattern as the pattern of points that make the three-dimensional volume of the sphere.

Similarly a twisted line, such as that in figure 25 (b), looping through three-dimensional space is a pattern, but is it a one-dimensional pattern or a three-dimensional pattern? To answer this question, we must consider how many irreducible facts constrain space to create a pattern. Each constraint reduces the freedom of location of symbols in the pattern. We must therefore consider not only dimensions of space, but also degrees of freedom when we consider the formatting of patterns (see box 37). Meanings hold information. Each variation of a pattern can represent a different meaning. Degrees of freedom measure the variability of patterns. Therefore the degrees of freedom of a symbol (pattern) represent its capacity to convey meanings to the physical world.

A line in three-dimensional space space, curved or straight, will have a different degree of freedom from a similar line in two-dimensional space space. For instance, a circle is a one-dimensional pattern. A circle in three-dimensional space space will have more degrees of freedom than a circle in two-dimensional space because it may be tilted and moved. Each orientation and position of the circle may represent a different meaning. Therefore, a circle in three-dimensional space space has a greater capacity for conveying meaning than one in two-dimensional space. In the same way, any shape, a line, a surface, or volume in three-dimensional space will have more degrees of freedom than a similarly shaped surface in two-dimensional space and will therefore have more potential (capacity) for conveying meaning.

Only flat surfaces may exist in two-dimensional space, but both flat and twisting surfaces can exist in three-dimensional space. Surfaces in three-dimensional space have more degrees of freedom and, in the role of symbols, can convey more information than surfaces in two-dimensional space space. Similarly, there can be patterns and shapes in higher-dimensional spaces, such as state spaces, that cannot exist in three dimensions. In general, the greater the dimensionality of the pattern, and the higher the dimensionality of the space that holds it, the greater its degrees of freedom will be, and the greater potential it will possess for conveying information hidden in meanings.

A three-dimensional surface could be used to convey the full shape of a mountain. A two-dimensional shape could convey only the profile of a cross-section of the mountain. If we wanted to convey more in two dimensions (like a photograph does), we would need additional parameters, such as color or shading. Thus, even if a photograph retains the two-dimensional character of the surface in physical space, it must add the dimensions to the pattern in state space to communicate the shape of the three-dimensional mountain, and commensurately increase the degrees of freedom of the pattern. (Even then this pattern would have fewer degrees of freedom and less information carrying capacity than the three-dimensional model did. The three-dimensional model can communicate the shape of the surface all around the mountain. The photograph merely conveys the shape of the mountain from a single, fixed viewpoint.)

Overall, shapes of patterns in space and time are described by at least three attributes:

- dimensionality of the shape
- dimensionality of the space that holds the shape
- degrees of freedom

The number of degrees of freedom can be computed for different shapes, and the capacity of a pattern to convey information will follow from its degrees of freedom (see the endnote on the "Measure of information" and box 37). The diverse patterns and symbols can pipe meanings from the metaworld of concept into the physical world of business, and each symbol can have a different information carrying capacity. The dimensionality of the space that contains the symbol and dimensionality of the symbol itself determine this capacity. Degrees of freedom are woven into the shape, its dimensionality, and that of the space that holds it. The dimensionality of a shape, the dimensionality of the space that holds the shape, and its degrees of freedom, are fundamental features of patterns and shapes that determine the state of a symbol (or pattern of symbols). These features determine the pattern's capacity to convey meaning. Dimensionality and degrees of freedom are distinct and different from features like association, delimitation, and separation (e.g. angular separation, linear separation, or other less common measures of separation[19] – see box 37).

(E) Boundaries of patterns

Consider the pattern in figure 51(a). It is a two-dimensional pattern in two-dimensional space (the plane of the paper). It has a boundary – the frame of figure 51(a) delimits the pattern. However, we can imagine that it goes on forever, spreading across the plane of this page in this book, with no beginning and no end in any direction. Had it gone on forever, it would have been an infinite pattern without any boundaries – it would have been an undelimited pattern. Therefore, patterns may be:

1 delimited (subject to a boundary – also termed a bounded pattern), or
2 unbounded (without boundaries)

Further, patterns may be:

1 finite or
2 infinite (extend infinitely through space or time).

We have discussed several examples of delimiters for finite patterns. It might appear that all finite patterns must be delimited, but that is a misconception – a finite pattern may or may not have boundaries. Consider the "hub and spoke" pattern in figure 51(a). Instead of being laid out on a flat sheet of paper, it might have been wrapped around the surface of a sphere as in figure 51(b). The extent of the pattern would still be finite – it is confined to the finite surface of a finite sphere – however, you could move in any direction around the curved two-dimensional surface of the sphere and not find a delimiter. The pattern in figure 51(b) has no boundary. This pattern is an example of a finite but undelimited pattern, and it demonstrates that finite patterns may or may not have boundaries.

A pattern is undelimited if you can traverse the pattern along connected points, in one direction, without ever finding a delimiter – even if you arrive back where you started.

[19] Measures of separation are metrics. The endnote on generalizing distance describes the basic rules that govern metrics.

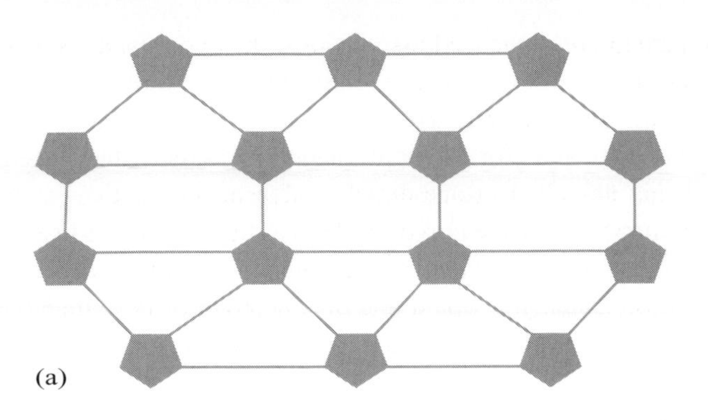

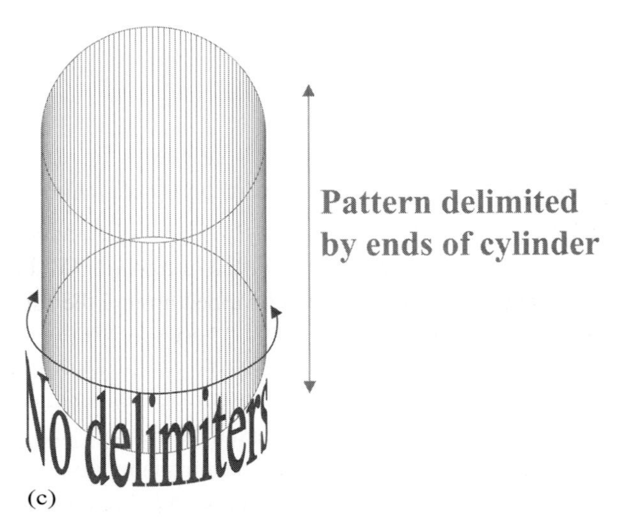

Figure 51 Examples of finite delimited and undelimited patterns
(a) Finite, delimited two-dimensional pattern
(b) Finite but undelimited two-dimensional pattern wrapped around the surface of a sphere
(c) Pattern of lines is delimited in one direction but not the other

If all associations in the pattern are sequenced associations, then you must always either traverse the pattern in the direction of the sequence, or always in the direction opposed to the sequence.

Like the other properties of shapes (for example, incomplete order), the pattern may possess the property of being delimited in one direction, but not in another: consider a pattern of vertical lines wrapped around the *curved* surface of the cylinder in figure 51(c). There are no delimiters if you traversed the pattern around the curved surface; but, if you move along the height of the cylinder, you would find that the ends of the cylinder delimit the pattern. Thus, this pattern is unbounded around the curved surface of the cylinder, but bounded along its height.

The notion of bounded versus unbounded patterns is a fundamental partitioning scheme for patterns that manifest the feature of "boundedness," or *delimitation* of a pattern. This partition tells us whether the pattern is delimited or not, and in what directions. The delimitation partition is different and distinct from the "finite versus infinite" partitioning scheme for patterns. Finiteness or infiniteness represents the *extent* of the pattern in space or time, whereas bounds represent the *existence of delimiters* in the pattern.

Of course, infinite patterns have no boundaries; but even infinite patterns may extend endlessly in one direction, but be finite in another. For example, imagine that the pattern in figure 51(a) extends endlessly along its length (to your right and left), but remains bounded and finite along its height (the top and bottom of figure 51(a)). The pattern would form an endless ribbon stretching away forever to your left and right, but the width of the ribbon would be the same as that of figure 51(a). This is an example of a pattern that is infinite along its length, but finite and delimited along its breadth.

Consider also a pattern that is infinite in one direction, bounded in another, and finite but unbounded in a third direction: had the cylinder in figure 51(c) stretched to infinity upwards, the pattern of lines would become infinite and undelimited upward, but would stay finite and delimited downward. The pattern would also stay finite and unbounded around the curved surface of the cylinder.

Like these variations of the pattern in figure 51(c), there are other patterns too that may be infinite in one (or more) directions, but finite, with or without bounds, in others. These examples serve to demonstrate that *the two key properties of patterns, delimitation and extent, can be different in different directions.*

It is also clear that *only sequenced patterns may have* ***start*** *and* ***end*** *delimiters*. Unsequenced patterns recognize no sequence; hence they do not distinguish a beginning from an end. Consider a disk delimited by a circle. We know that the circle is a boundary of the disk, but it is meaningless to say that one point in the circle (or one side of the circle) is the beginning and another is at the end of the disk. They could just as well be interchanged, if we did not care about the order of points in a path that traverses the disk.

Like the ranges that could be open or closed (see "Constraints on quantitative attributes", Chapter 3, section 2), delimited patterns in space may be open bounded or close bounded. Consider a pattern of points that coalesce into a two-dimensional disk. The law of location might permit the disk to touch the circumference, or it might allow the disk to approach infinitesimally close to the circle that delimits it, but never to actually touch it. If the disk touches its circumference, then the circumference is a closed bound for the disk; whereas, if it can only approach the circumference, the circumference is an open bound for the disk. A delimiter that describes a closed bound will be called a *close bounded delimiter*, or *closed delimiter*, whereas delimiters that define open bounds will be called an *open bounded delimiter*, or *open delimiter*.

The open-ended property of a pattern is actually a conjunction of two normalized meanings – a pattern of exclusion combined with a pattern of delimitation. An open-ended pattern is a delimited pattern (delimitation is a state of *pattern* that counts as one normalized irreducible fact) that excludes collocation of one or more points with its delimiter (a pattern of exclusion that counts as another normalized irreducible fact). This is why, in figure 53, *open bounded pattern* is a subtype with two parents – *pattern of exclusion* and *delimited pattern*. This subtype in the metamodel of knowledge is a bucket for normalizing facts common to open bounded patterns.

Two-dimensional patterns could be more complex than the open and closed ranges of Chapter 3, section 2. In the disk we just discussed, the law of location might bar the pattern from touching the circle only at specific points (which might also be a pattern). When we consider three dimensions, the pattern may not only be open at specific points on its delimiter(s), but also along specific lines and regions of delimiting surfaces. The complexity of delimiters and possible variants escalate very rapidly as the dimensionality of a pattern increases. We will review this in more detail when we extend the concept of *pattern*, from patterns in physical space to patterns in state space.

Box 37 Location in space

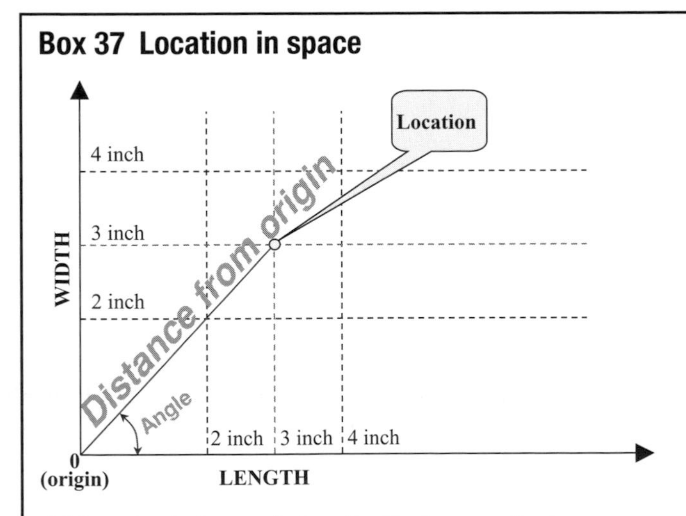

(a) Location in 2-space is described by two coordinates

Coordinates of locations and dimensionality of space: The figures above illustrate how a point (or symbol) in space can be located by its distance from a point of reference called the *origin*, and its angular separation from reference line(s) (axes) that pass through the origin. This method of locating a point is called the polar coordinate system.

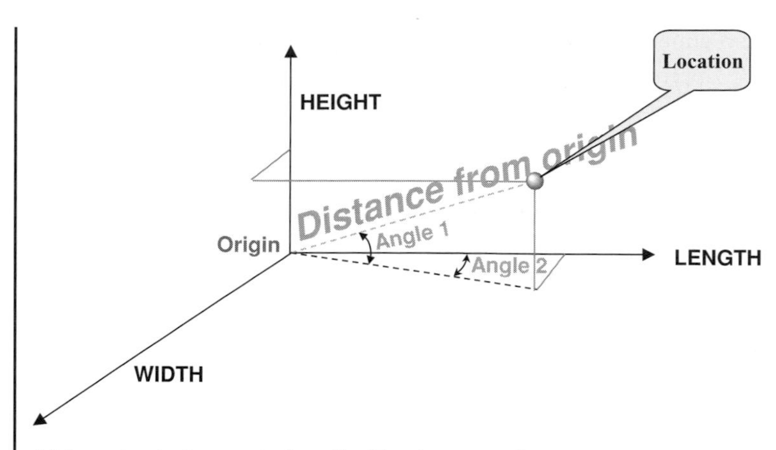

(b) Location in 3-space is described by three coordinates

The point could also be located as it was in figures 23 and 24. That method of locating a point is called the Cartesian coordinate system. Figures (a) and (b) illustrate both polar and Cartesian coordinate systems in two- and three-dimensional spaces. They also illustrate how only two coordinates suffice to locate a point in 2-space, whereas three coordinates are required to locate points in 3-space. It does not matter whether they are polar, Cartesian, or some other system of coordinates. The number of coordinates needed to locate a point in space depends only on its dimensionality. The meaning of location does not change, only the rule expression for describing it (the coordinate system) may vary. Location is the meaning, and the coordinate system its expression.[20]

Orientation and shape: When orientation does not matter to the identity of a pattern, it does not matter if the angles or distances from the origin in figures (a) and (b) change, as long as the *angles between constituent symbols* do not change (if these angles change, the pattern will be distorted). For example, the pattern may be rotated or moved without losing its identity. When orientation and position matter, either together or in combination, we will need to constrain parameters in Cartesian or polar coordinates appropriately.

Degrees of freedom: When three coordinates locate a point in 3-space, the point has three degrees of freedom. Consider a pair of points in 3-space. For each coordinate of one member of the pair, the other has three degrees of freedom. For each degree of freedom of one member, the other member has three degrees of freedom. As such, the pair has $3 \times 3 = 9$ degrees of freedom. The separation between points boils down to measuring differences between pairs of coordinates. Three coordinates locate each point. Therefore three differences are involved. Each difference is an irreducible fact. The difference between two locations in 3-space has $9 - 3 = 6$ degrees of freedom.[21] A line that connects two points in three-dimensional space involves a law that relates the three differences between coordinates. The law will determine the shape of the line –

[20] The polar and Cartesian coordinate systems are yet another example of two different rule expressions that have the same meaning – in this case, the location of a point in space.

whether it is straight, looping, or curved, and indeed how and where it bends if it does. This law is an irreducible fact. That takes away another degree of freedom. A line in 3-space has at least $6 - 1 = 5$ degrees of freedom – "at least," because the law could be qualified. The law might hold in a range, and then switch to another shape at another point. Each qualification will (a) increase degrees of freedom since it involves an additional point, which could move (where the new law kicks in), and (b) reduce a degree of freedom because the new law is an atomic rule.

Similar principles would apply to surfaces in three space, lines in two space or, in general, to shapes in spaces of any dimensionality. A shape is a pattern that has at least two attributes – dimensionality and degrees of freedom.[22] A pattern's degree of freedom is a measure of how much it can vary before it loses its identity. Each variation may express a different meaning. A pattern's degree of freedom is also a measure of its capacity to convey information.

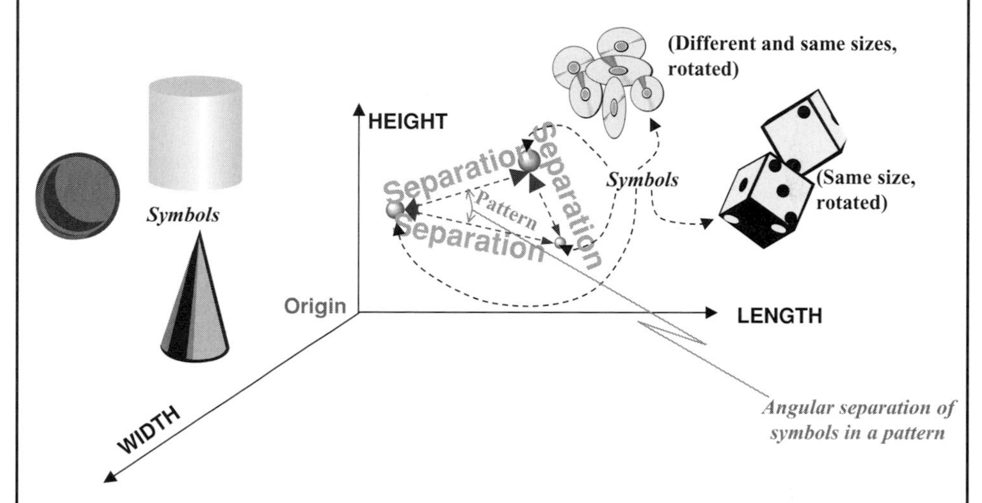

(c) Location in space: polar versus Cartesian coordinate systems

Which constraints on coordinates will preserve the orientation of a pattern of symbols, and which will affect a pattern's position? Would the shape of a pattern be preserved if we preserved the distance from the origin for each and *differences* in polar coordinate angles for each symbol? What would happen to the coordinates of each symbol if we rotated it in space while preserving its shape? Hint, use figure (a) of box 49.

[21] Any unspecified line, straight or curved, between two points in three-dimensional space, has 6 degrees of freedom. Three degrees of freedom are contributed by its ability to move along each of the three axes of space and three more degrees of freedom may be attributed to its ability to spin around the three axes without losing its identity as a one-dimensional shape that connects two points in 3-space.

[22] The information content of a message (or meaning) is a measure of the element of surprise in it (see the endnote on the measure of information). Each degree of freedom adds an element of surprise to a pattern. Thus the degrees of freedom in a pattern are a measure of its capacity to convey information. It is also a measure of its ability to represent meanings of different kinds.

Patterns in the examples above were simple, and the rules, obvious. This is not always so. Rules can be obtuse and patterns obscure. It might need considerable reflection before obtuse rules are understood and obscure patterns coalesce from the haze of possibilities. Children who have played with jigsaw puzzles and adults who have attempted to solve puzzles with missing numbers, words, and shapes in a series can understand how difficult and arcane these rules can be. However difficult these rules are, they must always align with the partitions and attributes we have discussed. They are:

- *Dimensionality*. Patterns in physical space can have up to three dimensions, and when we consider time, they can have at most four dimensions. (However, when we consider state space, there is no limit to the dimensionality of patterns.)
- *Dimensionality* of the space that holds the pattern.
- *Degrees of freedom* which determines its capacity to convey information. Degrees of freedom are related to dimensionality of the pattern, the space that holds it, delimitation, shape, and other constraints.

Partitions and subtypes

(A) ***Association partition.*** *Unsequenced versus sequenced patterns of association subtypes:*

(A1) rules that contain information on mere association, not sequences, of symbols in a pattern; versus

(A2) rules that contain information on sequences of symbols that make a pattern.

(B) ***Separation partition.*** *Distinction versus distance (separation) in space and time subtypes:*

(B1) rules that only distinguish one location in a pattern from another (in space or time or space and time);[23] versus

(B2) rules that contain information on ranking of physical separation between symbols (in space, time, or space and time) that make a pattern; versus

(B4) rules that contain information on physical distances between symbols (in space, time, or space and time) that make a pattern. (Subtype (B3) will be discussed further on in this section when we discuss patterns in state space.)

(C) ***Inclusivity partition.*** *Inclusion versus exclusion partition subtypes*:

(C1) rules that include items in a pattern; versus

(C2) rules that exclude items from a pattern.

(D) ***Extent partition.*** *Partition based on extent of a pattern subtypes*:

(D1) the pattern is infinite; versus

(D2) the pattern is finite.

(E) ***Delimitation partition.*** *Partition based on boundary of a pattern subtypes:*

(E1) undelimited pattern (that is, the pattern has no boundary); versus

(E2) delimited pattern (that is, the pattern has a boundary).

[23] The discrete metric in the endnote on generalized distances distinguishes between locations, but says nothing about separation.

Sometimes a single law of location may be subtyped differently when considering different dimensions of the physical world.[24] For instance, the law that did not distinguish between mirror images was one such example. It was an unsequenced pattern in one spatial dimension, but sequenced in others. The law that preserved shape, not size was another. It was a separation subtype when considering angular separation, but only a distinction subtype when considering linear distances between symbols.[25] The property of extent was yet another property of the same type. A pattern could be finite in one direction, but infinite in another. The property of delimitation was also similar. A pattern could be delimited in one direction, but not in another. In physical three-dimensional space, these directional variations are relatively easy to visualize. When we extend the concept into patterns in state spaces of higher dimensions, it becomes much harder. This is one of the principal challenges of pattern recognition, a topic of intense interest in expert systems and artificial intelligence. However, before we can understand pattern recognition, we will have to extend the concept of *pattern* from physical space and time into state space. To do so, we must understand what makes a pattern a pattern – the concept at the heart of all patterns.

Measures of similarity

The concept at the heart of a pattern is the resemblance between its constituents. Patterns are formed by including or excluding items based on some measure of similarity between them. This measure of similarity is the defining identity at the heart of the pattern.

Collections of symbols are patterns only if symbols in the pattern conform to a law. However, this definition is not adequate. The law, or rule, must be based on some criterion of similarity or dissimilarity between symbols before we can say that it defines a pattern. This criterion is the measure of similarity (or dissimilarity) of objects in the pattern. These objects are symbols when patterns are formats.

Distance in physical space and elapsed time are examples of measures of similarity of location in space or time. However, there may be other measures of similarity that map closeness in space and time, and *yet other* measures may map the proximity of states of objects and symbols in state space. A measure of similarity is a rule meaning.[26] However, it is a rule meaning subject to a constraint. Like distances in physical space, all measures of similarity are governed by the following rule, and must conform to it. We will call it the *golden rule of proximity metrics*:

- The measure of similarity between any pair of points in state space cannot exceed the sum of measures of similarity via intermediate pairs of points between them.[27]

  (When distance is the preferred measure of similarity, this rule asserts that the direct distance between a pair of points in state space must be less than, or equal to, the distance via intermediate points on an arbitrary path between the pair.)

[24] Topology deals with location and shape. See the introduction of [262], [264], and Chapter 11 of [314].

[25] Conformal mapping in mathematics preserves angles, but not separation between symbols. This is called *conformality*. When the pattern does not distinguish between orientations in space or size, but only cares about shape, the shapes are termed conformally equivalent.

[26] Like other rule meanings, measures of similarity may have multiple expressions (see box 33).

[27] Proximity between states is a metric. Measures of proximity must conform to the rules in the endnote on generalized distances.

So far we have discussed measuring similarities in position and shape in physical space or time. Position and shape in physical space and time represent only one aspect of the state of a symbol. Other parameters of state may have to be considered in determining the resemblance between two or more symbols. For example, musical notes widely dispersed in space and time may be considered similar when they have the same pitch or cadence. Similarly, in figure (c) of box 37, the dice and the cylinder may be considered close because their colors are close. In general, measures of resemblance will involve similarity (or dissimilarity) between states of an object (symbols too are objects). It is these measures that will determine inclusion, exclusion, and location of symbols (or objects) in patterns. Each pattern will have a measure of similarity. This measure is the identity of the (class of) pattern.

(An example of a pattern based on dissimilarity, or *exclusion*, is a collection of objects in which every object must colored differently. The measure of dissimilarity in this case is the identity of color. The pattern would not change if we asserted that the measure of similarity is the identity of color, and the rule that defines the pattern is a rule of exclusion, as opposed to a rule of inclusion. In both cases, the *meaning* of the rule is identical; only its expression is different – see box 33.)

Proximity metric

Measures of similarity are *rule meanings*[28] that determine closeness or resemblance of two or more symbols in state space.[29] The closer two or more points in state space are, the more the corresponding object instances will resemble each other. Therefore, a measure of similarity in state space can be called a *proximity metric*.

Like other measures, proximity metrics may be:

- Ratio scaled (the of ratios of differences between states are considered in creating a pattern): A cluster of points in space is a pattern. It is a pattern determined by mutual distances of points in it. Distance is a ratio scaled measure (ratios between distances are meaningful – for example, you can tell that one distance is half of another). This pattern has a ratio scaled measure of similarity.
- Difference scaled (the magnitudes of differences between states – not ratios of these magnitudes – are considered in describing the pattern). Consider an organizational hierarchy. It is a pattern of *differences* between levels in an organization. We can meaningfully tell how many levels one individual is *removed* from another in this hierarchy, but not what the *ratio* of this difference means. It demonstrates that the measure of similarity for this pattern is difference scaled in state space.
- Ordinally scaled (similarities between symbols are ranked when forming patterns).

  Consider the division of individuals into males and females. Individuals fall into a pattern of being male or female. We know that the gender difference between individuals of the same sex is less than that between individuals of different sexes, but not by how much (there is a difference, or there is not, and no difference is less than some difference). The

[28] Rule meanings may be expressed with one or *more* rule expressions (see box 33).

[29] Topology, in mathematics, studies rules that map one shape to another. It captures the notions of continuity, connectedness, and convergence. Chapter 11 of [314] introduces topology. Also see [262], [264], [255], [266], [267], and [278].

measure of similarity for this pattern is ordinally scaled. This is an example of a pattern with an ordinally scaled measure of similarity.

- Nominally scaled (when we know items exist in a pattern, but have no information on how similar they are – not even if they are collocated in state space).

  This boils down to knowing that a pattern exists, but not its basis. Measures of similarity of "patterns by decree" fall into this class. For example, in figure (c) of box 37, three points were decreed to be a pattern by an arbitrary rule. The rule merely told us that a pattern existed. If we did not care about the separation or location of those points, or whether we considered them distinct or the same symbol in the context of the pattern, this pattern would have had a nominal measure of similarity.

A nominal measure of similarity may also become a facility for modeling the uncertainty and incompleteness inherent in real-world information – a mechanism that lets us assert that a pattern certainly *exists*, only we do not know its basis – its parameters, meaning, or definition. It is an unknown pattern.

Note on terminology: A measure of similarity is different from the kind of measures discussed in section 2 of this chapter. A measure of similarity measures resemblance of patterns, whereas the measures in section 2 map meanings of magnitudes to numbers. *When the term "measure" is not qualified in this book, it will always mean the measure that maps a magnitude to a number*. Measures of similarity will always be qualified as such. They will also be called *similarity metrics* or *proximity metrics*. (See the endnote on metrics, metric spaces, and generalizing the concept of distance.)

Location in state space versus physical location

Position in physical space describes only the physical location of a symbol. Physical proximity is only one kind of measure, and physical location is only one aspect of state. The position in state space describes a symbol completely – its shape, intensity, color and all other properties relevant to its context and formatting domain. The closer two or more symbols are in state space, the more they resemble each other. If they are collocated, they are identical. Proximity in state space subsumes physical proximity, and extends to other measures of resemblance as well. For example, all sounds with a certain pitch could be considered close to each other, even if they are widely dispersed in space and time. Measures of similarity in state space are distances between points in state space. The measure of similarity is the supertype, and distance (separation) in physical space is only one of its subtypes – one that applies only to physical proximity of symbols.

Often, useful measures of similarity in state space will be intuitive, and will conform to our perception of reality via our five senses. However, it is not mandatory that every measure does so. Sometimes, measures of similarity may be obtuse, complex, and difficult to define.

Consider a pattern that consists of a set of three repeating notes that always have fixed *differences* in pitch between them. The note is an audible symbol. Its pitch determines the state of the note. Notes may or may not sound at fixed intervals, and may or may not always sound the same tone. Only the *differences* in pitch between notes will always be the same. Although they are different notes that sound at different intervals, the three still exhibit a pattern of behavior. This measure of similarity for the pattern may not be immediately evident to an observer listening to the tones.

The example above demonstrates that the proximity metric is the core of the *pattern*. It also demonstrates that the proximity metric is an attribute of *pattern*. The proximity metric is the attribute that measures the resemblance of symbols in a pattern with respect to the law that makes the pattern a pattern. The more a pattern's constituent symbols resemble each other, the closer they will be in state space. Therefore patterns are clusters in state space, in which mutual separation is determined by the pattern's *proximity metric*.

The location in physical space is merely a subtype of location in state space; clusters in physical space are only subtypes of clusters in state space. State space is richer than physical space. Therefore, to complete our understanding of pattern, we must understand clusters in state space.

Patterns in state space

State space can be much richer and more varied than physical space. Physical space is merely difference scaled,[30] whereas state space may be (box 17 in Chapter 2, section 2):

- ratio scaled (all axes are ratio scaled)
- difference scaled (all axes are difference scaled)
- ordinally scaled (all axes are ordinally scaled)
- nominally scaled (all axes are nominally scaled)
- mixed (different axes are scaled differently)

Furthermore, ordinally scaled state spaces may have:

- no natural origin[31] (as is the case with difference scaled space – see box 37), or
- a natural origin[32] (as is the case with ratio scaled space – see *Mr. Domain's secret* in Chapter 1, section 3 or the discussion of figure 67 in Chapter 4).

  For example, the natural origin of a serial number is the first item in a list. No item can have a serial number less than this. (The serial number usually starts with 1, but that is not mandated by its meaning. It could be any symbol from a set of ordered symbols – say the letter "A," or even a number 2. The only constraint on serial numbers in a list is that no serial number may rank below this minimum.)

The quantum of information intrinsic to the state of an object depends on the nature of its state space. The state of a pattern is rooted in the states of symbols that make the pattern. Each kind of proximity metric conveys a different quantum of information about what makes the pattern a pattern. The proximity metric is the heart of the pattern. It follows that all proximity metrics will not be meaningful in all kinds of spaces and the nature of state space will constrain the kinds of proximity metrics that can exist in it.

[30] Physical space is difference scaled because the origin of the frame of reference, from which coordinates of locations in space are measured, is chosen arbitrarily (see box 37). Ratio scaled measures require a natural zero – see the sixth golden rule of measurement in section 2. Like location in physical space, dates and moments in space–time are also difference scaled.

[31] A ranking scheme with no natural origin is called a *totally ordered* set in mathematics. Totally ordered sets have no lower bound. See the endnote on ordered sets and sequences.

[32] A ranking scheme with a natural origin is called a *well-ordered* set in mathematics. Well-ordered sets have a lower bound – the origin. See the endnote on ordered sets and sequences.

When the location of a symbol in state space is difference scaled, patterns based on proximity or distance in state space will follow the same rules and may be partitioned and subtyped in the same way as those in physical space. However, when location in state space is not difference scaled, rules must be amended as follows:

Nominally scaled state space:

- Patterns of separation that involve difference or ratio scaled measures cannot exist
- Open delimited patterns cannot occur in nominally scaled state space
- A new subtype, F1, based on *values* of coordinates and not merely on their mutual separation, may exist.

Ordinally scaled state space:

- Closed delimiters can always replace open delimiters without changing the pattern in any way. Therefore, open delimiters are redundant in ordinally scaled state space.
- Ordinally scaled state spaces may or may not have a natural origin (Chapter 4, section 3). If it does, a new subtype F1 will normalize rules about absolute ranks, not mere differences in rank.

Ratio scaled state space: A new subtype, F1, for patterns based on *magnitudes* of coordinates and not merely on their mutual separation, may exist.

The reasons for these differences flow from the information content of coordinates and measures of proximity that infuse meaning into each space as explained in the following paragraphs.

Nominally scaled state space

- *Meaningful measures of proximity between two or more points may be ordinally scaled, but coordinates of locations are nominally scaled.*[33]

Coordinates in nominally scaled spaces have just enough information to distinguish one location (state) from another. They convey no information on rank or magnitude – only that two different locations identify two different states.

Therefore *differences* between two or more points either exist, or they do not. That is all the information this kind of space conveys. Based on this information, we cannot tell the magnitude of differences between states (points), but we *can* tell which differences are *more* or *less* than other differences. This makes *differences* between states in nominally scaled state space an ordinally scaled measure.

In the parable of Jim and Jane in Chapter 1, section 3, Jim and Robert were male, whereas Jane was female. We could tell that the *difference* in gender between Jim and Jane was *greater* than the *difference* in gender between Jim and Robert, but not by how much. Thus we can tell which *distances* are greater than which in gender space. It demonstrates that *distances* between locations in nominally scaled spaces are ordinally scaled.

Since *differences* in nominally scaled state space have only two states, the state space for these *differences* may be ordinally scaled, but it is a special kind of ordinally scaled space: the state space of these *differences* (for instance, the state space of differences in

[33] Nominally scaled measures must be a discrete metric. See the endnote on generalized distances.

gender, not the gender itself) is limited to only two points, one of which is "nil," a value that indicates the absence of magnitude. Therefore we *can* say one *difference* is greater than another, although we cannot say the same about the individual *points* such as gender in a nominally scaled state space. We can extend this idea to cover the equivalence of two or more states, wherein we say that two or more states are distinct, but equivalent, in the sense that their distance (as measured by some proximity metric) in state space is nil. (See the discussion of pseudometric spaces in the endnote on generalized distances.)

- ***Partition B:*** *Fundamental subtypes B3 and B4 of figure 53 will be meaningless and cannot exist in nominally scaled space* because points in nominally scaled state space can be distinct, but have no magnitudes in relation to each other, and distances are either nil or not. *Only subtype B1 and B2 will exist.*
- ***Open delimited patterns:*** *Open delimited patterns cannot exist in nominally scaled space. The subtype is meaningless* because distances in nominally scaled state space have no information on sequence or magnitude, and hence ranges of values are meaningless.
- ***Partition F:*** *Subtype F1 of figure 53 may occur in nominally scaled spaces*. Subtype F1 normalizes laws of absolute location, not merely differences or distinctions of location in state space.

The values of nominally scaled attributes of objects determine their location in nominally scaled state space. Joint constraints, like those in figure 43, may be patterns that normalize rules about relationships between values of nominal states, not merely the proximity, of these states.

In the joint constraint illustrated by figure 43, the check was payable (a nominally scaled attribute of the check) if the CEO had signed it (another nominally scaled attribute of the check) and the CFO had signed it (a third nominally scaled attribute of the check). This is an example of a pattern, a joint constraint, of subtype F1 in the location partition of figure 53.

Ordinally scaled state space

- *Meaningful measures of proximity may be nominally, ordinally, or difference scaled in ordinally scaled space, but coordinates of locations are ordinally scaled.*

Measures of proximity may be nominally or ordinally scaled because the state of the object conveys information on classification and order. It may also be difference scaled because differences, not mere distinctions, between locations can be measured in terms of the number of ranks that separate one position from another.[34]

Consider ranks in the military. Military ranks are ordered values; we can tell which ranks are greater, lesser, or equivalent to which.[35] We can even tell how *many* ranks one rank is removed from another. *Differences* between ranks (not the ranks themselves) can be quantified and these are at *least* difference scaled.[36]

[34] See the theory of value difference functions in [211].

[35] Military ranks in a given branch (military service) are strictly hierarchical, but ranks across services like the army, the air force, and the navy may be different. However, each rank in a branch of the military has its equivalent in the other military services. Thus, military ranks convey information on which are greater, lesser or equal even across branches of military service. (Military ranks also serve as an example of points in a pseudometric state space – see the endnote on generalized distances.)

[36] See ordinal value theory in [211].

However, these *differences* between ranks are not ratio scaled because the *ratio of differences* between ranks is a meaningless quantity. This makes these *differences* in positions in an ordinally scaled space difference scaled, but not ratio scaled. (Contrast these differences with physical distances between points in physical space, where ratios of distances between pairs of points are also meaningful.)

Confused? Consider the following example: a private, a sergeant, and a major are three soldiers in the army. Assume that the sergeant is two ranks above the private, and the major is five ranks above the private. We can therefore infer that the *difference* between the sergeant and the major is $5 - 2 = 3$ ranks. However, it would be meaningless to say that a private who is promoted to sergeant has traveled 2/5ths of the way to becoming a major because we have no information on the *magnitude of gaps between ranks*. We can tell that the private has three ranks to go before (s)he becomes a major, whereas the sergeant has to move only two ranks up to become a major. The *difference* in military rank has meaning, but ratios of differences do not. It demonstrates that differences in rank are difference scaled, but not ratio scaled.

Patterns in ordinally scaled state space may flow from ordered hierarchies, or from magnitudes of *differences* between states. They can also flow from ordinally scaled proximity metrics.

The following example of an ordinal proximity metric demonstrates how patterns in ordinally scaled state space can flow from difference scaled proximity metrics, as well as from ordinally scaled proximity metrics. Consider the arrangement of an arbitrary set of military ranks in a hierarchy. It is an example of a pattern based on sequence in ordinally scaled state space. Let us assume that all ranks may not be represented in the repertoire of ranks thus arranged. We have no information on which ranks, or even how many ranks have been omitted from the pattern. There could be gaps between ranks and we have no way of telling where the gaps are, or how many levels (of military hierarchy) such gaps may span. It is a pattern based on an ordinally scaled proximity metric – we know which ranks are greater and lesser than which rank, but not by how many levels.

Now consider how a difference scaled proximity metric can give rise to a *difference scaled pattern in ordinal state space*. Consider an organizational structure that has a slot for every rank. The organizational *structure* that shows the hierarchy of military ranks is an example of one such pattern. It is a pattern based on differences in ordinally scaled state space: in this structure we not only know which ranks are adjacent to which, and also which are greater and lesser than which adjacent rank, but also by *how much* (in terms of numbers of ranks), because each level in the structure is separated from the level above or below it by one rank. Thus, the structure is a *sequenced* pattern based on a *difference scaled proximity metric*.

Similarly a more complicated rule that asserts that given a starting rank, every third rank from the starting rank must be a member of a group is a pattern based on *magnitudes of differences* between ranks. The groups fall into a pattern because they share a law. The law is based on *magnitudes of difference* because *magnitudes of separation* between ranks are a consideration. The pattern may or may not care about the hierarchy of ranks in the context of the *pattern* (i.e., the group may not be arranged in any sequence or hierarchy). If sequences or hierarchies are irrelevant in the context of the group, it will be a pattern of *unsequenced* association based on a *difference scaled proximity metric*. If hierarchies *are* relevant to arranging ranks *inside* the group, it will be a pattern of *sequenced* association based on a *difference scaled proximity metric*.

- ***Partition B:*** *Subtype B4 cannot occur in ordinal state space. Fundamental subtypes B1, B2, and B3 in figure 53 will exist.*

Since differences between locations can be measured quantitatively even in ordinal state space, subtypes B1, B2, and B3 convey meaningful information. On the other hand, subtype B4 is meaningless because coordinates in ordinal state spaces convey no information on ratios of differences between states.

- ***Patterns with open bounds:*** *Delimited patterns that are open at the delimiter can always be reduced to a pattern with closed bounds.*

The reasons are similar to those that described why open bounds on ranges of ordinal values can always be replaced by a closed bound without changing its meaning (see Chapter 3 under "Constraints on quantitative attributes").

- ***Partition F:*** *Subtype F1 of figure 53 may occur in ordinally scaled spaces with a natural origin* because the subtype involves absolute ranks (a kind of magnitude) of coordinates, not merely their distinction or difference. (This subtype is described in more detail under ratio scaled state space.)

A rule that states that the person who finishes a race first will be awarded a prize is a law based on absolute ranks. It is not concerned with differences between ranks, but only the absolute value of a rank in terms of its relationship with a natural origin. Similarly, a rule that asserts that the person who comes second will be awarded a consolation prize is another example of an atomic rule that is a law based on absolute ranks. These rules are normalized by patterns of subtype F1 in figure 53.

Difference scaled state space

- *Meaningful measures of proximity may be nominally, **ordinally**, difference, or ratio scaled.*

Consider physical space. It is difference scaled. The coordinates of a point in space (see figures in box 37) convey no information on any magnitude of the location by itself. They only convey information on magnitudes of *distances* between locations.

We can tell whether two or more *distances* are the same or different. We can also tell which distances are more or less than others, and by how much. We can even tell what the ratio of one distance, between one pair of points, is compared to another distance, between another pair of points. Therefore, nominal, ordinal, difference, and ratio scaled measures of proximity will all be valid in difference scaled space.

- ***Partition F:*** *Subtype F1 of figure 53 cannot occur in difference scaled space.*

Subtype F1 cannot occur because the subtype involves absolute magnitudes of coordinates, not merely their differences.

Both the polar and Cartesian coordinate systems in box 37 were examples of difference scaled coordinates. Fixing an arbitrary frame of reference, with an arbitrary origin, determined the coordinates. Naturally, these coordinates were also arbitrary. Their absolute magnitudes conveyed little by themselves. However, *differences* between coordinates were meaningful, and were ratio scaled.

- ***Partition B:*** Patterns of separation permitted in difference and ratio scaled state spaces are identical. All subtypes in partition B of figure 53 may occur because distances between points may be ratio, difference, ordinally or nominally scaled in both kinds of space.

(Remember that physical space is also difference scaled. If we use physical space as an analog of state space, it makes it much easier to visualize differences in difference and ratio scaled state spaces.) ***Patterns with open bounds****: Delimited patterns that are open at the delimiter can occur.*

The sole difference between physical space and difference scaled state space is that the state space may involve more than three spatial dimensions. Like open-ended delimited patterns in physical space, open-ended delimited patterns may occur in state space. The reasons have been discussed on page 184.

However, open delimiters in state space can be more complex than those in physical space because state space can have more dimensions than physical space. Consider a range. Only two points, its upper and lower bounds may be open. (See "Constraints on quantitative attributes" in Chapter 3, section 2.) The range was a pattern in one dimension. Consider a disk in two-dimensional space. A circle at its circumference delimits the disk. The disk may be allowed to get infinitesimally close to the circle, but might not be allowed to touch it. Thus the entire circle can be an open-ended delimiter for the disk.

On the other hand, the disk might be allowed to touch the circle at select points, or, conversely, might be allowed to touch the circle everywhere *except* at select points. These are two additional variants of open-ended delimiters.

In three dimensions, a surface could delimit a volume. There will be even more varieties of open-ended delimiters in greater in 3-space. Select points, lines, line segments, surfaces, and regions on delimited surfaces may be open ended.

In higher dimensions, patterns of openness can become even more complex. In general, any subspace or region of the delimiter may be open or closed.

Ratio scaled state space

- *Meaningful measures of proximity must be nominally, ordinally, difference or ratio scaled* because the state of the object conveys information on classification, order, magnitudes of separation, and absolute magnitudes of locations.
- We will need to add a new partition, Partition F, in this space to account for the fact that magnitudes of not just distances, but also locations (coordinates) are meaningful in ratio scaled space. Partition F distinguishes:

  Subtype F1: Patterns of absolute of location

  from

  Subtype F2: Patterns of separation

Subtype F2 includes the different kinds of patterns of separation we have discussed in partition B. It will normalize rules common to all patterns of separation. To understand subtype F1, consider figure (a) of box 37. Assume the vertical axis represents the pitch of an audible tone and the horizontal axis represents its volume (instead of width and length). Both *volume* and *pitch* are ratio scaled attributes of *tone*. The origin would anchor the natural zeros of volume and pitch. A state space like this is ratio scaled. The magnitudes of its coordinates *and* the magnitudes of distances between coordinates are meaningful in ratio scaled space.

Table 1 *Proximity metrics in spaces of different kinds*

	MEASURE OF PROXIMITY			
KIND OF SPACE	Nominal	Ordinal	Difference scaled	Ratio scaled
Nominally scaled	✓	✓ (only two values – nil and more than nil)		
Ordinally scaled	✓	✓	✓	
Difference scaled	✓	✓	✓	✓
Ratio scaled	✓	✓	✓	✓

The preferred frame of reference for ratio scaled state spaces naturally maps *absolute magnitudes* of coordinates to distance from the origin. For example, the state space of *tone* will map magnitudes of *pitch* and *volume* to distance from the origin.

Unlike this kind of space, the physical space in figure (a) of box 37 was difference scaled and the frame of reference was arbitrary. Both the distance of a point in space from the origin, and the angle it subtended at the origin, were quite arbitrary. For this reason, it would be meaningless to consider patterns based purely on magnitudes of any coordinates in physical space, polar, Cartesian, or other. Patterns in difference scaled state spaces must be based on magnitudes of *differences* between coordinates of two or more locations.[37]

Patterns in ratio scaled state spaces could flow from laws about absolute magnitudes of coordinates or their differences. For instance, a law could weave pitch and volume of tones into a pattern: it could relate the volume of a tone inversely to its pitch. Tones based on this law will form a pattern and be audible symbols. This is an example of subtype F1. Any law that related the pitch of a tone to its volume, independent of the occurrence of any other tone(s), would be an example of subtype F1 – a pattern of magnitudes of locations in state space, not merely magnitudes of differences between locations.

Laws could also relate differences between coordinates to absolute values of coordinates. Such laws will be subtypes of both F1 and F2 – terms that involve differences will inherit properties of *differences* between locations in state space, and properties that involve magnitudes of locations will inherit properties of *locations* in state space.

- *Patterns with open bounds: Open-ended delimiters can occur*. Just as open-ended delimiters could occur in difference scaled space, they can occur in ratio scaled spaces.

Table 1 summarizes the kinds of measures of proximity that may occur in each space. A check mark indicates that the measure of proximity is valid in the corresponding space.

Table 1 demonstrates that for:

Nominal measures of proximity: Ratio, difference ordinally, and nominally scaled attributes – i.e. state spaces – may all participate in nominal measures of proximity because they all convey information that distinguishes one state (location in state space) from another.

[37] Patterns in ratio scaled state spaces may be based on absolute values of coordinates or differences between coordinates.

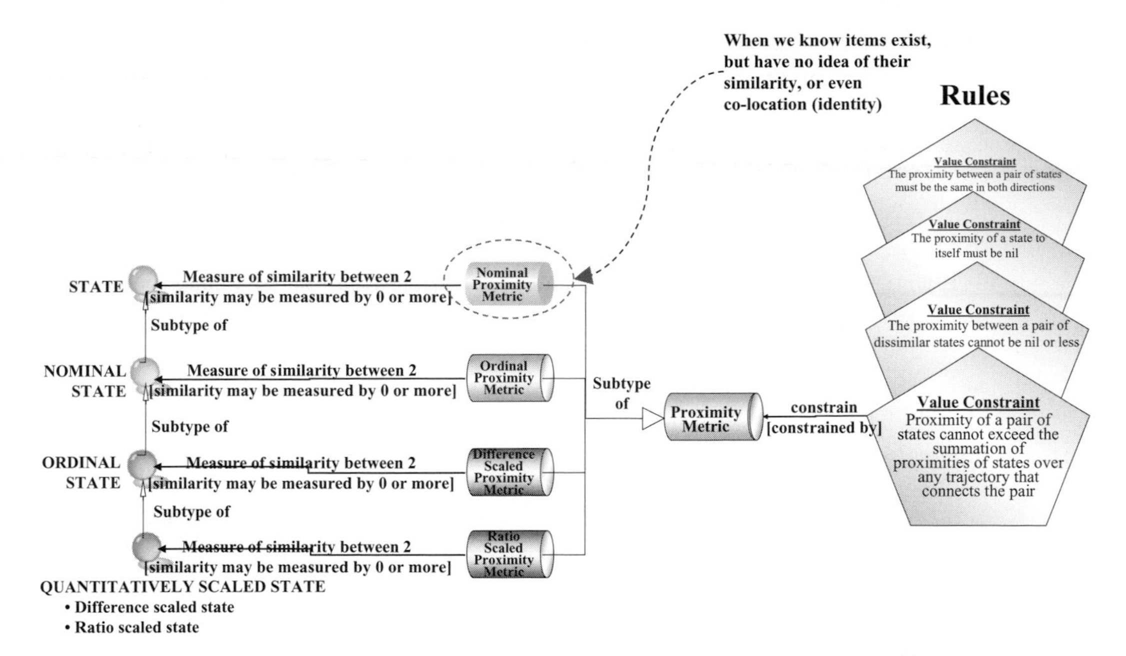

Figure 52 Metamodel of measure of similarity

Ordinal measures of proximity: Ratio, difference ordinally, and nominally scaled attributes – i.e. state spaces – may all participate in ordinal measures of proximity because they all convey information about ranking differences ("distances") between states (locations in state space).

Difference scaled measures of proximity: Ratio, difference, and ordinally scaled attributes may participate in difference scaled measures of proximity because they convey information on the magnitude of separation between pairs of states (locations in state space).

Ratio scaled measures of proximity: Only ratio and difference scaled attributes may participate in ratio scaled measures of proximity because they convey information on *ratios* between magnitudes of separation ("distance") between states.

Figure 52 represents these rules. It is the metamodel of *proximity metric.*

Figure 52 shows that the value constraint on the *proximity metric* is inherited by each kind of proximity metric in the figure (and is therefore normalized by the structure in the metamodel). However, the reason for arranging states in the hierarchy above is more elusive, but is key to the concept of similarity. It is evident that ordinal attribute values convey more information than nominal values. They *add* information on relative ranking of states to the information on *distinction* between states (conveyed by nominal values). Ordinal attribute values convey all the information nominal values do, *plus* information on ranks. Similarly, difference scaled attribute values *add* information on magnitudes of difference to information contained in ordinal values, and ratio scaled attribute values *add* information on ratios to information contained in difference scaled values.

In box 21, we understood, under *functional variation inheritance*, that it is simpler to build configurations by *adding* information. Subtypes should carry more, not less, information than their supertypes. The perspective in figure 52 satisfies this criterion. The *nominal state* in figure 52 inherits the nominal proximity metric from the set of all states, and adds the possibility of an ordinal proximity metric. Similarly, the *ordinal state* inherits the nominal proximity metric from the *nominal state*, and adds the possibility of a difference scaled proximity metric. Likewise, the *quantitatively scaled states* (difference and ratio scaled states) inherit the possibility of nominal, ordinal, and difference scaled proximity metrics from the *ordinal state*, and add the possibility of a ratio scaled proximity metric.

The subtyping hierarchy flows through to state space from the subtyping hierarchy of domains.The OAV model of box 27 shows that *state* is a collection of attribute value pairs. The discussion in this section highlights that these pairs are patterns of unsequenced association, as is the collection of these pairs that makes *state*. Figure 35 has the metamodel of *state*. This metamodel shows how attributes, and therefore states, map to domains of different kinds. It is through this metamodel, buried inside the *states* in figure 52, that the subtyping hierarchy of domains flows through to state space.

Based on the principle of defining subtypes by *adding* information, we could consider ordinal domains to be a subtype of nominal domains in which ranking information has been added, difference scaled domains to be a subtype of ordinal domains to which information on magnitudes of differences between values has been added, and ratio scaled domains to be a subtype of difference scaled domains to which information on ratios between values has been added. This hierarchy will then flow through to state spaces that map their meanings to these domains.

Mixed space

We know that state spaces often mix ratio, difference, ordinal, and nominal attributes. When this happens, what impact does it have on patterns? Partition B normalized information on separation in state space. Therefore, it is subtypes in this partition that were impacted by the nature of state space, and will be impacted when axes in state space are not all of the same kind – all nominally scaled, all ordinally scaled, all difference scaled, or all ratio scaled. In this book, spaces like these, with axes that convey different amounts of information, will be called *mixed* spaces.

Measures (metrics) of similarity in mixed space may involve combinations of ratio scaled, difference scaled, ordinally scaled, and nominally scaled values and terms. The scaling of the overall metric of similarity will be limited by the scaling of its arguments and terms. The argument or term that conveys the least information will limit the scaling of the overall metric as follows:

- If one or more nominally scaled values are involved, similarity metrics may be nominally or ordinally scaled and the rules for nominally scaled state spaces will apply.

  In terms of the metamodel in figure 52, this kind of state space will be a subtype of nominally scaled state space.

- If one or more ordinally scaled values are involved, but no nominally scaled values, similarity metrics may be nominally or ordinally scaled and the rules for ordinally scaled state spaces will apply.

  In terms of the metamodel in figure 52, this kind of state space will be a subtype of ordinally scaled state space.

Table 2 *Valid measures of proximity in mixed spaces*

KIND OF MIXED SPACE	MEASURE OF PROXIMITY			
	Nominal	Ordinal	Difference scaled	Ratio scaled
One or more nominally scaled axes	✓	✓ (only two values – nil and more than nil)		
No nominally scaled axes, one or more ordinally scaled axes	✓	✓	✓	
No nominally or ordinally scaled axes, one or more difference scaled axes	✓	✓	✓	✓
Only ratio scaled axes (not a mixed space)	✓	✓	✓	✓

- If one or more difference scaled values are involved, but no ordinal or nominal values, similarity metrics may be difference scaled, ordinally scaled, nominally scaled, or ratio scaled. The rules for difference scaled state spaces will apply.
  In terms of the metamodel in figure 52, this kind of state space will be a subtype of difference scaled state space.

Table 2 summarizes these rules.

The order of a pattern

Just as we can base patterns on closeness of positions in state space, there can be patterns of similarities between measures of similarity, which themselves show patterns of similarity and so on. We will call this property the *order of similarity*. Proximity of the first order would be the kind of proximity between components in a pattern that defines it as a pattern. Second order proximity will be the kind of proximity between measures of proximity that show that the measures of proximity themselves fall into a pattern and so on. We can have patterns of patterns of patterns and so on. This will be called the *order of a pattern*. Patterns can govern patterns, and table 1 can equally represent measures of similarity of any order: second-order measures of similarity for second-order patterns, or third-order measures of similarity for third-order patterns, and so on.[38]

[38] Table 1 also demonstrates that *differences* between states in nominally scaled state space will map to ordinal state space; *differences* between states in ordinally scaled state space will map to difference scaled state space; and *differences* between states in difference and ratio scaled state spaces will map to ratio scaled state space. We could then take differences between differences and apply the same logic and keep repeating the process. This rule is useful in pattern recognition, in the analysis of patterns of patterns, and in determining the kinds of proximity metrics that will be valid for higher-order patterns.

The metamodel of pattern

We all recognize several patterns and consider many of them to be simple. However, we find it hard to pin down the *concept* of a pattern – how would you define it? How would you normalize the rules that make a pattern a pattern? The concept is hard to pin down because it is abstract, but it is the concept at the heart of all knowledge. Every model in this book is a pattern. Every meaning is a pattern of information. Objects grow from patterns. The concepts in this chapter, the models in the universal perspective, and even the metamodel of knowledge, are all patterns. The metamodel of knowledge emerges from the metamodel of pattern and also contains it.[39]

> I am the Alpha and the Omega,
> The Beginning and the End,
> The first and the Last
> (John the Baptist in
> *Revelations, The Bible*)

The essence of a pattern

Patterns of symbols do not always express meanings, but they may. The interpretation of meaning boils down to targeting only meaningful patterns. This is the essence of a pattern.

All patterns possess only a finite capacity for conveying information. A pattern's capacity for conveying information is determined by its degrees of freedom. Patterns do not always fully utilize their capacity for representing meanings. Consider letters of the alphabet. Letters of the alphabet are shapes in two-dimensional space. They can be rotated in space and still retain their identity, but their orientation conveys no meaning. We are only interested in the one-dimensional delimited sequence of letters that spells words. These words convey meaning. Thus essential pattern is a one-dimensional sequence of letters.

Indeed each letter's orientation is usually constrained to be identical to those of its neighbors (each constraint involves the loss of degrees of freedom – one degree for each atomic rule). So how do we define the pattern that makes the word? If letters are not oriented identically, will it still be the same word? If words have different orientations, will we recognize the sentence? How much differences will we tolerate before the word or sentence loses its meaning? The answers to these questions will determine how we define words as patterns of letters and sentences as patterns of words. It all boils down to the law of location in state space, the heart of the pattern. Not all patterns convey meaning; the essence of a pattern is the law of location that involves shapes, states, and dimensions that do.

Just as letters and symbols do not utilize their full capacity to convey information, other symbols too may only utilize a few states, dimensions, and partitions to convey meanings. Consider a real life format – one made by nature, not the hand of man. The genome is the book of life. It consists of a set of coded instructions that are expressed as living organisms. A proteome is the collection of proteins, their structure and sequence, which make a complete living organism. The genome is nature's format for storing the information

[39] Patterns subsume sets much more. The *axiom of regularity* (see the endnote on the theory of categories) bars a set from containing itself, but patterns are not bound by this rule because they might not be sets. The algebra of objects subsumes and goes beyond the algebra of sets. See [171], [172], [173], [183], [184], [186], and [187].

content of a proteome. The instructions for creating a proteome are stored in a biological molecule called Deoxyribonucleic Acid, or DNA for short. The DNA is a pattern in three dimensions. It is like a twisted ladder consisting of two helixes joined by rungs; these rungs carry the information for making proteins that make a living organism.

Each rung of the twisted strand of DNA consists of a sequence of molecules called nucleotides. Each rung is made of two nucleotides. It is only this *sequence* of nucleotides and rungs, not the intricate three-dimensional shape of DNA, which carries the information that is eventually expressed as a living being.

The DNA molecule and its components form a pattern in 3-space. This pattern has information carrying capacity. However, its capacity is not fully utilized. The book of life only utilizes the sequencing aspect of this pattern to convey information. This is the essence of *that* pattern – the DNA format of life.

The essential pattern recognizes only a pattern of sequence (subtype A2 in figure 53). While interpreting and understanding the information in DNA, one need not care about the other partitions and subtypes in figure 53. One must recognize only the essential pattern subsumed in the complex three-dimensional, physical pattern of DNA.

Biochemical processes express this code as proteins. In proteins, the three-dimensional shape *and* the sequence of 20 amino acids carries significant information, whereas in a gene only the sequence, not the three-dimensional shape, of four nucleotide bases is significant. Thus, with reference to figure 53, the essential patterns of symbols in proteins involve interactions between three-dimensional patterns in physical space and subtype A2. Therefore the essence of the patterns in proteins involve both subtype A2, in figure 53, as well as the subtype (in the same figure) that represents three-dimensional patterns of locations in physical space.

The problem at the heart of pattern interpretation is to discover this essential pattern – the pattern's meaningful states and dimensions in a format. *Formats represent meaning*. They are symbols in space and time that we can perceive with our senses. *The essential pattern is the pattern within the format that actually conveys the meaning.*

A single meaning may map to many formats. Formats are not always printed numbers and letters. They could be graphs, pictures, and even patterns of molecules or patterns of magnetization on the disk of your computer. Like the essential patterns in DNA molecules, or in words and sentences, essential patterns in a format may be only a part of the physical arrangement of symbols that make the format. Usually the essential pattern is simpler because only a subset of states conveys meaning. (At most, all states of a pattern may be meaningful, but usually only some are.) This meaningful pattern is what we will call the *essential pattern* of symbols in a format.

Recognizing essential patterns is at the heart of disciplines like data mining, fingerprint recognition, genetics, bioinformatics, econometrics, and analytics. If pattern recognition is to recognize not just patterns, but meanings as well, it must recognize essential patterns – the patterns hidden within patterns of symbols.

To recognize essential patterns and normalize the rules that make them patterns, we must understand how different kinds of patterns normalize different kinds of rules. The following section is dedicated to understanding the universal atomic rules that distinguish patterns of different kinds.

Kinds of patterns

To recognize patterns, we must recognize the atomic rules every pattern normalizes. To identify these rules, we must start by recognizing the different kinds of patterns that flow from them. We have already discussed several properties of patterns in depth. They are (attributes that depend on direction in state space have been marked with a star (★) in the following list):

- ★ Association and sequence
- ★ Location
  - ★ Position
    - ★ Position relative to a natural origin (absolute position or state)
    - ★ Position relative to an arbitrary origin
  - ★ Cohesion and separation
- • Exclusion and inclusion
- • Order (order of governance of other patterns – the term has not been used as a synonym for sequence in this context)[40]
- • Dimensionality of the shape of the pattern in state space
- • Dimensionality of the pattern's state space
- • Degrees of freedom
  - ★ Extent
  - ★ Delimitation

Regardless of complexity or variety, all patterns will be governed by this list of fundamental characteristics that emerge from the metamodel of knowledge. Figure 53 illustrates how every pattern may be partitioned, regardless of its complexity or scale. In the partitions of figure 53, each subtype in partitions labeled with a star (★) will have attributes that describe the subtype's direction(s) in state space.

Partitions A, B, C, D, E, and F in figure 53 exist in the metamodel of knowledge only because each normalizes a different kind of information intrinsic to the meaning of pattern:

- Partition A normalizes information about position in a pattern – whether sequences are integral to the identity of the pattern or not.
- Partition B normalizes information about similarity of constituents in a pattern – the shapes and characteristics that form the criteria integral to the identity of the pattern.

  Patterns depend on measures of proximity like distances, angular separation, and other metrics of similarity. Proximity is calculated from the coordinates of constituent objects in state space. Subtypes in partition B normalize this aspect of similarity, a kind of clustering or *cohesion* between the pattern's constituents in state space. Partition A tells us which symbols are connected to which, and the directions of these connections (if any). It does not tell us how far these points are from each other. Partition B conveys this information. It tells us how dense or tightly clustered the pattern is.
- Partition C normalizes information about patterns of exclusion and inclusion – the (irreducible) fact that patterns may not only consist of symbols that occur together in a group, but also symbols that cannot (or do not) occur together.

[40] Patterns that govern patterns will be at the heart of intelligent agents (see box 36) that will continually adapt software to innovation in an intensely competitive, creative, and tumultuous environment driven by new ideas.

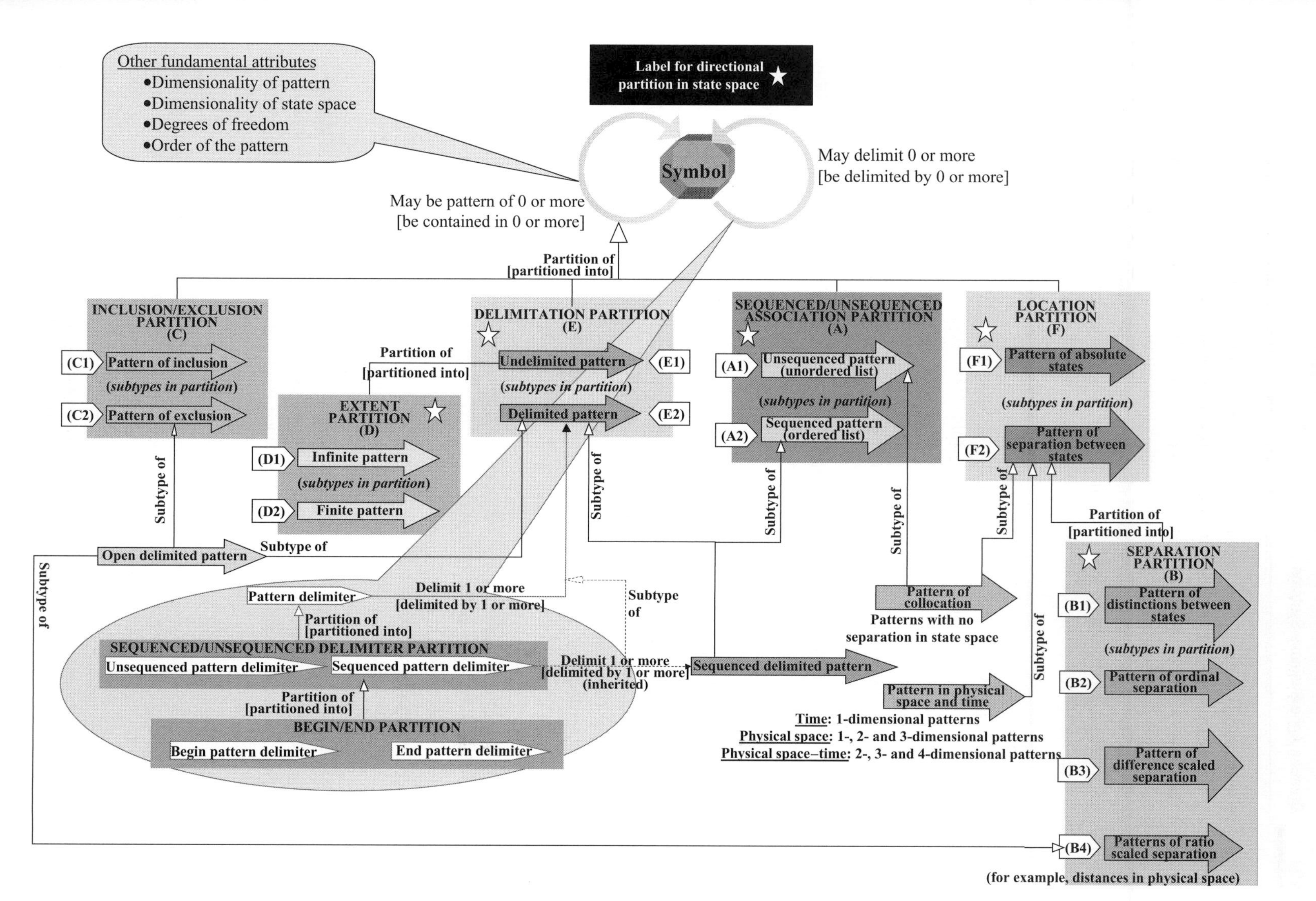

Figure 53 Kinds of patterns

- Partition D normalizes information about the extent of undelimited patterns. It discriminates between finite and infinite undelimited patterns (delimited patterns are always finite).
- Partition E normalizes information about delimitation.
- Partition F normalizes information about the location of a pattern's constituents in state space. Positions of a pattern's constituents in state space and their interrelationship are integral to the identity of the pattern. The rules that define a pattern determine how much leeway its constituents have in terms of changing their positions in state space before the pattern loses its identity and becomes something else.

  Both the coordinates of locations and separations between points are involved. Partition B, the partition for normalizing rules about separation and similarity between a pattern's constituents, is subsumed in partition F.
- Partitions of the delimiting relationship on symbols normalize information about pattern's (i.e. symbol's) role as a delimiter.

We have already discussed subtypes in each partition extensively. These characteristics can be summarized as follows for state spaces. They are similar to those we discussed for physical space, which is only one aspect of state space; the following subtypes extend the corresponding discussion of patterns in physical space to state space, which is a generalization that subsumes physical space – its meaning and information content:

(A) Association partition *(unsequenced versus sequenced patterns of association)*

Subtypes:

(A1) Rules that contain information on association, not sequences, of symbols in a pattern (mere association of a symbol with a pattern).

(A2) Rules that contain information on sequences of symbols that make a pattern.[41]

(B) Separation (proximity measurement) partition *(distinction versus separation versus magnitude of locations in state space)*

Subtypes:

(B1) Rules that only distinguish one location from another in state space.[42]

(B2) Rules that only *rank* separation in state space but not distance in terms of magnitudes of difference.

(B3) Rules about patterns of separation between symbols (in state space), in terms of their differences, not ratios of difference. (Separation may be in terms of angles and/or distances as well as other proximity metrics.)

(B4) Rules that contain information on magnitudes of separation between symbols, which makes a pattern in state space, in terms of differences and ratios of separation. (The normal concept of physical distance is an example of this kind of measure of separation.)

[41] The metamodel provides for the fact that a single meaning may be expressed very differently, and that formats may be translated to other formats subject to complex rules and constraints. These issues too are key concerns in disciplines such as bioinformatics, encryption and automated translation of natural languages. For instance, discovering and storing genetic information is an area of intense scientific and business interest. Data mining and pattern matching are others.

[42] The discrete metric described in the endnote on generalized distances applies when the law of location distinguishes between locations, but says nothing about separation.

(C) (Inclusion versus exclusion partition *(of positions in state space)*
Subtypes:
(C1) Rules that include items in a pattern.
(C2) Rules that exclude items from a pattern.
(D) Finiteness partition *(based on extent of a pattern)*
Subtypes:
(D1) Infinite pattern.
(D2) Finite pattern.
(E) Delimited versus undelimited pattern *(partition based on boundary of patterns)*
Subtypes:
(E1) Undelimited pattern – the pattern is unbounded and not delimited.
(E2) Delimited pattern – the pattern has a boundary and is delimited.
(F) Pattern of absolute versus relative location *(partition based on patterns of absolute location versus mutual separation of locations)*
Subtypes:
(F1) Rules that contain information on absolute locations (in state space) of symbols that make a pattern.
(F2) Pattern of separation between symbols in state space (i.e. pattern of similarity or dissimilarity.)

Figure 53 also shows interactions between subtypes across partitions, as follows:

- Only finite patterns may be delimited.
- Like the bounds in Chapter 3, section 2, a delimiter may define an *open limit only for patterns with ratio scaled proximity metrics.* ("Constraints on quantitative attributes" in Chapter 3, section 2 describes why unclosable open finite patterns can only occur in difference and ratio scaled spaces.)
- An open pattern is a pattern of exclusion – it excludes the entire delimiter, or a part of it (the part is region(s) or subspace(s) of the delimiter).
- The open delimited pattern in figure 53, being a subtype with three parents, represents all three interactions above. The supertypes of *open delimited pattern* are:
  Delimited pattern (naturally!)
  Pattern of exclusion
  Pattern of ratio scaled separation

Naturally, beginning and end delimiters will apply only to *sequenced patterns* (see "Constraints on ordinal attributes" in Chapter 3, section 2). Note that *beginning* and *end* are states of delimiters in sequenced patterns.

Furthermore, we understood that subtype A2 cannot exist when symbols in a pattern are collocated in state space. *Pattern of collocation* is thus a subtype of *unsequenced pattern.* Naturally, it is also a *pattern of separation* (separation *must* be zero, i.e. its constituents are must *not* be separated in state space). Therefore, *pattern of collocation* in figure 53 is a subtype with two parents. Its supertypes are *pattern of separation* and *unsequenced pattern.*

Patterns of constraint

You may have noticed the similarity between constraints and patterns. Yes, patterns are constraints. The law of location is a form of constraint; it tells us what locations are permitted in state space. Association is a constraint; it tells us what *is* associated with what, and *not* with what. The similarity metric is a constraint; it constrains the similarity of symbols in a pattern. The order of a pattern is a constraint on a constraint. Boundaries, extents, and dimensionality are all constraints on shapes, locations, states, and boundaries of symbols. Therefore, it is hardly surprising that the metamodel of pattern closely resembles the metamodel of constraint. Indeed, the law that defines a pattern is an aggregation of several kinds of constraints – one for each partition. Figure 54 demonstrates where each partition of figure 53 resides in the metamodel of *pattern*.

Figure 54 is the metamodel of pattern. Each subtype in figure 53 is marked in figure 54 by an arrow.

To understand figure 54, we must start by interpreting the different universal partitions of patterns in the context of *value constraint* (in Chapter 3, Section 2). Figure 53 has six different universal partitions of *pattern* and two different partitions of *pattern delimiter*. Figure 49 has the metamodel of *value constraint*. We will interpret figure 53 in the context of figure 49 to arrive at figure 54.

States are collections of attribute values. It follows that patterns of constraints will be collections of constraints like those in figure 49. Figure 49 was the metamodel of a single constraint. Figure 53 is the metamodel of the *aggregation* of constraints. Figure 53 also has a broader scope: the focus of figure 49 was on abstract meaning, not symbols (patterns) that could physically express those meanings. Figure 53 applies both to symbols and meanings. Symbols may be partitioned in ways values cannot be. After all, patterns of symbols normalize rules that values do not, and vice versa. We will examine these differences between figure 53 and figure 49 in this section.

Let us start with partitions that are common to both figure 53 and figure 49. Partition C, the inclusion/exclusion partition in figure 53 is identical to the inclusion/exclusion partition of figure 49. The upper and lower bounds in figure 49 are delimiters of range. *Range* in figure 49 was a one-dimensional region of sequenced state space. *Delimited pattern* in figure 53 generalizes the concept of *range* in figure 49. Therefore the "begin/end" partition of *sequenced pattern delimiter* in figure 53 generalizes and subsumes the upper/lower bound partition of figure 49.

Partition A, the partition that distinguishes sequenced from unsequenced patterns of association in figure 53, is missing from figure 49. It is missing because the value sets of Chapter 3 were sets, not lists (see box 30). Patterns on the other hand may be sets or lists; symbols in a pattern may be sequenced and repeated, and each repetition may count as a separate item that constitutes the pattern. As such, *pattern* subsumes sets, sequences, and lists – ordered or not. This is the fundamental difference between the *value set* in the metamodel of *value constraint* and *pattern* in the metamodel of knowledge. It is this difference that lies at the root of the other differences between the metamodel of *pattern* and the metamodel of *value constraint*.

The delimitation partition, partition E of figure 53, is missing in figure 49. This happens because a range is an ordered set of magnitudes of values. The set is not necessarily a set of discrete values. It may also be a continuum of values. There can be no finite but unbounded patterns of one-dimensional ranges (in the *value constraint* of figure 49). All undelimited ranges are infinitely large, and hence meaningless as ranges. Therefore, there was no need to

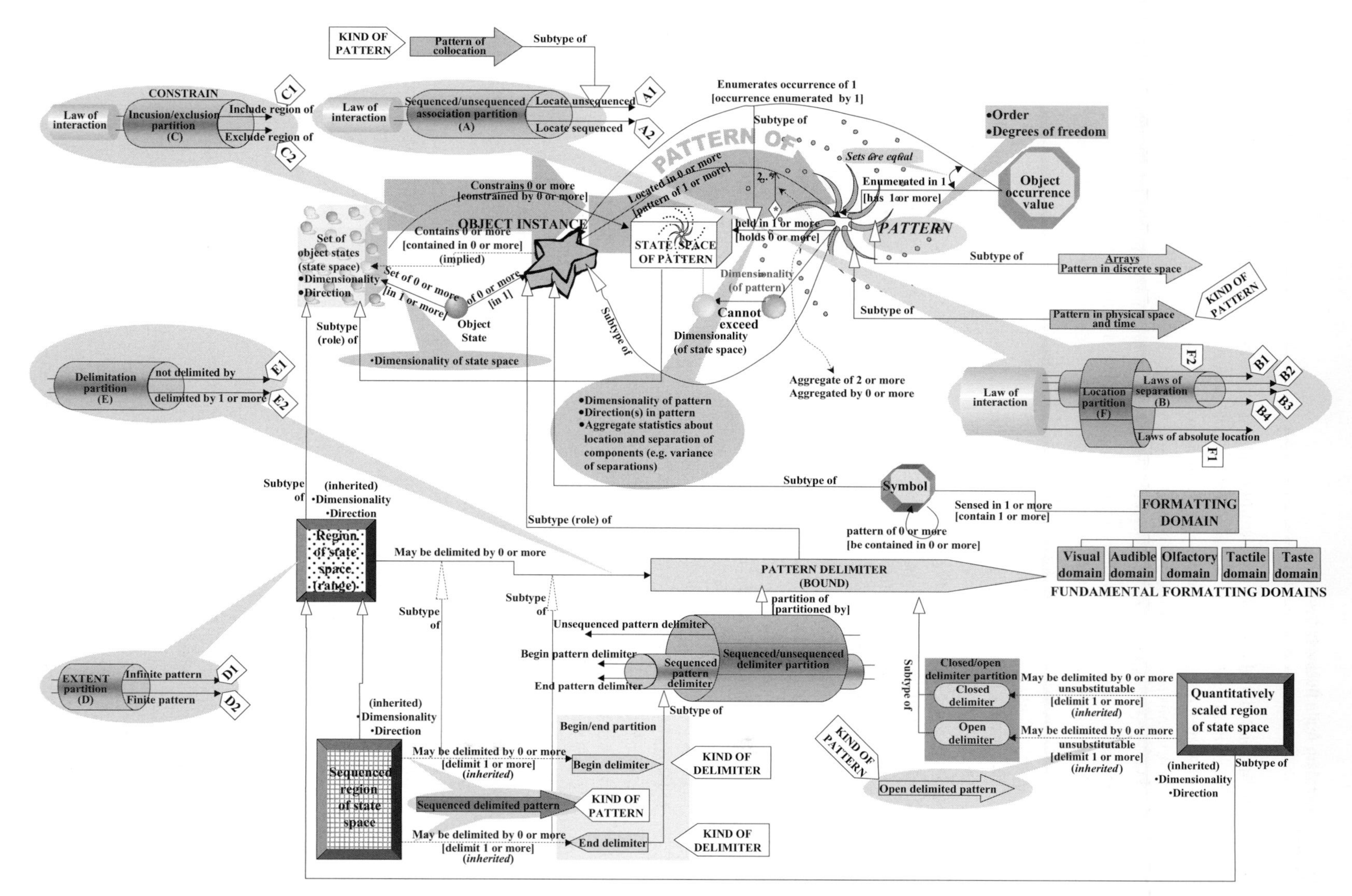

Figure 54 The metamodel of pattern is the source of universal properties, subtypes and partitions of pattern

distinguish delimited from undelimited ranges in figure 49 – it would have been meaningless and redundant. Partition E was irrelevant to the metamodel of value constraint in figure 49. In contrast, patterns of symbols may be finite and undelimited. Therefore, partition E is very relevant to the metamodel of pattern in figure 53 and figure 54.

The extent partition of figure 53 (partition D) is missing from figure 49. The extent partition of figure 53 partitions undelimited patterns based on their extent. It has been omitted from figure 49 because all undelimited ranges are infinite and meaningless in the context of *value constraint*. Therefore, like partition E, partition D would have been redundant and meaningless in figure 49.

The location partition of figure 53, partition F, is also missing from figure 49; hence its subpartition, the separation partition (partition B) is missing too. Both are subsumed in the *rule expression* of figure 49. That rule expression of figure 49 may be an expression that binds absolute locations into a law, like subtype F1 of figure 53, or be a law that merely constrains *separations* between locations, like subtype F2 of figure 53. The location and separation partitions of figure 53 (partitions F and B respectively) merely categorize and resolve rule expressions with greater granularity than figure 49 does.

In order to integrate the metamodel of *value constraint* in figure 49 with the patterns of figure 53, we must recognize that:

- *Pattern* generalizes the concept of set and list, sequenced or not. *Set* and *list* are subtypes of *pattern*.
- *Object* is a pattern described by its state. A symbol is a kind of object. The "law of location" in state space decrees how much play there can be in the state of a pattern before it is considered a different pattern. For example, a green ball may be considered the same pattern as a red ball because they are both colored spheres, or it might be considered a different pattern because the spheres have different colors. A pattern may also weave an abstract concept like the passage of time, or it could be a concrete symbol like the letter "w" with which "word" begins. The pattern is a shape, or a set of shapes in state space, and this shape (or a set of shapes) is also the very *identity* of the object or symbol that constitutes the pattern.
- *State* is a set (aggregation) of attribute values. Patterns constrain these values. Hence, pattern is an aggregation of value constraints. Aggregations have (and normalize) emergent properties that their constituents do not (see "Object class" in Chapter 2, section 1); the metamodel of *pattern* normalizes properties that the metamodel of *value constraint* does not (see Module 6 on our website).

Thus, the first step towards figure 54 is to replace the *value set* of figure 49 with *set of object states* in figure 54.

Next, we must recognize that this *set of states* may contain not only (some or all) states of a single object, but also states of several objects (which might belong to the same or different classes), all of which may influence a pattern simultaneously and jointly. The relationship between *set of object states* and *object instance* shows this. There is no injunction in figure 54 against *set of object states* containing states of different object instances in the same or different object classes, at the same or different times, or even the same object instance at different times. It is the state space of the aggregation of objects that make, constrain, or influence the pattern. State spaces of individual objects that make or influence the pattern will be subspaces of this state space in figure 54.

Region versus range

Like any state space, this space will have regions. Regions are more complex than ranges because they are multidimensional. Figure 51 demonstrates how finite patterns in space may be undelimited, whereas we know that finite ranges *must* be delimited. Also, several delimiters may delimit regions in different directions because regions may be aggregations of ranges. Like ranges, if sequences matter in some directions, regions may have *begin* and *end* delimiters (or both) in those directions. If the space is quantitatively scaled in certain directions, delimiters may be open or closed in those directions.

The delimitation relationship is a universal relationship, inherited by every kind of delimiter. This is similar to the partitioning of *bound* in figure 49. *Bound* in that figure was simpler, only because it did not have to recognize the multidimensional nature of state space, or the multiple ways constituents of patterns may be associated (or not) in sequences that make the pattern.

Location versus constraint

We only had to worry about permitted and forbidden values when we discussed *value constraint* in Chapter 3. The location of a value was automatically determined relative to other values in ordinally and quantitatively scaled space. However, the multidimensional nature of state space complicates the concept of location. We must consider inclusion and inclusion, i.e. what *must* go hand-in-hand, what *may* go hand-in-hand, and what *cannot* go together in multidimensional space, as well as *where* it must be in a pattern. We must consider forbidden, permitted, and mandatory *regions* of state space, not merely inclusion and exclusion constraints. This is the heart of the *pattern of* relationship in figure 50. Figure 54 illustrates how *pattern of* is a conjunction of *constrain* and *locate*.

Shaping and influencing patterns

Location in a pattern may be absolute or relative (discussed early in section 1). It is partitions of this *locate* relationship that normalize the meanings of different kinds of locations of a pattern's constituents. This partition distinguishes patterns of separation from patterns of absolute location and makes the pattern a pattern of normalized information.

In complex patterns, it may happen that objects that do not participate in the pattern nevertheless influence the location(s) of a pattern's constituents, or even their inclusion or exclusion from the pattern. Indeed, nothing bars a pattern's constituents from doing that too. *Thus participation in a pattern and influencing its shape are two different and independent roles of objects*. This is also the theoretical foundation of *context*, in which interpretations of the same meanings can change, depending on an external framework.

The rules that shape a pattern based on states of objects reside in the laws of interactions that are attached to the *locate* and *constrain* relationships of figure 54. They could be joint constraints of the kind we discussed in figure 48, except that they apply to *states* of objects, not single values. They could also be simple constraints about permitted and barred states. States of object instances are a *collection* of attribute values; therefore the laws of interaction in figure 54 are collections of value constraints like those in figure 48.

The state space of figure 54 may therefore be the aggregate state space of all objects that *influence* the location and existence of a pattern's constituents in the pattern. It may also be the aggregate state space of a pattern's *constituent* objects, or even the state space of the pattern itself. All these state spaces are subsumed in the concept of *state space*. The relationship between *state space* and *state* in figure 54 merely asserts that state spaces are sets of states (obviously!), and because states are states of object instances, it implies that it is states of these object instances that reside in state space (also obviously!), and so do these object instances (not so obvious – but remember it is the value, i.e., state, of the instance identifier that identifies an object instance). The instance identifier is the attribute that gives the object its identity. State spaces represent collections of attribute values. Hence the object instance may also be considered resident in state space.[43] This allows us to be as specific as we need to be to describe complex patterns. We can specify inclusion, exclusion, and location based on states of individual object instances as well as object classes[44] and relationships. Because patterns are objects, patterns have states, reside in state space, and can influence other patterns – even themselves.[45]

For example, it is common sense that isolating all tall men in one pattern excludes tall men from being present in other patterns. Thus it is also common sense that one pattern can influence another. To understand how objects outside a pattern may affect the state of a pattern, consider the requirement that all magnitudes larger than one million be expressed in exponential format, and all magnitudes less than zero be shown in red. The format is a symbol, a pattern. The magnitude is a number, a meaning. Both are also objects. It is clear that the magnitude of the number, a state of *number*, affects the color and format, a state of the symbol. Similarly, we could scale the two-dimensional image of a three-dimensional toy up or down to fit the size of the paper on advertising literature or, for on-line and television advertisements, the frame displayed on a screen. The image is the format, a two-dimensional pattern, influenced by another two-dimensional pattern in three-dimensional space, the surface of the toy.

It is also common sense that a pattern exists in the state space that holds it only because its constituents are located in that state space. Indeed, a pattern starts forming in this state space when two or more items create a pattern in it. Therefore, in figure 54, the relationship *held in*, between *pattern* and its *state space* is contingent on the relationship *located in* between *object instance* and *pattern*. In fact, *held in* is not just contingent on *located in*, it is the aggregation of *located in*. Directionality, orientation, and dimensionality of the pattern all emerge from relative locations of its components. Thus, directionality and dimensionality are emergent properties of the aggregate *held in* relationship.

Remember that directionality must not be confused with sequence. Directionality is direction in state space. Sequence is a property of the pattern – i.e. will the pattern be considered the same pattern or not if we do (or do not) care about the *sequence in which its constituents are associated* in certain directions in state space.

The directional properties of patterns flow from its state space. A pattern will always occupy a region of its state space, which may or may not be delimited in different directions,

[43] See discussions on figure 22 and figure 36.

[44] A collection of object instances is also an object instance (see Chapter 2, section 1).

[45] Explicitly asserting implied relationships in a model replicates knowledge instead of normalizing it. Implied relationships in figure 54 have been shown for clarity, but have been annotated for this reason (see Module V, section 1).

may extend more or less in different directions, with different separations between locations of constituent objects in different directions.

Directions and dimensions of state space flow from state space to patterns through the *located in* relationship of figure 54. We will leave it as an exercise for the interested reader to derive the relationship between the number of components and the potential dimensionality of a pattern. A pattern's dimensionality cannot exceed that of the state space that holds it, and is constrained by the number of components in it. These *value constraints* are components of knowledge that are naturally attached to all patterns. In this way, the metamodel in figure 54 is an expression of normalized common sense about patterns.

Object counts and statistical patterns

Because patterns may be multidimensional lists of objects, *pattern*, the aggregate object, normalizes information about how many object instances it contains. Indeed, because an object instance may be repeated in a pattern, *pattern* will also normalize the incidence of individual object instances in a pattern. Object counts, like the other properties of patterns, such as dimensionality, order, and degrees of freedom, and indeed any object that is a *value*, may be restricted by value constraints of the kind we discussed in Chapter 3. New patterns emerge when these restrictions are added to the universal framework in figure 54. Each such restriction is a component of knowledge, a *value constraint.*

Just as object counts are normalized by *pattern*, the aggregate object, so too is other statistical information about the pattern. This normalized information flows from the locations of a pattern's constituents in state space. Examples of statistical information that emerge from aggregation of locations of a pattern's constituents are: the average distance between a pattern's constituents in state space, the average coordinates of its constituents, the standard deviation of separations between constituents of the pattern, and other emergent statistical properties.[46] These attributes are derived by aggregating the *located in* relationship between *pattern* and its constituents in figure 54. The *held in* relationship between *pattern* and its state space in figure 54 is this aggregation of the *locate* relationship. These emergent properties of patterns emerge from subtypes (polymorphisms) of *held in.* The object count in a pattern is merely the number of *locate* relationships aggregated by *held in.* Therefore, it too is a polymorphism of *held in.* That is why, in figure 54, *enumerates* is a subtype of *held in.* The common thread that runs through the various subtypes of *held in* is the number of constituents of the pattern. This is the information *held in* normalizes. All statistical properties that emerge from subtypes of *held in* are emergent properties of *pattern*, the aggregate object.[47]

Arrays

Look at table 1 again. It is a pattern of permissions in state space. The pattern tells us what kind of proximity metrics may exist in which kind of state space. The row and column

[46] Averages and standard deviations assume a ratio scaled proximity metric. Ratio scaled proximity metrics are only possible in difference and ratio scaled state spaces (see table 1). There would be other kinds of statistics that would emerge from the aggregation of the locate relationship of figure 54 in nominally and ordinally scaled state spaces.

[47] Emergent properties may be derived from attributes of constituent objects (see Chapter 3, section 2).

headings of table 1 are axes that describe the state space of the pattern. The state space is not continuous like physical space, or the state space of figure 24. It is discrete like the space in figure (B) of box 17.

Tables are two-dimensional discrete patterns. Each cell of a table is a discrete point in a two-dimensional space. The cell may be empty, or may contain an object. The object may be a nominal value like that in table 1, an ordinal value like that in box 17, difference, or ratio scaled values, like the unit of measure conversion tables of box 5[48] Each cell may also contain a rule, a pattern, a picture, a sound, or symbol, all of which would count as nominally scaled values. These discrete state spaces are called *arrays.*

Tables are two-dimensional arrays. Discrete three-dimensional spaces would be three-dimensional arrays. Like state space, arrays could also have more dimensions. Just as three-dimensional arrays cannot be printed on the two dimensional plane of this paper, but its two-dimensional slices (cross sections) can, arrays of higher dimensions cannot be displayed in three-dimensional physical space, but their three-dimensional slices can. Indeed, their two-dimensional slices can even be displayed on this printed page.

Consider figure 22. It could be shown as a three-dimensional array – one dimension for object instance (a nominally scaled dimension, since object instances have no intrinsic magnitude or order), one for attributes (also nominally scaled), and one for time slice. (Time slices are ordinally scaled, because non-overlapping time intervals are intrinsically ordered from past to present to future, even if they are not equal, and each time slice is discrete because it is the *slice* – the *identity* of a region – not the continuum in *time*, we are considering.) Figure 55 has an array like this.

A slice of the array parallel to the plane of the paper would be a two-dimensional table of attributes values, a table commonly found in relational databases, and routinely displayed or printed in reports in response to queries from users. The person on the upper left-hand side of figure 55 has this perspective. It is this cross section that person A sees in figure 55.

Slices perpendicular to the plane of the paper would produce tables of object histories. For example, person B of figure 55 will see a cross section that is a slice perpendicular to the plane of the paper, but parallel to the side of this page (it slices through a single attribute, into the depth of the paper in figure 55). This is a historical table of values of a given attribute across all object instances. This is the cross section that the person in the right-hand side of figure is looking at.

Similarly, if we sliced a single object instance perpendicular to the plane of the paper (person C's viewpoint – a slice that is perpendicular to the plane of the page and parallel to its lower edge), we would get a two-dimensional historical table of attribute values for the object instance showing the path it has traced through time to evolve to its present state. The person at the bottom of figure 55 is looking at a cross section of this kind.

Just as cross sections of three-dimensional tables (and other solid objects) are two dimensional, shapes, arrays in higher dimensions can be sliced into lower-dimensional cross

[48] The contents of the conversion tables of box 5 were multiplication rules. However, we can normalize the information in the table by factoring the common act of multiplication out of the each cell of the table and making it common to the table (a pattern) so that the common rule is not repeated in each cell. Then each cell would be left with only a ratio scaled value.

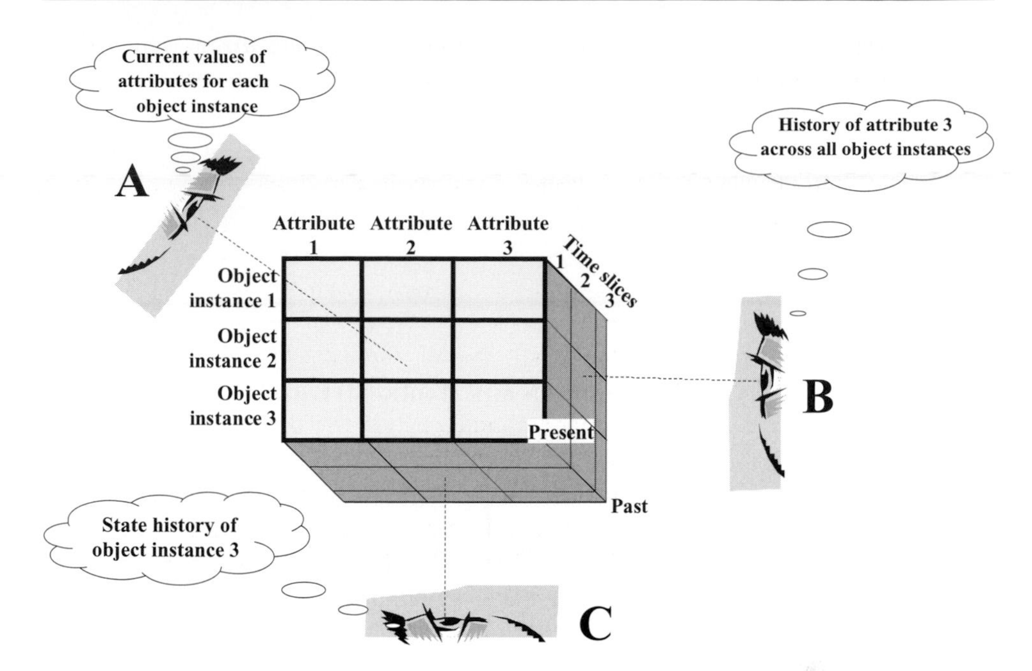

Figure 55 Perspectives of object – two-dimensional cross sections of a three dimensional array

sections. Most of us find it hard to visualize higher-dimensional shapes and arrays, let alone slicing and dicing them. An example might make it clearer:

Candu Compoot's Story – the tale of higher-dimensional arrays (on our website)

Candu Compoot's Story describes four- and higher-dimensional arrays in a parable with a business example. It shows how arrays need not always be patterns of concrete symbols, but could also be patterns of meanings. It demonstrates how lower-dimensional slices of higher-dimensional arrays may also be formatted as arrays with the following examples.

The *array* in figure 54 can be a symbol, an array of symbols or an array of meaning. Arrays that are formats are visual symbols like Candu's two-dimensional slice in table 3, or Candu's three-dimensional projections in figure 56. These symbols are subtypes with two parents – *pattern in physical space* and *array* – both of which are present in figure 54. Thus, as Candu Compoot's story demonstrates, figure 54 supports abstract arrays of meaning as well as formatting arrays, which are symbols we can see.

Patterns in physical space and time

Patterns in physical space and time have been shown as a subtype of patterns in state space in figure 54. We have discussed the reasons for this earlier in this section. Physical space and physical location are merely subtypes of state space and location in state space. Therefore,

Table 3 *A two-dimensional slice of Candu's nine-dimensional array*

	CONCERN FOR WORK LIFE BALANCE		
GROWTH PROSPECT	Poor	Average	Good
Poor	0%	1%	5%
Average	1%	2%	10%
Good	2%	3%	15%

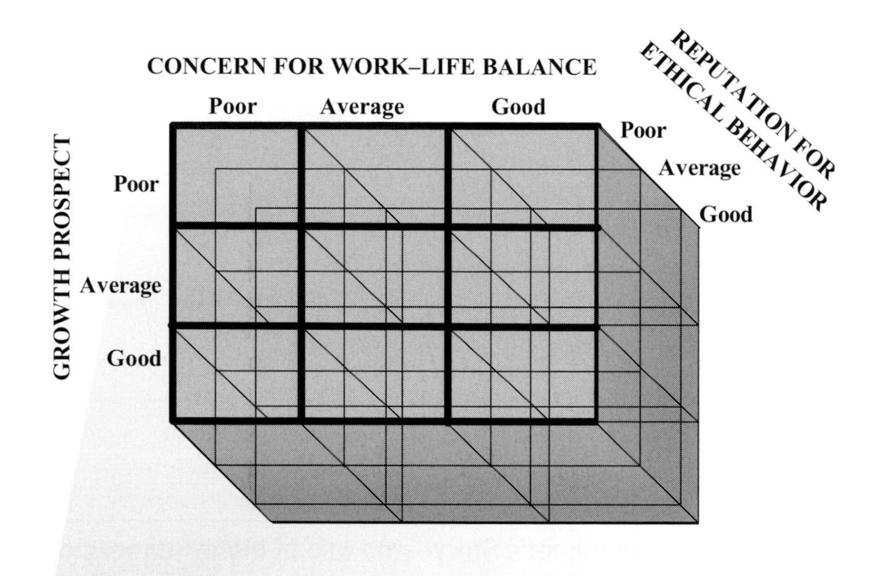

Figure 56 Candu Compoot's three-dimensional array

patterns in space and time are patterns that are located in physical space and time. In the language of figure 54, this flows from the fact that the *located in* relationship is linked to *object instance*, and patterns are a kind of object, which implies that *located in* may also be attached to *pattern* on one end, and, because physical space and time is a kind of state space, *located in* may be attached to physical space and time on the other end. In the language of figure 54, patterns may be located in physical space and time (remember inclusion polymorphism in box 21). Patterns in physical space and time that may be directly perceived by one or more of our five senses are symbols, which, associated with meaning, become formats.

(Discrete state space and arrays too have a similar relationship through inclusion polymorphism.)

Null space – patterns of everything, patterns of nothing

Those who have paid attention to the detail in figure 54 might find it strange that it asserts that the set of object states might be a set of *no* (0) object states. How can that be? How can the set be a set of object states if it contains neither objects nor states! It can be, because it is the empty set of box 19. Joined to an exclusion constraint, the empty set will assert that all objects are permitted to participate in the pattern, and attached to an inclusion constraint, it will assert that the pattern cannot exist – it is null. This is why state space without states is called *null space*. Null space tells us what *cannot* be. That too is an atomic rule (Module V, section 4 on our website discusses null space in more detail).

Patterns of symbols and patterns of objects

Patterns can be patterns of symbols in space, time, or state spaces, with attributes such as color, brightness, and pitch. They can also be patterns of pure information that constitute meanings. For example, consider the pattern of months in a year. A month is a concept, an abstract meaning we understand. It is not a physical symbol perceived with our senses. The eternal cycle of months is a finite but unbounded temporal pattern of meaning. We find no natural boundaries as we cycle through months. If we have not added the concept of a year, or the delimiters that tell us that a year begins in January and ends in December, there is no mark to tell us where the cycle begins and where it ends. We can choose any month equally as our starting point and the cycle will be eternal, although its extent will be 12 months.

On the other hand, consider the 12-month calendar year. It is delimited by two concepts: the month of January at the beginning, and the month of December at the end. We obtained the concept of "*year*" by adding these delimiters to the undelimited pattern that constituted the general meaning of eternity, the eternal passage of time. Not only is the resultant pattern a pure meaning, but its delimiters too are pure meanings, not symbols.

Consider the pattern we obtain by *combining* the linear concept of the passage of time from the past to the future with the cyclic concept of months. It is like the pattern in figure 51 (c) – the months, finite in number but repeated in a cycle without bounds, and eternity extending boundlessly and infinitely into the past and future. In the combined pattern, time has no bounds. The two concepts, the eternal passage of time and the eternal cycling of months are its components. These patterns of meaning are not formats, but may be *expressed* by formats. (As an exercise for the reader, what kind of pattern would we get if we combined all three, the concept of eternity, the delimited and finite concept of year, and the undelimited, but finite cycle of months?)

Consider the metamodels in this book. They are patterns of interlinked concepts. They are structures of meanings that engage each other through relationships (which are also meanings) to produce real-world behavior. They are patterns of meaning, not symbols. They are examples of patterns that are neither temporal nor spatial. The diagrams in this book represent these meanings. The diagrams are *symbols* that show how concepts engage each other to create new meanings. The diagrams are not the concepts they format; they merely represent them, and are only one of several possible expressions of these meanings. Thus, patterns need not always be patterns of symbols. Meanings are abstract patterns of information (see Module V, section 4 and Module VI on our website.) Figure 54 supports patterns of meaning, patterns of symbols, and also mixed patterns of meanings *and* symbols.

Patterns of symbols, by themselves, may be meaningless, like the patterns in figure 51, or they can express meanings. To express meaning, patterns of symbols must be associated with patterns of meaning. Meaningless patterns are not formats. Formats are meaningful patterns of symbols. Formats flow from the meeting of meaning and symbol – from the place where the metamodel of knowledge flows into the physical world of space, time, and perception. Figure 54 shows this meeting ground. The recursive *represent* relationship in figure 54 illustrates how patterns of meaning may meet patterns of symbols at infinitely many places, in an infinitude of ways. Therefore, it shows us that there are infinitely many ways of expressing abstract meanings with symbols, which thus become formats of meanings. (It also shows that meanings may represent other meanings, which implies that there are infinitely many ways of encoding not only symbols, but also meanings. We will discuss this in more detail in section 3 of this chapter.)

This fact, that a single meaning can be mapped to many different patterns of symbols, gives us the leeway to create, improve, and innovate information systems, as well as business processes. To seamlessly integrate business processes with the information systems that support them, we must seamlessly integrate the metamodel of patterns of meaning with the metamodel of patterns of symbols (as figure 54 does). It is a vital bridge and a key step in the metamodel of knowledge for aligning information systems with business processes and business processes with business requirements, which will keep agility and innovation in the forefront.

To understand how this can happen, we must understand the role, place, and context of the integrated metamodel of pattern framed by the metamodel of knowledge. This is what we will describe next.

The integrated metamodel of pattern

Box 36 and figures 50, 52, 53, and 54 collectively normalize the universal rules that we discussed about patterns of symbols. Figure 57 integrates this information. The structures on the top left-hand side of figure 57 are identical to those in figure 54, enhanced with the information in box 36. They not only provide a window into the *pattern of* relationship of figure 50, but also integrate the *represent* relationship in box 36 into the pattern. It is inherited by pattern from object (not shown to avoid clutter) and also by symbol (polymorphisms related to format and format conversion have been shown). In figure 54, we saw how the various partitions that normalize universal properties of patterns emerged from points in this structure. These partitions are on the top right-hand side. They are identical to those in figure 53. We have also seen, in figure 52, how properties of different kinds of proximity metrics lead to the different kinds of patterns in partition B. The lower right-hand side of figure 57 shows internal structure of each subtype in partition B in terms of its proximity metric. It is identical to figure 52.

We have discussed how patterns need not be confined to patterns of symbols only. They may be patterns of meaning, patterns of symbols, or may combine both – meanings associated with symbols. The last kind of pattern is *format*. *Object* in figure 57 generalizes the concept of *pattern* and supports the concept of *format*.

The window into *pattern of* also introduces an object that we did not have in figure 54. It is called *directional pattern in state space*. This object normalizes the rule that several partitions and subtypes in figure 53 are directional in state space. The subtypes in partitions labeled thus in figure 57 must all include their direction. It is a common feature of these subtypes that has been normalized by *directional pattern in state space*. This is why each is a subtype of *directional pattern in state space*.

The metamodel of knowledge in figure 57 not only provides for the fact that a single meaning may be expressed very differently, and that formats may be translated to other formats subject to complex rules and constraints, but also that knowledge may be expressed in any format, be it simple or complex – a printed report, a string of numbers, graphical multimedia, and multidimensional formats, or even molecular formats like those in DNA or the proteins that make living beings. All these formats supported by today's technology or waiting for their turn in tomorrow's, are enshrined in the metamodel of pattern. The integrated metamodel of pattern normalizes this knowledge.

> I saw Eternity the other night
> Like a great ring of pure and endless light,
> All calm, as it was bright, . . .
> Driven by Spheres
> Like a vast shadow moved, in which the world
> And all her train were hurl'd
>
> (Henry Vaughan, 17th century English poet in *The World*)

Pattern recognition and the metamodel of knowledge

Collections of symbols, values, and other objects are patterns only if they conform to a law. Pattern recognition is the discipline that discovers the law by analyzing a pattern.

Pattern recognition is of intense scientific and business interest with several applications in key areas such as information security and encryption, analytics, genetics and bioinformatics, credit risk assessment, robotics, weather forecasting, remote sensing, military reconnaissance, fingerprint recognition, face recognition, and biometrics.

Pattern recognition can be extremely complex and is an area of active and continuing research. The numbers of patterns was enormous when we considered even a few simple symbols in physical space. Many patterns can be exceedingly complex. When we consider patterns in state space, the possibilities explode.

Consider figure (c) of box 37. In physical space, we would consider only the shapes, sizes, orientation, and position of symbols to create patterns. The cylinder, cone, and ball might constitute a pattern based on their mutual proximity and size. Similarly, the die might

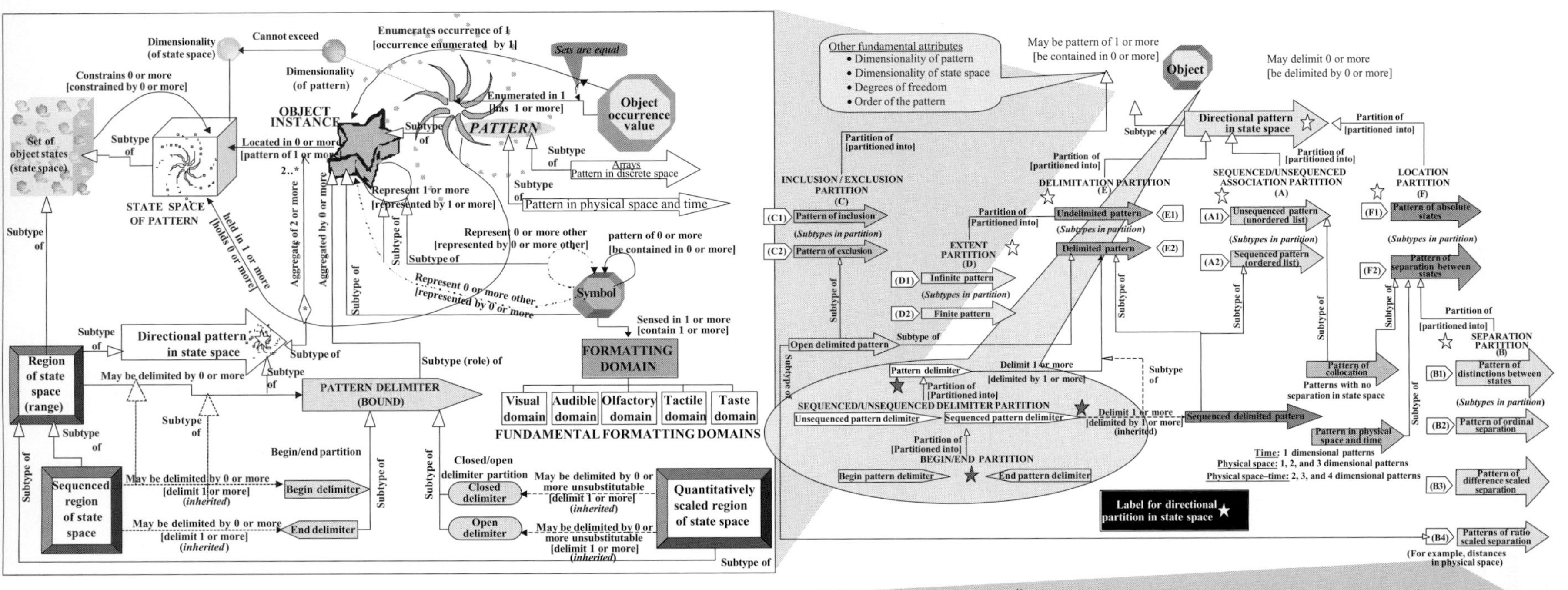

Figure 57 Integrated metamodel of pattern

be considered a part of a pattern based on the shape and proximity. The cluster of disks could be considered a pattern that flows from the same criteria. The three points are part of an arbitrary and *ad hoc* pattern by decree.

If we had to consider the state of the objects in figure (c) of box 37, we would have to add dimensions like color, weight, and reflectivity to the three spatial dimensions. The cone and the three points might become members of a pattern based on color, the die might go with the cylinder based on weight, the disks might go with the sphere and cone based on reflectivity, and so on. Possibilities and complexities compound themselves each time we add a dimension.

Just as some complex laws of location in physical space could be subtyped differently in different directions and dimensions of physical space (like the law that did not distinguish between mirror images), so too can some complex laws of location in state space morph into different subtypes in different directions and dimensions of state space. Finding these directions (if they exist) can be complex. When mixed state spaces are involved, laws in different subspaces may be constrained differently, governed by the different rules we have just discussed. This can compound the difficulty of an already complex problem. (Table 2 does give some guidance in terms of kinds of similarity metrics to look for.)

Compounding this difficulty even more is the fact that patterns are collections of symbols (or values and other objects). Therefore, the state space of a pattern is the collection of state spaces of constituent symbols (or objects – Module V, section 2 on our website discusses the states of aggregate objects in detail). The complexity of state space can rapidly become unmanageably large when we consider patterns.

Discovering patterns in this morass of possibilities in state space is the problem at the heart of pattern recognition. Patterns are knowledge, and the metamodel of knowledge is its fountainhead. Regardless of the complexity of the pattern, effective pattern recognition must leverage the fundamental attributes of patterns that emerge from the metamodel of knowledge. They are (the items marked with a "★" may depend on the direction of the pattern in state space):

★ Association and sequence
★ Location
  ★ Position
  ★ Cohesion and separation
- Exclusion and inclusion
- Order of the pattern (order of governance of other patterns)
- Dimensionality of the shape of the pattern in state space
- Dimensionality of its state space
- Degrees of freedom

★ Extent
★ Delimitation

The metamodel in figure 57 captures the universal rules that govern a pattern, regardless of how it is expressed – in what formats and what fundamental formatting domains – even mixed formatting domains. The metamodel is the container of common rules across formatting domains, whereas each fundamental formatting domain is the container of specific properties that the other domains do not have.

Every pattern recognition method must factor in these properties of patterns in order to recognize them, and to find which patterns resemble (or do not resemble) which. The footnote provides an overview of broad techniques that address each of them.[49]

Standards, language and patterns of patterns

As we have seen in this section, only some patterns are formats. Formats are *meaningful* patterns of symbols. Indeed, formats themselves can be arranged into systems, or patterns of formats. These are higher-order patterns or formats that govern formats. Language is one such higher order format of immense interest to business in the rapidly globalizing of man and machine.

Indeed, in the real world and the world of business, we often find systems of patterns – they are collections of patterns that govern other patterns like policies, strategies, language, and convention – they are higher-order patterns – patterns of patterns of vital interest to business. The metamodel in figure 57 addresses and unifies the whole. Of these higher-order patterns, language and standards for information exchange are vital to communication in the information-driven, relentlessly competitive, and paradoxically, the relentlessly collaborative global economy taking shape in the dawn of the new millennium. It might therefore be worthwhile to understand how language is framed by the metamodel of *pattern*.

Format, language, and semiotics

Language is a set of formats in visual domains (written script) and audible domains (speech). This set is a pattern in its own right – a pattern of unsequenced association. It is also an *ad hoc* association – *ad hoc* because the system is dictated by convention.

In each of the five fundamental formatting domains, there are potentially infinite numbers of symbols, and hence there are infinite numbers of formats. Standardization initiatives usually focus on selecting finite sets of formats and subtypes to facilitate communication. Language is one kind of standard (but not the only one).

Formatting rules and formatting domains are objects. Therefore, formatting rules as well as formatting domains can be generalized and subtyped. Thus, in the example of Chapter 3, section 3, where string length was expressed in English speech, we could have generalized

[49] Several statistical methods are used to find patterns in state space. They depend of the kind of proximity metric involved. It is impossible to exhaustively enumerate or describe all methods and their assumptions. However, some broad and basic methods are:

- *Ratio scaled proximity metrics*: principal component analysis, cluster analysis, rotational analysis, and regression analysis. Principal components analysis attempts to find the dimensionality of ratio scaled state space from observed states of symbols; cluster analysis tries to discover clusters of symbols in state space that are closer to other members of the cluster than members in other clusters; rotational analysis of different kinds tries to find directions in state space that measure variations in *extent* of the pattern; regression analysis tries to find relationships between states based on observed states of symbols.
- *Difference scaled proximity metrics*: multidimensional scaling and Kolmogorov–Smirnov statistics. Multidimensional scaling attempts to find the dimensionality of ordinally scaled state spaces from observed states of symbols; Kolmogorov–Smirnov statistics involve distances in ordinally scaled space.
- *Ordinally scaled proximity metrics*: discriminant analysis. Discriminant analysis attempts to allocate symbols to groups, based on a law that only says one group is *different* from another, from observed states of symbols.

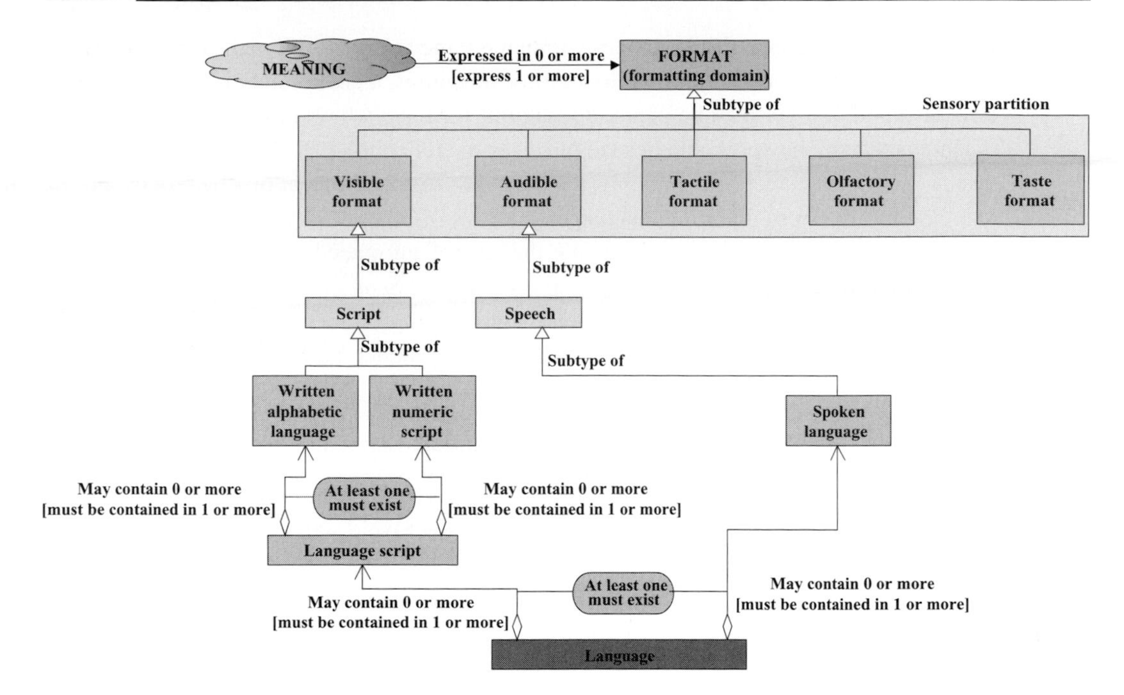

Figure 58 Structure of language

the format to read speech, instead of English speech. English speech would be a subtype of the more general speech domain, just as speech in other languages would be. Thus, the formatting rule, *speak*, would become a polymorphic relationship.[50] Depending on the language, the spoken word would be different for the same meaning.

Just as speech is a formatting domain, so are written symbols. Language is an aggregate object – a set of specific formatting domains – written words, scripts,[51] and spoken words. Figure 58 is a fragment of the metamodel of knowledge that represents these rules of aggregation.

Figure 58 recognizes that there may be primitive languages that have no written script, and there may be coded languages that are only written and have no spoken words. Figure 58 also recognizes that scripts and conventions may be reusable components across different languages. There are real world examples of this; in the state of Meghalaya in India, the spoken Khasi language is written in Roman script, the script of the English language. Similarly, in the English language, it is conventional to express numbers in both Arabic and Roman notations.

Figure 58 also recognizes that languages may be ambiguous, that there may be homonyms and synonyms. The relationship between meaning with format allows shows this: when a

[50] The polymorphic behavior of subtyped and composite relationships was also addressed in box 21 and Chapter 3, section 2, under "Implicit and intermediate rules".

[51] A script is the set of alphabet and numerals used in the written expression of a language. There are several different scripts in use today such as: the *Roman* script for most European languages, *Cyrillic* for Russian, *Greek*, *Kanji* for Japanese, and *Arabic*.

meaning is asserted in multiple formats within the same formatting domain, each format is a synonym. When a format expresses two or more meanings, it is a homonym. The metamodel also asserts that there may be meanings without formats, which implies the existence of meanings that are inexpressible in a language. Some reflection will show that this indeed is true for some languages. For example, the concept of infinite extent of a pattern cannot be expressed in most programming languages.

This link between meaning and format is the crux of automated language translation. The science of associating symbols with meanings is called *Semiotics*.[52] Semiotics is an issue of increasing importance in a world of rapidly globalizing business supported by culturally diverse peoples in equally diverse supply chains and product markets.

Format and format conversion – scope, size, and precision

A symbol is a pattern. It may also be a format. A format is a meaningful pattern we can sense directly with one or more of our five senses. We have discussed several examples of formats earlier in this book.

As we discussed under "Shaping and influencing patterns" (p. 213), interactions between objects can influence a pattern. Thus, the format of a value might be contingent not only on the state of the object it expresses, but also on states of other objects. Formats like this are context sensitive formats. Box 38 portrays several examples of context sensitive formats and format conversions.

A single meaning may be represented by many formats. Formats may even represent other formats. Just as formatting rules map meanings to formats, format translation rules map one format to another. Written words may be translated to spoken words and vice versa in any language. Just as *speak* is a polymorphic relationship, *translate* too is polymorphic for the same reasons. In a future world of global communication, a business person in New York may speak to a counterpart in Tokyo in English, but be heard in Japanese and vice versa, or a chemist might dictate an MSDS (Material Safety Data Sheet – a standard set of safety instructions for hazardous materials) in English in London, and it might be simultaneously printed in Arabic at one location and in French at another.

Formats need not always be characters like numerals and alphabets on our keyboards. In our discussion on patterns and formatting domains, we have seen how symbols can also be patterns that are pictures, graphics, even three-dimensional moving pictures or arrays, sounds, olfactory, taste, or tactile symbols.

These patterns might be patterns synchronized across formatting domains. The pitch of an audible note may tell the operator of an oven in a factory how hot the oven is, even as a screen displays the oven's temperature visually. In this context, both the sound and the numerals are formats of temperature – one audible and the other visual – because both are meaningful symbols. Together, they form a pattern, an audio-visual pattern. The audio-visual pattern too is a symbol and a format. It is a composite format – an aggregate object that normalizes rules about synchronizing its constituents (see "Pattern in physical space and time" (p. 217) for more information about patterns of this kind). Formats may be expressed in any medium

[52] [325] contains a lucid and non-mathematical introduction to semiotics.

we can perceive. The medium constrains the attributes of the format. As we have seen, some formats can even be arrangements of molecules in space. The only requirement of a symbol that is a format is that the number of states it will support must be commensurate with the quantum of information it represents. Otherwise it cannot represent the information fully (see box 38 and information carrying capacity under "Patterns of shapes: dimensions of freedom", p. 181). "A picture is worth a thousand words" may be a cliché, but it is a cliché based on the information carrying capacity of image formats, as opposed to character formats. No amount of written description can substitute the experience of actually watching the Mets play the Dodgers in Giants Stadium in New Jersey on a Sunday afternoon – even if you watch the game on television. The moving television picture, a format, cannot be converted into written words, another format, without losing some of its meaning.

Formats are symbols, and symbols are patterns. Therefore, formats inherit the emergent properties of patterns. Three emergent properties of patterns affect the behavior of format subtly but fundamentally. They are:

1 the information carrying capacity of the format;
2 the extent of the format; and
3 dimensionality of a format.

Information carrying capacity is fundamental because it has a profound impact on the *precision* with which a format expresses a meaning. Written words failed to convey the complete picture in the example above because their inherent capacity to convey information was far less than the capacity of a moving image. Extent is fundamental because it deals with the *scope and size* of the expression. Dimensionality is fundamental because it deals with the *shape and properties* of the formatting symbol. Of course, they are all interrelated.

Precision

The picture on the television screen lacked quality. It was grainier than in real life. It lacked the resolution and fine detail we would have seen with our own eyes had we been present at the scene. If we magnified the picture on the screen to real-life dimensions or beyond (perhaps with a projector or a screen larger than a house), this haziness or granularity would become even more apparent as our eyes sought the detail that would normally be visible at this level of magnification. The picture lacked this level of detail because its capacity to convey information was limited by the density of pixels on the TV screen and other technical factors. The picture did not have all the information our eyes would have seen, because it was short on information carrying capacity. It did not have the precision of our eyes. Similarly when we format written words or numbers, they may appear hazy or crisp. Precision is crisp and sharp.

Spoken words or numbers are audible formats. Like pictures, they may not be clear if the format lacks information carrying capacity. Have you compared the bell-like clarity of sound from a good compact disk with the tinny voice of a bullhorn? Most of us have experienced the crispness, clarity and richness of high-quality sound as well as the poverty and distortion endemic in low-quality sound. The fidelity of a format depends on its information carrying capacity. The information carrying capacity of a format normalizes precision.

When words, numbers, and pictures are truncated, they also lose information. However this loss (*if* it happens) flows from the extent of the formatting rule. Rules of truncation may or may not impact a format's information carrying capacity. When they do, they do so indirectly via their impact on the extent of the pattern. Truncation is not normalized by directly constraining the information carrying capacity of a pattern. Instead, it is normalized by directly constraining the pattern's extent. How and why this is so is our next topic.

Scope, size, and truncation

The image on the television screen was more limited than the real-life scene at the stadium or playing field. It cut off everything but the most important parts of the real-life image that the audience at the stadium would see. This reduced the *extent* of the object (the scene in physical space) that it was formatting and adversely affected our experience of the game. It limited the *scope* of the format. Limiting the scope also limited the *amount* of information that had to be formatted.

The picture (a format) was far smaller than the stadium and the playing field. Its extent was smaller than the extent of its scope – the scene it represented. The screen delimited the *size* of the format, but not its information content. (Most of us would have preferred a bigger screen. The closer the extent of the format is to the extent of its scope, the more "real" it would look to us – like an I-max movie looks real.) However, extent in physical space is only a part of the extent of *format*. Extent means much more than physical size and truncation of a symbol, because physical space is only a *part* of the state space of a format.

Consider a written word. It is a one-dimensional string of letters. Besides an extent in physical space (the plane of the paper), a written word has additional dimensions in state space. Points in an ordinal "serial number" or positional dimension determine the position of letters in a word. For example, in "WORD," the letter "W" occupies the first position on an axis that extends into the positional dimension. "O" occupies the second slot on this axis, "R," the third position, and "D" the fourth. Curbing this dimension will slice and dice the string of letters that make the word. An upper bound on this dimension (a direction in the state space of written words) will truncate a word. A lower bound would cut off letters at the beginning. Ranges with upper and lower bounds would slice up the string of letters into different segments depending on the upper and lower limits of the range.

The upper bound might limit the extent of the pattern in this direction of state space, but, if a word is shorter than its upper bound, it will not be truncated and will retain its accuracy as a format. It will lose accuracy (and information carrying capacity) only if the upper bound truncates it. Thus, bounds may limit the extent of a written word's state space, but may or may not impact its information carrying capacity. Similar arguments will hold for other formats and formatting domains (see box 38).

These were examples of how the *extent* normalizes rules of truncation of a format, but impacts its information content only if it:

1 Limits the scope of the format (limits the extent of the formatted object, not the formatting symbol); and
2 Reduces the state space of format to a level below that of its scope.[53]

[53] In a continuum, like in ratio or difference scaled state spaces (e.g. physical space and time), the number of possible locations (states) is infinitely large. Then, the *partial order* (see box 45) *on the set of difference and ratio scaled attribute values is dense* [208]. Cardinalities, or the relative sizes of these infinite sets, must then be

The information carrying capacity of the format determines its accuracy (precision) and extent determines scope, truncation, and size.

Dimensionality

Not counting the time dimension, the format of the moving picture was two dimensional. This too impacted our experience negatively by limiting the fidelity with which the two-dimensional moving picture could represent the three-dimensional information it represented (formatted). It distorted its shape. It also lost information because an object with more dimensions (the scene) was mapped to one with less (the flat picture).

Reducing the number of dimensions will not always reduce the information content of the format (box 38 elaborates on this). For example, an MRI (Magnetic Resonance Imaging) scan of a human body produces multiple two-dimensional images of cross sections of a human body, which taken together represents the three-dimensional image of the body and its internal organs. Computers can reconstitute the three-dimensional image of the body (with its internal organs) from the set of two-dimensional MRI scans. They can convert the two-dimensional format of an MRI scan into a three-dimensional format, because information was not lost in the scanning. Reducing the number of dimensions may change the shape, but not the fidelity of a format.

This was an example that demonstrated that the information carrying capacity of a format determines its accuracy (precision), and dimensionality, its shape. Indeed, dimensionality interacts with extent to determine shape. Dimensionality determines directions in space, and extent determines how directions are curbed to produce shape, an emergent property of the pattern. Together they shape a format as well as its scope.

Shapes extend not only in physical space, but also into state space. In Chapter 2, we discussed how different axes of state space represent different dimensions. Thus, the dimensions of a format also describe various properties inherited from its formatting domain. Convention, standards, and technology may act in tandem to limit the lawful state space of the format differently in different directions, and thereby define the shape of a format in state space (see the examples in box 38).

The polymorphic nature of format

Formats map values to symbols. Formats flow from the generic relationship in box 36, wherein one object may represent another. Thus, *format* is a subtype, a polymorphism of *represent*. The object being represented is a value, and that representing it is a symbol (see figure 60). A format may also be converted to another. That is another simple polymorphism of *represent* – the object represented and the object representing it are both symbols.

considered. The theory of transfinite numbers addresses this. *Ordinal* [212], *cardinal number* [206], *continuum hypothesis* [204], *countable* [202], and *countably infinite* [203] address cardinalities of infinitely large sets. In this book, it will suffice to assert that reducing the size of a physical image will not necessarily reduce its information carrying capacity. Contrast this with the truncation of words, which is a reduction of extent in a discretely scaled direction of state space. Reducing the extent of a discrete axis of state space reduces information carrying capacity.

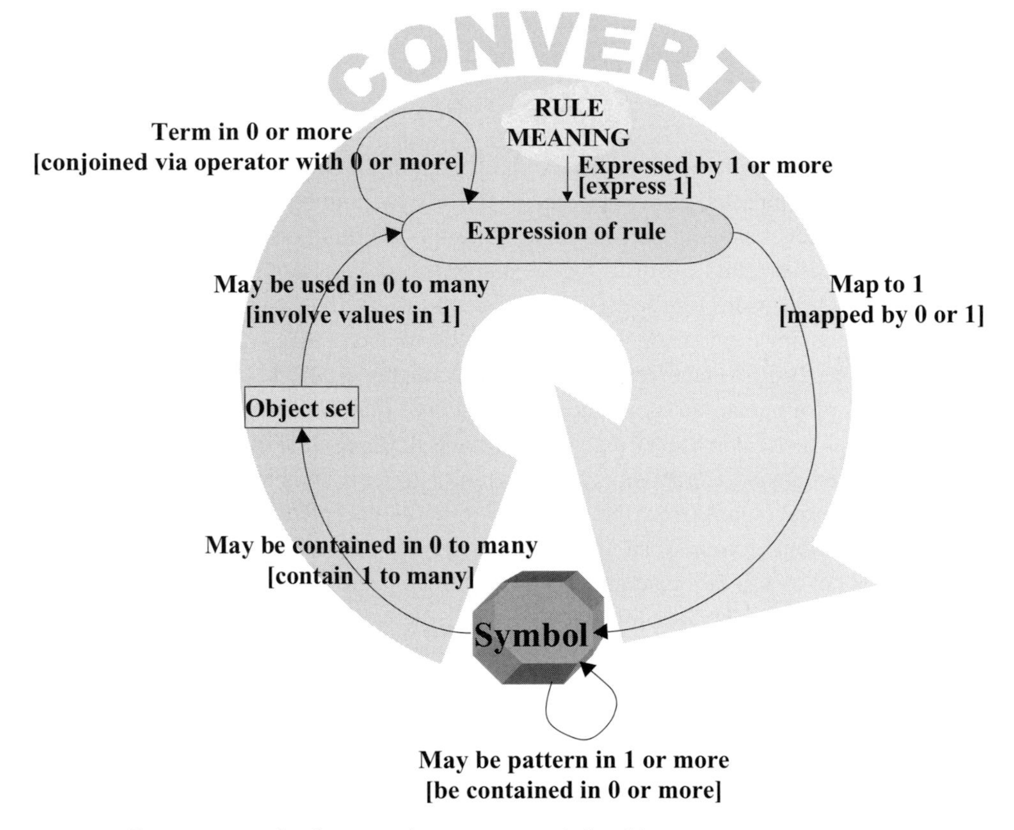

Figure 59 Format conversion is a recursive aggregate relationship

A symbol is a pattern. Interactions between objects can influence a pattern. As such, the format of a value might be contingent on other objects. For example, the size of a drawing may be automatically scaled up or down, depending on the size of the frame it will be displayed in; the color of text may be black, if the background is light, or white, if the background is dark. In these examples, the frame and the background are objects that influence the drawing and text that are displayed. Therefore, the frame and the background will be members of *object set*, and the drawing and text are the format, symbols that are influenced by these objects in *object set*.

The object sets in the figures of this section represent this kind of polymorphism. The object set in figure 60 contains the objects that influence the format. Thus, the map is polymorphic. Its parameters are members of object set.

Similarly, figure 61 is the metamodel of format conversion, and the object set has the context of the conversion – the objects and interactions that determine the state of the format after it is converted. For example, an audible tone may also be mapped to a waveform on an oscilloscope – a conversion from the audible to the visible formatting domain (see formatting domains under "Five fundamental formats", p. 172). Format conversion may also be constrained by physical devices used to support requisite formats. One kind of printer might support only black and white images, whereas another kind may support

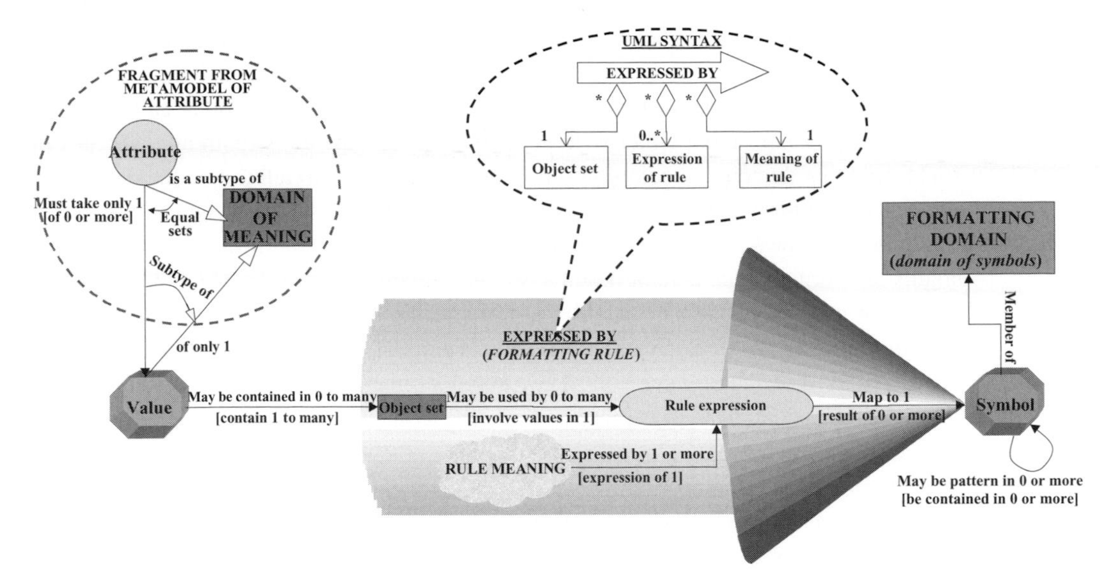

Figure 60 The metamodel of format maps values to symbols

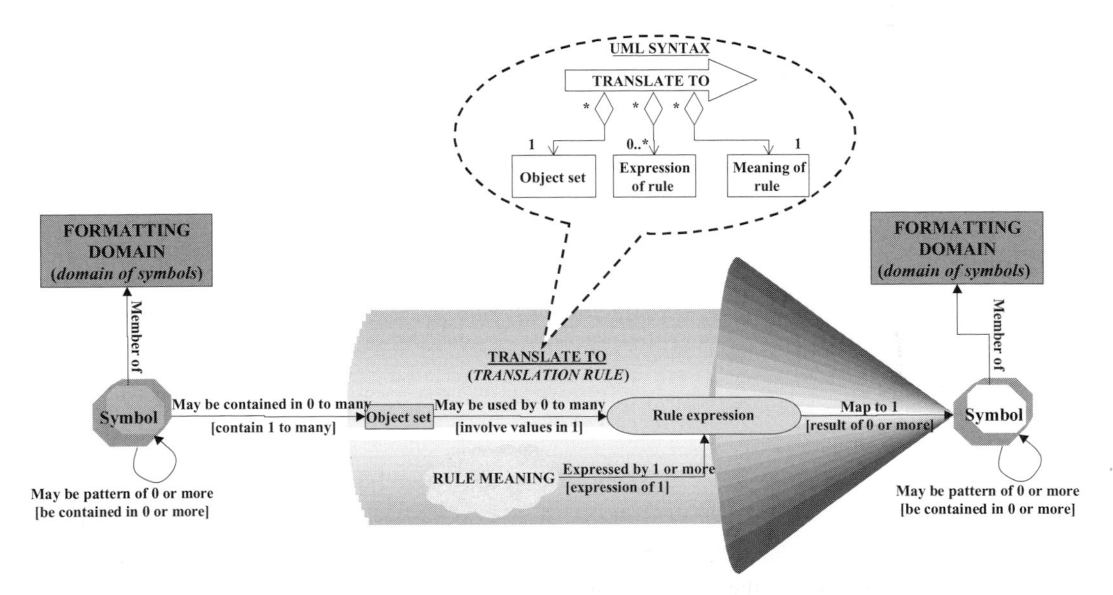

Figure 61 The metamodel of format conversion maps symbols to symbols

color. This is an integral part of business process automation. *Object set* also supports this kind of automatic context sensitive formatting and format conversion. The objects in this case would be the physical devices and their interactions. Constraints are parameters, and formats the results of these polymorphisms of *represent*.

Constraints may also go to the heart of the formatting domain, and bar or mandate expressions in specific fundamental formatting domains. An oscilloscope may not support the audible formatting domain and a speaker may not support the visual formatting domain. These are sweeping restrictions, from which automation could automatically determine what specific formats will suit which devices. For instance, it may be automatically inferred that a speaker cannot output either written script or printed pictures (both are subtypes of visual formatting domains) – common sense, but someone has to tell that to the computer!

Constraints on *format* could also mean higher-order constraints like *language*. For example, language can be a parameter that determines specific verbal (audio) and script (visual) domains in a format. A constraint might assert an "English only" rule. They may also bar, mandate, or merely permit specific subtypes of specific formatting domains. A publisher of a book might assert a "black and white only" rule for pictures. Constraints might even go to the heart of formatting symbols. They may bar, mandate, or merely permit, specific symbols within a domain. A chat room on the website may deem specific terms to describe ethnic, religious, or national groups unacceptable, and a business may mandate a logo on all written official communication.

Constraints could also be set on specific states of symbols, barring, mandating, or merely permitting specific regions of state space of a formatting symbol. For example, there might be size restrictions on the dimensions of a graphic. Constraints might even be attached to emergent states of the patterns that make the symbol, shaping it, mandating, barring or permitting its scope, size, texture, and information carrying capacity. These constraints too are parameters of *format* and its conversion – an aspect of the polymorphic map represented by figure 61 (see the examples in box 38). Thus, formats are one kind of bridge between business meaning and business process automation.

Formats (and format conversions) that are "aware" of states, constraints, and rule expressions attached to them will have the "intelligence" to change their character, depending on what they are mapping and by what rules.

Figure 62 describes a polymorphism of the *represent* relationship of box 36 that subsumes formatting and format conversion:

Objects may express or represent other objects. As highlighted in Module V on our website, higher-level processes can be implemented in different ways by alternative lower-level processes and tasks.

The broken lined arrow is a value constraint between the information carrying capacities of the object being represented and the object representing it. If we remove the value constraint (or weaken it – say, by limiting the difference between information carrying capacities), representation may still be possible, but with less precision once the information carrying capacity of the object that is representing the other object falls below that of the object it is representing. If we go on reducing the information carrying capacity of the object that is representing the other object, it will eventually become a mere token for the existence of the object it represents, like the diagramming symbols in this book are only tokens for the meanings you have been studying in it.[54] This notion is described in more detail in box 38.

[54] Eventually when the object doing the representation has no degrees of freedom left, it cannot be a format because it cannot represent a meaning. Constants are objects like this. They do not even have the freedom *not* to exist. They must exist, and can have only one value.

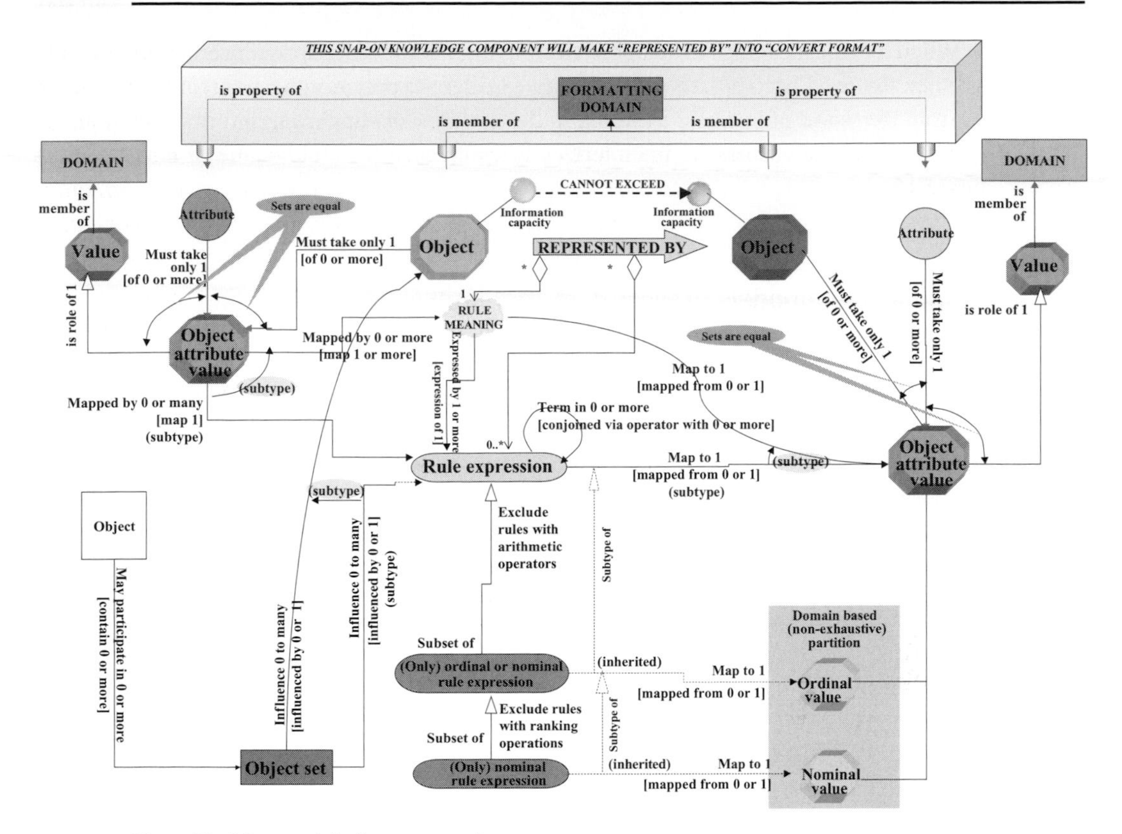

Figure 62 Metamodel of representation

In figure 62, when the object on the right is a symbol, the figure will become the metamodel of format – the additional detail behind figure 60. When both objects, that are being represented and that are representing it, are symbols, figure 62 will become the metamodel of format conversion – the additional detail behind figure 61. When only the object on the left is a symbol, figure 62 will become the metamodel for imputing values to a symbol. Thus figure 62 captures the polymorphic nature of *represent*, the relationship in box 36.

Different components of knowledge may be "snapped" into place and the behavior of the *represent* relationship will change commensurately to serve different ends. In the following description of figure 62, we emphasize format conversion, but remember the contents of figure 62 can play several roles, including the metamodel for encrypting symbols and/or meanings.

It is worth noting that maps between attributes and states that instantiate the metamodel in figure 62 may be between like domains as well as unlike domains. For example, a tone may be mapped to an identical, but louder tone. This kind of format translation is the component that supports the common act of adjusting the volume of an audio signal – something we do so often that we rarely even think about it. It is a translation between like formatting domains – audio domain to audio domain.

When it maps meanings to symbols, turning them into formats, or converts one format to another, figure 62 becomes a context sensitive bridge – a polymorphic transform – that takes us from the world of business meaning to the universe of supporting information systems; a bridge from the business to the interface rules layer of the architecture of knowledge in figure 15.

Box 38 Metamodels of format, format conversion, encryption, and formatting constraint (on our website)

Box 38 supplements the discussion in this section with deeper analysis and more sophisticated examples. It covers issues such as encryption, accuracy, fidelity of representation, and their relationship to the information carrying capacity of formats. Box 38 also describes how attributes of pattern, such as extent, directionality, dimensionality, delimiters, location, and proximity are inherited and relate to the behavior of formats and the represent relationship of box 36 and addresses the encryption of abstract meanings. Box 38 provides a more complete description of figure 62.

2 The meaning of units of measure

Values are meanings; symbols represent these meanings. Symbols are formats; formatting rules map meanings (values) to symbols. Values are abstract meanings; symbols are discrete, perceptible objects that occupy space and time. Many values are discrete, but some are continuous. A continuous region of state space contains an infinite number of distinct states. We cannot map infinite numbers of states to finite numbers of discrete symbols without losing information.[55] We need a continuum of symbols that matches the continuum of values we wish to represent. Otherwise, we will lose scope and precision when we map ratio and difference scaled state spaces to symbols.

Values in difference scaled space also carry information on magnitudes of differences between states. In ratio scaled space, they contain information on absolute magnitudes as well. Therefore, we need symbols that will not only match the continuum of values we must represent, but also match magnitudes and differences of the meanings we must represent. Numbers are symbols that satisfy all these criteria. Numbers, as well as their differences,[56] exist in a naturally ordered continuum. Numbers are also objects. They are objects that are meaningless by themselves, but acquire meaning only when they are associated with magnitudes of quantitative values in domains of meaning like length, money, weight, and others we have discussed in Chapter 1 and section 3 of this chapter. Unlike the formatting

[55] It is common sense that an object with a larger number of states cannot be completely represented by another object with a smaller number of states. *Accuracy of formats* and *encrypted information* in box 38 describes why formats lose precision in such maps.

[56] Differences between numbers are numbers too, and hence they inherit all properties of numbers.

symbols discussed so far, numbers are also abstractions. In this section, we will understand the difference between a number and its value.

Number versus value

Numbers are meaningless. If we say a room is 10 feet high, it has a meaning, but the number 10 is meaningless by itself. The meaning of the statement came from the meanings of a domain (length), an attribute (height), and an object (the room itself). It was borrowed from these abstract components of knowledge that engaged each other, like gears in a knowledge machine.

The domain (the length domain) told us that we were measuring a distance in physical space, the attribute (height) told us that the distance we were measuring was height, and finally the object (room) told us that it was the height of a room we were measuring. The value, a member of domain, carried information on the magnitude of the distance. However, we needed a symbol, a number, to express this value physically. We needed the number for *communication*. We can *experience* and understand the value without a number, but must have the number to *communicate* our experience.

The domain had meaning. It was also a class of meaningful values. This value was a member of this class. It was a specific value with meaning and magnitude (see figure 35). The number told us nothing by itself. It only borrowed meaning from the value it represented. We could have mapped the same meaning, or *value*, to a different number. It would not have changed the height of the room by one iota. Its *value* would not change, but the *number* that represented it would.

All the number did was that it *facilitated communication* of meaning, the value, provided we were consistent in how we *mapped* values to numbers (more on consistency to follow soon). The meaning of the number was borrowed from the engagement of an object and a domain through its attribute. The number, by itself, was meaningless.

The nature of numbers

Numbers are symbols, but they are also abstract concepts. Unless they are represented by perceptible symbols in physical space and time (like numerals), numbers cannot be perceived, and will remain abstract concepts like the values they represent.

To make ratio and difference scaled values tangible, they must be mapped to numbers, which, in turn, must be formatted by perceptible symbols. Only then can the meanings of ratio and difference scaled values become tangible information we can perceive. Two maps operate in tandem to manifest abstract ratio and difference scaled values, the concepts, into the concrete physical world of information.[57] Figure 63 shows this and figure 64 elaborates on it.

Just as a single value may be mapped to different numbers, the same numbers may map to different formats. These formats are different numerals and symbols in different scripts, languages, and systems. For example, the same number may be formatted by

[57] The two maps in tandem in figure 63 are mathematical *morphisms* that form a *composition* – see [186].

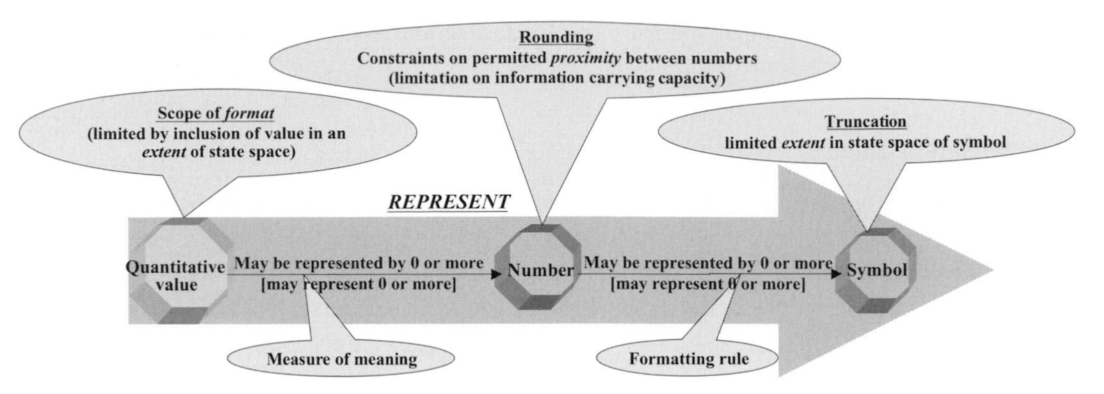

Figure 63 Two maps working in tandem map quantitative values to physical symbols

Arabic or Roman numerals. Even when Arabic numerals are used, they may be formatted as binary, octal, or decimal numbers.[58] Different formats do not change the number they represent any more than different numbers change the single meaning they may all express.

Just as the same number might be formatted differently by different symbols, the same sets of quantitative values may be mapped to different sets of numbers. Each such map is a *measure* of the value.[59]

Numbers normalize different formatting rules than do either values or symbols. Consider how the following formatting rules are normalized. They are also examples of polymorphic, context sensitive formatting rules:

1 **Rule:** "*Extremely large or extremely low values of temperature must be displayed in red and also voiced audibly to alert an operator of a furnace, regardless of the unit used to measure the temperature, such as Fahrenheit, Celsius, Kelvin, or any other number that displays the temperature.*"

  The behavior of format depends on the *value* of the temperature, not on the *number* displayed, nor on its *unit of measure*. This formatting rule is normalized by a relationship between *value* and *format*.

2 **Rule:** "*Extremely large or extremely small numbers must be in exponential format and those in between, in decimal format.*"

  The behavior of format depends on the *number*, not *value* or *unit of measure*. The format depends on *number* alone. It is a relationship between *number* and *format* that normalizes this rule.

3 **Rule:** "*All roman numerals are red.*"

[58] The endnote on number systems and radices describes how the same number, in the same script, may be written differently.

[59] The metamodel of measure is another polymorphism of the *represent* relationship in box 36. Figure 62 has the detail that box 36 omitted. If we replaced *object attribute value* on the right-hand side of figure 62, with *number*, *represent*, the relationship would become a *measure of value*, or *measure* in short. Figure 64 elaborates on figure 62 to show how units of measure emerge from the expression of a quantitative value, which is a polymorphism of *represent*.

Roman numerals are a visual *format*, a perceptible symbol that expresses a number. *Format* alone, not *number*, *value*, or *unit of measure*, normalizes this rule.

4 **Rule:** "*All temperatures in degrees celsius must be red.*"

The format depends on *unit of measure* alone, not *formatting symbols, number*, or *value*. A relationship between *unit of measure* and *format* normalizes this rule.

The pattern in number

The concept of number is not only an abstract symbol; it is also an abstract continuum, and a pattern in its own right.[60] The continuum is a pattern, a sequenced, one-dimensional, unbounded pattern of infinite extent. The origin of the continuum is the number zero. The proximity metric in this pattern of numbers is ratio scaled. It has all the emergent attributes of patterns, including information carrying capacity. Constraints on its information carrying capacity limit the precision with which it can express meanings, and are at the root of round numbers. Round numbers limit the precision with which numbers can express the values they represent. Figure 63 articulates this.

Round numbers versus truncated numbers

Rounding implies disallowing ranges of numbers. Only a set of discrete numbers is permitted in the resulting pattern. This set is based on the proximity of a number to its neighbor on either side. The continuum stops being a continuum. It becomes a set of discrete points instead.[61] When the expression of values is rounded up or down, a region, a *continuum* of quantitative values is mapped to a discrete, discontinuous pattern of numbers. Several values map to a single number. When values in a region, a continuum, all map to a single number, they must lose information.[62] The map loses fidelity in proportion to gaps between numbers. (It also violates the fourth rule of simple representation in box 38.) We have seen in section 1 and box 38, how truncation emerged from curbs on *extent* in the state space of symbols. We now can see that rounding emerges from curbs or the proximity metric of another pattern – a number in a continuum of numbers. Thus truncation emerges from *formatting symbol*, whereas rounding emerges from *formatting number*. Further, truncation emerges from constraints on *extent*, whereas rounding emerges from constraints on *proximity*. One kind of component of knowledge, a metaobject in our metamodel normalizes rules for rounding values and another normalizes rules of truncation. As such, rules for rounding values up or down have their origin at one place in the metamodel of knowledge, whereas rules of truncation have their origin at another (see figure 63).

[60] See [207], [202], [203], [206], [212], [213], [214], [215], [216], [217], [219], [220], [221], [222], [223], [224], [225], [226], [227], [228], [229], [230], and [231].

[61] Round numbers convey quantitative information on ratios and differences. Therefore they carry more information than ordinal numbers, which convey only sequence and distinction (see the first lattice of [218]). However round numbers convey less information than the real ("normal") number continuum–[222] because round numbers have no information on the gaps between them. Similarly the surreal number continuum has more information than the real number continuum – see [231].

[62] When two or more values map to a single number, the map between the domain of values and the domain of numbers is not *injective*, and hence loses information. See *morphisms* in the endnote on the theory of categories or items on the theory of categories, functors, and classes in the Bibliography.

Rounded numbers may also be truncated, and truncated numbers may be rounded. The two operations are independent, not mutually exclusive, but they do depend on sequence. (If they must be assembled into a subassembly of knowledge, the connective between them is like the "★" connective in the endnote on gluing objects together.)

Measures of meaning

Measures represent magnitudes. Formats were about symbols that represented meanings. Units of measure are about representing magnitudes. Values convey meaning. They also convey magnitude. However, not all meanings convey the same quantum of information about magnitude, nor are all arithmetic operations valid in all domains.

Consider nominal domains. The only information conveyed by values in nominal domains is that each is different and distinct, but they do not convey by how much these values differ, or which value has a greater or lesser magnitude. It follows that the entire information content of a nominal value can be expressed by a mere symbol alone. Hence formats will suffice in expressing all the information they convey and no arithmetic operations will be valid in their domains. Only operations that compare values to test that one value is distinct from another will exist in nominal domains.

Ordinal values convey more information on magnitude. They can tell us which values are greater, lesser, or of equal magnitudes compared to other values,[63] but cannot tell us by how much these magnitudes differ from each other. Therefore, their information content can be fully expressed by any set of symbols that convey information on sequences or ranks. There are many such symbol sets – numbers, the letters of the alphabet, and others. Alternatively, symbols will suffice, provided we impute a rank or sequence to the set of symbols. Formats will suffice to express all information about the measure of these values. Both tests for distinctness and sequencing (sorting) operations will exist in ordinal domains.

When we express naturally discrete magnitudes, like those in nominal or ordinal domains, formats suffice because each magnitude can be mapped to a discrete symbol. For example, it is *desirable* but not mandatory that the relative ordering of magnitudes in an ordinal domain must be consistent across formats (i.e., formats should be such that any natural sequencing of symbols should be consistent with magnitudes of ordinal values that map to them). For example, ranks could be mapped to whole numbers starting with 1, but we *could* use other symbols that have no natural order provided we associate a rank with each.[64]

[63] See properties of, ordinal numbers and ordered sets in [211], [212], [213], [214], [215], [216], and [217].

[64] Well-ordered sets that only differ in the "notation for their elements," i.e., their formats, are mathematically indistinguishable. Elements of the first set pair one-on-one with the elements of the second, so that if one element is smaller than another in the first set, then the partner of the first element is smaller than the partner of the second element in the second set, and vice versa. This kind of one-on-one correspondence is called an order isomorphism. Two well-ordered sets are always *order isomorphic*. Maps between ordinal values and numbers are *order isomorphic* and can be made order isomorphic to one and only one ordinal number. See isomorphism and order isomorphism in [212] and the endnote on the theory of categories. Also see [211], [213], [214], [215], [217], and [218].

However, with difference or ratio scaled domains, we will be dealing with a continuum of magnitudes[65] and we need a different kind of map to express magnitudes. Discrete symbols will not be sufficient.

Difference scaled domains convey more on magnitudes than ordinal or nominal domains do. Difference scaled values not only tell us which is greater or lesser, but also by how much. Unlike nominal and ordinal domains, magnitudes of values and differences in such domains can be a continuum, rather than a set of discrete points.

Difference scaled domains carry information on the magnitude of gaps between values in it.[66] Not only do comparison and sequencing operations have meaning in such domains, but so does subtraction. (However, difference scaled domains carry no information on ratios between values, and division has no meaning in these domains. For example, differences between dates are meaningful, but dividing one date by another is meaningless – see the parable of Metanesia in Chapter 1.)

Numbers are measures of magnitude. Moreover, numbers may be compared, have a natural sequence from small to large, and may be meaningfully subtracted to obtain the magnitudes of difference between them. Therefore, it is natural to map values in difference scaled domains to numbers to express their magnitudes.

Just as formats only *represented* nominal and ordinal meanings, numbers only represent *magnitudes* intrinsic to meaning. Just as several different maps can map a single meaning to symbols, a single magnitude can be mapped to different numbers. Each map will merely be a different measure of the intrinsic magnitude of *meaningful values* in a domain of *meaning*. The different numbers, which different measures map a single meaning to, are merely different representations of the magnitude latent in that meaning. Each measure is called a *unit of measure* (strictly speaking, the magnitude that is represented by the number 1 in each measure is its unit of measure, but in this book, we will call both the measure and its unit a unit of measure).

In short, the precision with which a quantitative value is translated to a number is a property of the unit of measure, a metaobject in the metamodel of knowledge. Based on the principle of subtyping by adding information, units of measure of greater precision are derived from subtypes of units of measure of lesser precision.

Moreover, in keeping with the nature of the intrinsic meaning of values being mapped, each measure must be internally consistent in expressing the relative ordering of magnitudes, distinctions between values and magnitudes of *differences* between values. We will call these the golden rules of measurement. In other words, the golden rules are:

[65] A continuum of values like those in ratio and difference scales domains is said to be mathematically *dense* – see [208]. The smallest unbounded totally ordered dense set is the set of rational numbers [220]. Therefore unconstrained quantitative domains are infinite sets of values that may be isomorphically mapped to at least the set of rational numbers (see isomorphic mapping in the endnote on the theory of categories). The continuum can be larger: it may involve all real [222], or even surreal numbers [231], or any totally ordered set of numbers of intermediate size. See [213]. Also see countability of members of a set ([202] and [203]), cardinality as a measure of the size of a set ([206] and [212]), and the continuum of magnitudes ([204] and [216]).

[66] The smallest totally ordered set that forms a continuum [204] (i.e. is dense [208]), is the set of rational numbers [220]. Other, larger sets ([212] and [206]) of totally ordered patterns [213], like the set of real numbers [222], p-adic numbers [230], hyperreal [225], and surreal numbers [231] may also represent difference and ratio scaled values.

1 *If values have different magnitudes, an internally consistent measure will not assign the same number to them.*[67]

However, the same number in different measures may represent different values. This is why the Mars Climate Orbiter crashed after a journey of over 416 million miles through interplanetary space.

> Friday, October 1, 1999
> "LOS ANGELES – A mix-up over metric and English measurements . . . caused the destruction of the $125 million Mars Climate Orbiter . . . last week . . . The spacecraft flew too close to Mars and is believed to have broken apart or burned up in the atmosphere. NASA said the English-vs.-metric mix-up . . . caused the navigation error." (THE ASSOCIATED PRESS (http://www.fas.org/mars/991001-mars01.html))

Someone forgot to convert units of measure! It led to the loss of $125 million, years of research, and one of the most advanced spacecraft that humans could build.[68] It happened because each measure is a different and independent map from intrinsic meanings of magnitudes in abstract domains of meanings to the domain of numbers. This component of knowledge was forgotten at NASA's peril.

Consider how different values may map to the same number in different measures. The temperature at which water turns to ice is 0° Celsius. 0° Fahrenheit on the other hand is much colder than freezing water. Celsius and Fahrenheit represent different measures of temperature, i.e. different maps between an intrinsic and naturally meaningful magnitude and the domain of numbers. Both temperatures map to the number zero, but do so in different (units of) measures.

2 *Conversely, given a measure of a domain, values with the same intrinsic magnitude will map to the same number.*[69]

Thus 0° Celsius will always mean the same temperature. Only the freezing point of water, and no other temperature, will map to the number zero, given that we are measuring temperature in °Celsius (obviously!). Of course, if we switch measures, the same temperature could map to different numbers. For example, the boiling point of water is 100° Celsius or 212° Fahrenheit. The Celsius measure maps the (magnitude of) temperature of boiling water to the number 100, whereas the Fahrenheit measure maps the same temperature to the number 212.

[67] The domain of values should map to the domain of numbers *injectively*. The second law is even more restrictive. It requires *bijective* mapping. See the endnote on the theory of categories.

[68] The climate orbiter was on a mission to study Mars weather and look for signs of water to determine if life could exist on Mars currently or in the past. Lockheed Martin Astronautics in Colorado submitted acceleration data in English units of pounds of force instead of the metric unit called newtons. NASA entered the numbers into a computer that assumed metric measurements. The numbers were used to find the force of thruster firings to adjust the orbiter's trajectory. "This is going to be the cautionary tale that is going to be embedded into introductions to the metric system in elementary school and high school and college physics till the end of time," said John Pike, director of space policy (*source*: Associated Press report, September 1999).

[69] The morphism between the domain of values and the domain of numbers must be *bijective*. See morphism and bijection in the endnote on the theory of categories.

3 *The relative ordering of magnitudes must be consistent across measures,*[70] i.e., sequencing of values in order of magnitude must be the same in all measures. For example, we know that the freezing point of water is a lower temperature than that of boiling water. If we map these (magnitudes of) temperature to numbers, the number for the freezing point of water must always be lower than that of boiling water in every measure. Thus, in Celsius 0 is less than 100, and in Fahrenheit, 32 is less than 212.

4 *Each measure must have a unit of magnitude for gaps between magnitudes that maps to the number "1."*

In (both ratio and) difference scaled domains, magnitudes of gaps between measures are also meaningful, and we must be in a position to compare these gaps consistently (within a given measure). Therefore, each measure must have a unit of magnitude for *gaps* between magnitudes that maps to the number "1." Thus 1° Celsius is a different magnitude from 1° Fahrenheit, but both are units of measure of *differences* of temperature.

5 *When two values are equal, their difference must map to the number zero* (naturally!).

Just as we needed a unit of measurement to measure differences between magnitudes, we need the number zero when there is no difference between two values, i.e. when values coincide.[71] Thus, whenever two values are equal, their *difference* must map to the number zero.

6 *No value can be said to be of an infinitesimally small magnitude.*

Values that naturally map to zero signal the *absence* of a property. For difference scaled values, there is no value that *naturally* maps to the number zero. Magnitudes of *gaps* between values may map to zero naturally to show that two or more values are coincident, but difference scaled *values* have no natural zero. Thus, this sixth condition might read: "It is not mandatory that a single value must map to the number zero across all measures, nor is it mandatory that measures of difference scaled domains must have a zero, and, if they do, the number zero has been arbitrarily imputed to a value in the domain (of meaning)."

For example, the length domain has a natural zero but not the domain of dates. We can conceive of two physical objects that touch each other, and hence the distance between them is zero, i.e. there is no distance between them. However, as we understood in the parable of Metanesia in section 3 of Chapter 1, some domains have no natural zero and the temperature domain was one such domain. The domain of dates is another. We can measure *differences* between dates (and times) in days, hours, minutes, or seconds, and can certainly say which dates come after (are greater than) which, but it is meaningless to talk about ratios between dates. We can certainly say that the gap between a pair of dates is twice that between another pair, but cannot meaningfully say that one date

[70] The theory of ordinal value functions has been lucidly described in [211]. Measures of a value (and also conversion rules between measures) must be *order isomorphic*. See order isomorphism in the endnote on the theory of categories. Also see [211], [216], and [217].

[71] Ordinal values too can be of equal magnitude (rank), but, if they are not, we have no information on how big the gap between them is. Thus, two ordinal values can map to the same *rank*, and it is implicit that the difference between them (their magnitudes) is zero. Ordinal values of the same rank are permitted in psuedometric state spaces (see the endnote on generalized distances). However, we do not need "1," the unit of measure, because measuring magnitudes of finite gaps between ranks is meaningless.

is twice another date. Midnight of January 1, 0 AD has been arbitrarily set to zero by convention. Other conventions in other parts of the world support other calendars with different imputed zeros for dates. For example, the Hindu, Muslim, and Jewish calendars, each imputes a different zero date to measure the timeline, and fix a number for a given year.

Ratio scaled values carry information on relative magnitudes in as well as the kind of information conveyed by difference scaled values. Ratios between ratio scaled values meaningfully compare how many *times* one value is more (or less) than another, a quality that difference scaled domains lack. Therefore, when we map magnitudes of ratio scaled values to numbers, the first five conditions for consistent measures would remain the same, but the sixth condition would read:

6 *A single value must map to the number zero across all measures because it represents the absence of a property*, not the absence of meaning of the property! It only says that the absolute value of the *magnitude* of the property is infinitesimally small.

The number zero means the same thing in all meaningful measures of ratio scaled domains. This is its natural zero. In the example above, when two objects touch each other, their separation will be zero in every possible units of measure – feet, inches, meters etc. – and even units of measure that are not invented yet.

Formats of units of measure

Formats of units of measure map magnitudes of values from domains of meaning to the domain of numbers. They are a bridge between the business rules layer and the implementation layer in figure 16. Business process design mandates that units of measure to be specified. Units of measure express and measure the *meanings* of quantities in *numbers*. However, we cannot represent these numbers in any recording system, manual, or automated, unless we assign symbols to them in one or more of the five formatting domains. Therefore, although each unit of measure *expresses* the meaning of a quantity (a single quantity may be expressed in many different units of measure), each unit of measure in turn must be expressed in a format (and possibly in many different formats) to make it perceptible to the real world of process automation, actors, and observers.

Formats are the symbols that express a value, and hence are physically sensed when we actually *perceive* a tangible expression. We hear "*you have mail*" when new mail exists, see a number colored red, or see it expressed in roman numerals.

In Chapter 3, section 3, we understood that the format of a nominal value would convey its entire information content, and be its full tangible expression ("Tangible expression": see Chapter 3, section 3).

Ordinal values have a rank. Their formats will also be their tangible expressions, and convey their full information content, provided discrete numbers (or another *totally ordered*[72] set of symbols) represent these ordinal values.

Unlike ordinal and nominal values, the format and value alone cannot convey all the information latent in difference and ratio scaled values. Both format and unit of measure

[72] See total order in [213], [214], [215] and the endnote on ordered sets and sequences.

are required to *tangibly* express a quantitative value. The unit of measure is not evident from a symbol that merely represents the value. To communicate full information about quantitative values, the (unit of) measure must also be identified and expressed tangibly.

Both the measure and the number need perceptible symbols (formats) that will physically represent them. The height of a building might be written down in a red decimal number (its format), but must be followed by a symbol like "feet" or "ft" to show that the number is measured in feet. Just like any other meaning, each unit of measure has potentially infinite numbers of formats that can give it perceptible form.

Of course, only a few of these formats for units of measure will be widely accepted conventional standards. For example, the US dollar is a unit of measure for the money domain. Conventional formats for it are *USD* and $. An inch is a unit of measure for length. Conventional formats for the unit of measure are *inch*, *in* and *″*.

Box 39 Full formats of values

The *full format* of a value is the set of formats that can express the *complete* meaning of the value. An ordered pair of symbols consisting of the format of value and the format of the unit of measure is the full format of a quantitative value. The pair is an aggregate object. It is also a pattern and a symbol on its own. The pairing sequence is usually determined by convention. For example, the written symbol for one dollar is "$1," whereas it is spoken as "One Dollar." When it is written down, the symbol for the unit of measure comes first, but when it is spoken aloud, the symbol (spoken word) for the value is said first.

Thus, quantitative values need *two* symbols to fully express their meaning, whereas qualitative values need only one symbol.[73]

The metamodel of units of measure

The metamodel of unit of measure resembles the metamodel of format (figure 60). The rule expression, however, is restricted to those expressions that conform to the six golden rules of measurement and numbers must be expressed in formats described earlier to obtain tangible expressions of quantitative values. The metamodel illustrates how (unit of) *measure* too must be represented by a symbol in the tangible expression of a *value* to convey the full meaning of a quantitative value.

The full format, as we understood under *formats of units of measure*, is a sequence. The sequence in which we arrange the constituents of a full format is dictated by convention, not logic. Under "Incomplete rules", in Chapter 3, section 2, we understood how structures like figure 64 let us express not only our knowledge, but also our ignorance with great precision by assigning "don't know" values to the right components. If the sequence of its constituent formats is unknown in a full format, we can still express all the information in the value that the format is expressing. The only item of information we lose will be convention –

[73] Figure 8 describes this basic difference between quantitative and qualitative values.

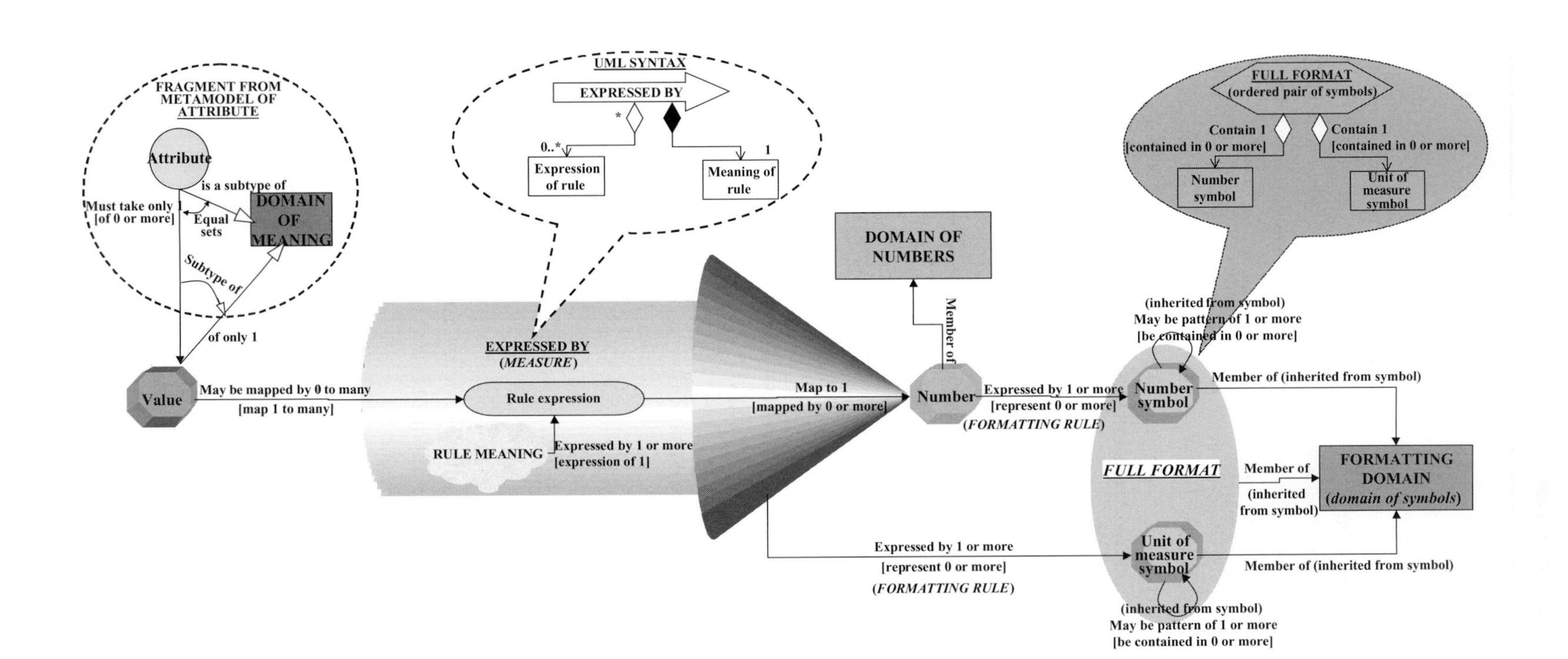

Figure 64 Metamodel of unit of measure

knowledge of how these symbols are *conventionally* sequenced, normalized by *full format*, the aggregate object in figure 64.

Of course, like any other format, the formats of *measure* and *number* are also patterns, and are bound by the architecture of patterns as well as the rules and constraints in box 38.

Consider how numbers are different and distinct from symbols in a format. Take an example of a complex rule. Assume that requirements dictate that all numbers in a certain range must be displayed in red, blinking digits. Outside this range they must be black digits. Moreover, the rule applies only to a *number*, not a *value*. Values may map to different numbers depending on the unit of measure, but the display changes only when a *number* falls within the range, regardless of what unit of measure has expressed it. Furthermore, the requirement states that it does not matter what kind of written script expresses the number. It could be in Roman numerals, Arabic, or Kanji (Chinese). Only the *number* matters to the format. The constraining relationship then would involve a set of numbers, not symbols or values, and the object set of figure 62 would become a *number set* instead. A number is a kind of object. The generalized object set can support rules like these – and even rules that are more complex, as we shall see in the following example.

Consider a new requirement. The requirement is that whenever distances are expressed in kilometers or meters instead of miles or feet, the symbol for the unit of measure (kilometers or meters) in the full format must be bold. The *object set* in figure 62 would then contain a set of measures for *length*, and the formatting rule would map specific measures to bold symbols. In this example, the measures are kilometers and meters. If their formats were the symbols "km" and "m" respectively, both "**km**" and "**m**" would be in **bold print like this**. In general, the expression of a rule could depend on any object or even interactions between objects of different kinds in different states. This is why *object set* influences *rule expression* in figure 62.

Measures like those in figure 64, formatting rules like those in figure 60, and Rule constraints like those in figure (c) of box 33 are all subtypes of the general *represent* relationship in box 36. The general relationship manifests itself as one of the three components of knowledge depending on the object that represents a meaning (the object on the upper right-hand side of figure 62 – a measure when the target is a number, a format when the target is a symbol and an encrypted meaning when the target is a value. Similarly, when both source and target objects are symbols, the generalized *represent* relationship is a format conversion rule. When the source is a symbol, but the target is not, it might be the key to encrypted information. The generalized *represent* relationship is polymorphic, and its parameters are:

1 The target object;
2 The contents of the object set involved with the rule expression – values, formats, numbers, and other objects; and
3 The object being represented.

Conversion between units of measure

Just as formats can be converted to other formats, units of measure of a domain can be converted to other units of measure *for the same domain*. Maps that convert between units of measure are similar to maps that convert one format to another. However, formats merely map meanings to symbols. Symbols can always be substituted for other symbols and there are few restrictions on mapping rules that convert formats to other formats. However, units of measure must satisfy the six golden rules we just discussed. Results of all units of measure conversions must preserve these rules, otherwise the result of the mapping is not a unit of measure. Therefore rules for converting between units of measure are restricted. They are special rules:[74]

1 *Units of measure can only be converted to other units of measure for the same domain.*
Units of measure map meanings to numbers. Domains are containers of meaning. Therefore each unit of measure applies to values in a single domain. When units of measure of a domain map to other units of measure, the resultant units of measure also must be units of measure for the *same* domain. For example, feet will always be a unit of measure of the length domain and no other; feet can never be converted to kilograms or dollars, but *can* be converted to centimeters.

2 *Units of measure for* **ratio scaled domains** *can be converted to another unit of measure for the same domain by multiplying every number in the unit of measure by a fixed, non-zero conversion factor.*
This kind of map between units of measure will preserve the six golden rules for ratio scaled domains across all units of measure. The conversion factor must be the same for every value. Otherwise it would distort the relative sizes of gaps between values. The rule also ensures that the same value maps to the number 0 in all units of measure. For example, to express length in inches, multiply length in feet by the fixed conversion factor 12.

Difference scaled values have no natural zero. As such, different units of measure need not all map to the same zero. Therefore, the second rule of conversion for units of measure of *difference scaled values* is less restrictive:

3 *Units of measure for difference scaled domains can be converted to another unit of measure for the same domain by multiplying every number in the unit of measure by a fixed, non-zero conversion factor. Even if we add (or subtract) a fixed number from the result, it will stay a unit of measure.*
This kind of map between units of measure will preserve the six golden rules for difference scaled domains across all units of measure. The multiplier must be the same for every value. Otherwise it would distort the relative sizes of gaps between values. Adding a fixed number to the result will only shift the zero of the new unit of measure. For example, to express temperature in Fahrenheit, multiply temperature in Celsius by the fixed conversion factor 1.8 and add 32 to the result.

Indeed, even if we merely shifted the zero and did not multiply values, we would still get a different unit of measure; one that is identical to the old unit, but with a different zero value (this is equivalent to the conversion factor being "1").

[74] Morphisms for converting between measures of the same value are called an *isometry* or isometric isomorphisms. See *isometry* and *isometric isomorphism* in [261]. See *morphism, isomorphism, surjection*, and *injection* in the endnote on the theory of categories.

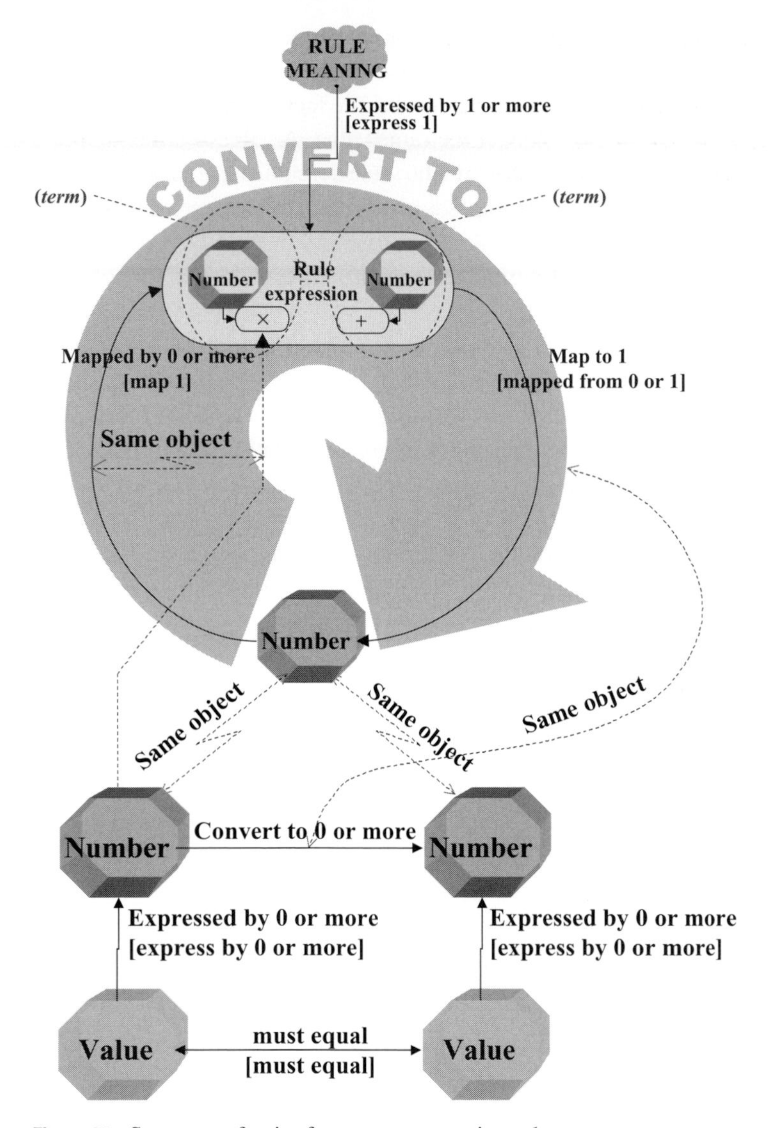

Figure 65 Structure of unit of measure conversion rules

The metamodel of measure conversion rules

Figure 65 is the metamodel of unit of measure conversion rules. It is a recursive relationship that resembles the format conversion rule of figure 59.

However the rule expression in figure 65 is more restrictive than the rule expression of figure 59; it must conform to rule 2 above and has two terms:

(A) The term on the left, inside the rule expression of figure 65, multiplies the number being converted by another number (the conversion factor)

(B) The term on the right merely adds a number (to account for zeros being different in different measures).

The lower half of figure 65 represents rule 1 for conversion between units of measure; the values are identical, but the numbers need not be. The equality relationship between the two value icons in figure 65 represents the irreducible fact that the number being converted and the result of the conversion express the same value. A value is a member of a domain; hence, conversion measures not only map to the same domain, but an instance of a conversion measure converts a number that expresses a *given* value to another number that expresses the *same* value. (Most examples of relationships so far in this book have been between object classes. The equality relationship between the value icons in figure 65 is an example of a relationship between object instances.)

Figure 65 also shows term (A), a joint constraint and quantitative relationship similar to figure 44. The relationship from *number* to *rule expression* in the upper half of figure 65 is actually the relationship between the number being converted (in the lower half of figure 65) and term (A) (inside the recursive loop). The relationship illustrated by the broken line from the number being converted to term (A) inside the rule expression makes this point.

The rule for converting temperature from Celsius to Fahrenheit is an instance of the unit of measure conversion rule in figure 65. The number in term (A) will be 1.8, and that in term (B) will be 32 in this case.

Box 40 Measure conversion (on our website)

Box 40 discusses how conversion rules between measures may be represented in a square matrix of the kind in box 5. It describes the general mathematical form of measure conversion rules and discusses how the metamodel of knowledge can derive new conversion rules from older rules. The box also discusses, with real life examples, how some conversion rules can be complex, and may even change over time, provided they are "order isomorphic", i.e. they conform to the first rule of measure conversion, and the golden rules of measurement (see the endnote on the mathematical theory of categories).

Another feature of a unit of measure conversion rule is that its inverse is automatically implied and fully determined by the conversion rule (inverse: see box 33). The inverse is not new information. For example, the rule for converting temperature from Fahrenheit to Celsius is determined by the rule for converting Celsius to Fahrenheit or vice versa.[75]

Of course, when ratio scaled values are involved, zeros in all measures will map to the same value and the rule expression will not have the second term. Alternatively, one could

[75] The rule for converting Celsius to Fahrenheit is $F = 1.8 \times C + 32$, where F is the temperature in Fahrenheit and C, the temperature in Celsius. The rule for converting Fahrenheit to Celsius can be derived from this. It is $C = (1/1.8) \times F - 32/1.8$, i.e. $C = (5/9) \times (F - 32)$. Thus the conversion rule from Fahrenheit to Celsius carries no *new* information. The two different rule expressions for converting from Fahrenheit to Celsius provide another example of the same meaning being expressed with different rule expressions.

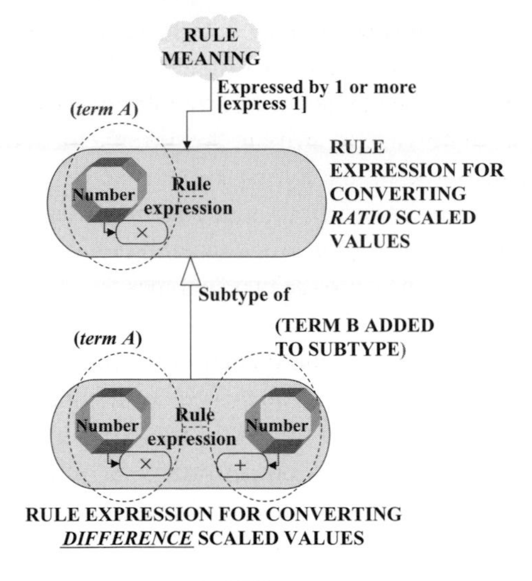

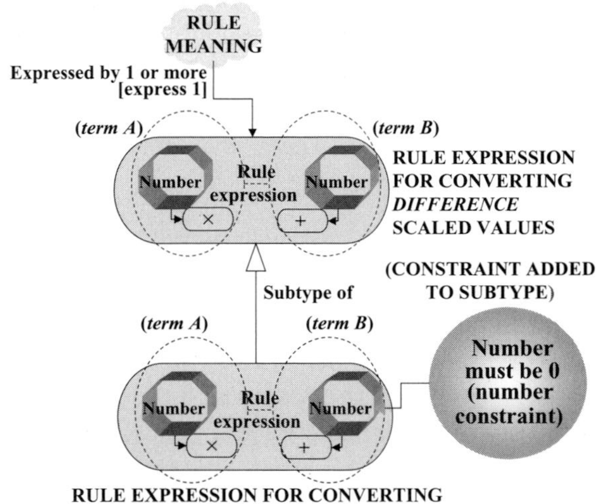

Figure 66 Conflicting subtypes?

say that the number in term (B) will be constrained to always equal zero for ratio scaled values.

Subtyping measure conversion rules

Consider the conversion rule from centimeters to meters, in box 5 again. Length is a ratio scaled domain. Centimeters and meters are two different measures of length. In the rule for converting centimeters to meters, the number in term (A) of figure 65 is 0.01 and term (B) is missing. On the other hand, we would be equally correct if we had said that the number in term (B) was 0.

Thus there are two ways we can subtype the rule expression in figure 65, by adding a term or adding a constraint. Both lead to the same *meaning* and map to the same number.

In figure 66(a), the rule expression for converting difference scaled units of measure is a subtype of the rule expression for converting ratio scaled units of measure. This is because we have added an extra component, a *term* (which is an item of information), to the formula for converting ratio scaled units of measure to arrive at the formula for converting difference scaled units of measure. In figure 66(b), the reverse is true. We have added an extra component, a *constraint* (also an item of information), to the formula for converting difference scaled units of measure to arrive at the formula for converting ratio scaled units of measure. However, they both cannot be true because the two subtypes are in conflict; in figure 66(a) the rule for conversion of difference scaled values is a subtype of the rule for converting ratio scaled values, whereas it is the reverse in figure 66(b).

This exemplifies how the object paradigm is sometimes an inadequate tool for modeling real-world meanings and behaviors. It is also an example of how blind subtyping in limited contexts can sometimes lead to results that conflict with interpretations in other contexts. To arrive at the right answer, we must interpret the information content and its underlying meaning of the rule – the pattern of information it represents.

To understand this answer, consider that the meaning of the conversion rule is the same. Only the rule *expressions*, not their meanings are different between figures 65(a) and (b): The rule expressions for converting ratio scaled measures in figures 65(a) and (b) satisfy both rules 1 and 2 for converting between measures of ratio scaled values (articulated in the section on conversion between units of measure). Thus, they convey the same meaning. Likewise, the rule expressions for converting difference scaled measures in figures 65(a) and (b) satisfy both conditions 1 and 2 for converting between measures of difference scaled values. They too convey the same meaning. The meaning of each object has not changed between the two perspectives in figures 65. Each is consistent and correct within its own perspective.

The reason for the ambiguity is that we lost information when we represented difference scaled values with numbers. This happened because we violated the first rule of simple representation (in box 38), which reads, "Each attribute of the object being represented will map to exactly one attribute of the object that represents it." Figure 66(b) is the correct interpretation, and the following gives the reasons why.

The confusion between subtypes in the conversion formula happened because two different meanings were attributed to the number zero. Section 3 will show that the nil value in difference scaled domains is "unknown." Conversely, the domain of numbers has a known nil value, the number zero, but no "unknown" value. To express values in difference scaled domains in numbers, we assigned the number zero to an arbitrarily chosen value in the difference scaled domain. This arbitrary value may be different for different units of measure. The additive term in figure 66(a) corrects for this difference between units of measure when converting between them. It is a mere computational procedure that does not add to the meaning of the value.

On the other hand, when we map the nil value in a ratio scaled domain (like length) to the number zero, we have added the information that the zero is meaningful, and represents the same value in all units of measure. It is a true constraint that adds information to the pattern.

Box 21 and Chapter 2, section 3 showed that it is simpler to build configurations of knowledge by adding information. Subtypes should carry more, not less, information than their supertypes. We understood that ratio scaled values convey more information than difference scaled values. The perspective in figure 66(b) satisfies this criterion. This is why it is the correct subtyping interpretation.

This interpretation is difficult to arrive at unambiguously, based on the object paradigm alone. This is why we must sunder meaning from its expression in the metamodel of knowledge. If we do not, we cannot normalize the information conveyed by the *meaning* (also see box 41, box 43 and the following section on conflicting subtypes). We do this in order to make our systems and processes more agile and adaptable. In Chapter 4, section 3 we will show how normalized meanings will help automated systems become more agile and more adaptable to new learning even as it occurs.

Conflicting subtypes – when objects are not enough

See our website for a more detailed technical discussion on conflicting subtypes and transmutation of meaning through polymorphisms in state space. This discussion is included in Box 41.

Box 41 Conflicting subtypes, state spaces, perspectives, and polymorphisms of metaobjects (on our website)

Box 41 elaborates on the right way of subtyping information in measure conversion rules based on their information content. It discusses the state space of rules and elaborates on the following figure to describe the right method of defining subtyping hierarchies for rules. It also shows how some of this intelligence and inference automatically flow from the metamodel of knowledge

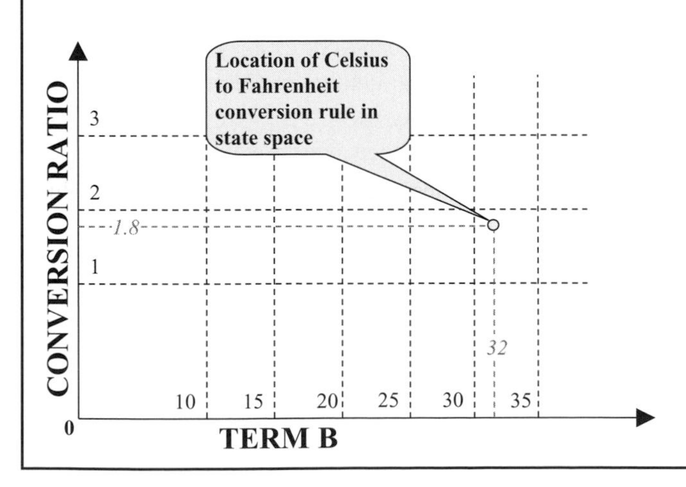

3 Domains of meaning and the metamodel of object

> Thus incorporeal spirits to smallest forms
> Reduce their shapes immense, and were at large,
> Through without number still, amidst the hall
> Of that Infernal Court
> (John Milton, *Paradise Lost*)

Meaning governs reality. Meaning conveys information. Meaning is also abstract and intangible. In this book, we have repeatedly wrestled with meanings and their multifaceted tangled representations in the real world. We have engaged, assembled, configured, and transfigured them into new meanings. Domains are the fountainheads of meaning. So what is a domain of meaning? This section will show us that meanings flow from the measurability of information, and domains are patterns of measurable information.

Domains are classes[76] of values. The class, an aggregate object, is a pattern[77] that lends meaning to its members. The members are values that convey information only about magnitude. The class says what it is a magnitude of. The *class* is the lowest common denominator of business meaning, and the *value* the lowest common denominator of the meaning of magnitudes, or intensities of the meaning conveyed by the *class*. *Thus, a domain is a class of immutable values based on a common meaning.*

The concept of "lowest common denominator of meaning" is best illustrated with an example. Take the concept of price. It has a meaning. The actual price at a point in time is the magnitude, or *value*, of price. Until we link it to an object (via an attribute) we do not know what it is a price *of*, nor what *kind* of price it is (list price, actual sale price, or quoted price). Price is a domain that contains potentially infinite numbers of values. It is a class of values – values with a meaning, but a meaning that is only completed by its context – the object and the attribute it is linked to (see Chapter 3, section 1). The basis for the class, its emergent property, is the common meaning.[78]

A class may count an infinite number of values among its members.[79] Some classes, like the class of difference or ratio scaled values such as price, may even count a continuum of values among its members, so that there is an infinite number of values between any two members of the class, however close they may be.[80]

We have seen how values may map to numbers and formats. The value of price can be mapped to a number. Each such map is a unit of measure. For example, the unit of measure

[76] Classes subsume sets. All sets are classes but not vice versa. See [171], [172] and [173] for differences between classes and sets.

[77] A domain is a pattern of values that is changeless and immutable. A domain does not change state. It only lends its meaning to attributes of objects (Chapter 3, section 1).

[78] ". . . a class is a collection of sets that can be unambiguously defined by a property that all its members share" – [172].

[79] See the size of a class in [172], [173], [202], [203], [206], and [212]. Non-mathematicians beware!

[80] The ordered set of values is said to be mathematically dense when there are infinite numbers of values between two distinct values in a set, regardless of how close the two values are to each other. The density of partial order in [208] describes the density of values in a domain.

of *price*, the domain, must describe the currency (US dollars, British pounds, Euros, etc.) as well as the number of units it is the price of (price per piece, per dozen, per gallon, or per square foot). Thus information content, the *meaning*, of this domain is actually derived from two other domains – the *money* domain and the *quantity* domain. This derivation involved a relationship, a division of every value in the money domain by every value in the quantity domain. This division relationship is a joint constraint between the price, money, and quantity domains (like the joint constraints in figure 43 and figure 44). As such, domains can even emerge from relationships between domains.

Box 33 describes these relationships. In box 33, we emphasized the value of an attribute. Now our emphasis is more generic. When the relationship is not specific to an attribute of a given object, but is between values in domains (like the relationship between price, money and quantity that we just discussed), it is the source of a new domain, a meaning that emerges by engaging the meanings of domains that it relates. A new meaning, a new irreducible fact, is therefore born of old meanings – also irreducible facts. All attributes that map to the new domain will inherit the relationship that engaged these facts (including constraints, joint or otherwise). It is this relationship that gave birth to the new domain. This is why domains are the wellspring of polymorphic behavior.

The polymorphic behavior of domains

Domains are the repository of common meaning – values, relationships, and behavior inherited by objects that instantiate them in different contexts. Relationships between domains will be polymorphic rules, because, based on their individual contexts, the attributes and objects that inherit them will "know" their behavior. For example, the price of yarn might be expressed in dollars per unit length, and the metamodel will "know" that the amount charged must equal price of the yarn multiplied by length of yarn sold, because the metamodel "knows" that the price domain emerged from a division relationship between the money domain and an amount domain (and multiplication is the inverse of division). It can even tell us that, when the length is measured in feet, we must multiply the price per foot by length sold in feet to arrive at the amount of money we must charge. Similarly, if land is priced in dollars per acre, the price domain "knows" that the amount charged must equal price multiplied by the area of land sold in acres. The price domain normalizes these rules derived from the relationship between the domains it emerged from.[81] If the Mars Climate Orbiter had "known" rules like these, the ship might not have crashed on the alien red deserts of Mars!

Domains may even be subtypes of other domains. For example, price, money, and quantity domains are all subtypes of the generic ratio scaled domain in figure 67. Subtypes inherit behavior and relationships (Chapter 2, section 3). This makes relationships between parent domains polymorphic. Indeed, just as inclusion polymorphism emerged from relationships between parent object classes (see box 21), so too might inclusion polymorphism emerge

[81] This normalization and inheritance of rules, or *morphisms*, that emerge from relationships between domains, is another reason why domains are classes, not mere sets. See [172], [173], [186], and the endnote on the theory of categories.

from relationships between parent domains. This polymorphic behavior will be inherited by all attributes that map to the domains involved.

In Chapter 3, section 2, we discussed relationships between attributes. We have just discussed how these relationships may be inherited from domains, and hence be polymorphisms based on relationships between domains. Domains may also normalize effects.[82] Effects change relationships between domains and object instances that map to them. These effects belong to the domain if they are common to attributes of all objects that draw on that domain. They then become another manifestation of polymorphic behavior: Take the color of a car. *Car* is an object class. The actual color ("value" of color) of an individual car is drawn from the *Color domain*. Domain is an object class too – it is a collection – the collection of all possible colors. Members of this class are instances of *color*. A specific (instance of) car has a relationship with a specific (instance of) color in the domain of colors. It is this (instance of) the relationship between the attribute and the domain that is manifested as the "*value*" of an attribute called car color (see figure 35 and figure 37).

We have also seen how other attributes of the same, or other objects, may map to the same domain. For example, chameleons have color too. Both cars and chameleons can change color. Cars change color when they are painted, and chameleons change color when the color of their ambiance changes. *Change color* is an *effect* that switches this relationship between (an instance of) an object and the color domain, from one color to another. The *change color effect* will be shared by all objects with attributes that map to the domain of colors. The *change color* effect belongs to the color domain, is normalized by it, and is inherited by all objects via attributes that map to the domain (specific objects may add special constraints, for example cars can be metallic gold, but not chameleons). This is the heart of polymorphism. Domains normalize polymorphic behavior.

The events that trigger the color change might differ for different objects. For cars, it might be a *painting* event and for chameleons, a change in ambiance. We could generalize all color changing events into a supertype called "trigger color change." The domain of color will normalize "trigger color change." *Trigger color change* will be different for different object classes. Indeed, some may have several different trigger color change events. Each is included in, i.e. is a subtype of trigger color change. The domain normalizes the generic effect and an object class normalizes each subtype. This is how domains and objects normalize polymorphism (see box 21).

Domains normalize generic effects. The effects normalized by a domain are inherited by all attributes that map to that domain. We have seen this in the example of the color change effect and the color domain, but what if an object's color is frozen – what if it cannot change color? If it cannot change color, it means there is no color change event associated with the object in the scope of the model. Although the colored object has no color change event, its potential to change color resides unrealized, but normalized in the color domain. Should a change in scope or an innovatively reengineered process add recoloring events to the object, the effect will be borrowed from the domain, and its potential instantiated by

[82] Effect – see Chapter 2, section 2.

the object (effect: Chapter 2, section 2). The object will thus naturally realize its recoloring potential.

All the model has to do to recognize new recoloring behavior is to link the new recoloring event to the object and make it a subtype of the generic recoloring event resident in the color domain. Inclusion polymorphism will take care of the rest. The new event will inherit the link between the generic color change event and the generic, "prefabricated" recoloring effect inherited from the color domain. It will change the color of the object when it occurs. Given this new learning, processes and systems assembled from these knowledge artifacts will mirror reality seamlessly and automatically.

Systems and business processes assembled from knowledge artifacts that recognize domains need no radical surgery to adapt to new effects. "Truths" in the real world are seldom absolute and often volatile. Recognizing domains can facilitate both resource and schedule compression for projects driven by the triple business imperatives of change, innovation, and survival.

Consider the volatility of real-world "truths." The state of knowledge, its configuration, is continually transfigured as we learn and innovate. For instance, take knowledge of gender. Gender is a nominally scaled domain. The gender of an individual, many believe, is frozen – a simple truth, but a universal truth it is not. So many species change gender that scientists have had to coin a new word for species like us that do not change – we are "gonochronistic" species (see the endnote on the question of gender).

Change is the only truth in this millennium of accelerated learning, unceasing new knowledge and increasingly volatile "truths." The question of gender is an example of how the metamodel of knowledge facilitates change by recognizing domains. In terms of the structure of information, species, the class of objects, acquires an attribute, gender, via a relationship between the object class and the gender domain (see Chapter 3, section 1, figure 34, box 27, and figure 35). A relationship between any object instance and a domain value can always switch. It is a rule about effects, intrinsic in the metamodel of knowledge, which we discussed in Chapter 2, and will discuss again here.

Until there was a need to recognize it, the potential for gender change, the gender change effect, lay hidden, but normalized in the gender domain. As the scope of our knowledge increased to cover greater and greater numbers of species, some species (classes of objects) instantiated the gender change effect and realized its potential. Then along came another change!

Many of us might have believed another "obvious" truth that turned out to be a lie – that an individual must have only one gender. We now know individuals in some species may have multiple genders and genders are not mutually exclusive in the same individuals of such hermaphroditic species (see the endnote on the question of gender).

The metamodel of knowledge lets knowledge artifacts flex easily to absorb these new "truths." We know two or more object classes may share several mutual relationships simultaneously. It is a part of the metamodel of knowledge.[83] We also know that attributes emerge from relationships between "normal" objects and domains, and that there could be

[83] [173] and [186] extend the concept and describe its implications.

more than one relationship between an object and a domain (Chapter 3, section 1). Thus, multiple attributes of an object class may map to the same domain. The new knowledge merely created a new configuration of knowledge – an additional relationship between the gender domain and some species, with a constraint that the two relationships cannot map to the same gender for the same individual (it would be strange indeed if two genders of an individual were both male or both female! – but, even if that happened, domains would easily flex to accommodate it). Our systems easily absorbed the new knowledge by reconfiguring components of knowledge.

Consider what domains imply for the metamodels of state and attribute in figure 35. Domains carries within them, not only values, but also their meaning and the potential for relationships with other objects. With the potential for lending its values and meaning to object instances through these relationships (see figure 35), the domain also carries the potential to switch relationships with its values, as well as the potential to recognize new attributes through new relationships with the domain. Objects can also gain new properties by recognizing new relationships with *values* in the domain.[84] Objects inherit these components from domains (via the subtyping relationship between *attribute* and *domain* in figure 35). The gender domain contains, within it, the potential for transformation of a hermaphrodite into a single sexed individual, or even of expanding the repertoire of genders beyond two should the need arise (and sure enough, it does!),[85] or of restricting a species to a single gender (via the value constraints of Chapter 3, section 2), as indeed it does (see the endnote on the whip tail lizard under the question of gender). It needs no extra labor to build in this potential for requirements unstated. The mere existence of the domain is enough. It carries within it the potential for requirements still unknown and presently unrecognized. We need only supply the parameters and conditions that will realize the hidden potential of *domain*.

Domains are containers of abstract common meanings, and it is through objects that meaning flows from abstract domains into the metamodel of knowledge, shaping it, seeping through relationships, creating patterns, and into the world of tangible things. It is thus that *meaning* orchestrates the tangible world of business.

Domains of information

So how do these meanings come to be? What is their root? Their root, like all the meanings we have discussed thus far, is information. Meanings carry information, and it is their information content that gives rise to subtypes, and it is the mutual engagement of their information content that creates new meanings – new repositories for atomic rules – repositories that grow in size, structure, and complexity, in step with the information they convey. To know the root of domains is to know the root of information. Let us start with domains that convey very little information. The principle of subtyping based on addition of information (see the discussion on figure 52) will add meaning to domains, a step at a time.

[84] Domains are mathematical categories. See [173].

[85] The need to expand the repertoire of genders does arise. There are species that have five or more genders! See the endnote on the question of gender.

The information in "value"

Arguably, the least information is carried by the mere fact that a domain exists. We may have no more information. Its meanings and values are all unknown. All it tells us is that it has three effects:

1. An effect that establishes a relationship between a domain and an object class. The relationship is also an object class. This (subtyping) relationship creates an attribute of the object (see figure 35).
2. An effect that establishes a relationship with an object instance and a single value in the domain. This effect assigns a value from the domain to an attribute of an object. It is an instance of the relationship class above.
3. An effect that switches the relationship in (2) to a different value. This effect is really a supertype of effect 2. Effect 2 is a special case in which the value switched was "null."

All three effects may have guard conditions (there are actually two effects because effect 2 is subsumed in effect 3). Module V, section 3 on our website elaborates on effects and guard conditions.

It is this unknown domain that lies at the root of all domains – a strange domain that knows only the potential to be, and has no knowledge of what is, or what may be.

Add a bare minimum of information about values in it – that different values are different and distinct. We still have no information on quantum of difference, nor do we know which values are greater or lesser than which. We only know that they are distinct. We have a nominal domain. We have added only just enough information to distinguish a nominal domain from the unknown generic domain. It is the bare minimum we could add, but it is enough to assert that nominal domains are subtypes of the unknown domain.

Add a little more information, just enough to say which values are more, less, or equal to which others. We still do not know by how much, but it is enough to distinguish an ordinal domain from a nominal domain. It is also enough to assert that ordinal domains are subtypes of nominal domains.

Now add the information on not only distinction and order of magnitudes of values, but also on the quantum of difference between them. It will not only distinguish difference scaled domains from ordinal domains, but will also assert that difference scaled domains are subtypes of ordinal domains based on the principle of adding information. Difference scaled domains convey all the information ordinal domains do, plus some.[86]

Consider the nominal domain again. In a sense, the mere existence of a value is a kind of magnitude, different, and distinct from non-existence. However, the absence of magnitude is not the same as the absence of the meaning of magnitude. It is the difference between nil and null values we discussed in Chapter 2, section 2. Add the nil value to the nominal domain. It is a new item of information, different from the sequencing information that had turned it into an ordinal domain. The domain now has a value that conveys the absence of magnitude. It is information. For example, the absence of illumination conveys darkness. The property of darkness may be conveyed by a magnitude, "nil," that shows that illumination is absent. This does not mean that the *meaning* of illumination is absent; only that illumination is

[86] [211] elaborates on the mathematics of ordinal domains and how they can grow into difference scaled domains with examples. See "Ordinal value theory" and "Value difference functions" in [211].

absent. The meaning is conveyed by the domain, and magnitude, by value. Even if meaning exists, the magnitude may be nil.

The fact that a domain can contain nil magnitudes is information, albeit different and independent of the kind of information we added to nominal domains to create ordinal or difference scaled domains. Thus, it is a subtype of nominal domains, but a subtype in a different partition.

For nominal domains, nil values do not really matter. Nominal domains merely distinguish between values, and this item of information only says that the nil value is known to be different and distinct from other values – no sweat – the nil value is just another value (see the footnote).[87] However, it has a more profound impact on other kinds of domains, as we will see next.

Jane's car color preference in Chapter 3, section 2 (under "Constraints on ordinal attributes") was an example of an ordinal domain with a nil value. Jane liked red and green cars, disliked black cars, but was neutral about white cars. Accordingly, car color preference was an ordinal domain with a natural origin (origin of a coordinate system – see box 37). The origin is anchored by the nil value, *the absence of preference*, or neutrality.

This kind of information on nil values, or a natural origin of ordinal domains is different from imposing a lower bound (by attaching a value constraint to the domain). A lower bound too is an item of information, but a different item of information from the nil value. For example, serial numbers have a natural lower bound, the first item. The serial number of the first item is different from the nil value,[88] which signals absence of magnitude. The nil value may not even impose a lower bound. Jane's car color preference, for example, had a nil value but was not constrained by it. She could like or dislike car colors (disliking a car color was akin to a negative liking, i.e. a "liking" that was less than nil), or be neutral about a car color, neither disliking nor liking it. This *neutrality* was the information the nil value added.

A nil magnitude may not be a lower bound for values in a domain, but it does signal the absence of magnitude. Conversely, a lower bound, when it exists, may not be the nil value. The nil value and lower bound are independent items of information that can be associated with domains, making them richer and more varied in the truth they contain and the values they express.

This also makes the question of "natural origin" of ordinal domains more complicated in a coordinate system such as that in box 37. If the domain carries both kinds of information, a natural lower bound as well as a nil value, we will have two different bases for choosing the natural origin of ordinal attributes that map to the domain. To unravel this knot, remember

[87] The nominal domain with the least information is the binary domain. It has only two values – existence and non-existence of magnitude, or merely the fact that two values are different (like the male and female genders). Only distinctions count in nominal domains, not degrees of magnitude. Ordinal binary domains are different. For example, the proximity of two values in a nominal domain is either nil or greater than nil (see table 2 and the discussion on patterns). A binary ordinal domain carries more information than a binary nominal domain. It carries two values *plus* information on their order. Binary domains lie at the heart of present day computing hardware and software. The paucity of information in binary domains is one reason for the many limitations at the heart of today's automated information systems.

[88] Ordinal domains with a lower bound are "well ordered," whereas ordinal values are "totally ordered." See the endnote on ordered sets and sequences, [212], [213], and [215].

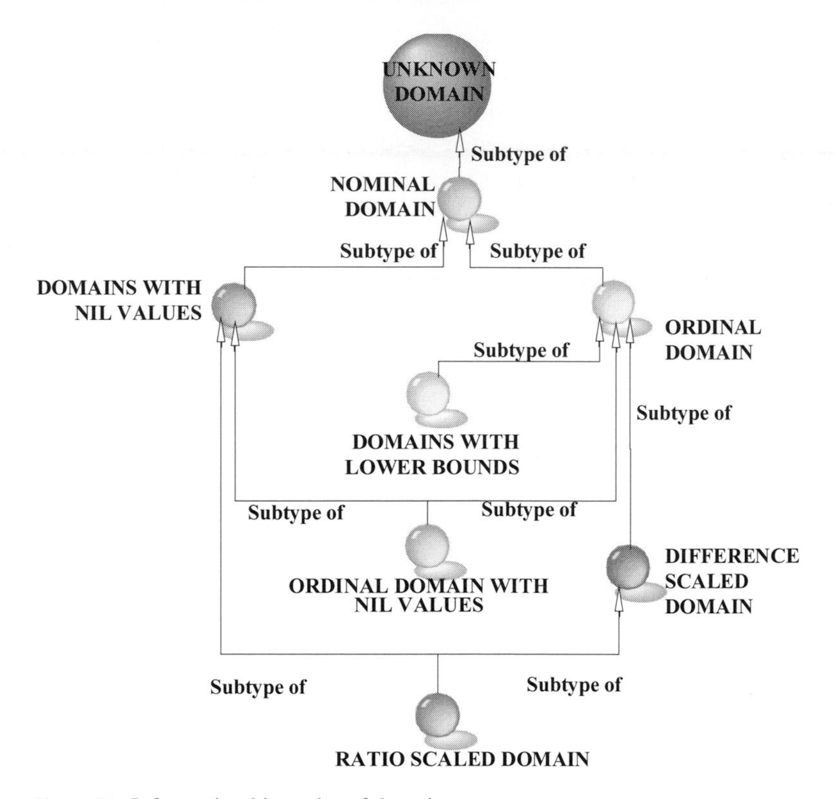

Figure 67 Information hierarchy of domains

that each kind of information has a different meaning, and hence each kind of origin will have a different meaning. It is difficult to represent two origins geometrically, but the two meanings exist in the real world, and care little about how we map them to geometry.

On quantitative domains, the impact of the nil value is even more profound. It changes a difference scaled domain to a ratio scaled domain.[89] A nil value in a difference scaled domain makes ratios between values meaningful. Thus, it changes the fundamental nature of the domain itself.

Figure 67 articulates this discussion on the hierarchy of domains.

Operations on values

The information content of a domain limits the quantum of meaning its values convey. It also limits *meaningful* operations on its values. The meaning of the domain constrains meaningful interactions between its values, but each domain is also a part of an information hierarchy that adds meaning a step at a time. Thus, each domain in the hierarchy of figure 67 inherits operations from its supertype(s), and adds its own, in step with the information it conveys (and also in step with the hierarchy of rule expressions in figure 48. That hierarchy was

[89] [199] describes the abstract algebra behind ratio scaled domains.

based on excluding information in steps, whereas in figure 67 we have included information in steps. That is why the hierarchy of figure 48 has been inverted in figure 67).

- ***Nominal domains:*** Only comparison of identity in terms of equality (or not) is valid in nominal domains. Effects normalized by the domain may assign or switch relationships between values and object instances. Other effects may establish relationships between the domain and the object class to create new properties, i.e. attributes and effects. Comparison of magnitudes and any arithmetic operations like addition, subtraction, multiplication, and division are meaningless.
- ***Ordinal domains:*** Comparison of values (ranks), both in terms of sequence and equality, are meaningful in ordinal domains.[90] Ordinal domains also inherit all operations and effects from the generic nominal domain.

  The concept of equality was inherited from the nominal domain, but comparison of sequence was information added. Arithmetic differences are meaningful in terms of gaps between ranks – the number of intermediate positions (ranks) between a pair of positions (ranks), but division and multiplication are meaningless (see ordinally scaled state space in section 1 under "Patterns").

  Arithmetic addition of a value in the domain itself is meaningless, but addition of *gaps* between values (ranks) is meaningful. The domain of gaps between ranks is a related, but different, domain derived from an ordinal domain (see table 1 and "Creating new domains from old" later in this section).[91]
- ***Difference scaled domains:*** Comparison of values in terms of equality, sequence, and closeness in a continuum[92] are all meaningful (inherited from ordinal domains). The generic difference scaled domain inherits all effects normalized by the generic ordinal domain, and adds information of its own. Arithmetic subtraction is meaningful (the meaning of subtraction was added to create difference scaled domains from ordinal domains), but not division or multiplication.[93] The gap between values is a related (ratio scaled) domain derived from a difference scaled domain (see table 1 and the discussion on difference scaled state space in section 1 under "Patterns"). Arithmetic addition of *gaps* is meaningful, just as it was for ordinal domains (inherited).
- ***Ratio scaled domains:*** All comparison and arithmetic operations are meaningful. Comparison and arithmetic subtraction were inherited from difference scaled domains. The operations on relationships are also inherited. The other arithmetic operations are added information, specific to ratio scaled domains.[94]

All arithmetic operations are defined in terms of algorithms on numbers. Therefore, implicit in the discussion on validity of operations on values in a domain, might seem to be the assumption that these values have been mapped to numbers before the operation is actually executed. However, it is the *concept*, the *meaning* of the operation that we are focusing on here, not the algorithm or *expression* that implements the meaning. For example, there are several algorithms that implement subtraction, division, and other arithmetic operations, but

[90] [211] contains a mathematical but lucid description of ordinal domains and their properties.

[91] These intuitive truths are backed by mathematical theory in [212], [213], and [215].

[92] Difference scaled domains are mathematically dense [208] and totally ordered [213].

[93] See value difference functions in [211].

[94] Quantitative domains may even contain surreal values [231] The smallest dense [208] totally ordered class is the class of rational numbers [220] and the largest is the class of surreal numbers. See [213] and [231].

all of them yield the same result and express the same meaning – subtraction is a measure of difference in terms of separation, addition is the inverse of subtraction; division is a relative measure of size in terms of ratios, and multiplication is the inverse of division. It was these *meanings* we added, one at a time – each meaning an item of information – to the domains in figure 67. We have not mapped these meanings to numbers or symbols yet. Numbers and symbols are different "snap-on" components (see sections 1 and 2).

The focus in this section has been *value*. We have described rules about values that domains normalize. We have also described how values express the intensity of the meaning conveyed by the domain. However we have not described what these values are intensities of. To understand this, our focus will now shift the meaning and the information *domain* normalizes as a *class*.

The prophet replied, "Surely the Creator cannot be described except by that which He has described Himself. And how should one describe that Creator whom the senses cannot perceive, imaginations cannot attain, thoughts cannot delimit and sight cannot encompass?" (Qur'an 112:1–2)

The information in domain

Thus far, the domains in this section have been bare. We have added information about how values behave, but little meaning to the domain itself. Domains normalize not only the bald mathematical behavior of values, but also imbue those values with meaning. These meanings flow from the physical world as well as the world of business. In the remainder of this section, we will add information a step at a time to the bald domains of figure 67, and meanings will flow from the rich tapestry of the business world, rather than the dry logic of mathematics.

At the most fundamental level, it is the physical world that lends meaning to six fundamental or primary domains (see box 42); these domains are the building blocks of other physical domains:

1 enumeration (ratio scaled)
2 mass (ratio scaled)
3 physical separation (ratio scaled)
4 date (difference scaled – includes date *and time* of occurrence)
5 electric charge (ratio scaled)
6 overall information content (ratio scaled)

Consider the meaning that enumeration lends to the bare ratio scaled domain. The availability of an item of inventory, say a car, might be counted in pieces. The Car is the object class and the number available, its attribute. The attribute maps to the enumeration domain, and its units of measure might be a single piece, dozens (of pieces), hundreds (of pieces) etc. Take the total number of cars sold. That is another attribute of car that maps to the same domain, and may be expressed with the same units of measure, and the same rules of conversion between units of measure. Take a different object class, say person. In Chapter 2, section 1 we saw how the class of all persons, an aggregate object, is also an instance of an object. The number of persons is an attribute of that aggregate object and it too maps to the enumeration domain. Hence, it may be expressed in the same units as the number of cars sold. The enumeration domain "knows" that it is a count. It "knows" this in addition to the bare rules and operations valid for ratio scaled domains and also in addition to the mathematical fact that it has a lower bound, nil.[95] The domain normalizes and lends additional meaning to the bald ratio scaled domain of figure 67. It adds information, and is therefore a subtype of that ratio scaled domain. In the same way, other domains too add meanings to the domains in figure 67, and hence are subtypes of domains to which they lend meaning.

Like the enumeration domain, the mass domain normalizes the common meaning of measures of mass. Likewise, the physical separation, or length domain, normalizes the common meaning of the measure of length in physical space.

Consider the only difference scaled attribute among the six basic physical domains – the date domain. We must distinguish it from the "time-gap" domain, also called "time domain," in brief. The date domain maps the time of occurrence of an event, whereas the temporal separation between two instantaneous events maps to the time domain. The time domain is derived from the date domain by arithmetic subtraction of pairs of values in the date domain. The date domain has no natural nil value.[96]

For example, it would be meaningless for a 40 year old father to claim his birthday is four times later than that of his ten year old daughter, but it would be meaningful to say that he is four times as old as his daughter (see the tale of the monster of Metanesia in Chapter 1, section 3) The time domain is a distinct ratio scaled domain, different from the difference scaled date domain (albeit related to it – we will examine how new domains emerge from other domains later in this section).

The overall information content domain, called information domain in brief, normalizes the meaning of the quantum of information (see the endnote on Shannon's information theory). It does not distinguish between individual items of information, but focuses instead on measuring the overall quantum of information in a pattern, the sum total contributed by all the structures and meanings that make the pattern. The information domain does not care about the individual meaning of an item, but it normalizes the common meaning of the measure of information. It also does not normalize the meaning of information quality. However, other subtypes of the information domain do so. We discuss them now.

Enumeration is a very special subtype of information. Enumeration is information about the population of constituents in an aggregate object. Enumeration applies only to the aggregate pattern, not to its constituent members (see "Object occurrence value" in figure 54 and figure 57). Object class, an aggregate, is at the core of the metamodel of knowledge. Enumeration is such a universal and frequently used form of information that we have listed it separately.

The information quality has four more fundamental subtypes. Each adds a unique meaning. Each is information about information. Each is frequently used, even if not as often

[95] We have deliberately not restricted the enumeration domain to positive integers. We would add this constraint to domains that enumerate indivisible objects like cars and people, but, if we were counting divisible objects like apples or oranges, we could count fractions of pieces. These fractions are all rational numbers [220]. Thus, the enumeration domain is constrained to the set of rational numbers. Other domains may be "larger," however, the enumeration domain is still a dense set [208], albeit the smallest set of values that can be dense [213] (roughly speaking, being dense is akin to forming a continuum of values).

[96] In cosmological terms, the date domain does have a natural origin, and hence may be ratio scaled. However, the origin of time is relevant only in the grand sweep of cosmic history (see the endnote on the natural zero of time and the references therein). Business can safely ignore this.

as enumeration. Each normalizes information quality, be it information about instances of objects, domains of meaning, or individual values in a domain:

- The completeness or *exhaustivity* domain:
  The completeness domain measures the degree of completion of information – for example, the extent to which a checklist might be complete. It might even tell us if a value has all the information it can about its meaning, sequencing, enumeration, validity, accuracy, and reliability.
- The validity domain:
  The validity domain measures meaningfulness in a context. Is the information really what you think it means – is the checklist for maintaining a Toyota Corolla valid for the Infiniti as well? Is 150% effort a meaningful value? Can the proportion of effort exceed 100%?
- The accuracy (or precision) domain:
  The accuracy domain describes the precision of information. What is the leeway for error when we estimate that a corporation's stock will earn 11 cents per share? If it returns 10.9 or 11.1 cents a share, was our estimate accurate? Was 10 cents inaccurate?
- The reliability (or *risk*) domain:
  The reliability domain measures consistency of information. How consistently do our forecasts have to be accurate to be considered reliable? If weather forecasts are correct only half the time, is the forecast reliable? What if forecasts are correct 90% of the time?

Sometimes, the four information domains listed above might appear chimerical. Consider again the checklist we discussed earlier. The enumeration domain counts the number of items in the list, and the completion domain measures if all items that must be included are, in fact, included. Let us examine four different states of knowledge about the list (not an exhaustive list – they are only illustrative of the chimerical nature of the completeness domain):

1. We may or may not know how many items the list has, but we may know that the list is either complete or incomplete.
2. We may or may not know how many items the list has, but we may know that the list is more complete than it was before, but not by how much.
3. We may or may not know how many items the list has, but we may know that the list is only 50% done.
4. We may or may not know the number of items in the list, and neither may we know if the list is complete or not.

In the first case, the value of completeness was "complete" or "incomplete," an ordinal measure with only two values (it is an ordinal, not nominal measure because we know that the degree of completeness is greater in a complete list than in an incomplete list. Therefore "complete" will rank above "incomplete" in terms of magnitude of completeness.) In the second instance, it was ordinal, with the potential for more than merely two ranks; if we had three lists, we could rank each in terms of how complete we think it is compared to the other two; if we had four lists, we could rank all four, and so on. In the third instance, it became one half, a ratio scaled measure. In the last instance, the value of completeness was "unknown."

We know that all domains count the "unknown" value among their members (Chapter 2, section 2), but it seems that the completeness domain is a chimera, sometimes binary, sometimes ordinal, and sometimes quantitatively scaled. This can happen to validity, accuracy, and reliability as well. It all depends on information content.

Information rich domains are subtypes of information poor parents. The converse is also true. Starved of information, the plush, ratio scaled information quality domains may become domains of sparse information that are not ratio scaled. We will elaborate on domains of information – enumeration, completeness, validity, accuracy, and reliability – later in this section.

In theory, every other physical domain may be derived from arithmetic operations, relationships, and subtypes of the primary physical domains we have just discussed (see box 42 for more information). Engineers and physical scientists have long used this principle to validate formulae and predict the behavior of physical systems based on incomplete empirical information.[97] In this book, we will not assemble every domain from fundamental domains. That approach is sometimes useful for engineering physical systems, but for business systems, it could easily lead to analysis paralysis (section 6.2 of Introduction). Instead, we will treat each domain as a domain that exists on its own right, and will only focus on its relationship with other domains when necessary.

Box 42 Domain analysis and primary physical domains (on our website)

Box 42 derives fundamental domains of business meanings from the domains in figure 67, and distinguishes these from secondary domains obtained by adding information to fundamental domains (also called *primary* or *base* domains).

It shows why *length*, *time lapse*, *mass*, and *information* are fundamental physical domains from which all physical meanings are derived. It adds *enumeration*, *completeness, accuracy, validity*, and *reliability* as subtypes of *information*. The box goes on to discuss the information content and derivation of secondary domains, coherent systems of measurement, shifting perspectives of primary domains, and how these issues are addressed by Buckingham's Pi theorem, Shannon's information theory, and dimensional analysis.

Fundamental domains also emerge from the subjective world of business. They too may combine with other domains – business, physical, or mixed, to create new domains. Two fundamental business domains reused frequently are:

7 Preference (ordinally scaled).

8 Economic value, also called utility or money (ratio scaled). This is the value a person or organization assigns to a resource or asset. It manifests itself in attributes when revenues, costs and funds of different kinds map to it.

Although preferences are usually discrete ranks and economic value is a ratio scaled continuum, economic value and preference are related. Economic value, being a ratio scaled domain, has more information than preference, an ordinal domain, and may legitimately

[97] Engineers call the fundamental physical domains fundamental dimensions; the study of domains that emerge from relationships between them is called dimensional analysis. Engineers use dimensional analysis to validate algebraic expressions of physical laws and to deal with situations in which the precise law is not known, but the variables the law ties together are. Readers interested in dimensional analysis may refer to box 42 and the references therein.

be considered a subtype of the preference domain (based on the principle of subtyping by adding information). The subtyping hierarchy between ordinal and ratio scaled domains in figure 67 covers this, and is inherited by this pair of domains. However it is more specific than the general hierarchy of figure 67. Information (intermediate preferences) was added to the *preference* domain, not to just any ratio scaled domain, until it became dense[98] enough to be considered a continuum. Information on a nil value (indifference) was also added to the preference domain. Only then did the preference domain turn into the ratio scaled economic value domain.[99] This was an example of inclusion polymorphism (see box 21) between domains and this polymorphism will be inherited by all attributes that map to these domains.

Economic value is also an example of how all ratio scaled domains may not be equally flush with information. The information content of the economic value domain lies somewhere between the sparse information of the ordinal preference domain, and the rich certainty of "hard" information inherent in physical ratio scaled domains. It is harder to be objectively certain about the economic value, or degree of satisfaction in a barter or trade, than it is to be certain about the length of a room or the mass of a spacecraft. This uncertainty is inherent in the economic value domain. It degrades the reliability *potential* of values in the domain. "Potential" is the key word here. We could have used an elastic tape measure to measure the length of the room, and our measurement would have been unreliable, but that unreliability would have come from the process of measurement – not lack of information in the length domain itself. That process would not have realized the full information potential of the length domain. On the other hand, the unreliability of the economic value domain stems from the domain itself. It just does not have the reliability potential of the "harder" physical domains because it has less information than these "hard" physical domains. Thus, all ratio scaled domains are not equally rich in information. These differences stem from differences in business meaning.[100]

Subtypes have more information than their parents.[101] Every primary domain is a subtype of the "bald" domains of figure 67. Primary domains add business meaning to bare the mathematical logic of their parents. In the list of primary domains above, each primary domain has its parent next to it in parenthesis. The preference domain is a subtype of an ordinal domain.

Each primary domain may, in turn, be subtyped. Each subtype adds new meanings to meanings already present in its parent. Sometimes these meanings are mathematical, but more often they are meanings from the world of business – meanings normalized by the domain. For instance, the preference domain may be used with or without the nil value attached. That is a mathematical meaning added to support real-world behavior.

On the other hand, a domain can masquerade as different domains in different roles only because of the way businesses use it. For example, the ordinal domain may reflect an

[98] [208] describes the mathematical concept of *dense* sets.

[99] [211] elaborates on the mathematical relationship between, and the properties of, preference and economic value.

[100] See "softness" of information in box 46, softness of the economic value in box 45.

[101] By the principle of subtyping by adding information a subtype domain inherits the information in its supertype(s) and adds special information to this inherited component to create a new meaning.

organizational hierarchy or a hierarchy of serial numbers. These domains are only subtypes of the same bald ordinal domain of figure 67, with a lower bound added, being used in two different ways. Business meaning was included when we specified the *kind* of hierarchy we mean – serial number, reporting hierarchy, hierarchy of titles etc. Thus, based on the principle of subtyping by adding information, it became a subtype of the bald ordinal domain of figure 67. Its use added business meaning.

Besides preference and economic value, there are several subjective domains that we cannot, strictly speaking, "measure" objectively because they are subjective sensations or concepts. An exhaustive list of primary subjective domains that can be building blocks for making other domains has not yet been developed. It is an area of continued research.[102] Similarly, no exhaustive list of primary domains in the non-physical sciences exists today. The physical, or "hard" sciences deal mainly with information rich quantitative domains.[103] The others need nominal and ordinal domains more often. The primary domains we know do not have this focus.

As we drain primary domains of information, they lose business meaning until they reach the root of knowledge – the great divide between the known and the unknown. At this boundary between the known and unknown, we find the bald domains of figure 67 – pale ghosts of business meaning, leached of all knowledge save the meanings of magnitude, distinction, and difference.

For us, it will suffice to know that primary domains exist, although many are still unknown. But it is even more important to understand what a domain is, to recognize it in a business model, and to use it to normalize information when necessary – even if the domain is bald and its meaning is still barren.

Creating new domains from old

So far, we have understood the information content of primary domains. We now know what meanings they normalize. In this subsection, we will understand how meanings flow from primary to secondary domains to create new meanings. We will discuss how domains engage each other, like the gears of a knowledge machine, to create new configurations of knowledge – new atomic rules, as well as new domains, those abstract containers of meaning that will anchor and normalize these new rules built upon the old.

We have already seen one way new domains come to be. They can inherit information from the old – sometimes from more than one parent – and add meanings of their own. We get the first rule of secondary domains:

[102] Subjective domains are a largely undeveloped area of research. The list of primary subjective domains is unknown. Unlike the list of primary physical domains published in 1954, no broadly agreed upon list of primary non-physical domains exists. There are no standards. See [measurement] *In psychology* under "Measurement theory", *Macropedia*, volume 23, page 795 of [336].

[103] The *emphasis* of the physical sciences is on ratio scaled domains flush with information, but ordinal domains also emerge from purely physical concepts. For example, hardness, a physical property of materials, is measured on the basis of what materials will, or will not scratch which others – a purely ordinal scale, based on a pecking order – the ability to scratch another. This example demonstrates that ordinal and nominal domains can also be useful in the physical sciences.

Rule 1: Adding meaning to a domain creates a new domain. Meanings added may be new or inherited. If the new domain includes meaning(s) inherited from other domains, it is a subtype of the domain(s) it was created from (examples under figure 67).

Domains engage each other via special kinds of relationships and it is these relationships that add the meanings that give birth to new domains. It is relationships between domains that normalize meanings intrinsic to the new domains they create. These meanings are the joint constraints we discussed in Chapter 3, section 2 and box 33. In Chapter 3, our focus was on attributes, now it is on domains. The same constraints, attached to primary domains instead of attributes, create new domains. Secondary domains may also engage each other to create new domains, but in terms of *rule meaning* (box 33), it boils down to engagement of primary domains (box 42).

Ratio scaled domains are the richest in information content, and nominal domains are the poorest. All arithmetic operations are meaningful for ratio scaled domains, and none applies to nominal domains. We will start with ratio scaled domains and work our way down to nominal domains.

Ratio scaled domains

As we discussed earlier in this section, all arithmetic operations are valid in ratio scaled domains, hence all magnitude constraints (Chapter 3, section 2) may meaningfully engage ratio scaled domains. In general, magnitude constraints may engage attributes or domains. Ratio scaled domains come packaged with addition, subtraction, multiplication, and division operators (as well as others built on these basic building blocks). Magnitude constraints attached to this domain and its subtypes may use these operations.

When magnitude constraints engage domains instead of attributes, new subtypes of ratio scaled domain emerge. Their meaning is normalized by the magnitude constraint. For instance, take *area*. All measures of area, be it the area of a room or the footprint of your computer, map to the area domain. Area is a secondary ratio scaled domain that emerges by engaging the primary length domain. The length domain is joined to itself with the multiplication operator to obtain the area domain. The meaning of area resides in the structure – a recursive multiplication relationship on the length domain that mutually multiplies each pair of values in the length domain once.[104] The class of values thus obtained is a new domain, the *area domain*.

Next consider the *volume domain*. Each triplet of values in the length domain is mutually multiplied to yield a new domain of values, the volume domain. You could also think of it as the area domain multiplied by the length domain. This is an example of two domains engaging via an arithmetic operator to create a third. It is also an example of how it all boils down to engagement of primary domains.

Similarly, cost (or price) per unit area is a domain obtained by dividing the economic value domain by the area domain. Prices of land, tinplate, fabric, and many other object classes would map to this domain. Speed is obtained from the length and time domains the same way. It is the distance covered per unit time.

[104] *Fundamental group* in [193] describes looping relationships like these in mathematical terms.

On the other hand, for ratio scaled domains, subtraction and addition of values within a domain map back to a subtype of the same domain. Difference in speed is still speed, albeit relative speed. Difference in length is still length, as is addition of length.

To understand why these domains are subtypes of the domains added or subtracted, remember that domains are objects, and objects are patterns (see figure 31). Therefore, domains are also patterns. They are patterns of values. Patterns are aggregate objects that have several universal attributes in common (see figure 53, figure 54, figure 57, and "Patterns" in section 1). One universal attribute of a pattern is its degrees of freedom. The degrees of freedom of a sum is less than that of its summands – if a sum is fixed (i.e. the sum has no freedom), there is a wide choice of patterns of values that may add up to the sum. However, the converse is not true; we have no choice of sums if we freeze the summands.[105]

It follows that the pattern of values being summed has more degrees of freedom than the pattern of sums. Therefore, the two patterns are similar, but the sum is more restrictive. The sum has more information than the values that were summed. The extra information is the curbs on its freedom – the summation operation. Based on the principle of subtyping by addition of information, the domain of sums is a subtype of the domain of summands. The domain of sums contains exactly the same values, and has exactly the same effects and operations as the domain of summands; therefore, it is a subclass of that domain.

A similar argument holds for subtraction. The rules for creating new ratio scaled domains from old are:

Rule 2: Any multiplication or division operation on values in ratio scaled domains creates a new ratio scaled domain. Multiplication and division operations may be between values in the same or different domains.

Rule 3: Addition and subtraction operations on values in the same ratio scaled domains map back to a subtype of the same domain. Addition and subtraction are permitted between parent and subtype domains.

For example, you cannot add apples and oranges, but you can add apples and fruit, and the result will map back to the class of fruit, not apple. It will actually be a subclass of fruit that has a little more information than the class of fruit had – we now know it is a class of fruit that contains information about apples. If we ignore or lose this information, we can say that the result mapped back to the parent domain.

Let us look at the structure of information that let the metamodel of knowledge apply rule 3 to infer this: the class of fruit mapped to the enumeration domain and acquired the enumeration attribute (see figure 35). Subclasses apple and orange inherited the attribute from fruit. Therefore, the "piece" of the enumeration domain inherited by fruit, its enumeration attribute, is a supertype of the same attribute for apple and orange.[106] By the laws established by rule 3, the enumeration of apples and fruit may be added, and it will map back to fruit. The enumeration of oranges

[105] We have no choice of sums if we freeze the pattern of summands, only a singe sum will fit the pattern.

[106] Paradox: fruits may be counted by adding counts of apples and oranges. If the domain of sums is a subtype of the domain of summands, why is the count of fruit a supertype of the count of apples or oranges rather than the other way around? This apparent contradiction is resolved in box 43.

and fruit may also be added. That too will map back to fruit. However we cannot add the numbers of apples to the numbers of oranges, because neither apple nor orange is a subclass of the other.

The metamodel thus created the "intelligence" to not only count fruit, but also to count fruit even as it counts apples or oranges. It also "knows" that the converse is not true, that counting fruit does not translate into counting its subtypes, apples, oranges, or whatever. The ratio scaled domain normalized this knowledge, and the enumeration domain inherited it. We did not have to "program" these injunctions and instructions multiple times for each requirement for counting fruit of different kinds. The counts were intrinsic and available, framed by rules specified once for the domain. This reasoning applied not only to fruit and its enumeration, but to any additive operations on any ratio scaled attribute of any objects and subclasses. It was knowledge normalized by a domain.

(It was also parametric polymorphism. In programming terms, the parameter, apple, orange, or fruit was "passed" to the "count" function, and counted, but it was also an enhanced form of parametric polymorphism. "Count," the function, "knew" that counting apples or oranges counted fruit, but not vice-versa.)

Thus rule 3 leads to rule 4.

Rule 4: Addition and subtraction operations across values in different ratio scaled domains have no meaning. "Different domains" is used in the sense that neither is even a subtype of the other.

Box 43 The principle of subtyping by adding information

The principle of subtyping by adding information asserts that a subtype object class has more information than its supertype class(es). Subtypes share the information in their common supertype(s) and add information of their own. Creating subtypes by adding attributes (Chapter 2, section 3) was just one instance of this principle. Business meanings, relationships, and constraints also add information.

You might ask why the enumeration of fruit is not a subtype of the enumeration of apples and oranges instead of the other way round. After all the sum of numbers of apples and oranges adds up to the numbers of fruit. Thus, if the domain of sums is a subtype of the domain of summands, why is the count of fruit a supertype of the count of apples or oranges rather than the other way around? The reason is that we are counting fruit. Just as fruit added business meaning, information, to the bald enumeration domain and thus made enumeration of fruit a subtype of the enumeration domain, apple and orange added mutually exclusive business meanings to fruit. This added information made the count of apples, as well as the count of oranges subtypes of their common parent, the count of fruit. Counts of apples and oranges are not bald counts. We know what we are counting. They have emerged from a relationship between an object apple (or orange), and the domain of enumeration (see figure 35). Although counts of apples and oranges add to the count of fruit, they contain more information than the count of fruit. Each is a count of a specific kind of fruit; each is a subtype of the general count of fruit.

Fruit has the freedom to be an apple or an orange, but an apple or an orange must be what it is. The count of fruit has more freedom, i.e., contains less information than the count of either apples or oranges. Therefore, the count of fruit is the common parent of both the count of apples and the count of oranges. When mathematical operations and business meaning conflict, about which object is a subtype and which a supertype based on the principle of subtyping by adding information, business meaning always wins. Follow this simple rule when in doubt and you will not go wrong.

In abstract terms, think of the object as a pattern of information. Parts of the pattern may be shared with other patterns. This is shared information. However, the pattern extends beyond the portion that is identical to other patterns. These extensions add information and give the pattern its unique identity. Thus, the pattern may be conceived as a shared part (the supertype), plus extensions (the added information). The composite of the two are the subtype.

A pattern with fewer degrees of freedom has a greater burden of information than a similar pattern with more freedom. For example, a straight line is a pattern of two points – its ends – and a rule about how they are connected. The line may be of any length. The pattern will not lose its identity. The rules that make the pattern a pattern also give it the freedom to retain its identity.

If we restricted the length of the line, we would add information. The pattern would lose some of its freedom. The restricted pattern will be a subtype of the unrestricted pattern. This is the principle of subtyping by adding information. The pattern with more information is always a subtype of the pattern with common information when information is shared by two or more patterns.

Subtype versus subset: the difference between subsetting and subtyping operations
Subsetting (box 19) can be a different and distinct operation from subtyping. The *principle of subtyping by adding information* makes this clear. Consider the metamodel of rule expression in figure 48. The rule expression in that figure was a collection of all possible rule expressions. We *excluded* all rules with arithmetic operators from the set of rules to arrive at the collection (set) of nominal and ordinal rule expressions. It was a subset of the set of all rule expressions at the top of the hierarchy (the hierarchy has been reproduced in figure A). From this subset, we *excluded* all ordinal rule expressions to arrive at the set of nominal rule expressions at the bottom of the hierarchy. The set of nominal rule expressions was a subset of a subset.

On the other hand, consider the subtyping hierarchy in figure B. It is based on information content. It is similar to the hierarchy of domains in figure 67. It inverts the subtyping hierarchy of figure A (and figure 48) because a nominal (Boolean) rule expression only conveys classification information, an ordinal rule expression contains information on relative ranks – which result is larger than which, but not by how much, whereas a quantitative rule expression with arithmetic operations conveys information on relative and absolute magnitudes. Naturally, if you can rank a result, you can also classify it on that basis, but not vice versa. Similarly, if you know by *how much* one result exceeds another, you can rank and classify it, but not necessarily the other way around. A nominal rule expression conveys less information than an ordinal rule expression, which in turn conveys less information than a quantitative rule expression with arithmetic operations.

(Note that occurrence relationships between objects, the "normal" relationships we have discussed thus far, are also instances of nominal rule expressions.)

The nominal rule expression normalizes classification information, to which the ordinal rule expression adds ranking information (which it normalizes, even as it inherits classification information from its nominal parent). A quantitative rule expression normalizes and adds information on quantified magnitudes, not just their relative ranks. It inherits ranking and classification information from its ordinal parent. Based on the principle of adding information, the class of quantitative rule expressions is a subtype of the class of ordinal rule expressions, which in turn is a subtype of the class of the class of nominal rule expressions.

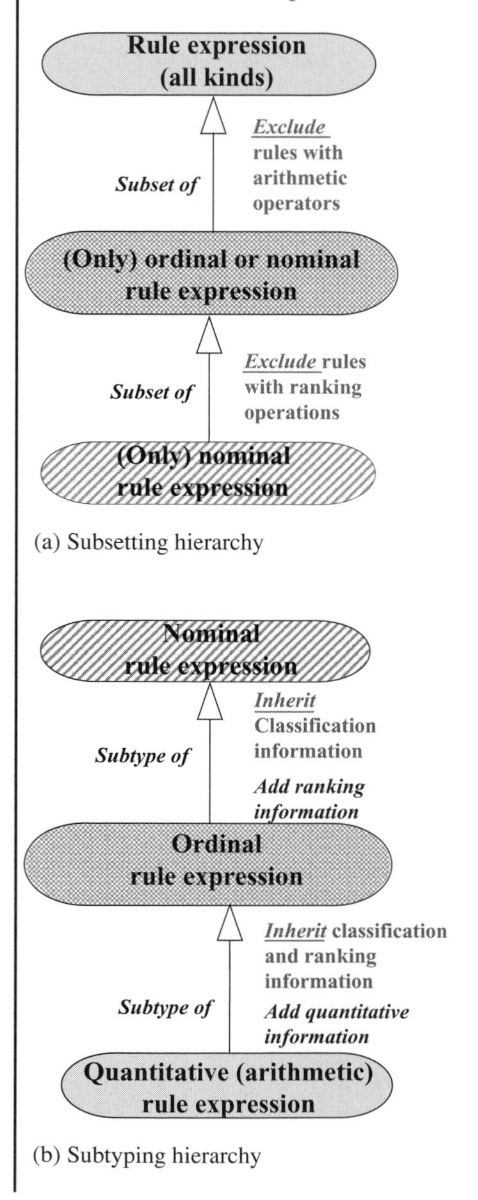

The hierarchy of rule expressions was an example of a subsetting hierarchy that is different from the corresponding subtyping hierarchy. A subset may be carved out of a set based on any criterion. Subsetting and subtyping hierarchies will coincide only when subsets are carved out of sets based on specialized versus common information content.

To understand how these rules translate into practice, consider the speed domain. Distance covered per unit time is speed, and dividing the length domain by the time domain creates the speed domain (rule 2). Subtracting one speed from another is still speed (rule 3). It gives us the relative speed of one object with respect to another. If you are in a car doing 55 mph on a highway, and the car in front of you is doing 70, it will pull away from you at a speed of $70 - 55 = 15$ mph. Change in speed continues to be speed. Adding two speeds is also change of speed and hence no new domain is created (rule 3 again). If you threw a ball towards the other car at 5 mph from your car traveling at 55 mph, the speed of the ball, as seen by an observer standing still on the highway will be 60 mph (because your speed [55 mph] + the speed of the ball relative to you [5 mph] = 60 mph).

In contrast, division creates a new domain. The acceleration you feel is your change in speed per unit time and acceleration is a new domain (rule 2 again). This is equivalent to dividing the length domain by the time domain twice – the first time to obtain the speed domain and the second time to obtain the acceleration domain.[107] Similarly the "jerk" you feel is the rate at which acceleration changes with respect to time. The "jerk domain" is obtained from the length and time domains, by dividing the length domain three times by the time domain.

On the other hand, if you tried to add length and time domains, it would be a meaningless arithmetic operation. If you added speed to jerk, it would be equally meaningless. Rule 4, in everyday language, is nothing but the proverbial injunction against mixing apples and oranges.

Metrics and the domain of quotients

Quotients are ratio scaled. The domain of quotients consists of the quotient of every possible pair, or tuple (see box 19), of values in a ratio scaled domain. Obviously, the quotient depends on which value is the divisor and which the dividend, just as the order of values in a tuple can distinguish one tuple from another.

Naturally, dividends, divisors, and quotients are all ratio scaled. The domain of quotients is a subtype of the general ratio scaled domain of figure 67.[108] It is a subtype based on the existence of a division operator that connects it to the general domain of ratio scaled values.

Moreover, the enumeration domain is a subtype of ratio scaled domains with a lower bound of nil. The domain of quotients of the enumeration domain (the domain of *enumeration quotients*) is therefore a subtype of the domain of quotients that has a natural lower bound of nil. It also has a null value for the reasons in box 47 under figure (b).

[107] The theory of mathematical groups is the basis for generating secondary domains from primary domains. See group theory in the bibliography, especially the fundamental group in [193].

[108] The domain of quotients is a subtype of the general ratio scaled domain just as the domain of sums was a subtype of the general ratio scaled domain.

The division operation, a magnitude constraint, joins the meanings of the divisor and dividend to create a new meaning and a new domain. Often these domains are the basis for performance metrics because they are domains of ratio scaled rates – rates that can be interpreted as ratio scaled domains of growth and influence, as we will see next.

Domains of growth

Quotients are rates, and are frequently used in performance measurement. For example, the ratio of last year's sales to this year, the ratio of prices last year to those this year, and growth rates in terms of percentages, all map to the domain of quotients.

We have seen how the domain of gaps is ratio scaled for ratio and difference scaled domains (see table 2 and the discussion later in this section). Quotients of domains of gaps with other ratio scaled domains measure rates of change.

For example:

- *Change* in temperature, divided by the time it took to change temperature is the rate at which an object is heating up or cooling down.
- *Change* in position, divided by the time it took to change position is speed, the rate of change of position with respect to time.
- *Change* in revenues, divided by the total number of sales people is the marginal rate at which revenues change with respect to the size of the sales force.
- *Change* in annual revenues over last year, divided by last year's revenues is the rate of change of annual revenues with respect to itself.
- *Change* in annual revenues over the last five years, divided by the time gap of five years is the annual rate of change of revenues with respect to time.
- *Change* in number of service calls divided by number of customers is the rate growth rate of service calls with respect to the customer base, whereas the same change in number of service calls divided by the length of the time period they were received in, is the temporal growth rate of service calls.

Performance metrics frequently map to the domain of quotients, and growth rates are a special subtype of the domain of quotients in which the domain of gaps is divided by another domain to obtain a domain of growth. Some growth rates in the examples above are temporal; others are not, yet they all are rates of change. The examples above show us that growth rates need not always be temporal rates of growth. The change does not always mean temporal change, and the meanings of growth rates are as diverse as the diverse domains of ratio scaled properties. However, a rate of change is always a quotient – a quotient that maps to a domain described by the arithmetic division of a ratio scaled domain of gaps by another ratio scaled domain.[109]

Domains of proportions

Proportions are a special kind of quotient. The domain of proportions is a special subtype of the domain of quotients (and therefore also a subtype of ratio scaled domains). The domain

[109] Gaps in difference scaled domains are ratio scaled. A valid divisor must always be ratio scaled. Therefore, growth rates in difference scaled domains are also ratio scaled domains.

of proportions has upper and lower bounds, as well as special rules attached to it. These bounds and rules are inherited by all attributes that represent proportions.

For example, the proportion of female employees in an organization cannot exceed 100%, moreover the sum of proportions of male and female employees must equal 100%; the proportion of sales to each industry group can never exceed 100% of total sales, moreover the sum of proportions over all industries must equal 100%; the area of a room can never exceed 100% of the area of the building that contains it and so on. These examples were specific instances of the general rules that surround proportions. Let us examine the general rule – the structure of information in the metamodel of knowledge – from which these rules spring forth.

Consider first the relationship between an object class and the object instances that constitute it. Each object instance has attributes, relationships, and effects. The object class is an aggregate object. It too normalizes attributes, relationships, and effects. The information the class normalizes is different from the information its instances normalize, but they are related. To begin with, every object class has an enumeration attribute. It is the population of instances at any given moment. Every object class also has a sum for every ratio scaled attribute of an object instance that belongs to it.

Glass pane may be an object class, and every instance of glass pane will have mass, volume, and area. Derived from these ratio scaled attributes will be the total mass, the total volume, and the total area of the population of glass panes. These totals are ratio scaled attributes of the *aggregate object*, the *class* of glass panes. The sum of every attribute is a class level attribute, derived from its instances. There will be one sum for each ratio scaled attribute of the object instance, and each sum will map to the domain of sums.

Moreover, if the object is partitioned, each subtype in the partition, the aggregate object, will inherit exactly the same attributes as its parent, but *values* of these sums may be different, because the instances of the parent object class have been divided up and allocated separately to each subtype, and their populations are different. However, if we summed the population attribute of each subtype (the aggregate object), the result would equal the population of the parent. This rule would apply equally to all attributes of the subtype that are sums of attributes of its instances. Indeed, the rule will not only apply to each of the attributes the subtype inherits from its parent, but also to attributes it adds to its parent.

There are four fundamental reasons why the domain of proportions is different from other domains of quotients. Two reasons stem from relationships between instances and the classes they belong to, and two others from relationships between aggregate objects – the object classes that were partitioned, and their subtypes in those partitions. Their root cause is the relationship between the information normalized by an object instance and that normalized by the object class:

(a) *Population*: Every aggregate object (the object class in this case) has a population attribute. The population enumerates its members (common sense!).
(b) *Sum*: Given a ratio scaled attribute of an object instance, corresponding object class(es) will have an attribute that sums up the instance level attribute over all instances of the class. This sum is an attribute of the aggregate object, the class (also common sense!).
(c) *Relationship between the parent object class and its subtypes*:

(i) If populations of individual subtypes in a partition are summed, the sum will equal the population of the parent object.

(ii) This will be equally true for each derived attribute in (b) also, if the instance level attributes are all nil or greater (or nil or less).

There are several primary domains like length, enumeration, area, information content, and mass that have a lower bound of nil. The rule can apply to attributes that map to these domains and several kinds of secondary domains built from them. The rule can be useful in validating information for domains like these.

If these values can be either negative or positive, some object instances may have negative valued attributes, and others may have positive valued attributes, so that the net effect after summing them up might be that the class level attribute of the parent is actually less than that of its subtypes. In fact, if the negative values balance positive values exactly, their class level sum might even be nil.

However, if we take the sum of absolute values, i.e. if we ignore the sign of negative values and treat them as if they were greater than nil, then, summed across subtypes in an exhaustive partition, the result must equal the value of the same attribute of the parent (common sense once again – remember that these subtypes are mutually exclusive). This rule can then be used to check for valid information.

Indeed, since the partition represents the collection of subtypes in it, this relationship between the parent and its subtypes is a relationship between the partition and the object class partitioned. The partition normalizes this rule, common to each subtype in it. Each subtype in the partition inherits the relationship from its partition.

The following examples illustrate rule (c):

- The sum of the population of male persons and the population of female persons is identical to the population of *person*, the object class that was partitioned into male and female persons – an example of rule (c) as it applies to population.
- The sum of weights of all males is an attribute of the object class male person, and the sum of weights of all females is an attribute of the class of female person. The weight of all male persons (a class level attribute of the object class male person) plus the weight of all female persons (a class level attribute of the object class female person) equals the weight of all persons (the same class level attribute of their parent object class, person) – another example of rule (c).

In both examples, the *gender partition* partitioned *person*, the object class, into its subtypes *male person* and *female person*. Class level attributes of these subtypes were summed over the gender partition.

(d) *Proportion*: A proportion is the quotient obtained by dividing a class level attribute of a subtype, such as population, total weight, or total volume (obtained from the summation of absolute values of corresponding instance level attributes), by the corresponding attribute of the parent object.

Like the relationship in (c), this rule applies equally to all subtypes in a partition. It is a shared rule normalized by the partition, and inherited by each subtype in the partition. Examples are:

- The proportion of the males in the population equals the population of male persons (an attribute of subtype *male person)* divided by the population of *person*, the object class that was partitioned by the *gender partition* into *male person* and *female person*.

- The sum of weights of all steel balls (an attribute of object class *steel ball*) divided by the sum of weights of object class *ball*, the parent object class, equals the proportion contributed by steel balls to the weight of balls.

The partition normalizes both rules (c) and (d). Each is a relationship between an object class and its partition(s), from which each subtype in each partition inherits each rule. If we talk of the population of males, we automatically imply a population of females; if we talk of the total volume of green glass panes, we automatically imply the total volume of red and purple glass panes, and so on. As we will soon see, the implications are profound for automation.

A proportion can, at most, be 100% of the sum. Values in the domain of proportions lie between *nil* and *total*, both inclusive. This constraint is shared by every proportion of every attribute of every subtype in every partition of every object. It is inherited from the domain of proportions. Moreover, given an attribute, the sum of proportions across all subtypes in an exhaustive partition must equal 100%. If the partition is not exhaustive, it may be less. These rules too spring from relationships between the *object class* and *partition*, and are inherited by subtypes in a partition.

The number, 100%, is only a unit of measure. We have mapped *total*, the highest value in the domain of proportions to 100, and *nil*, the lowest value to the number zero. We could have also mapped *total* to the number one, in which case numbers between zero and one would express all proportions. Mapping the range of values between nil and total to the range of numbers between zero and one, or zero and 100 are both accepted convention. Each is a different unit of measure. Indeed, even if it is unconventional, we could validly map the range of values between nil and total to any range of numbers. Each such range will represent a different unit of measure for the same domain of proportions (however, the nil should always map to zero – see the footnote).[110]

(The format of proportions may be pie charts, tables of numbers, graphs, or any other symbol. It does not matter which as long as it is a symbol that represents a proportion in a unit of measure.)

These considerations lead to additional rules:

(e) *Range of values*: Values in the domain of proportions lie between *nil* and *total*.

(f) *Sum of proportions*: The sum of proportions in an exhaustive partition must equal *total*. In a non-exhaustive partition, it may be less than total, but cannot exceed it.

(g) *Injunction against arithmetic addition of proportions*: A proportion is a ratio of a class level attribute of a subtype, to the same attribute of its parent. The subtypes all belong

[110] Mapping the nil value to zero simplifies arithmetic because zero mirrors the natural properties of nil in ratio scaled domains. Adding nil to another value results in the same value, just as adding zero to any number results in the same number; similarly, multiplying any ratio scaled value by nil results in nil, just as multiplying any number by zero results in zero; dividing nil by another ratio scaled value results in nil, just as dividing zero by another number results in zero. Dividing nil by nil yields null. It maps to no value, ratio scaled or otherwise in any domain (unless we include the null value in the unknown domain of figure 67, which every domain inherits – see Chapter 2, section 2). This is why it is good practice to map nil to zero in every unit of measure (see Chapter 4, section 2).

to the same partition, and are mutually exclusive. It follows that proportions for a given attribute of a subtype in a partition may be meaningfully added. Across partitions, addition of proportions has no meaning.

We can meaningfully add the proportion of high income people to the proportion of middle high income people in a town; we can also add the proportion of children and the proportion of senior citizens in the town, but adding the proportion of high income people to the proportion of senior citizens is meaningless.

The object class person was partitioned by income in one partition, and by age in another. *High income person* and *middle income person* were subtypes of *person* in the income partition. The population of high income persons was an attribute of the subtype *high income person* and the population of middle income persons was an attribute of the subtype *middle income person*, both of the income partition. Their proportions could be added.

Child and senior citizen were subtypes of person in the age partition. The population of children was an attribute of subtype *child* and the population of senior citizens was an attribute of the subtype *senior citizen*, both of the income partition. Their proportions could also be added.

However, the proportion of high income persons and the proportion of senior citizens were proportions in different partitions – the income partition and age partition respectively. The sum of these proportions is meaningless.

The meaning of a proportion emerges by mutually engaging the meanings of an attribute, a partition, and an object class. As we have seen, proportions spring from this division relationship between a partition and its object class. We also know that we can add like with like (rule 3). Proportions of a given attribute across a single partition all map to the same (subtype of) domain of proportions. However, proportions of different attributes, or proportions in different partitions do not have the same meaning. They cannot be added.

Just as apples and oranges could not be added together, even though both were subtypes of a common class, *Fruit* (see the discussion of rule 3), *proportions of different attributes, or proportions in different partitions cannot be added together because each has a different business meaning not shared with the other.*

Rule 5 consolidates these rules normalized by the domain of proportions. They are inherited from the domain of proportions by every attribute that is a proportion. These rules and relationships are timeless and immutable. The rules belong to domains and exist in domains, immutable and silent. They need no other reason to be, neither time nor event – they just are. Attributes only use them:

Rule 5: Proportions are ratio scaled attributes of aggregate objects that are subtypes of a parent object class. All proportions conform to the following rules:

(a) *Population*: Every aggregate object has a population attribute. The population enumerates its members.

(b) *A sum of a ratio scaled attribute over all instances in a class*: Given a ratio scaled attribute of an object instance, corresponding object class(es) will have an attribute that sums up absolute values of the instance level attribute over all instances of the class. This sum is a class level, not instance level, attribute.

(c) *Sum of class level attributes in a partition*:
 (i) The sum of populations of individual subtypes in an exhaustive partition will equal the population of the parent object. In an inexhaustive partition, the sum may be less, but cannot exceed the population of the parent object.
 (ii) The sum of class level attributes – attributes that are sums of absolute values of corresponding instance level attributes – summed across all subtypes in an exhaustive partition, will equal the value of the class level attribute of the parent object. In an inexhaustive partition, the sum may be less, but may not exceed the class level attribute of the corresponding parent object.

 Since the partition represents the collection of subtypes in it, the partition normalizes these relationships. They are between the parent and the partition. Each subtype in the partition inherits them.

(d) *Proportion*: A proportion is a quotient obtained by dividing the population or another class level attribute of a subtype (obtained from the summation of absolute values of corresponding instance level attributes) by the corresponding attribute of the parent object.

 Like the relationship in (c), this rule applies equally to all subtypes in a partition. It is a shared rule normalized by the partition, and inherited by each subtype in the partition.

(e) *Range*: Values in any domain of proportions lie between *nil* and *total*.
(f) *Sum of proportions*: The sum of proportions in an exhaustive partition must equal *total*. In non-exhaustive partitions, the sum of proportions cannot exceed *total*.
(g) *An injunction against arithmetic addition of proportions*: Proportions for a given attribute of subtypes in a partition may be meaningfully added. Across partitions, or across different attributes, addition of proportions has no meaning.

The automation of these rules is obvious and the benefits evident:

1 Given the attributes of an object instance, processes and information systems based on the metamodel of knowledge will "know" that class level summations automatically exist. Users do not even have to explicitly call and store sums. Class level sums of these ratio scaled attributes implicitly exist, and any reference to the total for the class automatically uses the operation inherited from the domain of sums.[111]
2 The metamodel of knowledge "knows" that a partition automatically and implicitly bears proportions. Declaring partitioning criteria and subtypes automatically implies allocating proportions of each ratio scaled attribute to subtypes.[112] The domain "knows" the meaning of proportion, and every ratio scaled attribute of subtypes in a partition inherits this meaning, along with constraints and validity

[111] Sums exist implicitly; the operation is inherited from the domain of sums by every ratio scaled attribute, but when the computer executes the code, it is a technical design issue beyond the scope of this book.
[112] When the computer physically calculates and stores specific proportions it is a technical decision beyond the scope of this book.

criteria that this meaning carries (rule 5(e), (f), and (g)). Some benefits that flow from "knowing" are:

- Users need only refer to one or more proportions of one or more attributes in one of more partitions, and proportions will automatically be available. It will not have to be separately "programmed" each time. Each proportion will also automatically carry rule 5 with it.
- Actors inputting or importing information need only know which items are proportions in what partitions, and the metamodel of knowledge will automatically validate that they conform to rules 5(e) and (f).
- If all proportions but one in an exhaustive partition are available, and we know that the attribute in question is always nil or greater (or nil or less), rule 5(f) tells the process how to fill it in. This knowledge flows automatically to every process that involves proportions. It is inherited from the domain of proportions, not "programmed" each time it is required.

Indeed, the partition, by its very nature, brings rule 5 with it. Until it is needed, the relationships in rule 5 lie unused and unrecognized, hidden within the domain of proportions, waiting to spring forth when a requirement instantiates a proportion.

Difference scaled domains

Difference scaled domains are like ratio scaled domains, except they lack the nil value (see the discussion on figure 67). Therefore:

Rule 6: Division of one difference scaled value by another is meaningless, and hence so is multiplication of one difference scaled value by another (see *operations on values* in the discussion of figure 67).

Difference scaled domains have no information on ratios, i.e. they have lost the meaning of division (and hence also its inverse, multiplication – all because the domain has no nil value).

> Note the difference between "nil" and "null" values. The nil value conveys the absence of magnitude, whereas the null value conveys absence of meaning. See the discussion of figure (b) of box 47 for a detailed discussion with examples.

However, difference scaled domains do have information about *equality* of values and have information on relative magnitudes of gaps between values in a continuum.[113] The domain knows when two values are equal, and therefore, knows when the gap is zero. This is why gaps are ratio scaled, whereas the original domain was difference scaled – gaps between values are not only quantified, but could also be nil. The domain of gaps has a natural zero. Hence, subtracting a difference scaled value from another in the same domain creates a ratio scaled domain.

[113] Ratio scaled and difference scaled domains are mathematically dense [208]. [208] describes the continuum of values in these domains.

Let us examine the natural structure of information in the domain of gaps. In recognizing the magnitude of gaps, we recognized the difference scaled quality of the original domain; and in recognizing that gaps can be closed, we recognized the nil value. Thus, the domain of gaps was a subtype of the original domain with a nil value. That made it a ratio scaled subtype. The subtyping relationship in figure 67 was inherited via inclusion polymorphism to create the ratio scaled domain of gaps of a difference scaled domain.

Therefore, for difference scaled domains, rule 3 must be made more explicit as follows:

Rule 7: Subtracting pairs of values in a difference scaled domain will create a new domain. The new domain will be a subtype of the original domain. If the original domain was difference scaled, the new domain will be ratio scaled. It is the domain of intervals, or gaps between pairs of values, obtained by attaching a nil value to the original domain. It is also a subtype of the original domain

Rule 7 does not mention addition. The missing nil value impacts the validity of rule 3 in difference scaled domains. Rules of arithmetic addition also change. To understand these changes, think of difference scaled domains as domains of quantitative information in which the relative position of a hypothetical nil value is unknown (the domain has no information on nil). This does not impact arithmetic subtraction, because this hypothetical nil value must have mapped to some unknown number, and, as long as we subtracted numbers in the same units of measure, the subtraction operation cancelled it out, reducing it to zero. However, when we add values, we will be adding up and double counting the number that maps to the unknown nil. We do not know what numbers we have mapped to the unknown nil value, so we cannot meaningfully add values together in difference scaled domains. The results of addition will all map to "unknown." Rule 8 follows:

Rule 8: Addition of values in the same difference scaled domains is meaningless by itself (because it would double count the unknown number assigned to the hypothetical nil value). However, arithmetic expressions that do not distort the hypothetical nil value are meaningful. For example, taking the arithmetic mean of two or more values is meaningful, even though it involves addition of values in the domain. When the values are summed together, the unknown number assigned to the hypothetical nil value is multiply counted – once for each value – but that sum is also divided by the number of values added, which corrects the distortion. All such operations that do not distort the hypothetical nil value map back to the same domain. The validity of an arithmetic expression in the difference scaled domain depends on the entire arithmetic expression, not the validity of its terms in isolation. The entire expression must be considered in toto.

Rule 8 also implies that rule 4 must be made more explicit, because gaps between values in a difference scaled domain may be added and subtracted from values in their parent domains:

Rule 9: Values in the domain of gaps may be meaningfully added to, or subtracted from, values in the domain it was generated from. The results of the operation will map to the latter domain. This does not contradict rule 3 or rule 4 because the domain of gaps is a subtype of the original domain (albeit a ratio scaled subtype). Addition of values between subtype domains and their parent domains will be meaningful provided the hypothetical nil value is not distorted. Addition and subtraction operations with values in other domains (with this exception) have no meaning.

Box 44 Domain analysis for components of knowledge

The rules of traditional dimensional analysis (see box 42) are simpler. Traditional dimensional analysis recognizes neither difference scaled domains, nor partitions. Therefore rules 4, 5, 6, and 7 are ignored by traditional dimensional analysis. For this reason, traditional dimensional analysis need not consider modifying rule 4 for difference scaled domains. That reduces the number of rules we must consider, and makes those we must, much simpler. To get around the complexity created by difference scaled domains, traditional dimensional analysis recognizes only ratio scaled primary domains. It was the *time gap* domain, not the *date domain* that was identified as a primary domain by the Tenth General Conference on Weights and Measures mentioned in box 42. This is the correct approach for design and analysis of physical systems, because the time *lapse* between events, rather the actual time of occurrence, is important for physical laws and the machines that depend on them. However, that is not true for business systems. Business rules may involve both time lapse between events, as well as the actual date and time of occurrence. Therefore we must modify the rules of domain analysis for physical systems to support engineering of business knowledge. Engineering of business knowledge must recognize difference scaled domains, and add Rules 4, 5, 6, 7, 8, and 9 to the repertoire of domain analysis.

Ordinally scaled domains

The only arithmetic operation valid for ordinal domains is subtraction of ranks within the same domain.[114] Under patterns, we understood that subtraction of pairs of ordinal values will create a difference scaled domain (see table 1 and ordinally scaled state space in section 1). Therefore, rule 7 must be modified for ordinal domains and combined with rule 9 as follows:

Rule 7: Mutually subtracting pairs of values in difference or ordinally scaled domains will create a new domain. The new domain will be a subtype of the domain it was made from. The new domain is the domain of intervals, or gaps between pairs of values, obtained by attaching a nil value to the original domain. Values in the domain of gaps may be meaningfully added to, or subtracted from, values in the domain it was generated from. The results of the operation will map to the latter domain. If the original domain was:

(a) difference scaled, the new domain of intervals will be ratio scaled;

(b) ordinally scaled, the new domain of intervals will be difference scaled.

Consider the impact of this "prefabricated knowledge" on business processes that must flex. Let us return to the example of military ranks in section 1, under "Patterns in state space." Military ranks in a given branch (military service) are strictly hierarchical, but ranks across services like the army, air force, and navy may be different. However, each rank in a branch of the military has its equivalent in the other military services. Thus

[114] Besides subtraction, other arithmetic operations have parallels in ordinal domains with natural lower bounds. See addition, multiplication, reachability, countability, and other operations in [212].

military ranks convey information which are greater, lesser, or equal across branches of military service.[115] The ranks in a branch of service are all subclasses of the general class called *military rank*.

Assume we introduced a new rank in the army – inserted it between two ranks. The metamodel of knowledge would "know" the pecking order of the rank, and hence it would intrinsically and automatically become an ordinal position in (i.e. be inserted into) the general hierarchy of military ranks. Thus all ranks in the other services would automatically and intrinsically "know" their relative position in relation to the new rank in the army. Interservice protocols and rules that involve the pecking order of ranks would automatically apply to the new rank, and all other ranks would "know" how to behave. Not a line of additional code need be written, nor any new protocols created (the military does have the option of redefining rank equivalence across services if needed, but that is another matter, and would involve switching relationships between values, a "prefabricated" effect inherited from the "unknown" domain, ready for use when required – see the discussion on figure 67).

The class, *military rank*, inherited this "intelligence" about relative hierarchical positions and the behavior of supertypes from the generalized ordinal domain, added the information it already had about protocols between hierarchical differences in positions, and thus the formal military protocols between other ranks and the new rank were automatically available to each service.

Nominally scaled domains

A nominal domain has no information on magnitude, besides the fact that a value either exists or does not. Therefore, it can have no information on magnitude constraints of the kind in Chapter 3, section 2. However, it may participate in joint constraints with other domains.

Joint constraints (Chapter 3, section 2), linked to attributes, limited the state space of objects. Attached to domains, they create new domains. Indeed, any kinds of domains, in any numbers, may be mutually related by a joint constraint. We will examine this next.

Domains by association

A joint constraint is a law that binds attributes together. It is a law that binds values by mere association, and, on that basis, it constrains the state space of objects. When it binds values in domains together, it creates new domains, a domain of association – a domain that is generated by the law that created it, a domain from which the law may be inherited by all attributes that map to it.

Consider a relationship between the temperature and preference domains. It creates the *temperature preference* domain. An object class will provide the context for temperature preference. If the object class is, say, *furnace*, an attribute *furnace temperature preference*, of *furnace*, will map to the temperature preference domain. Had the object class been *swimming pool*, the attribute would have read *swimming pool temperature preference* instead. The new domain, temperature preference, is in the "wiring" of the relationship between the

[115] Military ranks are an example of points in a pseudometric state space – see the endnote on generalized distances.

temperature and preference domains, i.e. it is normalized by the relationship. Indeed, the relationship *is* the new domain.

We not only related two different domains to create a new domain, but we also related two different *kinds* of domains. The temperature domain was difference scaled (see section 2), whereas the preference domain was ordinally scaled. The new domain was also ordinally scaled. When all domains had sequencing information, the information content of the domain with the least information determined how values in the new domain were scaled.[116]

On the other hand, consider Jane's color preference. It was an attribute that mapped to the color preference domain, a domain created by the junction of a nominal color domain and the ordinal preference domain. The resultant color preference domain was ordinally scaled. In this case, it was the richer of the two domains, the ordinally scaled preference domain, which determined the scaling of the domain that emerged from the junction of domains.

Based on the above, we obtain the tenth rule of secondary domains:

Rule 10: A relationship between domains creates a new domain. The scaling of a domain created thus is:

(a) at least identical to the scaling of the participating domain with least information, when all domains thus joined contain sequencing information (we will not lose information if we assume it is identically scaled to the scaling of the participating domain with the least information);

(b) ordinal, when one or more domains thus joined is ordinally scaled (a special case of rule 10(a) above);

(c) nominal only when all domains thus joined are nominally scaled.

The key to rules 10(a) through (c) lies in information content. Why is the information content of a domain derived by joining ordinal, difference, and ratio scaled domains of different kinds restricted to that of the domain with least information? The answer is simple – it is not. Its information content is actually somewhere in between.[117]

Think of temperature preference again. It is common sense that there will be little difference in preference between temperatures very close to each other. That is information, and it is normalized by, and resident in, the temperature preference domain. However, the preference domain is an ordinally scaled domain of discrete values and the temperature domain is a continuum of difference scaled values.[118] As we consider temperatures closer and closer to each other, differences between temperature preferences too get less and less. When preferences are marginally different, it is difficult to tell them apart. As differences between preferences shrink, distinctions become a matter of chance.[119] The metamodel of

[116] Unless all domains in the combination are nominally scaled, the new domain will carry sequencing information. It will be a partial order (see box 46). [217] elaborates on box 45. Bijection and countability are key to relationships between values in domains. See [202] and [203].

[117] The Cartesian product of any domain with sequencing information, with any other domain, will create a partially ordered domain. Box 45 and [217] and [218] discuss partial order.

[118] [202], [165], [204], [206], and [212] discuss mapping discrete ordered values to a continuum.

[119] [211] and the references therein discuss mapping ordinal domains to other kinds of domains.

knowledge in this book is deterministic. It ignores chance. That is the information we have lost (see box 46). This is the reason behind rule 10(a).

In the same way, when any domain with information on sequencing of values is related to a nominal domain, the composite keeps this information intact. This is the reason for rule 10(b).

Of course, when only nominal domains join, there is no information on sequencing of values, and the junction is also a nominally scaled domain. This leads to rule 10(c).

Box 45 Partial order, fuzzy meaning, and the scaling of derived domains (on our website)

Take a pair of points, any pair. If you can find a path from one point to the other such that the value increases along one of the coordinates and does not simultaneously decrease for others as you travel along the path, then the pair is partially ordered.

For more details, please visit Box 45 on our website. The latter page also contains details of how these three concepts relate to the business environment.

Consider how domains of association help us normalize knowledge and isolate change. Take two "normal" temporal objects, *car* and *person*, and two domains, *color* and *preference*. The objects can change state under the pressure of events and the tides of time. The domains are forever still. A person's car color preference is an irreducible fact normalized by a four-way relationship, between *person, car, color*, and *preference*. This four-way relationship can change state. Our intent is to isolate and encapsulate the impact of temporal changes such as changes of state. For this, we will use derived domains of association and derived objects.

A relationship between objects is a derived object. A relationship between domains is a derived domain. Color preference is a relationship between two domains *color* and *preference*. It gives us a new domain, *color preference*.

The four-way relationship, *person's car color preference* has several equally correct expressions. To start with, it is equivalent to a three-way relationship between the object classes *car* and *person*, and the *color preference* domain (for the benefit of mathematically inclined readers, the relationship is a Cartesian product, and the two Cartesian products are equivalent because Cartesian products are associative mathematical operations – see box 47 and the endnote on associative operators under the mathematical theory of categories).

We can go one step further and reduce *person's car color preference* to a two-way relationship between an object, a relationship between *person* and *car*, and a domain, the *color preference* domain. *Person's car color preference* will be an attribute of the relationship between the *person* and *car*. The attribute maps to the immutable *color preference* domain. Event-driven changes of state are now isolated in the relationship between *person* and *car*. The color preference domain is temporally stateless and forever still.

Information quality – domains of information about information

The behavior of domains is determined by their information content. Implicit in every domain, as well as every association between domains, is its association with the domain of information. This association, a relationship between domains, has a profound effect. The information domain is ratio scaled, and information rich. The others may not be as flush with information. An association of domains is also a domain. All domains involved in an association of domains determine the information content of their mutual relationship. A domain is a pattern of information. Implicit in the fabric of every domain lie the following properties of information. You could think of each as emerging from the junction of every domain with the domain of information, or more precisely with one of its subtypes:

- enumeration
- accuracy
- reliability (risk)
- completeness
- validity

Each has a unique meaning and a unique impact. Each adds information to their common parent – the domain of information. We will examine the distinct meaning and impact of each.

Enumeration

Enumeration is a count. It is always ratio scaled, but there are subtleties we must be aware of. Consider the relationship between enumeration and a nominally scaled domain like the gender domain. The relationship is *gender enumeration* – a count of distinct values of genders. It is a domain. When we consider the completeness or incompleteness of the numbers of genders, we are, strictly speaking, considering the completeness of the *gender enumeration* domain, not the *gender domain*. The completeness of the gender domain determines the completeness of *all* the information the domain conveys.

If we know there are five genders (see the endnote on the question of gender), but not what these genders are, our information about the *gender enumeration domain* is complete, but information about the *gender domain* remains incomplete. Similarly, the accuracy of gender enumeration is not the same as the gender enumeration domain. In the first instance, we mean the accuracy of our information on the *number* of genders, and in the second instance, we mean accuracy of information on the *meaning* of gender. Similar arguments can be made for reliability and validity of information as well.

Indeed, the same arguments may be made for any domain and any object. Every object, be it a domain, a relationship, or any other component of knowledge, will map to the domains of information above. Therefore, intrinsic in any domain, or any object, is the concept of population. Objects are patterns (see Chapter 2, section 5 and the architecture of pattern in section 1, and the count of its constituents is a universal property of all patterns (Object occurrence value in figure 54 and figure 57). In an object, it is the number of instances that make the class; in a domain, it is the number of values that make the domain.

That population cannot be negative is a value constraint, a natural component of knowledge, attached to the population domain. All populations inherit it. However, populations do not always have to be whole numbers. Consider the population of cheese cubes. You could consider a half cube, or a quarter, or indeed any fractional size. Therefore, restraining populations to whole numbers is a constraint attached to the *attribute* that maps to the population (enumeration) domain, *not the population domain itself*. Populations of all aggregate objects must not inherit it.[120]

The concept of population is easy to understand when the number of values in a domain is finite, as in the gender domain. It is harder to compare or comprehend infinitely large populations of values such as those found in ratio and difference scaled domains. Domains are patterns of values. The extent of the pattern, the proximity and population of its values are all interdependent. The mathematical concept of cardinality (the size of a class) subsumes both finite and infinite counts.[121] Discussion of infinite cardinalities and their comparison is beyond the scope of this book.[122] It will suffice to understand that the size of a class is a universal property that flows from the concept of pattern, and the number of constituents – its population – is a universal property of all domains, object classes, and patterns ("Object occurrence value" in figure 54 and figure 57).

A relationship between domains too is an object class – a very special kind of class – a domain. *The population of a relationship between domains* is the sum total of all possible ways individual values across related domains may be associated in the tuple (see box 19 and the section on complex domains). In a system with finite states, only finite numbers of instances might be physically instantiated at a time, but the possibilities are immense, even infinitely large. It is this population that lends meaning to enumeration, and it applies universally to all domains, all objects, and all relationships. We arrive at the following law for the size or enumeration of values in a domain:

- Relationships between domains are also domains. Every domain has a size. The size of a relationship between domains cannot be less than the size of the smallest domain in the relationship. It may be larger. Sizes of domains that emerge from relationships between domains of known size are known. If the size of any domain that participates in the relationship is unknown, the size of the relationship will also be unknown.

[120] An initial state (default value) is meaningless for a domain; domains do not have temporal states. However, populations are restricted to whole numbers often enough to make it convenient to declare that the default, for the enumeration domain, is an electronic repository of knowledge artifacts. This is a design artifice. This default would be inherited by all enumeration domains in the repository, and then could be overridden for exceptions.

[121] Populations may be restricted to fractions (quotients of natural numbers or integers) – i.e. rational numbers), or of real numbers, p-adic numbers etc. Rational numbers form a continuum; real numbers, a denser continuum; other kinds of numbers extend the continuum in different ways (see [219], [220], [221], [222], [223], [224], [225], [226], [227], [228], [229], [230] and [231]). However, present-day computers cannot handle the continuum because they cannot accommodate the concept of infinitely large or infinitesimally small values. They approximate them with discrete sets of numbers (see [209] and [210]. It is always possible to find a rational number arbitrarily close to an irrational number in the continuum of numbers.) Therefore, any constraints that discriminate between infinite cardinalities or infinitesimally small differences are moot, and so are constraints that try to distinguish fractions from other kinds of numbers. However, infinitesimally small differences can impact how we model reality. For example, chaotic ([292], [293], [323]) or stochastic ([305], [310], [312]), phenomena could be conceived as stemming from infinitesimally small differences in initial or intermediate states of ratio scaled trajectories in state space. [204], [206], [207], [212], [219], [221], [222], [230], and [231] discuss infinitesimally small differences and differences between infinitely large numbers in a continuum.

[122] [202], [203], [206], [208], and [212] discusses the cardinality of classes.

Between known and unknown populations runs the whole gamut of possibilities based on how much information we have about sizes of domains – we may have information on ratios of populations (even if they are infinitely large); we may have information on differences in population between domains, but not ratios; we may have even less information, just barely enough to arrange domains in order of increasing population; and we may not even be able to do that if we have scarcely enough information to know that different domains have different populations, but cannot say which are larger than others. If we do not even have that, we eventually meet the unknown domain at the line that divides knowledge from ignorance.

Accuracy

The precision of information maps to the accuracy domain. Precision measures proximity. For example, a manufacturer of car parts might make a shaft to specification. Let us assume the length of the shaft has been specified, as well as the permitted variance in its length that will be tolerated. The manufacturer specified tolerance, or permitted imprecision, in terms of its proximity to the desired length of the shaft – a value. Similarly, the precision of difference scaled values is described by its closeness to a desired standard. The precision of an ordinally scaled expression may also be expressed in terms of permitted imprecision, except that in this case imprecision is measured in terms of *differences* in rank – does the rule expression map to the rank it should, or will it be wrong? If it is wrong, how large is the error – one rank, two ranks, three ranks, or more? Domains are patterns, and precision is a proximity metric. The proximity metric is a universal property of patterns. It is a property inherited by domains. Precision is only one manifestation of the proximity metric – how close must a pair of values be before we accept them as virtually identical, or at least acceptably close? This is the question precision answers. That is why precision is the proximity metric of section 1.

Precision therefore conforms to the laws in table 2. These laws are simple; their reasons, complex. Both are described under proximity metrics in section 1.[123] Thus, we arrive at the law of precision:

- The precision of a domain will conform to the rules for proximity metrics in table 2. The domain may be a simple domain, or a complex domain, described by an association of domains. The scalability of domains in the association may be of different kinds (nominal, ordinal, difference, or ratio scales).

The right-most metric in each row of table 2 conveys the most information for each kind of state space. A greater degree of precision is not possible, but our standards may be less precise. In the example of the shaft, the requirement may have merely asked that the manufactured shaft be shorter than a standard shaft, or even that it only be different. Both will be valid requirements. On the other hand, a requirement that demands that the shaft satisfy the quality inspector to the extent of 80 units of satisfaction or more is meaningless because preferences (satisfaction levels) are ordinally scaled. Table 2 articulates all these rules.

[123] The state space of any domain that emerges from a relationship between precision and any other domain will be a partial order (see box 45). Table 2 makes this explicit.

In terms of the structure of information, precision of information in any domain is rooted in a relationship between the domain in question and the precision domain. To demonstrate this, we will return to the example of the shaft being manufactured for a car. The car manufacturer specified the tolerance or requisite precision for shaft length. The manufactured shaft will meet the specification subject to a degree of accuracy in terms of its length. The shaft length maps to the length domain. The length domain, in this case, provided the context for accuracy. Length precision is a complex domain – a secondary domain that is a relationship between the accuracy domain and the length domain.

The length domain is ratio scaled. Therefore, the law of precision merely states that length precision may also be ratio scaled. Thus, engineering tolerances are ratio scaled measures. On the other hand, if accuracy had been related to the gender domain, it would be a binary, ordinally scaled domain such as the similarity metric in nominally scaled state spaces – we can only be right or wrong when we identify a person's gender (it might be common sense, but someone must tell the computer that!). The domain normalizes this knowledge.

If we were to assess the precision of the preference domain, precision would be expressed in terms of how many ranks of preference we may err by. That is a difference scaled measure as we discussed under patterns in section 1. The precision of a difference scaled domain like temperature will be ratio scaled for similar reasons – the gaps are ratio scaled, and precision measures error in terms of gaps.

Similarly, precision could relate to other primary or secondary domains like time, mass, volume, area, and others – even to the information content domain itself – to create new secondary domains, all framed by the law of precision. The law of precision is information. It is intelligence encapsulated and normalized in the information domain – or, more precisely, in its subtype, the precision domain.

Reliability (risk)

Reliability is a measure of the consistency of information. Whether the same information always has the same meaning and the same degree of accuracy. If it is hard for me to decide whether I prefer red to blue, my choice of color will not always be consistent and information on my color preference will be unreliable. Measurement of reliability often involves measurement of chance and likelihood. Chance is a ratio scaled measure (technically it is called probability).

If a measurement is always accurate, or if we are completely certain that it is accurate and there is no chance of being wrong, it is completely reliable – its reliability is *total*. If it is always inaccurate, it is completely unreliable – its reliability is *nil*. Values of reliability may fall anywhere between these two extremes – total and nil – because the certainty of being right (or wrong!) may be anywhere in between.

Total reliability may be expressed as 100% reliability, or in fractional terms, as a probability of one. Intermediate degrees of reliability will then be expressed in fractional probabilities, and complete unreliability, the nil value will map to the number zero. If we are right only one half the time about gender, the reliability of gender information may be expressed as 1/2, or as 50%. The units, or numbers that express the probability is a matter of choice – a unit of measure for the probability domain (0.5 and 1/2 are not different numbers, but the same number expressed in different formats. On the other hand, 50 and 1/2 *are* different numbers. The unit of measure is different in the second case, whereas it is the same in the first case – the difference between 1/2 and 0.5 is only one of format).

The domain itself is ratio scaled, bounded below by the nil value, and above by the *total* value. It is a proportion. The law of reliability follows naturally from the logic of information:

- The reliability of information in a domain is a quality of information shared by all domains. It is potentially ratio scaled with an upper bound. A given value of reliability of information means the same in every domain. It is a measure of confidence in the accuracy of the value – the chance of being correct.

Reliability is measured by probability. In plain English, reliability is a measure of the uncertainty about information. It is a property of information shared by all information – information in domains and information in objects that inherit it through their attributes – from domains.

Like the other domains we have discussed, measures of reliability may be scaled back if we do not have enough information on consistency of data. It might be difference scaled, wherein we can merely articulate differences in consistency (the chance of being accurate within requisite limits); it might be ordinally scaled, whereby we can only articulate the degree of risk in ordinal terms, such as "very risky," "somewhat risky," "quite reliable," but have no information on the quantum of difference between these positions, or it might be nominally scaled, so that we are only aware that different categories of risk exist, but not which is more or less. As information gets sparser, we start approaching the unknown domain.

In a purely deterministic model, risk is not considered. Everything is either true of false, with total confidence. There is no chance of being wrong, no shades of gray in a world of black and white – truth and falsehood with nothing in between.

Our metamodel is deterministic. Therefore, technically we have no room for reliability or the lack of it. However, recognizing the reliability (risk) domain compensates for this deficiency. We can express risk, when it is important to do so, with ratio, difference, ordinally scaled, or ratio scaled domains of risk – all subtypes of the unknown risk domain.

Until recognized, the potential of the risk domain lies buried, hidden within every domain, waiting to be realized in attributes and objects when and if needed. The information domain is its root, and the reliability domain normalizes this information.

Reliability lends its meaning to every domain and every value through the universal relationship every domain has with the reliability domain, a relationship every domain inherits from the relationship that the information domain enjoys with the domain at the root of all domains – the unknown domain – a domain at the border between knowledge and its absence.

In our deterministic model of knowledge, the domain of reliability is a hook, a portal into the shadowy shifting world of chance, where opposites can overlap and coexist, and absolute truths cease to be. Only their shadows mingle in a half world where the observer and observation lose their identity to become one pattern – a pattern of probability. Reliability is a universal property of all domains of values, and a natural bridge, a portal into a world where chance and chaos can rule supreme.

Completeness (exhaustivity)

Completeness tells us how exhaustive our information is. Consider the completeness of a list: we may know just enough to understand that some items are missing from a list. We do not know how many, or what proportion. When this happens, completeness maps to a nominal domain, like the exhaustivity attribute of partitions we discussed in Chapter 2, section 3.

Sometimes we may be able to distinguish degrees of completeness, but not the differences between them. For example, we might know that a project is closer to its end than it was before, but not how much closer. Then completeness will map to an ordinal domain. We could even know how much work of the project has progressed, but not the total amount finished to date. We cannot say what proportion of work is finished. Then completion maps to the difference scaled domain. Of course, we may have the processes in place to let us measure the total work required, and the proportion finished. Then completion maps to the ratio scaled domain. The business meaning of completeness stays the same, but the its scaling depends on the amount of information provided by its context – the relationship between the completion domain and the object it is the completeness *of*.

We had touched on the context of completion in our discussion of enumeration. It is a common mistake to confuse the completeness of enumeration with the completeness of all information conveyed by the class. The count of its members is only one item of the information conveyed by a class. It *contributes* to the information content of a class, but other items do so as well. Completeness of information will depend on the total information conveyed by a class. Domains are classes of values. Objects inherit their information from domains. Completeness of information in a domain must be based on its total information content – sequencing information, meanings, constraints, operations, and effects resident in the domain. In other words, completeness measures the *proportion* of the total information content that is known and available. This leads to the law of exhaustivity. It is similar to the law of reliability because both reliability and completeness are proportions:

- The completeness of information about a domain is a measure of information known versus that still unknown. Specifically, it is the ratio of known information to the total information content of the domain (see the endnote on the quantum of information). A ratio scaled domain with an upper bound thus normalizes it. A given value of completeness means the same for every domain. It is a measure of how much domain knowledge we have extracted or realized, compared to what the entire domain can tell us.[124]

Completeness is potentially ratio scaled. When we have incomplete information, the measure of completeness may slip backwards to difference, ordinal, and nominal values, as we have described earlier on, or even become "unknown" (as happens frequently in business). Measuring information content is not simple, and business often ignores it. The cost of not knowing adds risk. Analyzing domains helps reduce risk. Obtaining information uses resources. It costs money. Judiciously addressing the risk of incomplete information can sharpen management of risk.

[124] Calculating the information content or the degrees of freedom of a pattern can be complex. The section on patterns, the endnote on the measure of information, and the references therein discuss these concepts.

Validity

Validity tells us whether we are interpreting information correctly. It is different from reliability or accuracy. For example, you may have an imprecise thermometer, and your measurement of temperature may not be accurate, but it will still be temperature you are measuring. Therefore the measurement is valid, albeit inaccurate. On the other hand, if you used a speedometer to infer temperature readings, the information is invalid. You are simply not measuring temperature, even if you say so!

Validity will often be nominally scaled like it was in the example above. An assertion will either be valid or invalid. However, like completeness, the scaling of validity may also be ordinal, difference, or ratio scaled, depending on the information provided by its context. We often measure properties by proxy: driving our car, we actually infer our speed by the deflection of the needle on our speedometer. Greater speed maps to proportionately greater deflection and vice-versa. The deflection and speed are completely correlated, so the validity of our information is total. In the real world, different items may be correlated to a lesser or greater degree; the stronger the correlation, the greater the validity of the proxy measurement.

Validity depends on correlation, a measure of the confidence with which we can say that values move in lock step. It does not matter how much one value changes when the other(s) does, it only matters that it does. Validity is always between at least two things, an object and its context(s), and, like reliability, its value is limited within a range.[125]

In terms of information structure, the validity domain always links to a domain that is the intersection of two or more domains or object classes. Validity never stands alone. The law of validity asserts:

- Validity is a ratio scaled measure that is the assessment of the intensity of a relationship between objects. It may be nil, total, or in between. It assesses the chance of values in one object class changing lock-step with values in others. In a deterministic model, a rule is either valid or it is not. Only the nil and total values are entertained.

Relationships are also object classes. They can provide the context of validity, even if they are relationships between different time slices of the same object instance, or relationships across values in the same domain. The validity of an assertion that relates the current price of corporate stock with past values makes sense only within a specific time frame. The current stock price is an object, and so is the collection of past values, and the time frame. It is the validity of this three-way relationship that is in question.

Similarly, the validity of the assertion that values in the length domain cannot be negative makes sense only because it is a relationship between a value in the length domain and nil length, and other values in the length domain. Validity is meaningless for a value without a context, and it is also meaningless for a domain or a value unless it is the validity of information in a relationship *across* domains or values (remember relationships between domains are domains too). *The only requirement is that validity always must involve two or more objects.*

Like the other ratio scaled measures of information, the validity of a specific relationship may slip from being ratio scaled to being difference scaled, ordinally scaled, nominally

[125] [311] discusses correlation of ratio, difference, ordinally, and nominally scaled information.

scaled, and even "unknown" as information content about the *existence* of the relationship becomes sparser and sparser. However, the domain of validity, the class that normalizes the concept remains ratio scaled; only the scaling information in its relationship with an attribute or another domain slips. It is appropriate to say that the validity of a rule is unknown, or that the validity of one law is different from another, or even that the validity of one relationship is greater than another, without being able to say how much different or how much greater. It all depends on information content.

Think of validity as a domain of intensity from which instance identifiers like those in figure (b) of box 27, figure 35, and figure 36 draw their meanings. As an abstract concept, think of validity as a domain of information that measures the intensity of relationships between values, or the intensity of meanings of relationships between object classes.

Our metamodel ignores chance, hence for us a rule is ether valid or invalid and an association either exists or does not. Those two ends of the domain of validity – *nil* and *total* – are the only possibilities in a purely deterministic model: total validity is an absolute truth, and total invalidity an exception. *Exceptions are our hook into the world of chance and chaos.*

Assumptions about what is permitted and what is not lets us test for the validity of our businesses processes and information systems. Perspectives are patterns of objects and relationships (see Chapter 2, section 4), and provide the context that makes validity in a perspective valid. New learning can change perspectives, information systems, and business processes (see box 49). Our metamodel must embrace change.

Assumptions about validity are hidden in every domain and perspective. They flow from the validity domain that normalizes the concept of validity. It is a universal property of information. It is a ratio scaled domain like reliability, that lies hidden in the metamodel of knowledge at the trijunction where chance meets certainty and ignorance, ready to test and be tested should the need arise.

Validity, like reliability, is another portal through which we can connect our deterministic metamodel to the shadowy world of chance – a world of shifting shades of existence, mingling in strange patterns of probability – a world beyond our scope.

Completeness, validity, accuracy, and enumeration are different and independent items of information about information. Items from which the eleventh law of domains comes forth:

Rule 11: Every domain inherits the following properties from the domain of information:

(a) Every aggregate object, including every domain, is a pattern that has an attribute derived from the ratio scaled enumeration domain. It is a count of the number of members in it. The count may even be infinitely large.

(b) Every domain has one or more domain(s) of proximity metrics associated with it. The domain of the proximity metric is a subtype of the domain it is associated with, and measures the accuracy of values in the domain. The scaling of domains of proximity metrics conform to table 2 (indeed, the domain of gaps is this domain for ratio and difference scaled domains).

(c) Implicit in every value of every domain is the potentially ratio scaled property of reliability. The property of reliability of values in a domain springs from an association (a Cartesian product – see box 47 or *complex domains*) between the domain in question and the domain of reliability. Values of reliability range from *nil* to *total*. The reliability domain articulates the consistency of meanings. In a purely deterministic model, reliability may only be nil or total.

(d) Every object maps to the ratio scaled completeness domain. This domain measures the proportion of information in the object that has been realized. Values of completeness range between *nil* and *total*.
(e) Every relationship maps to the ratio scaled validity domain. It is a universal attribute of relationships. Values of validity range from *nil* to *total*. The validity domain articulates the meaningfulness of the relationship. In a purely deterministic model, validity of the relationship is either nil or total.
(f) Specific relationships with these domains of information may be nominally, ordinally, difference, or ratio scaled, depending on the information content of the relationship.

Indeed, validity, reliability, completeness, accuracy, and enumeration can even relate to each other, and, recursively, also to themselves. Thus, the completeness of enumeration, the reliability of validity and even the reliability of reliability are mutual relationships between these domains that are also valid domains of information, and so are higher-order domains of information, such as the reliability of reliability of reliability, or the validity of completeness of enumeration.

Box 46 "Softness" of information

We have often referred to "soft" information, such as those in non-physical domains, even ratio scaled domains such as economic value. What is soft information and how does it relate to information content? Consider the interaction between accuracy, reliability, and enumeration, all subtypes of the domain of information. Enumeration tells us how many values the domain contains (even domains with infinitely many values, as we saw in our discussion on cardinality). Accuracy is a measure of proximity between a pair of values, and reliability is our confidence in being accurate – the chance of being close enough to a target value with requisite accuracy. Each is information, and contributes to the overall information content of the specification. Given these facts, let us consider Jane's color preference domain again.

Assume Jane can only discriminate between colors she likes, colors she is indifferent to, and colors she dislikes. The domain has three values. That is the information enumeration contributes to the overall information content of the domain. Assume Jim, in the tale of Metanesia in Chapter 1, forced her to discuss colors she only disliked a little or liked a little, versus those she disliked a lot or liked a lot. He forced additional values into the domain – values that do not really exist. Unless Jane becomes more discriminating about he color preferences, the quantum of information in the domain will not change. The accuracy with which she can discriminate between colors still stays the same. She can only discriminate neutrality from dislike, and neutrality from liking for a color (and of course dislike from liking. This domain is ordinal, so we also know that the difference between liking and dislike for a color is larger than the distance between neutrality and either position.) If she is forced to discriminate between colors she likes only a little versus those she likes a lot, when she truly cannot, her responses will be random. In terms of our recent discussion on reliability, her responses will not be *reliable*. The domain will become soft and uncertain. The overall information content determines overall "softness" and hence the kind of scalability the domain has. If we try to impute a higher degree of scalability to a domain – i.e. make a nominal domain ordinal, or an ordinal domain difference scaled without increasing its intrinsic information content – it "melts" and becomes "soft" – i.e. values in it become less and less reliable.

On the other hand, if we reduce the size – the number of values – in a domain that is reliable, we will lose information. We will be able to reliably discriminate between the values that are left, but we could be even more accurate and still be reliable. The domain becomes "grainy."

Thus, if Jane became a connoisseur of colors, and could make very fine distinctions in her preferences between subtle shades of color, we could increase the number of preference values in the color preference domain and still be able to distinguish between preferences reliably. If Jane continues to grow ever-more discriminating, the number of values in her color preference domain will keep increasing and the requisite proximity between points that she can reliably discriminate between will keep decreasing, until, for practical purposes, she can discriminate between infinitesimally close values in a continuum. The color preference domain would have gathered enough information from the enumeration, accuracy, and reliability domains to become a difference scaled domain. In fact, because she can discriminate between like, dislike, and *neutrality*, this domain has a natural zero – it has become a ratio scaled domain like the money domain. This is how the interaction between subtypes of information adds to information content, and this is how the information content of a domain changes the very nature of its scalable behavior

In real life, there are no absolutes. Jane would never be able to make infinitesimally fine distinctions between her preferences. As colors and preferences got closer and closer, the chance of her consistently making the distinction would become less than certain. The smaller the difference became, the less her chance of being consistent would become. The domain would get softer and softer. Our metamodel is deterministic. We recognize only certainty in a world of black and white – choices are either always consistent or always inconsistent in such a world. To reconcile uncertain reality with the non-existent world of absolute certainty, we will have to decide how much uncertainty and inconsistency we will tolerate before we declare that Jane cannot meaningfully tell the difference because the values in question are too close – the values are virtually identical and *we will declare that they are indeed identical*. This is how the cardinality of a domain emerges from its information content. This is also how enumeration, accuracy, and reliability all contribute to information content.

The property of validity tells us that it is indeed Jane's color preference we mean, and not some other quality, such as the shape of colored objects presented to her, or their texture or odor that she might be confusing with preference for their color. This too is a matter of chance, but, in our idealized black and white world, it either happens or does not. Together, validity enumeration, accuracy, and reliability determine the information content of the domain, and information content determines its "softness" and scalability. If we are willing to tolerate less reliability, we can increase the number of values (and decrease gaps between them, i.e. increase potential accuracy requirements for measurements in the domain). If we are willing to live with less accuracy, we can increase reliability, and will recognize larger gaps between values. These gaps measure unreliability of information on differences – the meaninglessness of proximity with commensurate levels of reliability in a world where black and white coexist in shades of gray – a world where domains may be as hard or soft as the information they convey – a world without absolute certainty or absolute meaning.

The information domains focus on risk – the risk of imperfect information. Incomplete, invalid, inaccurate unreliable, and uncounted information all contribute to business risk – the risk rooted in bad information, blinkered vision, unsubstantiated assumptions, or assumptions belied. The information domains help us sharpen management of risk.

Complex domains and mixed meanings

Domains are value objects with timeless states.[126] We have just seen how some are classes of single values, while others are associations that cannot be neatly reduced to one value. The latter are relationships between domains. These relationships are also domains. They are classes of ordered lists, or *tuples* (see box 47 and box 19), wherein members of the class are lists of values, and members of each such list are the values of domains that have been associated by the relationship.

The color preference domain was one such domain. The color preference domain was different from other secondary domains like "money per unit area" we have discussed thus far. Single, unjoined values were members of domains like "money per unit area" even though its values were derived from other domains. The relationship (division in this case) reduced the values it joined to a single magnitude. The color preference domain was different. Its members are *pairs* – pairs of values. One member of the pair is color, and the other is preference. The *pair* is color preference.

We will call domains that are classes of single values *simple domains*, whereas those that are classes of tuples, *complex domains*.

A simple domain may describe a complex pattern of the kind in figure (b) of box 47 when we consider its derivation from other domains via a magnitude constraint, but by itself (if we ignore its derivation) it is a simple one-dimensional pattern such as values of area, volume, or temperature. Complex domains, on the other hand, are always multidimensional patterns. They are multidimensional patterns because each value is a tuple. Members of the tuple have not "melted" into a single value through a magnitude or joint constraint. Thus each member of the tuple preserves its separate identity, as illustrated in figure (a) of box 47 and figure (c) of box 19. This is why the dimensionality of the pattern must equal the size of each tuple – the number of members it has.[127] Domains like color preference, and even color and physical space (see box 49) are complex domains.

[126] Domains have states, but these states do not change in response to temporal events. States of domains are their timeless relationships, subtypes, and constraints (for example, in our discussion of proportions). Domains are classes and classes are categories, which may be subtyped based on information conveyed by relationships and constraints (see [173]).

[127] Each tuple in this domain, a class of tuples, is the same size (see box 19). The dimensionality of the pattern equals the cardinality of each tuple – the number of members it has.

Thus, domains are value objects because they are classes of values that share a common meaning. The domain is the object class, and the value the object instance. Together they are the value object.

Domains are immutable and changeless value objects. Domains are immutable and changeless because they are fields of pure values. They convey immutable and changeless meanings – values that flow to objects through attributes to lend them their meaning.

Domains are strange shadowy objects that contain only the potential to be. Other objects realize this potential when they map to specific values in domains. When domains relate to other objects, they are manifested as attributes. When domains relate to other domains, they create new domains – domains that are also value objects pregnant with meaning, and laden with possibilities – each possibility a value, and each effect a requirement. Each has the potential to be, waiting to be mated with its context; a context provided by an object that then realizes this potential.

Box 47 Domains, relationships, and the Cartesian product (on our website)

Box 47 on our website provides more detail on how the Cartesian product discussed in box 19 can serve as the basis for relationships between domains, and how these relationships create new domains, which might be nominally, ordinally, difference, or ratio scaled. The examples also demonstrate how different mathematical operators might join domains to create new domains. It discusses the difference between null and nil values, and how some domains, assembled from domains without null values, may have "holes," expressed by the null value as a result of the joining. The following figures have been reproduced here, and are discussed in detail on our website.

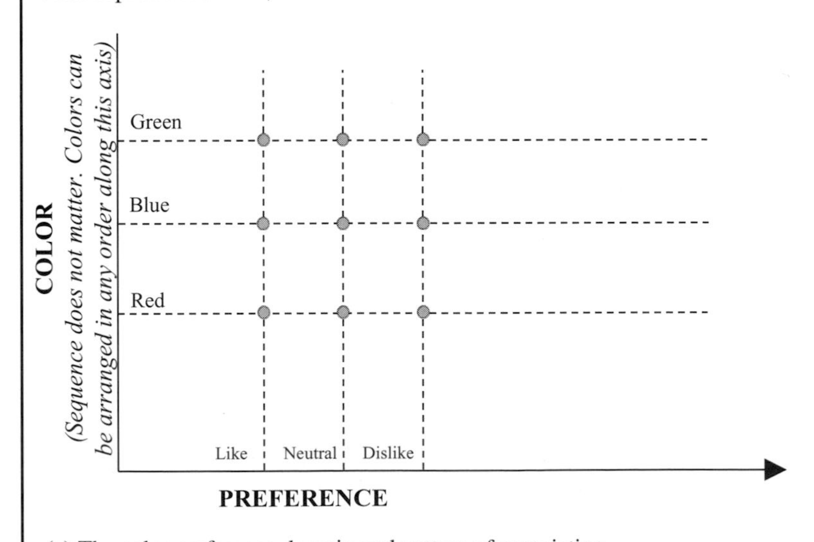

(a) The color preference domain and pattern of association

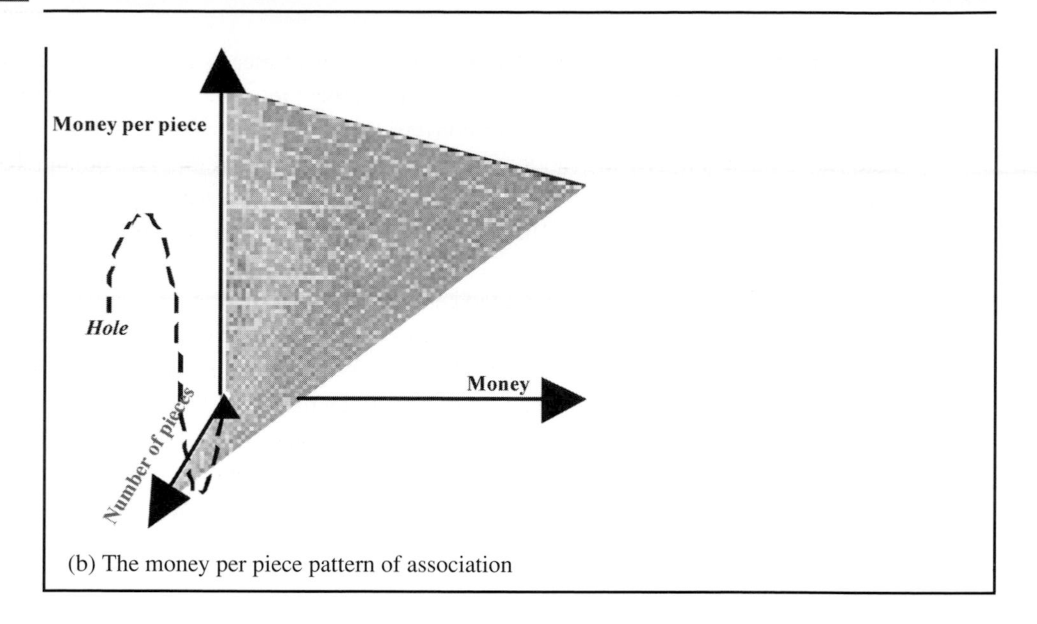

(b) The money per piece pattern of association

The risk and benefit of domain analysis

The primary benefit of domain analysis is that we can reduce the risk of unintentionally replicating knowledge scattered throughout our systems and processes. Domains are the most fundamental components of knowledge and behavior that are shared by virtually all information. Knowledge that belongs to a domain is common knowledge, shared by attributes, objects, and processes throughout organizations and value chains, regardless of local practices, perceptions, and measurement units. If we ignore this and put the knowledge that belongs to domains into attributes instead, we will replicate knowledge and risk chaos under the twin imperative of rapid and continual change.

Normalizing the behavior of domains and managing them as components of shared electronic knowledge can positively impact processes and the information systems they support. It can help improve information quality. It can assist in sharing of knowledge across complex processes and in obtaining more flexible systems and processes. It can also facilitate integrated and robust responses to change and help embrace innovation. It can automate design, development, and maintenance of information systems, and it can also help rebuild and integrate old knowledge into new configurations. Let us see how.

Reusing knowledge – building upon the old

We have seen how new meanings emerge from old, and analyzed the rules for assembling new domains from older components. Now we will take each rule, one at a time, and see the impact it can have on improving reliability of processes and systems while reducing time and cost of eliciting requirements and developing or modifying systems and business processes. Often the benefits of these rules will only flow if we automate domain analysis. The task may be too complex and too labor intensive to do manually.

Usually these benefits will stem from sharing a unit of measure and conversion information. Domains also share other rules such as bounds, ranges, formats etc., which will also be normalized along with rules for converting between them This too can help facilitate change at a greater pace and lower cost. Also keep in mind that all domains contain not only the effects that update values of attributes, but they also carry within them the effect that creates new attributes. They can help us create new attributes and refactor knowledge, as new learning forces us to embrace change. Domains can help reconfigure what we know already, as we learn new things and get new ideas. They can help preserve, conserve, and reuse our knowledge assets – the common learning shared across organizations and people:

Rule 1: Adding meaning to a domain creates a new domain. Meanings added may be new, or inherited. If the new domain includes meaning(s) inherited from other domains, it is a subtype of the domain(s) it was created from and:

(a) The new domain will inherit unit-of-measure information from its parent domain and add information of its own.

(b) If the nil value was included in the new domain, the units of measure of the old domain will be inherited with their zeros reset to coincide with the new nil value.

The nil and zero should coincide to reflect the mathematical properties of nil. Otherwise simple arithmetic with numbers in that unit of measure becomes difficult. The number zero mirrors the natural properties of nil. Adding nil to another value results in the same value, just as adding zero to any number results in the same number; similarly, multiplying any ratio scaled value by nil results in nil, just as multiplying any number by zero results in zero; dividing nil by another ratio scaled value results in nil, just as dividing zero by another number results in zero; dividing any non-nil ratio scaled value by nil results in an infinitely large value, just as division by zero does, and dividing nil by nil yields null. It maps to no value, ratio scaled or otherwise in any domain (unless we include the null value in the unknown domain of figure 67, from which every domain inherits it. See null value in Chapter 2, section 2 and "Null versus nil value" in box 47). This is why it is good practice to map nil to zero in every unit of measure.

If a hitherto unknown natural nil or lower bound is discovered (for example, by a natural law), rule 1(b) can also automatically establish new, "natural" units of measure by realigning hitherto arbitrary zeros of older units of measure with the newly discovered nil value or lower bound, while keeping differences between values intact.

Based on the principle of subtyping by adding information, the new units of measure will be a subtype of the corresponding units of measure they were derived from. It is a form of inclusion polymorphism we discussed in box 21.

When the natural lower bound of temperature was discovered, and it happened to be the nil value (it is approximately negative 273.15° Celsius),[128]

[128] See the endnote on the natural zero of temperature and references therein.

absolute temperature continued to use the units of measure of difference scaled temperature, with the caveat that zero of the older Celsius scale was realigned to match the nil value of temperature, and the older units of measure were all bounded below. The Celsius scale was reused to create a new Kelvin scale. The Kelvin scale was identical to the Celsius scale, with the exception that 0° was set at the newly discovered, natural nil value of temperature.

Given this new information, and older conversion rules from other units of measure to Celsius, a system assembled from knowledge artifacts could automatically derive conversion rules from all older units of measure to the new Kelvin scale. Indeed, knowledge artifacts can even derive the Kelvin scale from the Celsius scale by applying rule 1(b): since the new domain is a ratio scaled subtype of the old difference scaled domain, it will inherit the old units of measure, but will move the zero value to coincide with the newly discovered nil value. These revised units of measure will be subtypes of the older units of measure (see the end of box 41). Knowledge artifacts could also infer, from rule 9, that temperature differences computed in Celsius may be added to temperature readings in Kelvin or Celsius, but not in Fahrenheit.

This kind of sophisticated reasoning simply does not exist in most software development tools currently available. It must be manually programmed. The metamodel of knowledge naturally supplies these components of knowledge. Anything else would be unnatural, difficult, and tangled reasoning from this perspective. We have started unraveling the knot and normalizing the natural rules of business. Many reside in domains within the metamodel of knowledge.

Knowledge artifacts will save redundant requirements definition and design and development time, as well as resources; because knowledge will be normalized, neither the new scales, nor conversion rules will need to be manually formulated. New measures and conversion rules will be inferred by knowledge artifacts from the behavior of domains and knowledge of older units of measure and conversion ratios already held by information systems and business processes. It will facilitate change while promoting lower cost and resource requirements.

Rule 2: Any multiplication or division operation on values in a ratio scaled domain creates a new ratio scaled domain. Multiplication and division operations may be between values in the same or different domains. *Units of measure of the new ratio scaled domain will be expressed in terms of the same operations on units of measure of the domains it was created from, as will conversion ratios between units of measure of the new domain.*

Multiplying length three times derives volume. Thus, if length is measured in inches, volume will be measured in cubic inches, and if length is measured in feet, volume may be measured in cubic feet. The rule for converting from feet to inches is "multiply by 12," therefore the corresponding rule for converting from cubic feet to cubic inches is to multiply by 1728 ($12 \times 12 \times 12 = 1728$).

Similarly, specific density is mass divided by volume. Therefore, if mass is measured in ounces, and volume in inches, specific density will be measured in ounces per cubic inch, and if mass is measured in pounds and volume in cubic foot, specific density will also be measured in pounds per cubic foot. Moreover, because the rule for converting pounds to ounces is "multiply by 16", the rule for converting pounds per cubic foot to ounces per cubic inch is to multiply by "0.01252" ($16/(12 \times 12 \times 12) = 0.01252$).

The rules for conversion of units of measure of the secondary domain were derived automatically from its primary domains. These requirements were not specified independently. They were not designed and developed separately for each attribute that mapped to these domains. This is an example of how processes and systems assembled from knowledge artifacts can automatically adapt to new inputs based on old knowledge.

Rule 3: Addition and subtraction operations on values in the same ratio scaled domains map back to a subtype of the same domain. Addition and subtraction are permitted between parent and subtype domains. The subtype domain will inherit all units of measure and conversion and formatting choices from its parent domain. We have already discussed the benefits of this rule under "Creating new domains from old".

With this rule, we will also know which additions and subtractions imply which others, and also what may not be added or subtracted from what, even if they have similar meanings. The examples under "Creating new domains from old" demonstrated how intelligence on which additions and subtractions implied which others was inferred from domains, along with injunctions against adding subtypes of a common parent. This kind of inference, especially if automated, can promote quality and reduce resource requirements, while simultaneously compressing requirements definition and software development schedules; all under the pressure of rapidly shifting scope, as new business processes address new objects, recognize new events, create new behavior, and enhance older object classification schemes.

Rule 4: Addition and subtraction operations between values in different ratio scaled domains have no meaning if one is not a subtype of the other. *This rule can facilitate quality assurance of requirements by raising warning flags about mixing apples and oranges.*

Rule 5: Proportions are ratio scaled attributes of aggregate objects that are subtypes of a parent object class. All proportions conform to the following rules:

(a) *Population*: Every aggregate object has a population attribute. The population enumerates its members. Its units of measure are enumeration units.

Various metrics will automatically be implied and will exist under this rule. If we have an object called insurance claim, and it has a state, *unsettled*, information on numbers of unsettled insurance claims is automatically implied. No additional "programming" will be needed each time a demand for similar information arises. *Each enumeration will not have to be "programmed" separately. Partitions may be defined "on-the-fly" based on attributes and states of object classes. Counts of object instances for each subtype will be automatically implied by the partition. This can anticipate requirements, cut time, cost, and resources, even as demand for new measurements, controls, and processes picks up steam.*

(b) *A sum of ratio scaled attributes over all instances in a class*: Given a ratio scaled attribute of an object instance, corresponding object class(es) will have an attribute that sums up the instance level attribute over all instances of the class. This sum is a class level, not instance level, attribute. The units of measure of the class level attribute will be identical to that of the instance level attribute.

The benefit is similar to that for 5(a) above. Totals will be automatically implied and may be addressed without repeatedly programming them each time we have a requirement for a new total or a new ratio scaled attribute.[129] If the insurance claim object class has an attribute called claim amount in the example above, the total claim amount for unsettled insurance claims will automatically exist.

(c) *Sum of class level attributes in a partition*:

(i) The sum of populations of individual subtypes in an exhaustive partition will equal the population of the parent object. In an inexhaustive partition, the sum may be less, but cannot exceed the population of the parent object. The units of measure of the sum will be inherited from the enumeration domain.

(ii) The sum of class level attributes – attributes that are sums of absolute values of corresponding instance level attributes – summed across all subtypes in an exhaustive partition, will equal the value of the class level attribute of the parent object. In an inexhaustive partition the sum may be less, but may not exceed the class level attribute of the corresponding parent object. The units of measure of the sum will be inherited from the domain of the summed attribute.

Since the partition represents the collection of subtypes in it, the partition normalizes these relationships. They are between the parent and the partition. Each subtype in the partition inherits them.

This rule can be used for validating information or filling in missing information. It can help enhance quality of processes and information. Sums will not have to be repeatedly mapped to the same domains, nor will conversion between units of measure have to be specified separately for each sum. This can help reduce resource requirements and compress schedules under the pressures of continual and never-ending change.

(d) *Proportions*: The sum of populations of individual subtypes in a partition will equal the population of the parent object in an exhaustive partition. In

[129] Existence of a total need not imply physical computation and storage of each total for every ratio scaled attribute of every object and its subtype.

a non-exhaustive partition, their sum cannot exceed the population of the parent. For each subtype (subclass) within a partition, there will also be one ratio/difference scaled class level attributes for each ratio/difference scaled instance level attribute. The value of this subclass level ratio/difference scaled attribute will be the sum of absolute values of the corresponding instance level attribute in the subtype. In an exhaustive partition, this subclass level value, summed across all subclasses within the partition, will equal the value of the corresponding class level attribute of the parent object that has was originally partitioned into subclasses. On the other hand, in a non-exhaustive partition, this sum (of subclass level values of the attribute in question), might be less than, or equal to the value of the corresponding attribute of the parent class, but may never exceed it. Since the partition represents the collection of subtypes in it, this relationship is between the parent and the partition. Each subtype in the partition inherits it. The result of a proportions calculation is independent of the units of measure used to express the divisor (or dividend – both divisor and dividend may be expressed with the same unit of measure), provided the divisor and dividend are expressed in the same units of measure. *The existence of proportions is automatically implied by the existence of an attribute and partition of an object class. Both may be defined when required, even "on the fly" (see the effects shared by all domains in the discussion of figure* 67*). The business leverage from this rule is similar to the benefits of rules 5(b) and (c).*

(e) *Range*: Values in any domain of proportions lie between *nil* and *total*. The unit of measure of a proportion should assign the nil value to the number zero,[130] and the total value to another number. The default unit of measure could be%. That will assign the number 100 (%) to *total*. *Data validation need not be "programmed" each time a proportion is required. It is implied and can be automatic.*

(f) *Sum of proportions*: The sum of proportions in an exhaustive partition must equal *total*. In non-exhaustive partitions, the sum of proportions cannot exceed *total*. The units of measure of the sum of proportions are the same as units of measure of proportions. *The benefits are similar to 5(e). Also, if only one value is missing, rule 5(f) can automatically fill it in, reducing the data entry/acquisition effort and making the process more user friendly.*

(g) *An injunction against arithmetic addition of proportions*: Proportions for a given attribute of a subtype in a partition may be meaningfully added. Across partitions, or across different attributes, addition of proportions has no meaning. *The benefits are similar to the benefits derived from the injunction against mixing apples and oranges in rule 3.*

[130] The nil value added to another value has no effect. The arithmetic product of the nil value with any other value, save infinity, is also nil. The number zero mirrors these properties. If a unit of measure maps nil to another number, addition and multiplication will become more complex. This is why the number zero should be assigned to nil by all units of measure.

Rule 6: Division of one difference scaled value by another is meaningless, so is multiplication of one difference scaled value by another (see "Operations on values" in the discussion of figure 67). *The benefit is similar to that of rule 4.*

Rule 7: Mutually subtracting pairs of values in difference or ordinally scaled domains will create a new domain. The new domain is the domain of intervals, or gaps between pairs of values, obtained by attaching a nil value to the original domain. It is a subtype of the domain it was created from. *The units of measure of the domain of gaps will be inherited from the domain it was created from.* If the original domain was:

(a) Difference scaled, the new domain of intervals will be ratio scaled;
(b) Ordinally scaled, the new domain of intervals will be difference scaled.

All attributes that map to these domains, and all differences will inherit these rules as well as rules for converting between units of measure. The rules for ratio and difference scaled domains, respectively, will automatically apply to all differences, tolerances, and protocols every time a difference is referred to or calculated. These rules will not have to be specified separately for each attribute and domain. This will facilitate both cost and schedule compression.

Rule 8: Addition of values in the same difference scaled domains is meaningless by itself. It may have meaning in expressions that do not bias the result by changing the arbitrary zero value of a given unit of measure (e.g., by adding, subtracting or multiplying it multiple times), in which case it maps back to the same domain and the same unit of measure. The entire arithmetic expression must be considered in toto.

Adding two temperatures in degrees Fahrenheit is meaningless by itself, but it could be a term in a meaningful rule expression like an average calculation. It is meaningful in an average calculation because the arbitrary constant may be summed multiple times as the sum of values being averaged is computed, but it is also divided an equal number of times, so that the arbitrary zero is not distorted. The average, of course will map back to the temperature domain in the same unit of measure. *The benefit of this is similar to rule 4, albeit it is a far more complex validation of requirements.*

Rule 9: Values in the domain of gaps may be meaningfully added to, or subtracted from, values in the domain they were generated from. The results of the operation will map to the latter domain, *in the same units of measure.* Addition of values between subtype domains and their parent domains will be meaningful, provided the hypothetical nil value is not distorted. Addition and subtraction operations with values in other domains, with this exception, have no meaning. (*Combines benefits of rules 4 and 7*).

Rule 10: A relationship (Cartesian product) between domains creates a new domain. The Cartesian product may involve a single domain or several domains. The scaling of a domain created thus is:

(a) At least the same as the scaling of the participating domain with least information, when all domains thus joined contain sequencing information. We can assume that the new domain is scaled the same as the participating domain with the least information without attributing non-existent information to the domain. If the domains were all quantitatively scaled, the unit of measure of the new domain, a complex value object, will be the arithmetic product of units of measure of domains associated by the Cartesian product.[131]

The three-dimensional physical space is the Cartesian product of the length domain three times (see box 49), and the measure of physical space is volume. The unit of measure of volume is obtained from the arithmetic product of the unit of measure of the length multiplied by itself three times. *This is an example of how rule 10 can facilitate automatic derivation or validation of measures for complex value objects.*

(b) Ordinal, when one or more domains joined in this manner is ordinally scaled (a special case of rule 9(a) above).

(c) Nominal only when all domains joined in this manner are nominally scaled.

Rule 11: Every domain inherits the following properties from the domain of information:

(a) A count of its members is an attribute of every aggregate object, including every domain and object class. It maps to the ratio scaled enumeration domain.

(b) Every domain has one or more domain(s) of proximity metrics associated with it. The domain of the proximity metric is a subtype of the domain it is associated with, and measures the accuracy of values in the domain. The scaling of domains of proximity metrics conforms to table 2 (indeed, the domain of gaps is this domain for ratio and difference scaled domains).

(c) Every object, including domains and their values, has the ratio scaled attribute of reliability. Values of reliability range from *nil* to *total.* The reliability domain articulates the consistency of meanings. It maps to the domain of proportions. In a purely deterministic model, reliability may only be nil or total.

(d) Every object maps to the ratio scaled completeness domain. This domain measures the proportion of information in the object that has been realized. Values of completeness range between *nil* and *total.* Completion maps to the domain of proportions.

(e) Every relationship maps to the ratio scaled validity domain. It is a universal attribute of relationships. Values of validity range from *nil* to *total.* The validity domain articulates the meaningfulness of a relationship (rule) between objects. In a purely deterministic model, validity is either Nil or Total. Validity maps to the domain of proportions.

(f) Specific relationships with these domains of information may be nominally, ordinally, difference, or ratio scaled, depending on the information content of the relationship.

These rules, judiciously used, can sharpen management of risk and provide ports to stochastic metamodels beyond the scope of this book. Benefits have been discussed under "Information quality – domains of information about information".

[131] The mathematical basis of this assertion lies in the logic of categories. See [173] and product category in [236].

All objects do not always realize the full potential hidden in domains. However, this potential is the key to flexibility and adaptation under the immense pressures of rapid and continual change – change driven by ceaseless learning and unrelenting innovation. The potential of information domains may be realized as new learning and new processes pave new ways. The immense and unrealized potential of information lies latent and waiting in the metamodel of knowledge – the potential to embrace change as it happens.

The most frequently used domains

Normalizing atomic rules in domains is important, especially because innumerable attributes inherit these rules, relationships, and effects. For example, the natural upper and lower bounds, units of measure, units of measure conversion rules, and effects normalized by a domain will be inherited by all attributes that map to it, and will need be defined only once. Domains used most frequently depend on both the kind of industry and the kind of application involved. A manufacturing process may use the temperature domain frequently, and electrical or electronic applications might use electric charge and electric current (the flow rate of electric charge) frequently, but the insurance or financial services industry may not. However, common processes consistently use a number of domains frequently across all areas of business.

Frequently used primary domains

The most frequently used domains are the primary domains. They are used both alone and in combination. In combination with other primary domains, or even themselves (for instance length combines with itself to make area or volume), they become secondary domains. Secondary domains are used virtually everywhere. ***Enumeration***, ***information, date***, ***mass***, ***length***, ***economic value*** (i.e., money), and its weaker counterpart, ***preference***, are primary domains used most often in business.

Enumeration domain: The enumeration domain is especially important. Every aggregate object maps to it. Aggregate objects are also instances of objects (Chapter 2, section 1). An object class is an aggregate object; it is the bedrock on which reuse of knowledge is founded. The single most important attribute of every aggregate object is the number of objects in it. That defines the concept of aggregation. Enumeration of its members is inherited from the aggregate object by every instance of object class. Its members define the pattern that lends an aggregate object its meaning; the count of its members maps to the enumeration domain.

Information domains: Management of risk is at the heart of every business. The information domains – risk, validity, accuracy, and completeness identify risk and facilitate its management. Indeed, they are intrinsic to information and inherent in business processes. When automation does not recognize them explicitly, decision makers recognize them implicitly. These domains are the key to process quality.

Date: The date domain involves not just calendar dates, but also time of occurrence. Both concepts are subsumed into values of date in the date domain. Whether we choose to express these values as purely calendar dates, or also specify the hour, the minute, the second, and parts of a second, is merely a question of precision of expression, a property of unit of measure (section 2).

Date and time are at the heart of business performance. Virtually all business transactions are sensitive to dates and times. It is pertinent to every state change and time slice in figure 22. Naturally, every business known and unknown uses it frequently. *Date* is a very frequently used domain.

Mass and length: Almost all fundamental physical aspects of physical products or resources of businesses stem from mass and length. Mass and length are often used directly, or are hidden in secondary domains such as space, density, depletion rates, and others. Their frequent use is natural, considering that the physical universe frames all business.

The money domain: The economic value domain is not only the heart of business, but it also has an interesting twist. It conveys "softer", more uncertain information than "harder" engineering information such as mass or length. The domain is ratio scaled, but its information content is less than that of its ratio scaled physical counterparts. We have discussed this in box 46. Also, unlike its physical counterparts, its unit of measure is non-stationary (i.e., changes over time – see box 40). If you expressed a value, such as a pay scale, in US dollars ten years ago, it will not equate to its expression in US dollars today, and, even if you converted ten year old dollar values to present day dollar values, the conversion would intrinsically lack the reliability (certainty of being correct) that, say, converting from feet to inches has (nations that have experienced hyperinflation know the severity of this problem). It is a fact we must live with, because the domain just does not have the requisite information. Its information content is somewhere in the nether zone between the information content of a physical ratio scaled domain and that of an ordinal domain that has a nil value.

However, rules for converting money from one unit of measure to another conform to the conversion rules for ratio scaled domains. Whether we convert one currency to another in terms of their current values, or convert a measure of currency past to the measure of the same currency today, we multiply by a conversion ratio. Converting between currencies is called an exchange rate; converting between different times is called an index. Each is a conversion ratio, and both change state. Conceptually, they are identical – they belong to the same class of objects.

Of course, exchange rates between currencies are more volatile than currency indices. Exchange rates can change with every tick of the market. However, that does not change the basic intent of either exchange rates or indices. Both convert one unit of measure of money to another. The exchange rate matrix may be considered an object that changes its state in step with the market, and may be used to track equivalence of economic value across currencies at any given time, but knowledge artifacts can do more. With the rule in box 40 (under "How many conversion rules?"), a knowledge artifact can even automatically combine exchange rates and indices to convert between currencies *across* different periods of time. The knowledge lies normalized in the money domain, ready to be automatically configured and used when required.

Gender and other classification domains: At least one domain based on the non-physical sciences also finds widespread use – the *gender domain*. We discussed gender at the beginning of section 3. The gender domain is a nominally scaled biological domain. More nominal domains emerge from biological, social, economic, and other classifications. Nominal domains are impossible to list exhaustively at present. Many are chimerical, continually emerging, merging, vanishing, growing into ordinal or quantitative domains, or otherwise changing their meanings (see box 47, box 49 and "Measurement of meaning – a paradox of perspectives" later in this chapter).

Frequently used secondary domains

Secondary domains, we know, are derived from primary domains. They are configurations of primary domains that have mutually engaged with each other. Some secondary domains are used frequently in business. We could consider "prefabricating" these knowledge artifacts in our electronic repository of meaning:

- The time domain, the domain of differences between pairs of values in the date domain, is a frequently used secondary domain. It is ratio scaled. Attributes such as ages of objects and durations of processes map to it.
- Physical space is a complex, difference scaled, partially ordered domain that is frequently used. It follows that the following secondary domains, both measures of physical space, will also be frequently used:
  - Volume
  - Area
- Rate domains of various kinds are used frequently. A rate domain is a secondary domain of quotients created by the arithmetic division of one ratio scaled domain by another:
  - Money rates are used very often in business. Money per item (economic value per enumerated item), per unit area, per unit volume, per unit length, per unit mass, and per unit time are all examples of rate domains. Attributes such as unit costs, unit prices, revenues, and burn rates of project portfolios map to domains of money rates.
  - Rates of change (growth) with respect to time are also frequently used.
  - Proportions are frequently used in business performance measurement and quantitative metrics of various kinds

The risk of domain analysis

Domain analysis has its benefits. It also has risks. The risks flow from two practical limitations:

1. *The risk of analysis paralysis:* as knowledge becomes complex, and numbers of secondary domains grow in step with the complexity of business rules, the onset of analysis paralysis can be rapid. Automation may control this risk somewhat.
2. *The risk of incomplete information:* not all primary domains have been identified as primary domains. Only the list of physical primary domains is complete. Secondary domains will be impossible to identify as secondary domains if primary domains are unknown. We will not be able to derive their properties from simpler and fewer parts. Behavior might be unintentionally replicated.

Problem 1 is the smaller of the two problems. The larger problem is that all primary domains are not identified as primary domains.

What do we lose if we ignore domain analysis? We risk inadvertently replicating domains and the knowledge held by them – behavior, relationships, values, and measures. If we cannot coordinate and synchronize processes under the relentless impact of change, innovation, and new learning – a task facilitated by domain analysis – we will risk loss of integrity of information on behavior and measurement. We will risk loss of quality. Completely ignoring domain analysis is risky.

How large is our risk? It depends. Primary domains are reused most often – especially if we include their use in common secondary domains like space, time lapse and growth rates – but relationships between primary and secondary domains are not always intuitively clear. In box 49 we will see that value objects can be complex – the color domain, that seemed a simple nominally scaled domain at first, blossoms before our eyes into a rich and complex structure as we added information to it. Colors have many expressions, but only one meaning. It is not always easy to see the underlying unity of meaning behind these diverse expressions because their meaning lies buried in abstraction – abstract information but concrete behavior.

Meanings can be complex, even if they are only meanings of classes of values. Despite complexity, we have to normalize meanings across diverse and complex processes when demand for quality is stringent, and our operations are global and diverse, or simply too large to manage intuitively.

If we create meanings in secondary domains, completely ignoring their relationships with primary domains, we risk replicating domains and their meanings inadvertently. The same domain may be manifested in different expressions. If domains are replicated, we will, without knowing it, also replicate knowledge – rules and meanings buried in our business processes and systems. Coordination of change can become difficult, and the impact, under the pressure of continual unrelenting change, can be catastrophic.

Whether the effect of unintentionally replicated rules is minor or catastrophic depends on how frequently we reuse these domains, how frequently we change them, and how important the objects that map to them are to the businesses they support.

A practical approach is to pick the "low hanging fruit" and move on – to use what we know, the best we can, pass over the low impact items, and reduce the risk of replicating domain information for difficult high impact items to the best of our ability. In this, we have help – fortunately many primary domains are known. The solution is to apply a heuristic approach – an approach that may not always be correct, but one that will use the primary information we have, and will simultaneously not get bogged down in analysis paralysis.

We would rarely reduce temperature to its primary components, but we might analyze biometric signatures closely. We should attempt to reuse secondary domains as much as possible, and tie them to primary domains in key areas. The intent must be to ensure that we do not replicate key domains when we create new ones. Often, this will be intuitively obvious. It is the best we can do given the state of the art today. The need to curb analysis paralysis takes precedence over the need for perfection. Some shortcuts can help.

Shortcuts

We have an exhaustive list of primary physical domains, but not of non-physical domains. Secondary domains are not always easy to identify. The following heuristic rules can help bring order to the chaos of domain analysis. They are essentially prefabricated domains or prefabricated rules about how to classify domains so that we can often skip or shorten the procedure in box 34:

- An instance identifier always maps to a nominal domain. Skip domain analysis for these (each is a separate "attribute" derived from a different nominal domain). Use the universal perspective and the usual rules of data normalization to normalize the information they carry.
- The common codes of an organization usually map to nominal or ordinal domains. Common codes might be integrated if they map to the same domain.
- Common codes are often nominal domains.
- We should check for multiple components in the code structure to see if the code maps to a complex secondary domain (the benefits of rule 10 will follow only if you can analyze domains of association in terms of their constituents). The universal perspective can help. You can also use the conventional data normalization techniques.
- We should check for a natural ordering sequence (partial order), a natural nil value or bounds to determine if a common code is ordinally scaled.
- Many attributes will map directly to known primary domains. Check that first.
- The money domain is a ratio scaled primary domain. Rates, such as revenues and prices that involve money are secondary ratio scaled domains.
- We should avoid reducing temperature or sensory information in the five fundamental formatting domains of section 1 to primary domains for most applications.
- Many objects will have audit attributes
  - The process, person, event, rule, reason, and automation that caused a state to change (for each slice of the object in figure 22). Skip domain analysis for these attributes.
  - Date and time stamps.
- Dates, times, and date–time all map to the date domain.
- Pairs of dates, times, and date–time stamps may be subtracted. The results all map to the ratio scaled time domain.
- Each rate points to a ratio scaled domain. Rate domains are always secondary domains. The numerator and denominator point to different domains. Each may be a primary domain, or another secondary domain that may reduce to a primary domain on further analysis.

Domains and value sets – states of domain

The values in the value set of Chapter 3, section 2, were all drawn from the same domain. Therefore, some might argue that value sets are domains, sometimes finite, and sometimes not. This argument is false because a value set can change its state: values may jump in and out of value sets in response to events, and the lawful state space of objects will change in step.[132] Even partitions can change state between exclusion and inclusion partitions in this

[132] Ron Ross, in the chapter on calculators in [294], distinguishes between constraints that test for integrity (type 1 calculators) versus those that change values (type 2 calculators). At the end of box 33, we have also discussed how equations flow from the properties of partitions and value sets – two components of knowledge.

elaborate choreography of state space, an event that orchestrates real-world systems and processes.

As we have seen, domains may have relationships, constraints, and rules attached to them. They do have states, but these are not temporal states. These states just are. The flow of time and the tide of events pass them by. Neither time, nor event are aware of these states, nor are domains aware of time or event. Domains are timeless, eternal, and still. They are meanings – sometimes meanings structured from other meanings. Domains are the meanings that lend meaning to states, temporal objects, and time itself.

Value sets are subsets, carved out of domains. Without these "larger" domains, there would be no value sets, and, conversely, these "large" domains are value sets shorn of temporal states. Domains are still; value sets are not.

The *subsets* of domain, the immutable class of values, are mutable and have states, but the larger class, the "ocean" of values they draw upon, the domain itself, is changeless and immutable. Value sets are therefore subtypes of domains – domains, or pieces of domains, with temporal information.

(A value set can consist of any combination of values in a single domain. Technically, the class of value sets is the power set of a domain – see box 19.)

What about domains that grow over time, acquiring new values in step with new knowledge, like the gender domain we discussed earlier in section 3, or the color domain in box 49? How can we say that they are still and timeless?

The domains were always there. It was only the state of our knowledge about them that changed. "Domain" is a concept that helps normalize non-temporal behavior. If we represented domains with knowledge artifacts stored in an electronic repository, the knowledge artifact would change state, not the concept itself. Domains have no birthday. The knowledge artifact does. Domains have no change day. The knowledge artifact does. Domains might only have a recognition day, a state of knowledge about knowledge of a domain – a relationship between a domain and those who use it.

It is the state of a relationship, a relationship between a domain and the class of actors, not of the domain itself. This relationship is a bridge between business meaning and its automation. When we step into the realm of knowledge artifacts, we cross this boundary.

As our knowledge grows, we can add values, constraints, relationships, and other information to the knowledge artifact – information that was always there, but information we did not know. The artifact will change state, and the information will automatically flow to all value sets and attributes that map to the domain.

In box 49, we might have started with a model of the color domain, a knowledge artifact with three nominal values (one for each primary color). All colors in our processes would then be restricted to only these three. On learning about ratio scaled luminosity, we could create a new ratio scaled color luminosity domain, a domain of association between the color domain and the luminosity domain. The repertoire of colors now available would expand. The secondary domain of colors always existed – they were always there – we only made the information available to processes that are assembled from the repository of knowledge artifacts afterwards. If this physical process of assembly of systems is dynamic, processes will flex instantly and recognize the new colors. Otherwise, we might have to reassemble them again. It depends on the design of the electronic repository of knowledge artifacts. It is a technical design issue beyond the scope of this discussion. The domain was always there, eternal, even if the repository did not have it.

Even the relationship that gives birth to value sets is a timeless mathematical concept that only articulates that a domain may have subsets. *It is the subsets that are mutable, and have mutable states*, even if they are subsets that span the entire domain (i.e., in terms of box 19, they are not *proper subsets*). Domains are the seeds from which inclusion and exclusion sets, as well as attributes and objects grow. It is through value sets, attributes, and temporal objects with histories that domains influence the elaborate choreography of events and states. In a world where movement, meaning, and location come together in the fabric of space and time, domains are the timeless "ocean" of stateless meaning – the silent still, where movement meets meaning.

The metamodel of domain

> Before their eyes in sudden view appear
> The secrets of the hoary deep, a dark . . .
> Without dimension; where length, breadth, and highth,
> And time and place are lost; where eldest Night
> And Chaos, ancestors of Nature, hold
> Eternal anarchy. . .
> (John Milton in *Paradise Lost, Book II*)

Figure 68 is the metamodel of domain. It expands on the hierarchy of domains in figure 67 and includes the primary domains of business meaning. The primary domains bridge the gulf between the abstract domains in the metamodel and their instantiation in the physical world of business. Meanings are patterns of information – simple, complex, and small and large patterns – patterns made from the information in primary domains. All business meaning is, in the ultimate analysis, assembled from primary domains subject to the rules in box 48. (To complete this model, we should also add the strong and weak forces in physics to the list of primary domains in Figure 68, then it would support all conceivable meanings and technologies: past, present, and future – see box 42).

Figure 68 shows how *preference* is an ordinal domain with a nil value – a point of neutrality that signals an absence of preference – a value that suggests neither liking nor disliking. It just says "no preference."

Figure 68 also shows how the money, or economic value, domain is a subtype of the preference domain. The subtyping relationship between economic value and preference is inherited from the generic ratio scaled domain. The generic ratio scaled domain is a subtype of the generic difference scaled domain, which, in turn, is a subtype of the general ordinal domain. This makes the generic ratio scaled domain also a subtype of the generic ordinal domain. The *money* and *preference* domains merely inherit that subtyping relationship. The money domain being a subtype of preference also demonstrates how inclusion polymorphism emerges as subtypes inheriting the subtyping relationship itself from their supertypes.

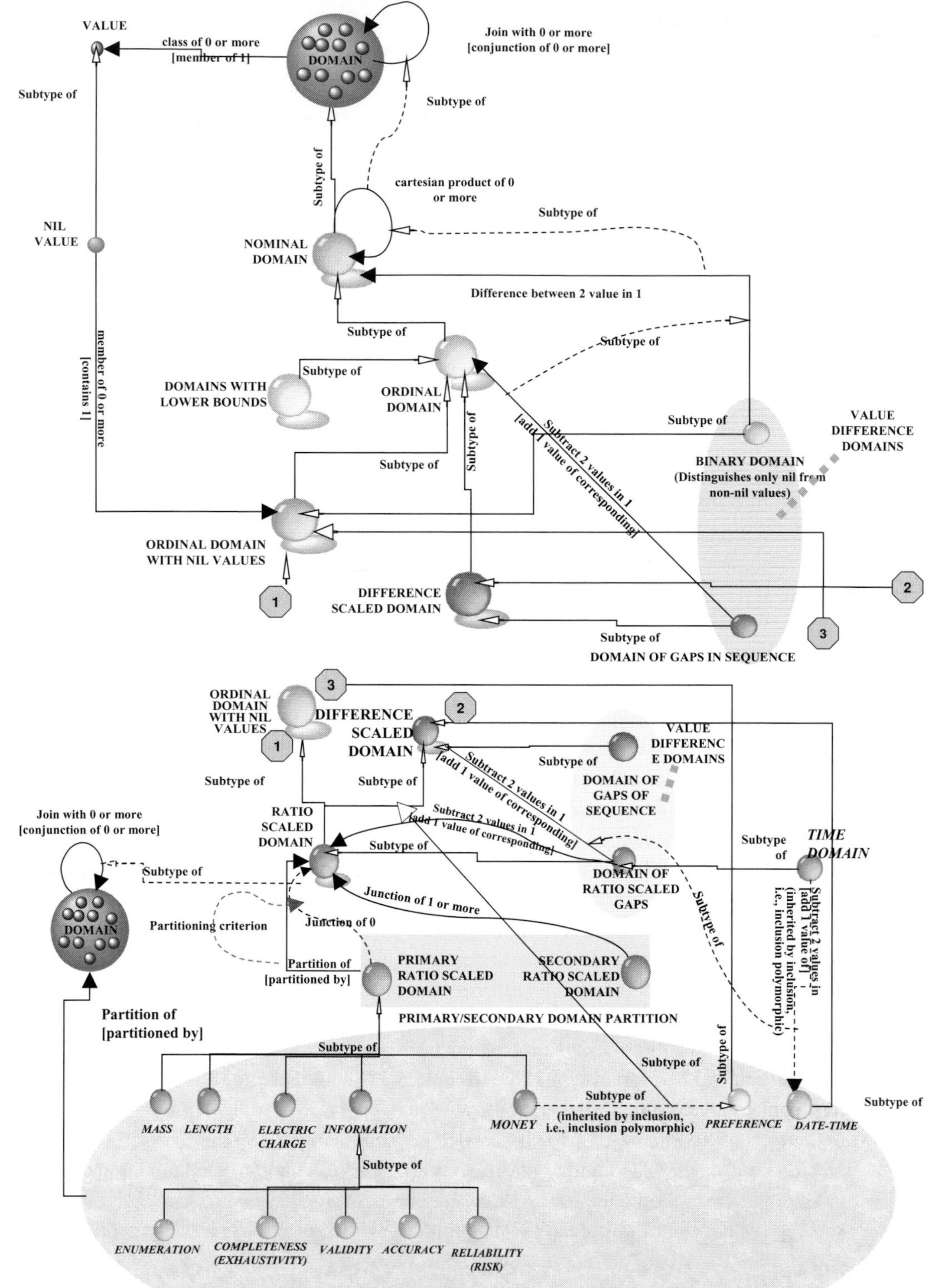

Figure 68 The metamodel of domain

Figure 68 also articulates the behavior of the value difference domain – the domain of gaps between values in a domain. It conforms to the proximity metric of table 2 – gaps between ordinally scaled values form an ordinally scaled binary domain. Gaps between ordinal values are difference scaled, and gaps between difference scaled values are ratio scaled, as are gaps between ratio scaled values.

The time-lapse domain (labeled time domain in figure 68) emerges from differences in date–time values. Figure 68 illustrates how this is also an instance of inclusion polymorphism. The arithmetic subtraction operator is inherited from the general difference scaled domain by the difference scaled date domain. Arithmetic differences between dates map to the time-lapse domain, which is a subtype of the general domain of gaps in difference scaled domains.

Figure 68 also illustrates how the arithmetic subtraction operator is a subtype of the more general "difference" operator, which, in turn, is a special kind of connective[133] between values in a domain, as is the Cartesian product of domains.

All domains in figure 68 inherit the Cartesian product and the general connective from the "unknown domain." The Cartesian product, like the difference operator, is a subtype of the general connective operator between domains. Indeed, all operators that connect domains to create new domains – whether simple domains like unit price or complex domains like color preference – are subtypes of the general connective operator. The general connective articulates the fact that new domains may be built from old. Only the primary domains do not owe their existence to connectives (not counting the subtyping connective). They are not made from other domains; rather they *make* secondary domains.

As information is added to domains, specific subtypes of the general connective such as arithmetic operations, comparison of values and others emerge in step with the fullness of information in each kind of domain. Thus it is not just values that change behavior in step with the information content of the domain; operators too march lock step with it (see the rules in box 48). Operations and values are information, and each adds meaning to the domain. Figure 68 articulates this.

Notice the cardinality ratio of the relationship between value and the unknown domain on the upper left-hand side of figure 68. It has a lower bound of zero. How can we have a domain of *no values*? We can have a domain with no values because it is the null space we discussed under patterns. A domain is null if it consists of only null values; it is the empty set of box 19. The unknown domain embraces both the null domain as well as domains with other values. On the other hand, unlike the domain of null values, domains with nil values must contain at least one value – the nil value.

Towards the bottom of figure 68, we find domains of information quality. These domains are universal parameters of information quality. They sharpen management of risk and frame quality assurance of information systems and business processes. Each is potentially ratio scaled, but may lose some scaling information in its relationships with other domains (or even in recursive relationships with itself) and become difference scaled, ordinally scaled, nominally scaled, or simply "unknown" in different contexts.

[133] Connective: see the endnote on gluing objects together.

All domains are classes of values. They are value objects. Value objects may only be assembled from other value objects, and their assembly is subject to rules. These rules are the nub of the metamodel in figure 68. We have discussed them at length. Many are too complex to represent graphically.[134] They are consolidated in box 48.

Taken together, domains and values represent the lowest common denominator of immutable meaning. Meanings that can neither change nor shift nor die with the shifting sands of time, space, or context – meanings that just are. They are the meanings that create other meanings. They are immutable.

Box 48 Domain rules ready reckoner

Rule 0: Measurability of domains:

(a) Nominal domains only carry information on distinctions between values.

(b) Ordinal domains rank values in terms of their magnitudes, but contain no information on absolute magnitudes or differences in magnitude between values.

(c) Difference scaled domains convey information on the magnitude of differences between values, but no information on a nil value (and hence no information on absolute magnitudes of values).

(d) Ratio scaled domains convey information on absolute magnitudes of values. This includes information on a nil magnitude.

Rule 1: Adding meaning to a domain creates a new domain. Meanings added may be new, or inherited. If the new domain includes meaning(s) inherited from other domains, it is a subtype of the domain(s) it was created from and:

(a) the new domain will inherit unit-of-measure information from its parent domain and add information of its own;

(b) if the nil value was included in the new domain, the units of measure of the old domain will be inherited with their zeros reset to coincide with the new nil value.

Rule 2: Any multiplication or division operation on values in a ratio scaled domain creates a new ratio scaled domain. Multiplication and division operations may be between values in the same or different domains. Units of measure of the new ratio scaled domain will be expressed in terms of the same operations on units of measure of the domains it was created from, as will conversion ratios between units of measure of the new domain. Units of measure and conversion ratios for the new domain may be derived from its constituent domains (and ultimately its constituent primary domains).

Rule 3: Addition and subtraction operations on values in the same ratio scaled domains map back to a subtype of the same domain. Addition and subtraction are permitted between parent and subtype domains. The subtype domain will

[134] The entity relationship diagramming technique we have been using in this book is not robust enough to handle that level of complexity without becoming too cluttered to clearly articulate what it must about assembly of domains from other domains. It is best said mathematically, or said in words. In this book, it is said in words.

inherit all units of measure, conversion, and formatting choices from its parent domain.

Rule 4: Addition and subtraction operations between values in different ratio scaled domains have no meaning, if one is not a subtype of the other.

Rule 5: Proportions are ratio scaled attributes of aggregate objects that are subtypes of a parent object class. All proportions conform to the following rules:

(a) *Population*: Every aggregate object has a population attribute. The population enumerates its members. Its units of measure are units of enumeration.

(b) *A sum of a ratio scaled attribute over all instances in a class*: Given a ratio scaled attribute of an object instance, corresponding object class(es) will have an attribute that sums up the instance level attribute over all instances of the class. This sum is a class level, not instance level, attribute. Class level totals will automatically be implied by the existence of each instance level ratio scaled attribute and the units of measure of the class level attribute will be identical to that of the instance level attribute.

(c) *Sum of class level attributes in a partition*:

(i) The sum of populations of individual subtypes in an exhaustive partition will equal the population of the parent object. In a non-exhaustive partition, the sum may be less, but cannot exceed the population of the parent object. The units of measure of the sum will be inherited from the enumeration domain.

(ii) Adding to (subtracting from) the population of a subtype will always *automatically* add to (subtract from) the population of its supertypes, but not necessarily vice versa (the partition may not be exhaustive, and, even if it is, we will not know which subtype(s) has increased (or decreased) its population(s) based on increases (or decreases) in the population of supertypes alone).

(iii) The sum of class level attributes – attributes that are sums of absolute values of corresponding instance level attributes – summed across all subtypes in an exhaustive partition, will equal the value of the class level attribute of the parent object. In an inexhaustive partition, the sum may be less but may not exceed the class level attribute of the corresponding parent object. The units of measure of the sum will be inherited from the domain of the summed attribute.

Since the partition represents the collection of subtypes in it, the partition normalizes these relationships. They are between the parent and the partition. Each subtype in the partition inherits them.

(d) *Proportions*: The sum of populations of individual subtypes in a partition will equal the population of the parent object in an exhaustive partition; in a non-exhaustive partition their sum cannot exceed the population of the parent. The sum of class level attributes, attributes that are sums of absolute values of corresponding instance level attributes, summed across all subtypes in an exhaustive partition, will equal the value of the class level attribute of the partitioned object; in a non-exhaustive partition, their sum cannot exceed the population of the parent. Since the partition represents the collection of subtypes in it, this relationship is between the parent and the partition. Each subtype in the partition inherits it.

The result of a proportions calculation is independent of the units of measure used to express the divisor (or dividend – both divisor and dividend may be expressed with the same unit of measure), provided the divisor and dividend are expressed in the same units of measure. The existence of proportions is automatically implied by the existence of an attribute and partition of an object class.

(e) *Range*: Values in any domain of proportions lie between *nil* and *total*. The unit of measure of a proportion should assign the nil value to the number zero, and the total value to a larger number.

(f) *Sum of proportions*: The sum of proportions in an exhaustive partition must equal *total*. In non-exhaustive partitions, the sum of proportions cannot exceed *total*. The units of measure of the sum of proportions are the same as units of measure of proportions (rule 3 implies this).

(g) *An injunction against arithmetic addition of proportions*: Proportions for a given attribute of a subtype in a partition may be meaningfully added. Across partitions, or across different attributes, addition of proportions has no meaning.

Rule 6: Division of one difference scaled value by another is meaningless, and hence so is multiplication of one difference scaled value by another.

Rule 7: Mutually subtracting pairs of values in a difference or ordinally scaled domain will create a new domain. The new domain is the domain of intervals, or gaps between pairs of values, obtained by attaching a nil value to the original domain. It is a subtype of the domain it was created from. The units of measure of the domain of gaps will be inherited from the domain it was created from (as required by rule 3). If the original domain was:

(a) difference scaled, the new domain of intervals will be ratio scaled;

(b) ordinally scaled, the new domain of intervals will be difference scaled.

Rule 8: Addition of values in the same difference scaled domain is meaningless by itself. It may have meaning in expressions that do not bias the result by changing the arbitrary zero value of a given unit of measure (e.g., by adding, subtracting or multiplying it multiple times), in which case it maps back to the same domain and the same unit of measure. The entire arithmetic expression must be considered in toto.

Rule 9: Values in the domain of gaps may be meaningfully added to, or subtracted from, values in the domain it was generated from. The results of the operation will map to the latter domain, in the same units of measure. Addition of values between subtype domains and their parent domains will be meaningful provided the hypothetical nil value is not distorted. Addition and subtraction operations with values in other domains, with this exception, have no meaning.

Rule 10: A relationship (Cartesian product) between domains creates a new domain. The Cartesian product may involve a single domain or several domains. The scaling of a domain created thus is:

(a) At least, the same as the scaling of the participating domain with least information, when all domains thus joined contain sequencing information. We can assume that the new domain is scaled the same as the participating domain with the least information without attributing non-existent information to the domain. If the domains were all quantitatively scaled, the unit of measure of the new domain, a complex value object, will be the arithmetic product of units of measure of domains associated by the Cartesian product.

(b) Ordinal, when one or more domains thus joined is ordinally scaled (a special case of rule 10(a) above).

(c) Nominal, only when all domains thus joined are nominally scaled.

Rule 11: Every domain inherits the following properties from the domain of information:

(a) A count of its members is an attribute of every aggregate object, including every domain and object class. It maps to the ratio scaled enumeration domain.

(b) Every domain has one or more domain(s) of proximity metrics associated with it. The domain of the proximity metric is a subtype of the domain it is associated with, and measures the accuracy of values in the domain. The scaling of domains of proximity metrics conforms to table 2.

(c) Every object, including domains and their values, has the ratio scaled attribute of reliability. Values of reliability range from *nil* to *total*. The reliability domain articulates the consistency of meanings. It maps to the domain of proportions. In a purely deterministic model, reliability may only be nil, or total.

(d) Every object maps to the ratio scaled completeness domain. This domain measures the proportion of information in the object that has been realized. Values of completeness range between *nil* and *total*. Completion maps to the domain of proportions.

(e) Every relationship maps to the ratio scaled validity domain. It is a universal attribute of relationships. Values of validity range from *nil* to *total*. The validity domain articulates the meaningfulness of a relationship (rule) between objects. In a purely deterministic model validity is either nil, or total. Validity maps to the domain of proportions.

(f) Specific relationships with these domains of information may be nominally, ordinally, difference, or ratio scaled, depending on the information content of the relationship.

Measurement of meaning – a paradox of perspectives

> What is the substance, where of are you made,
> That millions of strange shadows on you tend?
> Since every one hath, every on a shade,
> And you, but one, can every shadow lend.
> (William Shakespeare, *Sonnets LIII*)

Domains are objects, even if they are only value objects. Are domains, like other objects, also subject to the chimera of perspective?[135] The answer is a resounding "yes," and yet, paradoxically, domains also unify meanings. Domains might be chimeras, but unify meanings they must, because domains are immutable and timeless – they are meanings that cut across many perspectives. Let us see how domains can be chimerical and yet bring constancy to the mingling shadows of meanings that flitter across perspectives without count.

Domains in perspective

First, let us understand what perspective means to a domain. In Chapter 2, section 4 we saw how meanings can change, fracture, dissolve, and merge between perspectives. Domains are pure information, immutable, eternal, and without change – the seeds from which other objects, attributes, and entire structures of meanings crystallize, grow, and engage each other in ever-expanding circles to create new possibilities and new meanings – complex structures that model a world as subtle as it is rich. Despite being pure meanings, perspective impacts domains. Let us see how even pure meanings can be fickle shades, flittering between a riot of perspectives. We will start with a simple domain – unit price.

Domains of viewpoints

Unit price is a simple secondary domain – a quotient. It is the amount we pay per piece of a product we buy. It has two items of information – the amount paid and pieces bought – or should the two items be unit price and amount paid? Perhaps it should be neither. We can infer the pieces bought from the amount paid and unit price. Therefore, it might be correct to assume that amount paid and unit price are the two independent meanings we seek. Any two items, of the three – unit price, amount paid, or the number of pieces – implies the third. Which two items are primary items of information, independent in meaning, and which item is dependent on the other two? It all depends on a point of view. There are no right answers and no wrong answers. It is a matter of perspective. They are all interdependent.

We resolved this question in box 42 with the artifice of primary domains. Unit price is a secondary domain, in which primary domains have met and joined. Box 42 showed how it was all a matter of pure information. We were free to choose the identity of these primary domains, but not their number. Their number depended on information content – the "amount of meaning" that was intrinsic to the real world and the rules that frame it. We arbitrarily standardized which domains we would consider primary and which secondary. We nailed the chimera of perspective by decree. It was a decree built on common understanding, a unified perspective of the reality that binds us all. Primary domains help us anchor all

[135] Chapter 2, section 4 discusses the problem of perspective.

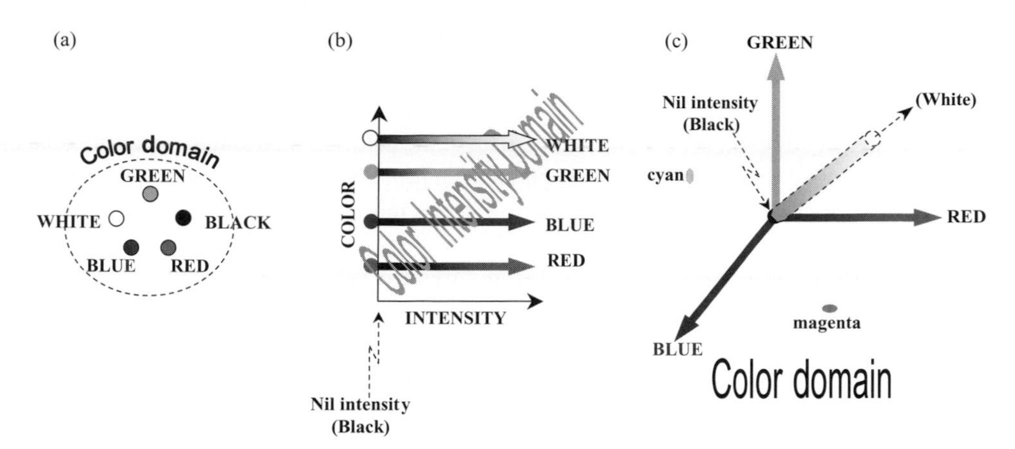

Figure 69 Shifting perspectives of color

perspectives. These *domains* are primary only because we have declared them so. This *perspective* of domains is primary, only because we have declared it is so. It could just as easily have been some other set of domains. That is the first nail we have driven into the chimera of perspective.

Embracing scope creep

Let us take another step into the mist of crystallizing meaning. In the last example, the quantum of information was constant. It helped nail down the primary domains. What if information content itself shifted like quicksand, as it does when new learning forces us to embrace change – when perspectives are fickle because information is? Consider color.

Assume that it has only struck us now that different colors exist. We neither know, nor care that some colors can be more similar than others. Mere distinction of colors will suffice in our model. Color may be important to some objects. We can map their color attribute to the color domain – a mere nominal domain of finite extent (cardinality), with a limited number of colors we can distinguish. This is the color domain of figure 69(a).

Scope shifts. We realize that color depends on illumination. Illumination is ratio scaled. Colors get brighter as the intensity of light increases. They get darker as light dims, until, as light fades, all colors darken till they turn black. One color becomes indistinguishable from another in dim light. Our old perspective of color does not fit our newfound knowledge. We must revise our perspective. That is easy. We add the illumination domain.

We can now create a new domain, a Cartesian product, of the nominally scaled color domain and the new illumination domain (figure 69 (b)). It is the color intensity domain – a mixed space – the intensity axis is ratio scaled, and the color axis is nominally scaled. The sequence of colors along the vertical axis of figure 69 (b) is irrelevant – it only matters that all the colors we need are there.

When the intensity of the color of an object matters, we map the attribute to the new color intensity domain. This domain has inherited all prefabricated effects such as color switching and changing illumination from its constituents. Given a color, similarities between shades of the color are ratio scaled (or ordinally scaled with a natural zero at worst). Across colors, it follows the laws for nominally scaled state spaces. We can tell which shades are close and which are not. If we need a new hue, say, purple, we can drop it into the "bag" in figure 69(a). All shades of purple will automatically become available to all objects because the color intensity domain is the Cartesian product of the color domain and the intensity domain, and the color domain now recognizes purple. Our processes and information systems, assembled from knowledge artifacts, can easily embrace this change, but new learning again sweeps the concept away.

We discover there is no limit to color. A riot of colors can exist. Mixing different colors may make new and different colors. Colors that have never before existed may be created on the fly. Old knowledge fails. We must adapt to change once again. We can take the Cartesian product of the colors in the color intensity domain we just created to create a new domain that will let us mix colors to obtain new colors – but, alas, reality confounds us once more. There are different mixtures that yield the same end result – a given color can be made in different ways. Moreover, we can tell some hues are closer than others, just as we could tell which shades of a *given hue* were similar earlier.

We recognize that the color domain has less information than we are imputing to it. Some colors we are treating as independent items of information (independent colors) might be mixtures themselves (see box 49). We must reduce the dimensionality of the pattern (its extent will still be infinite because illumination has no upper bound). We find we need only three colors (how do we know, and why only three? – see box 49). We can combine various intensities of these three colors in different proportions to create every possible shade of color.

Which three shall we use? We can declare red, green, and blue to be primary color domains (this is also aligned with convention). It is easy to tell that they are distinct colors, different from each other, but it is hard to say how different, or even if they are equally close or far. On the other hand, pink is obviously partway between red and white – it is a pale shade of red, and it is easy to tell that black and white are opposites, with shades of gray in between.

We could have chosen magenta, cyan, and yellow as our primary colors, instead of red, green, and blue. That would change the axes, the basis of the color domain,[136] but its information content would be exactly the same, and it would contain exactly the same colors. Tilting the axes would not create a new space, nor make a new meaning; only coordinates of points, their expressions, would change. Tilting the axes in figure 69(c) might show the color domain in a new perspective, but the *meaning* of color would stay the same because no new information was added to the meaning of color – just as the *meaning* of unit price would not change if we changed units of measure.

Changing units of measure is like stretching or compressing the axes we use to express or measure meanings in state space. Moving the origin is like mapping the nil value to a non-zero number.[137] Neither changes the meaning of the domain. In the same way, tilting the axes of state space cannot change its meaning. It is only another perspective of the same meaning.

[136] [260] and [261] discuss the bases of state space.

[137] Moving the origin is like mapping the nil value to a non-zero number for ratio scaled state spaces. For domains with no known nil values, it is like mapping to a different arbitrary reference point. See Chapter 4, section 2 and the discussion of rule 1(b) under "The risk of domain analysis".

Stretching (or compressing) the axes of state space and shifting the origin apply to both simple and complex domains. Tilting axes only matters to complex domains. Complex domains normalize complex behavior such as rotation of axes and other kinds of operations such as bending or distorting axes in other ways. Many of these distorted axes are merely secondary domains derived from those that the original axes represented (see box 49). They merely express the same meaning in different terms – a new rule expression, not a new rule meaning, or a new domain – merely another perspective of the same meaning.

Some of these distortions might even impact the universal attributes we have discussed earlier under patterns, such as dimensionality, extent, separation, sequence, and direction in state space, but if no new information is added, no distortion will create new meanings. It will only create a new perspective based on different secondary (derived) domains – a mere restatement of meaning – a new expression (see box 42).

If we start by expressing complex domains in terms of primary domains and then build in its other perspectives based on secondary domains, we have derived from the original axes, it is easy to keep track of equivalence of meaning. However, if we just have a bunch of complex domains, their equivalence may not always be intuitively obvious (remember the temperature domain – it is not immediately obvious that it is derived from other domains).[138] This is the problem at the heart of the problem of perspective – a topic we discussed in Chapter 2, section 4. Domains are abstract values, and, in complex domains, the problem of perspective can become especially acute.

Given a set of static points in space, their relative positions will not change in different coordinate systems, but expressions of their coordinates will (see box 37 and box 49). The expression will depend on the coordinate system – its origin as well as the orientation of the frame of reference. Each coordinate system will be a perspective of the same underlying meaning – the same domain.

The conversion between these coordinate systems will not lose or gain information if they conform to the golden rules of measurement in section 2. The conversion rule may be like the rule in figure (a) of box 33. This kind of rule generalizes the second special rule of conversion between units of measure (in section 2). The generalized rule can apply to complex domains. Indeed, the rule expression may even be like that of figure (b) of box 33 – actually an entire set of rule expressions like figure (b) of box 33 – one for each degree of freedom, or axis, of the state space in question, so that old coordinates are transformed to new expressions, but the information content of the pattern of values is preserved. This will present a new perspective – one that expresses the same meaning in terms of secondary domains derived from domains represented by the original axes (see box 49).

If we do not know the equivalence of coordinate systems beforehand, we will need to know properties of various kinds of operations that distort the axes of state space, as well as pattern recognition algorithms that can use this information to discover the equivalence of patterns in different coordinate systems. An exhaustive discussion of these operations and

[138] [290] and the references in the endnote on Shannon's information theory show how temperature is derived from primary physical domains.

their properties is beyond the scope of this book.[139] It will suffice if readers understand that the same meaning may be expressed in several ways, and even though their equivalence may not be obvious, *conversion between perspectives is similar to converting between units of measure* – ***provided information content does not change***.

Conversions between perspectives of this kind merely generalize unit of measure conversion rules – unit of measure conversion involves compressing or stretching individual axes, whereas conversion between perspectives of a complex domain involves secondary domains that bend, tilt and warp the original axes in state space.

In our discussion of patterns under "Patterns of symbols, patterns of objects" in section 1, the pattern of months and years could represent the time domain accurately and reliably, because it had a *larger* information carrying capacity than the concept of time. Time is a one-dimensional, infinite, and unbounded pattern in information space, whereas the pattern of months and years was more like the pattern in figure 51 (c),[140] a pattern with more freedom, more dimensions, and a larger capacity to convey information than a mere line with neither beginning nor end. "Smearing" and "bending" the linear concept of time into the infinitely long cylindrical surface in figure 51(c) obtained this expression of time.

We see that when information content is constant, we can switch between perspectives, and a common meaning will be their constant anchor, but when information content changes, domains can give birth to new domains – domains built upon the old – as they did when the color domain unfolded in all its variety and fullness. Attributes that map to new domains will automatically embrace new behavior – both new attributes, as well as old attributes that are remapped, manually or automatically, will adapt to change. Processes and information systems will change their behavior in step with the knowledge artifacts they were assembled from.

When knowledge artifacts change state, knowledge keeps pace – business knowledge embedded in business processes facilitated by automation. That is where return on investment in knowledge lies – change at the speed of thought.

Box 49 The information content of domains – new learning and changing perspectives – an example in color (on our website)

Box 49 has a more sophisticated, partly mathematical discussion on domains of information, absorption of new learning, and changing of perspective. It show how the perspective of the color domain can shift and change state in step with new learning and cognition. Box 49 also discusses how information can be kept normalized as perspectives shift. This box includes a discussion of physical space as a domain and shows how this depends on the meanings we derive as we add or subtract information.

[139] See the references on topology and abstract algebra in the bibliography, the endnote on lambda calculus and the Church–Rosser theorem. Volume 19 of [336] also addresses operations that map between values and may be used to determine the equivalence of expressions: see "Algebraic geometry" (pp. 951–958), "Abstract geometries" (pp. 969–970), and "Topology" (pp. 977–998).

[140] "Patterns of symbols", "Patterns of objects" in section 1 describes how this pattern is assembled from patterns of information that describe the cycle of months and the passage of years.

Meanings that represent meanings

"The Eagle has landed!" – words pregnant with meaning; words that bridged a quarter million mile gulf between worlds at the speed of light to make history. They told the world that the first manned ship to the moon had arrived. The words had one literal and unexciting meaning, but that is not what they expressed. The phrase was a metaphor for another meaning.

Metaphors are common in every natural language. Meanings can represent meanings. Box 36 said they could (see the metamodel in box 36). Both the object represented, and that representing it, may be meanings. They could even be domains.

Sometimes meanings are deliberately encrypted this way. Then it is not a problem – *if* you have the key. If you do not, the result can be either hilarious or disastrous depending on your perspective. Even if you have the cues (key), the second meaning must convey at least as much information as the meaning it represents, otherwise it might clip and distort the meaning it represents. Metaphors will then be imprecise, and words may mislead.[141]

Meanings can be complex, even if they are only collections of values with common meanings; even meanings of words. Have you, dear reader, ever had a thought that you found hard to express in words? Have you ever heard somebody describe a complex idea, and not understood it? A written word might be a format, a spoken word might be a format, but the concept in the word is a meaning. It could be a complex meaning, a complex domain of mixed meaning. Many are only roughly aligned – perspectives of domains that share less than all their axes, or only roughly align their axes. This is terribly important when we translate expressions from one language to another, even simple words; even common words like "Peace."

"Peace" shares a common meaning – absence of conflict across most languages, but nuances are different. For example, in the languages of Northern India, the word that translates most accurately into the English word Peace is "shanti." "Shanti" also mixes absence of conflict with ideas of harmony, tranquility, and the spirit of universal unity of all creation. On the other hand, the phrase "non-violence" has embedded within it nuances of struggle and conflict! Words are labels for meanings. Many meanings are complex domains. Synonyms, especially across cultures, may not mean exactly the same when they label complex domains. Similar domains may not share all their axes, and, even if they do, perspectives may not all be perfectly aligned. Each is a perspective of meaning. The problem becomes intractable because we have no well-understood and widely accepted standard primary domains to fall back on where human perception and mental constructs are involved. The distinction between homonym and synonym starts to blur as we leave the bedrock of deterministic meaning behind.

Although most cross-cultural concepts share components of meaning, each can be associations of meanings unique to the language or culture in question. Unbeknown, and insidiously, differences in nuance, configurations of meaning in complex domains, have seeped into the collective consciousness of people around the globe – people being woven ever

[141] The recent discussion on months and years representing the time domain demonstrated how meanings can represent other meanings with precision and reliability provided they have a *larger* information carrying capacity than the meanings they represent.

more tightly into the whirling web of global commerce – people spinning a web of global knowledge around a hub of rapid fire, even instantaneous, global communication.

Precise communication of complex ideas will be the key to competitive advantage in the rising riptide of a global economy driven by knowledge, information, and customer value. In this environment of surging, rapid-fire global communication of complex information, matched by rapid, finely targeted, and complex global responses, risks will rise when perspectives across cultures do not translate easily. Even without consciously knowing it, meanings – complex state spaces labeled by words – have different axes in different languages and cultures. The more translation automata "know" these differences, the more accurate translations will be, and the more it will mitigate the surging risk of poor communication.

Whether culture creates meaning or language creates culture is a topic of continuing debate among experts in the field. However, most agree that each influences the other – at least a little.[142] The true tyranny of words is that it shapes the very thoughts that frame them – the thoughts of men and women whence meanings are created, labeled, and communicated, as dear reader, we are communicating with you.

Fortunately, there is a universal perspective; common understanding can prevail. Common understanding dwarfs differences for all but the most complex nuances of meaning. Ultimately it boils down to the metamodel in figure 68, the common basis of all understanding, alien, human, or unknown from which all meanings are derived.

As we descend the hierarchy in figure 68, domains become flush with meaning. At the line that separates business meaning from the bland rules of mathematics, we find the domains of primary meaning – the seeds from which all concepts grow – physical, business, and cultural. Below them, we find nuances of meaning, secondary domains, and perspectives without count, but all founded on common understanding – elemental meanings – the meanings of primary domains. If we change direction and start ascending the hierarchy, we lose information. Domains gradually lose measurability and meaning until chance again rules supreme, and finally, at the gray border of null space, we meet the unknown domain, a pale ghost, in which meaning, measurement, observer, and observed, all lose their identity and become one unknown pattern.

> Behold a wonder! They but now who seemed
> In bigness to surpass Earth's giant sons,
> Now less than smallest dwarfs, in a narrow room
> Throng numberless.
> (John Milton, *Paradise Lost*, Book I)

4 Storing abstract meaning

Domains are intangible classes of abstract magnitudes – magnitudes shorn of numbers or formats, but imbued with meaning. In box 49, we saw that domains can be complex. The

[142] Experiments have confirmed that cognition and thought are affected, to a limited extent, by language and culture. See [291].

color domain, that seemed a simple nominally scaled domain at first blossomed into a rich and complex structure as we added information to it. Colors had many expressions, but only one meaning. It was not always easy to see the underlying unity of meaning behind these diverse expressions, because the meaning lay buried in abstract information. How may we store such abstract concepts, let alone assemble, configure, and manipulate them electronically? To use them, we must store them, and to store them, we must represent them – represent them physically.

A simple solution would be to assign standard formats and codes to all domains, along with standard units of measure to quantitative domains, and then to store the structure in an electronic repository. All other representations can then map to the standard, which will be the hub of their common meaning.

The repository may also store maps that show which meanings represent which others, even which meanings are assembled from others, in what perspectives. Perspectives too, can point to a standard perspective, a hub around which other perspectives revolve – a hub that expresses meanings in terms of standard primary domains.

Perspective was a model and an object (Chapter 2, section 4). Every value object (domain) defines a state space, a meaning. Different coordinate systems may describe this state space and the points within it (box 37 and box 49). Each coordinate system is only one of many possible ways of describing the space it frames: only one perspective among many of a single meaning.

The frame of reference for locating a point in a complex domain is also an object, as is the set of relationships it has with other frames of reference in other domains. They are all components of *perspective*, the object. We can always map coordinates in one frame of reference to those in another, and one set of axes in one perspective to another set of axes in another perspective, if both coordinate systems frame a common meaning (see box 49).

True rule meaning is an abstraction that subsumes all its expressions, and focuses on pure information. It is independent of the coordinate system, and is the substance of *domain*, simple or complex. Complex domains may have complex nuances – meanings that are subtle variations of a shared meaning; nuances that have added or deleted meanings. To model these nuances, remember that meanings are dimensions in state space, and you can use the principle of subtyping by adding information to define nuances and perspectives that share common meanings.[143] Shared axes will be the basis of meanings at the top of subtyping hierarchies. Subtle nuances will be at the bottom of the pile (see the examples in Module V, section 4 on our website).

Thus, the expression of time in the example under "Patterns of symbols, patterns of objects" was a subtype in which the concept of a monthly cycle was added to the concept of a timeline. The supplementary materials in Module V on our website describe how the principle of parsimony (in the endnotes) determines the level of abstraction for a meaning

[143] A topos is a mathematical category that is internally complete and consistent within its own laws. See [173] and [183]. Each perspective is a topos. The topos of perspectives is also a topos, and contains rules for mapping one perspective to another. Topoii are objects as are topoii of topoii. See [178]. Different models may be considered points in a topos of models, and different perspectives may be points in a topos of perspectives. Some perspectives (models) in a topos may be subtypes of others and thus inherit shared information. See topos theory in [175], [176], [181], [183], [184], [185], and [263].

that is right for a given purpose. A purpose is also a pattern of information and a meaning. Module V shows, with business examples, why the right subtype is the pattern with the least information payload required to represent the irreducible fact that matches a process with a purpose (which is also an irreducible fact).

It ultimately depends on joining objects – meanings of aggregations forged by relationships and compositions both simple and complex, compositions that orchestrate meaning beyond the sums of their parts – the meaning of knowledge.

Relationships are the glue, and aggregate objects the fount, from which object classes, attributes, domains, and processes all emerge in their multitudes of shapes and nuances. They are the objects from which the metamodel of knowledge flows. It follows that the next step towards the integrated model of knowledge must be the study of relationships, and their manifestations in processes and aggregate objects. This is described in the supplementary modules on our website. Those modules show us that fundamental meanings and properties of the real world, like process, location, containment, and the power of reason all flow from the nexus between patterns of abstract information and their associations.

The abacus was a simple calculating machine used in ancient Rome. That simple machine was to modern computers, what our computers will be to the infinitely more complex computing machines of the future. These machines will operate on the plane of meanings. Thus, although this book has ended, it is a new beginning. As T. S. Eliot, the poet who pioneered modern verse once said:

> What we call the beginning is often the end.
> And to make an End is to make a Beginning.
> The End is where we start from

Beginings have no end. The automation of tomorrow will rise from the begining and the end of the nexus of meaning.

Appendix: Key shared components of knowledge described at the nexus of meaning in this book and its supplementary modules

Activity (and other) costs (Module V, section 3).

Aggregate object. A collection (Module V, section 2). A **composition** is a structured aggregate.

Array (Chapter 4, section 1).

Assemble. A polymorphism of *process* and the *part of* relationship (Module V, section 4). Assemble emerged from a process that made an item a part of an aggregate in step with the flow of time. Similarly *disassembly* cuts the relationship between an aggregate and its parts, so that the part does not remain a part of the aggregate after disassembly has occurred. Thus **disassemble** is also a process, but it is a polymorphism of the *exclude* relationship (near the top of figure 116). Polymorphisms of *disassemble* will tell us how an aggregate is picked apart – explosively, all at once, or in steps – perhaps even one item at a time.

Attribute. A kind of object property that is also a subtype of *domain*. It is a relationship between an object class and a subtype of a domain that consists of a single value at any given time (Chapter 3, section 2).

Beginning and **ending moments** of an event (both are subtypes of *moment*).

Borel object. A generalization of the concept of *array*, useful for categorization and segmentation of objects and state spaces – a power set of values, or an infinitely large power set of ranges (see Module V, section 1).

Bounds (Chapter 3, section 2).

Capacity. A kind of cardinality constraint (see Module V, section 1).

Cardinality (the "size" of a class. See *enumeration domain* in Chapter 4, section 3. *Cardinality* is a supertype of *enumeration*).

Composed of. A subtype of *consist of* (Module V, section 2). In figure 116, its inverse has been labeled *component of*.

Consist of. The inverse of *part of* and a subtype of *locate* (Module V, sections 2, 4, figure 114 and figure 116).

Constraint (generic). A generic *constraint* is a generalized *meaning*, synonymous with *object property* (Module VI). *Rule constraint* and *value constraint* (Chapter 3, section 2) are special subtypes of this generic constraint.

Contain. A supertype of *consist of*, and a subtype of *locate* (Module V, sections 2, 4, figure 114, and figure 116).

Cycle time. The time interval from the start to the end of a process (Module V, section 3 – cycle time is a subtype of *event*).

Delimiters (Chapter 4, section 1).

Domain (introduced in Chapter 1, section 3.2, detailed in Chapter 4, section 3). A domain is a class of values. The class may contain finite or infinite numbers of distinct values and lends its members a common meaning, such as "length." The meaning of *qualitative* measurement is encapsulated in nominal and ordinal domains: *nominal* domains only distinguish between values; *ordinal* domains add information on sequences. The meaning of *quantitative* measurement is encapsulated in difference and ratio scaled domains: difference scaled domains add information on magnitudes; ratio scaled domains add information on ratios and the concept of nil magnitude. The metamodel of knowledge infers that quantitative values must be expressed in units of measure, of which it may have several (Chapter 4, section 2). Domains are arranged in a subtyping hierarchy shown in figures 67 and 68 The most elementary business and physical meanings start with *primary domains*: *Enumeration* (ratio scaled), *Mass* (ratio scaled), *physical separation* (ratio scaled), *date/time* (difference scaled – includes date and time of occurrence), *electric charge* (ratio scaled), overall *Information* content (ratio scaled), and *Preference* (ordinal). *Secondary domains* are derived from primary domains as polymorphisms, or from relationships between domains. A few frequently used secondary domains are *time lapse*, domains of *information quality* (*validity*, that we are measuring the right thing; *reliability*, that the measurement is always consistent; *completeness* and *accuracy*, that the measurement is unbiased), *Economic value added* (ratio scaled polymorphism of *preference*), various domains of proportions, various domains of change/growth and *gender*. The *cardinality* of a domain is a measure of its size, which might be infinite. A *dense* domain has an infinite number of values between any ordered pair of values (for example, a difference scaled domain like temperature, or a ratio scaled domain like mass).

Effect is a kind (subtype) of process that changes the state of a single object. It is not always a *business* process, but effects always map directly to computer systems processes (Chapter 2, section 2 under "events, effects, and actions", and Module V, section 3 under "Transforming business processes into effects of events in crossing the chasm". Also see figure 109. An effect is a subtype of *object property* in the same partition as *attribute*).

Efficiency and **productivity** of processes (Module V, section 3).

Essence (of a pattern) is the information that gives the pattern its identity and distinguishes it from other similar patterns. It is closely tied to the **freedom** the pattern has to be that pattern. The meaning of "*essential*" is derived from "**essence**," and the meaning of "**freedom**" is derived from the degrees of freedom of a pattern (Chapter 4, section 1 under "Pattern").

Event. A time interval (introduced in Chapter 1, section 2, described in more detail in Module V, section 3).

Exception process (polymorphism of *process*). Processes triggered when constraints are violated. Exception processes are polymorphisms of *process* in a different partition from input and output processes. Thus there may be exception processes for inputs, outputs, and transformations (Module V, section 3 under the "Risk management transform" under "Crossing the chasm" (Module V also discusses exception management patterns in that section).

The **expression** of a rule (box 33 and figure 117). A meaning may have many expressions. Each expression is a perspective of that meaning. Therefore *expression* and *perspective* are identical. *Expression* is the result of *express* (*expression of* and *express* are synonyms; their inverse is *expressed by*). *Express* is a polymorphism of the subtyping relationship (as is "*instance of*") (Module V, section 4). *Expression*, an object, is identical to *expressed by*, its defining relationship; the information conveyed (and hence meaning) is identical (see Module VI, section 2 and the endnote on functional programming).

Extent (Chapter 4, section 1).

Feature. Any property of an object – an attribute, relationship, effect, or constraint. See object property (box 10; Module V, section 3 and Module VI, section 3 expand on the description of feature/object property in box 10 and figure 32).

Format (introduced in Chapter 1, section 3, detailed in Chapter 4, section 1).

Freedom (degree of) (Chapter 4, section 1).

Governance and **non-stationarity**. Applies to constraints, patterns, and processes. Non-stationarity is the property in which features and parameters change over time; governance sets parameters and features. ***Governing processes*** are processes that set parameters of processes (Chapter 3, section 2; Chapter 4, section 1; and Module V, section 3 on our website). Governance processes often depend on tracking and exception processes to govern – another commonly used theme in business.

Idempotent relationship (Module V, section 1).

Inclusion and **exclusion sets.** Mutually exclusive subtypes of *partition* (see figure 39 under "Constraints on nominal attributes").

Incorporation. A subtype of *consist of*, wherein the object loses its identity as a member of a separate class of objects. It becomes a subtype.

Instance of. A different polymorphism of the subtyping relationship in the same partition as *express* (see Module V, section 4).

Intransitive relationship. When a composition of relationships disallows the existence of another relationship (Module V, section 1).

Joint constraints. When a value is constrained by an *interaction* between multiple objects. Joint constraint is a polymorphism of *value constraint*; it is a relationship of a higher order, with more information in its rule expression and meaning (Chapter 3, section 2).

Language (Chapter 4, section 1).

List of. A subtype of *consist of* (Module V, section 4 and figure 116).

Load balancing of processes (Module V, section 3).

Location (*locate*) and **origin** (Module V, sections 2, 4, figure 114 and figure 116).

Location, containment, part of/consists of, subset of/superset of, subtype of/supertype of (Module V, sections 2 and 4).

Magnitude constraints. Restricts the magnitude of a difference or ratio scaled value. Based on the principle of adding information, a magnitude constraint is a polymorphism of *value constraint*. Joint constraints and magnitude constraints are subtypes in different, independent partitions of value constraint, so a constraint could simultaneously be both (Chapter 3, section 2).

Meaning. Meanings are polymorphisms (Chapter 4, section 1, "Pattern"). They are patterns of abstract information. Meanings include the meaning of a rule, as opposed to its expression (box 33 and Module VI, figure 117). Indeed, this is the inchoate universal object. Polymorphisms of *meaning* carve object instances and object classes from the primal metaobject (Module V, section 4, Module VI, sections 1 and 2).

Metaobject. A generic and inchoate ***instance*** of an object. All objects are subtypes of this primal object (introduced in Chapter 2, discussed in Module VI).

Moment. An event of nil duration (Module V, section 3) and hence a subtype of event (Module V, section 4).

Mutability. Substitutability of one object by another (Module V, section 1).

Name, and its subtypes, ***synonym***, ***homonym***, ***alias***, and ***concept id*** (Chapter 2, section 4).

Number. Number is an expression of *quantitative value*, and therefore a subtype of both *expression* and *quantitative value* (Chapter 4, section 2). Also, note that *format* is a kind of *expression of* value in symbolic form (Chapter 4, section 1). This makes *format* a subtype with two parents, *value* and *symbol* (the relationship *expression of /express* is a polymorphism of the subtyping relationship (Module V, section 4).

Object class. (A subtype of an aggregate object. A ***list*** is also a different subtype of an aggregate object in this partition (Chapter 2, section 5 and Module V, sections 2 and 4).

Object instance (Chapter 2, section 1).

Object partition. A criterion for dividing an object class into mutually exclusive subtypes. A partition may be *exhaustive* (the subtypes in the partition collectively cover all possible members of the partitioned class) or *inexhaustive* (the subtypes do not cover all possible members of the partitioned class) (Chapter 2, section 3, *Object partitions and role modeling*).

Object property. Attributes, relationships, effects of events, and constraints associated with the object (box 10. Also described formally in Module VI, section 2; see figure 117).

Observation, inquiry, and reporting. Processes that are polymorphisms of a generic "inquiry" process, which changes the state of the object queried/observed to "queried/ observed", and may or may not change it in other ways (Module V, section 3, box 54).

Pattern (Chapter 4, section 1). This is the root of the metamodel of knowledge. All its components are polymorphisms of *pattern*; an object instance is also a kind of pattern – a meaningful pattern (Module VI and figure 31).

Perspective is a classification scheme. It is expressed in a network of objects and relationships. It is also a *composition* (Module V, section 1, under "Compositions of relationships"). Compositions are also subtypes of relationships. A *composition* is also a synonym for *expression*. Therefore *perspective* is the same as *composition*, which is a subtype of *relationship* (Module V, sections 1, 4 and Module VI, section 2).

Pick. A polymorphism of *process* and the *instance of* relationship (Module V, section 4). *Pick*, the polymorphism, may also have subordinate polymorphisms. For instance, one polymorphism may pick a single item out of a collection or assembly of items, whereas another might pick a class of similar items out of that collection of parts, and yet another polymorphism could pick a batch of similar or dissimilar parts out of the collection.

Planned. Intended state. A universal state applicable to all objects. **Plan** is a polymorphism of **purpose** (Module V, section 3).

Polymorphism. Synonym for subtype (box 21 and Chapter 3, section 2, Chapter 4, section 3; Module V, section 4 and Module VI).

Precision (Chapter 4, section 1) is a synonym for ***accuracy***, and **exhaustiveness** is a synonym for ***completeness***. Note that less-precise and less-complete patterns convey less information than their more-precise or more-complete counterparts. Therefore the more-precise or more-complete pattern is a subtype of its less-precise or less-complete counterpart.

Process. A subtype of two parents – event and relationship (see Module V, section 3 and "Processes, events and temporal relationships"). Processes *use* resources to *produce* products (Module V, section 3). *Process* inherits the features of *relationship*, and combines them with temporal information from *event*, such as cycle time. Combined with temporal information from *event*, these features inherited from *relationship* acquire new characteristics such as: *temporal succession*, *productivity*, *reversibility*, *temporal mutability* – the time dependence of mutability between objects; *temporal order* – how far back into history does a process reach to articulate rules about a change of state at present; *temporal degree* – repeatability and concurrency; for idempotent relationships: the number of times a process loops back to the same product, or reuses the same resource. A ***reporting process*** changes the state of an object from unknown to a known value. An ***inquiry*** changes the state of an object from *unknown* to *observed*. It may or may not change other features that constitute the overall state of the object.

Process owner (various kinds (subtypes) in Module V, section 3 and "Process ownership".

Product (Module V, section 3).

Proximity metric measures similarity. May also measure distance, which is a polymorphism of similarity (Chapter 4, section 1. See also the endnote on generalizing the concept of distance).

Purpose or **goal** (Module V, section 3).

Ranges (ranges are subtypes of twin parents – *sequenced pattern* and *value set* (see Chapter 3, section 2).

Recursive relationship (Module V, section 1).

Relationship is an interaction. It is a polymorphism of a *list*, which in turn is a polymorphism of *aggregate object*. (Module V, section 4, also figures 31 and 116).

Representation. A polymorphism of expression (Chapter 4, section 1).

Resource (Module V, section 3).

Resource life. A temporal polymorphism of capacity; when time is added to the meaning of capacity, the capacity to engage with objects will change over time. When the capacity decreases, we might conceive of an, "unknown" process that has engaged the capacity of an object. The "unknown" process starts "consuming" it, or diminishing its capacity for engagement. If the decline is precipitous at a particular point in time after the resource is created, that interval may be considered the life of the object. **Resource consumption** is a polymorphism of *resource life*, in which the capacity of a resource to engage is diminished over time by a known process. If a process changes the state of a resource, it is considered consumed, and the changed resource is a product (it could be a work product, a waste product, or a by-product (Module V, section 3).

Reversibility and **reversion** (of processes) (Module V, section 3). Reversion is a process that is the inverse of another process – it restores the original states of all involved objects, i.e. undoes the effects of the reversed process.

Rule constraint. A rule that constrains a nominal, ordinal, or ratio scaled *value*; a kind of **constraint** (Chapter 3, section 2).

Saga. A process with no definite end, which is also a supertype of a process with a definite end. An **endless saga** is a polymorphism of saga, in which it is definitely known that the process will not end (Module V, sections 3, 4; figure 116).

Size. A polymorphism of *Capacity* (see Module V, section 4).

State, **state space**, **trajectory in state space**, and **set of possible trajectories in state space** are all subtypes of *aggregate object*. The last two are also ***compositions***. A composition is a subtype of *aggregate object* (Module V, section 2) Thus *trajectory in state space* and *set* *of possible trajectories in state space* are actually subtypes of *composition*, and therefore a subtype of *aggregate object*, once removed.

Subtype and **Supertype.** Both subtypes of *object class* (in the same, exhaustive partition) (Chapter 2).

Subtyping relationship (a kind of relationship). See *incorporation* (Module V, sections 2 and 4).

Supply chains and **demand chains** (polymorphisms of *process*) (see "Supply and demand chains").

Symbols (Chapter 4, section 1).

Symmetry (Module V, section 1. Note that processes cannot be symmetric; they incorporate information on the flow of time, which is asymmetrical).

Temporal succession. Sequence in time; a supertype of *process* and subtype of *relative location* (Module V, sections 2, 3, and 4).

Tracking process. A process obtained by infusing temporal information into the proximity metric. It is a polymorphism of the proximity metric and *event.*

Transformation, **input**, and **output processes** (subtypes of *process*). Transformation processes use resources to create products. Input processes convey resources to transformation processes and output processes convey products from transformation processes. They are all polymorphisms of *process*, and every business process consists of all three – input, transformation, and output process – assembled in tandem (Module V, section 3; Chapter 4 section 3).

Transitive Relationship. When a set of relationships implies another, the implied relationship is ***transitive*** with respect to the others. In a **transitive triad** of relationships, any two relationships in the triad imply the third (Module V, sections 1 and 3).

Truncation slices a pattern into a part. *Truncate* relates an object to its truncation. A truncated pattern conveys less information than the pattern that was truncated. It is therefore a supertype of the original pattern, and the inverse of *truncate* is a polymorphism of the subtyping relationship (Chapter 4 section 1).

Unit of measure (Introduced in Chapter 1, section 3.2, detailed in Chapter 4, section 2).

Universal perspective is a subtype of *perspective.*

Use. The defining relationship between a process and its resources. The input process is a polymorphism of "Use" (Module V, sections 3, 4).

Value constraints. A kind (subtype) of *rule constraint* in which specific values are permitted or excluded – box 28 (Chapter 3, section 2).

Value. Encapsulates the concept of existence and measurability. It may convey distinctness, an ordered sequence, a magnitude, the absence of magnitude (the *nil value*), infinite magnitude, the absence of meaning (the *null value*), the concepts of "all," "any," and "unknown"

(Chapter 2, section 3, Chapter 4, section 3; under "Added value" in Module V, section 3, Module V, section 4; and box 51).

Value sets. A collection of values at a point in time (Chapter 3, section 2, figure 40).

View (Chapter 2, section 5).

Table A1 *More meanings from the nexus of knowledge in Module VIII*

THEME (OBJECT)	DEFINITION	TYPICAL POLYMORPHISMS (EXAMPLES)	TOKEN FEATURES
ACYCLIC PATH (subtype of *path*)	A path without loops; a path that cannot loop back to a node if it is always traversed in one direction; a path in which any given node may be traversed at most once when the path is negotiated in a single direction (although the path may converge on the same node along two or more different associations)	• The directional topology of a network of rivers and tributaries, possibly flowing around islands • A radio broadcast. • The topology of a one-way communications network of repeater stations • The directional topology of a supply chain that forbids return of goods	• Longest distance from a starting node in terms of the largest possible number of nodes that must be traversed to reach it • Longest distance to an ending node in terms of the largest possible number of nodes that must be traversed to reach it
ADDRESS (subtype of *format* and *location*)	Formatted information for locating a place	• Mailing label • Telephone directory, address • Postal address, e-mail address • Grid locations on a map	• Location (inherited from *locate*) • Line number • Text, style • Language • Map coordinates
AFFIRMATION (subtype and state of *negotiation*)	An event that asserts the concurrence of the parties in a *negotiation* to the terms under discussion	• An exchange in which the parties to the sale concur on its terms and conditions	• Start time (inherited from *event*) • End time (inherited from *task*) • Terms and conditions concurred upon (a subset and select state of all terms and conditions being

Table A1 *(cont.)*

THEME (OBJECT)	DEFINITION	TYPICAL POLYMORPHISMS (EXAMPLES)	TOKEN FEATURES
			negotiated; terms and conditions are inherited from *agreement* via *negotiation*)
AGREEMENT (subtype of *meeting*)	An arrangement that is negotiated or accepted by two or more parties to a meeting	• A sale • Insurance policy • Warranty • Marriage	• Terms and conditions • Parties • State (potential, planned, being negotiated, affirmed, bound, etc.)
ASSET (role of *information*, *document*, *physical object* [*construction*, *equipment* etc.], *organization*, *event* [*agreements*, *projects* etc.], *place* [land, electromagnetic frequencies etc.], *fund*)	A tangible or intangible item of value owned by a person or organization; an owned resource	• Construction • Network element • Facility • Right • Product • Accounts receivable	• Ownership • Proportion of ownership • Type of ownership • Value
BASELINE (subtype of *resource*)	A reference state that is used as a basis for comparison	• The first agreed upon project plan • Standard • Reference item	• Baseline status for a resource • Items it is a baseline for
BUSINESS PRODUCT (subtype of *asset*)	Assets positioned in markets to define the corporation's business (those assets that will be sold rented or offered in the normal course of business to generate income)	• Product–service offering • Service • Planned product • Withdrawn product	• List price • Purpose • Market positioning (intended product-market)
BUSINESS PRODUCT OWNERSHIP (an inclusion polymorphism of *resource ownership*)	The fact of ownership of a right, resource, service or product positioned for sale, lease, lending or use in the marketplace in order to trade it for other resources	• Joint ownership of a property meant for sale • Ownership of shares	• Owner • Owned product

(cont.)

Table A1 (*cont.*)

THEME (OBJECT)	DEFINITION	TYPICAL POLYMORPHISMS (EXAMPLES)	TOKEN FEATURES
CALENDAR (a sequential composition of events; also a polymorphism of event)	A sequence of time slots	• Gregorian calendar • Jewish calendar • Meeting calendar • Holiday schedule • Production schedule	• Start time (inherited from event) • Time slots
CHANGING PERSPECTIVE (polymorphism of *interpretation*)	Mapping meanings from one perspective to another	• Incoming payment to outgoing payment • Owner to person/organization • Invoice to payment, document and information	• Object correspondence • Accuracy (inherited from *information*) • Validity (inherited from *information*)
COLLABORATION	A set of mutually supportive goals	• A supply chain • Intermediate work products used to produce a final product that meets a goal • The set of processes that are intermediate steps in meeting a goal or satisfying a need	• Membership of the collaborative aggregation
COLLABORATOR (a role of a person or organization relative to another person or organization)	Persons or organizations assigned mutually supportive objectives	• Owners of processes in a supply chain	• Membership of a collaboration
COMPETITOR (a role of a person or organization relative to another person or organization)	Persons or organizations assigned mutually exclusive, competing objectives	• Opponent • Rival	• Participation in competition • Person/organizations who are competitors
CONFIRMATION (subtype of *affirmation*; a state)	An event that binds parties to the terms of an agreement	• A sale event	• Parties bound • Terms of conditions they are bound to within the agreement
CONFLICT	A set of mutually exclusive goals	• Competition • War • A case in a court of law	• Conflicting goals or processes

Table A1 *(cont.)*

THEME (OBJECT)	DEFINITION	TYPICAL POLYMORPHISMS (EXAMPLES)	TOKEN FEATURES
CONSIGNMENT (a possible aggregation of one or more resources; polymorphism of *resource*)	A resource that is picked up from one place and dropped off at another on a conveyance	• Mail addressed to an individual or organization • Road consignment to a particular place • Air consignment • E-mail to an individual • Payload with a single target • Passenger in a vehicle	• Consigned quantity • Mode of transportation
CONSTRUCTION (subtype of *physical object* and *place*)	An immobile construction used to support or service business activities	• Building • Room • Living unit • Bridge • Tunnel • Factory • Parking lot	• Floor space • Geographical location (polymorphism of physical location)
CREDENTIAL (subtype of qualification)	Qualification issued by a person or organization that is the basis of entitlement to rights or privileges or the basis for confidence, belief or credit	• Degree • License • Certification, • Authentication • Permit	• Permission, • Expiry date/time • Exceptions exemptions, and limitations • Issuing authority • Owner or certified resource
CUSTOMER (polymorphism of *person/organization*)	A person or organization that is the transferee an actual or potential *transfer of possession event*	• Shopper • Person or organization that buys resources from another in order to manufacture its products • A client	• The relationship to the potential or actual transfer of possession event that casts a person or organization in the role of a customer • Requirements
DOCUMENT	A collection of information formatted in a medium	• Form • Letter • Recording • Check	• Medium • Language

(cont.)

Table A1 (*cont.*)

THEME (OBJECT)	DEFINITION	TYPICAL POLYMORPHISMS (EXAMPLES)	TOKEN FEATURES
ELABORATION (polymorphism of the subtyping relationship)	Furnishing of detail about an object or meaning previously described in less detail	• Description of the meaning of a word • Explanation	• Elaborated item
EMPLOYEE (polymorphism of *supplier* and *person*)	A person bound to an organization by an employment agreement to furnish personal time to the employing organization	• Manager of a department • Sales representative	• Salary • Frequency of payment • Work hours
EMPLOYER (polymorphism of *customer*)	A person or organization that hires an *employee*	• A corporation • Government	• The employment relationship with one or more employees
ENERGY (medium of information that normalizes the fact of physical location; a bridge between information space and physical space)	The capacity for performing physical work	• Heat • Light • Kinetic energy of an object in motion • Gravitational energy of an object lifted against the force of gravity	• Quantum • Form (kind of energy) • Physical location
EQUIPMENT (subtype of *physical object*)	A tangible tool	• Machine • Vehicle • Software	• Function • Usage
EVENT	Something that takes place in a time slot; a significant occurrence in a time slot or a moment in time; a happening in a time slot	• Phone call • Accident • Payment • Customer order	• Start time • End time (optional)
EXCHANGE (polymorphism of *task*)	Swap; to transfer a resource, and to receive a resource in return	• A barter • The exchange of goods for funds in a sale	• Resource(s) exchanged • Objects exchanging resources • Place of exchange

Table A1 (*cont.*)

THEME (OBJECT)	DEFINITION	TYPICAL POLYMORPHISMS (EXAMPLES)	TOKEN FEATURES
FEATURE	An object property or constraint	• Product feature • Insurance coverage/exclusion offerings • Telecommunication USOC codes • Equipment capabilities	• Color • Capacity • Speed • Boundary
FEATURE GROUP	A set of features	• Set of services that go with a purchased product	• Features in the group • Relationship with requirements • Relationship with goals • Relationship with product-market
FORMAT (polymorphism of *proxy*; a symbol that is a proxy for *information*)	A symbol that may be sensed	• Printed letters • Sound • An image • Odor • Tactile symbols like Braille codes	• Form or shape of the symbol • Physical or relative location of the symbol • Composition of the symbol
GENERALIZATION (polymorphism of *interpretation*)	The common aspect shared across several specific aspects	• Generalization of a solution to address several classes of problems • Generic concept • A class of products	• Classes subsumed • Exhaustivity • Exceptions
GOAL or PURPOSE (polymorphism of *information*)	An intention, an aim or objective	• The objective of a business plan • The intended destination of a journey	• The objects (such as a person, organization or process) that have the goal
GUIDELINE (polymorphism of *issue*)	A course or method of action based on specified conditions to guide or determine decisions	• Regulation • Supplier guidelines • Underwriting guidelines • Policy • Instructions	• Authority • Purpose or goal • State indicators: Proposed Planned Filed

(*cont.*)

Table A1 (*cont.*)

THEME (OBJECT)	DEFINITION	TYPICAL POLYMORPHISMS (EXAMPLES)	TOKEN FEATURES
			Approved Rejected Endorsed Unendorsed Violated (normalized by the relationship between an event and the guideline)
INFORMATION	Knowledge or intelligence about a concept or meaning	• Market need • Message • Risk factor	• Validity • Accuracy • Reliability • Amount
INTERMEDIARY (an inclusion polymorphism of *mediator*)	A person or organization in the role of a mediator between other people or organizations	• Value added reseller	• Persons or organizations mediated
INTERPRETATION (polymorphism of *location* of information)	An ascription of a particular meaning	• Interpretation of a law • Interpretation of a meaning in a context (for example, "time flies" might mean that time moves swiftly, or that flies must be timed in some activity performed by them)	• Context of interpretation
ISSUE (polymorphism of *information*)	A subject of concern; information related to the achievement of one or more goals	• Non-availability of requirements for a project • Shortage or excessive resources • Lack of coordination • Delays in schedule • Deviation from plan	• Goals involved • Priority • States Open Closed resolved Unresolved Irresolvable Reopened
LEGAL AGREEMENT (subtype of *agreement*)	An arrangement that is recognized by law	• Credit card agreement	• Covering law

Table A1 (*cont.*)

THEME (OBJECT)	DEFINITION	TYPICAL POLYMORPHISMS (EXAMPLES)	TOKEN FEATURES
MARKET (Borel object defined on the transfer of possession relationship)	A class of *market places* which may, or may not consider business products as a classification parameter	• The generation-X market • Baby boomers market • Healthcare market • Stock market • Futures market • Auction • Residential real estate market on the West Coast • The mainland China market	• Characteristics of objects involved in the transfer of possession event
MARKET NEED (subtype of information)	Intelligence about requirements of a market segment	• Preferences • Product use	• Relationship between issues, problems, and requirements • Preferred features • States of *market need* Satisfied Unsatisfied
MARKET PLACE (polymorphism of *meeting ground*)	An actual or potential meeting ground where transfer of possession agreements may occur	• A seven-eleven store • Auction website • Telemarketing call • A mall • The stock exchange	• Classes of buyers, sellers, and business products traded
MARKET SEGMENT (Borel object defined on product transfer/usage agreement)	A category of actual or potential product transfer/usage agreements	• Line of business • Class of potential customers • Class of products • A geographical footprint • Class of customers, for a class of products in a geographical footprint	• Potential value • Profitability • Boundaries or limits on values or ranges of parameters of the market that define the market segment
MEDIATOR (polymorphism of *node*)	A resource that connects resources in a *structure*	• Reseller • Router in a network • Power distributor	• Connected nodes • Terms of mediation

(*cont.*)

Table A1 (*cont.*)

THEME (OBJECT)	DEFINITION	TYPICAL POLYMORPHISMS (EXAMPLES)	TOKEN FEATURES
MEDIUM	A *class* of places that imposes constraints on formativeness or format	• The electromagnetic spectrum • Paper • Air	• Formatting rules • Permitted formats • Impermissible formats
MEETING (a kind of event)	A gathering of two or more people or organizations for a time period	• A Christmas party • A conference • A joint product design event • An interview	• Start time (inherited from *event*) • Person/organizations meeting (must be at least two, may be more)
MEETING GROUND (polymorphism of *place*)	A place of exchange	• Market place • A telephone network • Internet chat room • Fair • Shop • Mall	• Object classes involved in an actual or potential exchange • Contents of a meeting ground (inherited from *place*)
MESSAGE (polymorphism of *information* and *shipment*)	Information or signal in transit between nodes	• Radio broadcast • Memo or letter • E-mail message • Telephone conversation	• Source (inherited from *shipment*) • Destination (inherited from *shipment*) • Content (inherited from *shipment*; in this case the content may only be some subtype of *information*)
NEGOTIATION (polymorphism of *agreement* and *information exchange*, a kind of *resource exchange* in which the item exchanged is pure information)	A meeting in which attempts are made to reach an agreement through discussion and/or compromise	• Negotiating a sale • Negotiating the settlement of a conflict • Negotiating a collaboration	• Start time (inherited from *event*) • End time (inherited from *task*) • Negotiating parties (inherited from *meeting*) • Purpose of *negotiation* • (Proposed) terms and conditions (inherited from *agreement*)

Table A1 *(cont.)*

THEME (OBJECT)	DEFINITION	TYPICAL POLYMORPHISMS (EXAMPLES)	TOKEN FEATURES
NETWORK (subtype of two parents: *structure* and *node*)	An association of resources	• A telecommunications network • A trellis • A set of linked documents or web pages • A network of people who exchange information • A network of roads linking geographical places and facilities	• Capacity • Footprint
NODE (polymorphism of *resource*)	A resource associated with itself, or another resource in a structure	• LAN node • Telephone switch • Place on travel itinerary • Start or end of task in project • Position in organizational chart • Account in chart of accounts	• Associated node • Capacity
ORGANIZATION (subtype of *person/organization*)	An association of people (it could be an empty association)	• Joint venture • Controlled organization • Bank • Clearing house • Industry evaluation organization • Task force • Project team • Department • Community	• Organizational charter • Mission (an organization might be controlled and operated by one or more persons who are its members and have a common mission, purpose and responsibility)
ORGANIZATIONAL STRUCTURE (an inclusion polymorphism of *path tree*)	A path tree in which the nodes are people or organizations	• Management hierarchy • Hierarchy of supervisory roles	• Level in organizational hierarchy

(cont.)

Table A1 (*cont.*)

THEME (OBJECT)	DEFINITION	TYPICAL POLYMORPHISMS (EXAMPLES)	TOKEN FEATURES
PATH (subtype of *structure*)	The continuous series of positions that are assumed in any motion or progression	• Directional topology of a flight plan	• Predecessor node • Successor node • Direction • Directional capacity
PATH TREE (subtype of *path* and *tree structure*)	A directional *tree structure*	• Organizational hierarchy • Reporting structure in which an individual must report to only one other	• Level number • Subordinate roles • Supervisor
PAYMENT (subtype of event)	Actual or potential transfer of money from one fund to another	• Incoming • Outgoing payment	• Amount • Currency • Payer • Payee
PAYMENT INSTRUMENT (subtype of asset)	An asset that is actually or potentially transferred to make payment	• Funds • Property	• Value
PERSON (subtype of *person/organization* and *physical object*)	A human being	• Employee • Spouse • Male person • Female person • Child	• Gender
PERSON/ ORGANIZATION	Any individual or organization that has an invested interest, stake, or business dealing with the enterprise	• Collaborator • Competitor • Applicant • Beneficiary • Broker/distributor • Payer • Payee • Customer • Vendor • Participant • Legal entity	• Date of birth/creation or appointment in role
PHYSICAL OBJECT (polymorphism of *energy*)	A tangible object detectable by our physical senses or instruments	• Vehicle • Equipment	• Weight • Volume • Physical shape • Physical footprint (polymorphism of physical location)

Table A1 *(cont.)*

THEME (OBJECT)	DEFINITION	TYPICAL POLYMORPHISMS (EXAMPLES)	TOKEN FEATURES
PHYSICAL PLACE (subtype of *place*)	Contiguous (or disconnected) location(s) within boundaries or points in physical space, where physical objects or energy may be located	• Continent • Location of a bridge • Ports of call • The surface of a ball • The floor space inside a room • Interplanetary or interstellar space	• Physical area • Perimeter • Physical length • Volume • Zoning • Time zone • Zip code • Surface area • Latitude (for geographical place only) • Longitude (for geographical place only) • CPFR's global location number (GLN) (The CPFR model is described under *supply and demand chains* at our website. • CPFR's Duns plus 4 code
PLAN (subtype of *goal*)	An intended state with or without intended processes and state transitions to achieve the goal	• Delivery schedule • Strategic plan • Sales targets • Estimated production • Maintenance schedule	• A planned state • State transitions from current to planned state • Processes with or without the process map to achieve the goal
PLACE	An object that contains, locates or conveys information, energy, events, material objects, organizations or people. Contiguous or disconnected location(s) where information, energy, events or physical objects, organizations or people may be found.	• Internet bulletin board • Part of electromagnetic spectrum • Country, city, zone. • Contour • Ports of call • State space • Pattern	• Coordinates • Web URL (universal resource locator, or web page address) • Address • Contents

(cont.)

Table A1 (*cont.*)

THEME (OBJECT)	DEFINITION	TYPICAL POLYMORPHISMS (EXAMPLES)	TOKEN FEATURES
PRODUCT-MARKET (polymorphism of *market*)	A market that includes characteristics of products and services traded in it	• Market for soap among teenagers • Market for pharmaceutical products • Market for two-wheeled vehicles in China	• Characteristics of products exchanged in the market
PROJECT, TASK (subtype of two parents: *event* and *path*)	A clearly defined piece of work that consumes or references resources to create or alter resources or their relationships; usually the responsibility of a person or organization	• Project • Litigation • Negotiation • Service call	• Responsibility • Resources • Cost
PROXY (polymorphism of *mediator*)	A resource that represents another resource in a structure	• Agent • Distributor • Format • Encrypted message • A map • Floor plan • A photograph • A scale model	• Item represented (inherited from *mediator*) • Terms of representation (constraints, scale etc.; polymorphism of *terms of mediation*) • Resource represented to (inherited from mediator)
QUALIFICATION (subtype of information)	Information that provides the basis for confidence or belief	• Skill • Experience • Permission	• Relationship with qualified resource
REGULATION (subtype of guideline)	A mandatory rule	• Stock trading regulation • Mandatory instructions	• State indicators: Enforced Unenforced
RENEGOTIATION (polymorphism of *negotiation*)	A negotiation that references a prior negotiation and its terms and conditions	• Renegotiating the terms of collaboration	• Prior negotiation • Terms and conditions under renegotiation

Table A1 (*cont.*)

THEME (OBJECT)	DEFINITION	TYPICAL POLYMORPHISMS (EXAMPLES)	TOKEN FEATURES
REPRESENTATIVE (an inclusion polymorphism of *proxy*)	A person or organization in the role of a proxy	• Person/organization with the power of attorney for another • Spokesperson • Congressman	• Resource represented (inherited from *proxy*) • Context of representation (polymorphism of *terms of representation*)
REQUIREMENT (polymorphism of *issue*)	An articulated need	• Market need	• Level of satisfaction • Validity (inherited from information) • Accuracy (inherited from information) • Reliability (inherited from information) • Priority (inherited from issue)
RESOURCE (role of *information*, *document*, *physical object*, *place*, *person*, *organization*, *event*, *fund* [or their interrelationships and aggregations])	Real-world objects or concepts that may be altered, consumed, referenced or created by tasks or processes	• Work product • By product • Node • Consumable • Catalyst • Facilitator	• Description • Relationship with task or place in a structure
RESOURCE CALENDAR (subtype of *calendar*)	A set of relationships between event(s) and resource(s)	• Maintenance schedule • Financial calendar • Meeting calendar, holiday schedule • Production schedule	• Capacity booked • Capacity left • Capacity used
RESOURCE OWNERSHIP (relationship between an asset and the person/organization who owns it)	The fact of owning a resource as a property	• Property ownership • Ownership of a baseball team • Ownership of a television show • Ownership of a right such as a copyright or patent	• Owning person/ organization (*asset owner*) • Owned asset • Proportion owned

(*cont.*)

Table A1 (*cont.*)

THEME (OBJECT)	DEFINITION	TYPICAL POLYMORPHISMS (EXAMPLES)	TOKEN FEATURES
RESOURCE RETURN TASK (polymorphism of *reversal*, in which no substitutions are permitted)	To take back a resource to where it came from	• Return of a person from a business trip to the exact point the journey started from • Return of a borrowed book to the library it was borrowed from • Reversion of an individual to the *same* rank or level he or she had held previously in a hierarchy	• Place or position returned from (inherited from *return*); *substitutions barred.* • Place or position returned to (inherited from *return*); *substitutions barred* • Resources being returned (e.g. *book*, *person*); *substitutions barred*
RESOURCE TRANSFER (polymorphism of *task*)	The transfer of resources between resources	• Delivery of mail • The process for feeding resources to a mechanism that will transform them • Output process from a machine after it has worked on the resource fed to it • Transportation of passengers from one airport to another	• Source • Destination • Resource (object) being moved
RETRACTION (polymorphism of *negotiation* and *reversion* – reversion; see box 30)	The reversion of an affirmation or proposal	• Retraction of intent to buy a home	• Proposed terms and conditions retracted
RETURN (polymorphism of idempotent *relationship*, a kind of *structure*; included in the metamodel of knowledge as shown in figure 2.5)	To go back to the same place; the last leg of an idempotent composition	• Return from a trip • Return of a borrowed item • Reversion to a rank or level in a hierarchy	• Place or position returned from • Place or position returned to

Table A1 *(cont.)*

THEME (OBJECT)	DEFINITION	TYPICAL POLYMORPHISMS (EXAMPLES)	TOKEN FEATURES
RETURN EVENT (polymorphism of *resource transfer*, a kind of *task*, and *return*, which is a kind of *structure*; the event has no information on what it is returning, or the equivalence of places being returned to or resources being returned)	An event that takes back to a place	• A return trip, the leg of a journey that returns an individual to his or her starting point • The event or process that restores of a borrowed item to its owner • The process that reverts an individual to a rank or level he or she held previously in an organizational hierarchy	• Place or position returned from (inherited from *return*) • Place or position returned to (inherited from *return*) • Start time (inherited from *event* via *task*) • End time (inherited from *task*)
REVERSAL (polymorphism of *return event*)	A return to a condition deemed equivalent to its former condition	• The event or process that restores of a borrowed item or an item of *equal value* to its owner • The process that reverts an individual to a rank or level *equivalent* to one he or she held previously in an organizational hierarchy	• Substituted item(s) (could be an aggregation) • Substitute item(s) (could be an aggregation) • Place or position returned from (inherited from *return event*) The place that is its *equivalent* (the equivalent could also be the same place). • Place or position returned to (inherited from *return event*) The place that is its *equivalent* (the equivalent could also be the same place). • Start time (inherited from *event* via *task*) • End time (inherited from *task*)

(cont.)

Table A1 (*cont.*)

THEME (OBJECT)	DEFINITION	TYPICAL POLYMORPHISMS (EXAMPLES)	TOKEN FEATURES
REVOCATION (polymorphism of *retraction*)	An event that reverts a confirmed agreement to an unconfirmed state	• Revocation of a treaty between nations	• The confirmed agreement being revoked
RISK (polymorphism of *issue*)	A hazard	• Risk of loss • Risk of fire • Risk of unforeseen exceptions	• Goal involved • Probability
ROUTE (a subtype of two parents: *path* and *node*)	A sequence of points visited	• A network of rivers and tributaries, possibly flowing around islands • Travel itinerary • Routing of materials in the standard operating procedure for manufacturing an item • Route plan • Supply chain	• Sequence number • Resource id
SHIPMENT (an aggregation of one or more consignments)	A particular cargo that is sent from one place to another on a conveyance. Each consignment in the shipment may be dropped off (and picked up) at a different destination (or source).	• Mail in the mail van • Road shipment with several drop-off points • Air shipment • E-mail broadcast to several individuals • Network transmission • Payload with multiple targets • Group of passengers being transported by a vehicle	• Shipment quantity • Consignments in the shipment
SKILL (subtype of qualification)	Qualification of a person that provides the basis for confidence or belief for executing a task	• Languages known • Technical ability	• Relationship with person/organization and task type
SOFTWARE (subtype of *document* and *equipment*)	Program code to elicit specific responses from equipment	• Switching software • Numerically controlled machine program	• Medium (inherited from *document*) • Language (inherited from *document*)

Table A1 (*cont.*)

THEME (OBJECT)	DEFINITION	TYPICAL POLYMORPHISMS (EXAMPLES)	TOKEN FEATURES
		• Computer program/operating system	• Function (inherited from *equipment*) • Usage (inherited from *equipment*) • Instruction (inherited from *regulation* via *document*)
SPECIALIZATION (polymorphism of *interpretation* and the inverse of *generalization*)	An adaptation of a broader concept to a particular niche	• Customizing a service to fit the needs of a specific customer	• Non-standard components or custom patterns not shared with other members of the class
STRATEGIC AGGREGATION OF GOALS	An aggregation of conflicting and collaborating goals	• A set of competitive targets along with the goal of collaborating with a competitor to promote common interests of the industry	• Goals in the aggregation and the fact of their mutual support or mutual exclusion
STRATEGIC PERSON/ ORGANIZATION (a role of a person or organization relative to another person or organization)	Persons or organizations assigned mutually exclusive *and* mutually supportive objectives	• A person or organization in a strategic relationship that includes both competitive and collaborative goals	• Other members of the strategic relationship • Their interests and goals
STRUCTURE	A set of associations	• Topology of a telecommunications network • Topology of a trestle • Topology of a trellis • Topology of linkages between web pages	• Linked objects • Capacity of link between objects
SUPPLIER or VENDOR (polymorphism of *person/organization*)	A person or organization that owns a right or resource and is a potential or actual conceder of ownership in an actual or potential transfer of possession event	• Supplier of components for the manufacture of a car • Real estate developer • Car distributor • Internet services provider • Temp agency	• Resource owned • The vendor's business products

(*cont.*)

Table A1 (*cont.*)

THEME (OBJECT)	DEFINITION	TYPICAL POLYMORPHISMS (EXAMPLES)	TOKEN FEATURES
TERMS AND CONDITIONS (subtype of guideline)	A guideline associated with a negotiation or agreement which establishes and (potentially or actually) binds a party to a set of constraints on resources and their relationships	• Right • Terms and conditions of sale • Terms and conditions of use • Terms and conditions of employment • Credit card terms and conditions • Insurance terms and conditions • Settlement terms and conditions • Order terms and conditions	• Relationship with one or more resources, parties and negotiations or agreements
TRACKING PROCESS (polymorphism of *process*)	A process in which the state of one object is compared with the states of others	• Deviation from flight plan	• Quantum of difference between baseline and state(s) of tracked object(s)
TRANSFER OF POSESSION (subtype of *negotiation* – a task. A transfer of possession may be under negotiation, a successful negotiation such as a confirmed/affirmed agreement, or even a failed negotiation that did not end in agreement)	An agreement or negotiation to transfer ownership, or permit use of a product (including services)	• Sale • Lease • Rental • Gifting	• Price • Start time (inherited from *event*) • End time (inherited from *task*. The duration of the event is often negligible; *start time* and *end time* might coincide and be subsumed by *time of occurrence*; both *start time* and *end time* are polymorphisms of *time of occurrence*) • Terms and conditions (inherited from negotiation)

Table A1 *(cont.)*

THEME (OBJECT)	DEFINITION	TYPICAL POLYMORPHISMS (EXAMPLES)	TOKEN FEATURES
TRANSPORTATION (a sequential composition of one or more *resource transfer* events; a subtype of *resource transfer*)	Movement from one place to another with possibly multiple drops	• Transportation of cargo • Transportation of passengers in a vehicle • Movement of a frame from one web page to another • Conveying a message • Movement of mail • Transfer of an individual from one organization to another	• Pick up points (polymorphism of *source*) • Drop off points (polymorphism of *destination*) • Resource (object) being moved (inherited from resource transfer)
TREE STRUCTURE (subtype of *structure*)	A branching topology without closed loops, which may be traversed in at least one direction without converging on the same node along two different associations, and has at least one node that cannot be traversed forward (termination node(s)), and at least one other that cannot be traversed backwards (starting node(s))	• A hub with spokes • The topology of a hierarchical communications network	• Position in a hierarchy
VIRTUAL PLACE (subtype of *place*)	A non-physical object that contains information	• Web page • Frequency spectrum	• Contents (only information, formatted or not) • Location

Bibliography

Papers

Intelligent agents

1 Mark Nissen, Professor BA248D, Naval Postgraduate School, Monterey, CA. Telecommunications and Distributed Processing. Intelligent Agents: A Technology and Business Application Analysis, November 30, 1995 at http://web.nps.navy.mil/~menissen/

2 W. Shen and D. H. Norrie of Division of Manufacturing Engineering, The University of Calgary. Agent-Based Systems for Intelligent Manufacturing: A State-of-the-Art Survey, *International Journal of Knowledge and Information Systems*, 1(2), 129–156, 1999 at http://imsg.enme.ucalgary.ca/publication/abm.htm

3 Cetus Team. Distributed Objects and Components: Mobile Agents, 2001/03/17, 1996–2000 at http://www.cetus-links.org/oo_mobile_agents.html

Business Process (re)engineering and e-commerce

4 B. de Vries, J. P. van Leeuwen, and H. H. Achten of Eindhoven University of Technology, The Netherlands. Design Studio of the Future, 1997 at http://www.ds.arch.tue.nl/Research/publications/bauke/ CIBW78_97.htm

(Describes structures of physical object, feature, activity and application of virtual reality to engineering design.)

5 Craig Standing, School of Management Information Systems, Edith Cowan University, Joondalup, Western Australia. Managing and Developing Internet Commerce Systems with ICDM, 1999 at http://www.vuw.ac.nz/acis99/Papers/PaperStanding-048.pdf

(Perspective of the full BPR process – strategic planning through process design and rollout. Focuses on differences between business processes in a traditional vs. collaborative e-commerce environment.)

6 William J. Kettinger, James T. C. Teng, and Subashish Guha, Business Process Change: A Study of Methodologies, Techniques, and Tools, Appendices 4 and 5, in *MISQ Archivist*, March 1997 at http://129.252.51.247/bpr/aa-4.htm

(Alphabetical list of major business process re-engineering techniques and tools with brief descriptions. You may have to access the paper from the MISQ Archivist site at http://www.misq.org/archivist/ home.html)

7 Activity Based Costing and Management from QPR Software at http://www.qpronline.com/abc/activity_based_intro.html

(You may have to go there via http://www.qpronline.com)

8 The (US) Department of Defense, 12/15/94. Framework for Managing Process Improvement.

9 Ellen Gottesdiener, President, EBG Consulting, Inc. OO Methodologies: Process and Product Patterns, EBG Consulting, Inc., SIGS Publications, published in *Component Strategies*, 1(5), 1998 at http://www.ebgconsulting.com/OOmethodsArticleCSmag.html

10 Wilfred van der Vegte, Assistant Professor, Delft University of Technology presented at EDIProd Conference, October 14, 2000, Dychow, Poland. Reflections on Artifact Related Process Modeling, October 20, 2000 at http://www.ediprod.uz.zgora.pl/files/ediprod2000.html, http://dutoce.io.tudelft.nl/%7Ewilfred/WFvdVegte-EDIProd2000.htm, http://www.sdpsnet.org/journals/vol6-2/vegte1.pdf, http://dutoce.io.tudelft.nl/~wilfred/

(Summary and assessment of different process modeling techniques and a process classification scheme.)

Ontologies and component re-use projects

11 Jose Vasconcelos, Department of Computer Science, University of York, UK and Multimedia Resource Center, University of Fernando Pessoa, Portugal; Chris Kimble, Department of Computer Science, University of York, UK; Feliz Gouveia, Multimedia Resource Center, University of Fernando Pessoa, Portugal; Daniel Kudenko, Department of Computer Science, University of York, UK. A Group Memory System for Corporate Knowledge Management: An Ontological Approach, September 2000 at http://www-users.cs.york.ac.uk/~kimble/research/ ECKM-2000-paper.pdf

12 Peter Green of Department of Commerce, University of Queensland, Australia and Michael Roseman of School of Information Systems, Queensland Institute of Technology, Australia. Ontological Analysis of Integrated Process Modeling: Some Initial Insights, a paper presented in the Proceedings of the Australian Conference on Information Systems (ACIS 2000), Brisbane, Australia, December 6–8, 2000.

(Evaluates ARIS against BWW criteria.)

13 Michael Rosemann of Queensland University of Technology, School of Information Systems and Peter Green of University of Queensland, Department of Commerce in the Proceedings of the Information Systems Foundations Workshop on Ontology, Semiotics and Practice 1999. Enhancing the Process of Ontological Analysis – The "Who Cares" Dimension at http://www.comp.mq.edu.au/isf99/Rosemann.htm

(A discussion of the BWW model applied to facets and information systems analysis and design.)
(Knowledge reuse algebras and test beds for techniques.)

14 C. N. G. (Kit) Dampney and M. S. J Johnson, Department of Computing, Macquarie University in Proceedings of the Information Systems Foundations Workshop on Ontology, Semiotics and Practice, 1999. An Information Theory Formalization and the BWW Ontology at http://www.comp.mq.edu.au/isf99/DampneyJohnson.htm

(Bunge Wand Weber (BWW) framework – rigorous algebra for testing the completeness of techniques/ontologies re business rule expression.)

15 Andreas L. Opdahl and Brian Henderson-Sellers of School of Computing Sciences, University of Technology, Sydney in Proceedings of the Information Systems Foundations Workshop on Ontology, Semiotics and Practice 1999. Evaluating and Improving OO Modeling Languages Using the BWW-Model at http://www.comp.mq.edu.au/isf99/Opdahl.htm

16 Glynn Winskel and Mogels Nielsen, Computer Science Department, Aarhus University, Denmark. Categories, in Concurrency, 1997. See abstract at http://www.brics.dk/upd/EP/97/WN_CC/EP-97-WN_CC.bib, https://booktrade.cambridge.org/catalogue.asp?isbn = 0521580579

(A comprehensive process algebra based on category theory and functors.)

17 David Rowe and John Leaney, Computer Systems Engineering, School of Electrical Engineering, University of Technology, Sydney. Evaluating Evolvability of Computer Based Systems Architectures – An Ontological Approach in IEEE International Conference on Engineering of Computer-Based Systems (ECBS Workshop 1997) at http://csdl2.computer.org/persagen/DLAbsToc.jsp?resourcePath = /dl/proceedings/&toc = comp/proceedings/ecbs/1997/7889/00/7889toc.xml&DOI = 10.1109/ECBS.1997.581903

(Applies BWW to systems evolution trajectories and architecture.)

18 John Mylopoulos, University of Toronto. Information Modeling in the Time of Revolution *Information Systems* 23 (3–4), June 1998.

(Compares various well-known reuse, modeling and knowledge representation algebras.)

19 Julieanne van Zyl and Dan Corbett, School of Computer and Information Science, University of South Australia. Framework for Comparing Methods for Using or Reusing Multiple Ontologies in an Application, a paper presented in the Proceedings of the 8th International Conference on Conceptual Structures, Darmstadt, Germany, August, 2000.

(Also lists and compares several major ontology and reuse projects/frameworks.)

20 Urban Nulden, Department of Informatics, Göteborg University, Sweden. The Why, What, and How of Reuse in Software Development at the 20th Information Systems Research seminar in 1997 at Scandinavia, Hankø, Norway at http://staff.cs.utu.fi/IRIS/y/1997.htm

(Translates the set-theoretic BWW framework to a more easily understood metamodel and compares various modeling algebras in terms of BWW criteria. Also evaluates the BWW framework itself.)

21 Andreas L Opdahl, Department of Information Science, University of Bergen. A Comparison of Four Families of Multi-Perspective Problem Analysis Methods, technical paper from project Technology Inc., 1998 at http://web.archive.org/web/20011218043935/www.project.com/pubs/papers.html.

(Analyzes the nature of multiple perspectives in BWW ontology for information systems and identifies principal differences between structured analysis, object-oriented analysis, faceted analysis, and viewpoints-based analysis.)

22 Yair Wand, Management Information Systems, Faculty of Commerce and Business Administration, The University of British Columbia, Canada and Richard Y. Wang, Sloan School of Management, Massachusetts Institute of Technology, Cambridge, MA. Business: Anchoring Data Quality Dimensions in Ontological Foundations, 1994 at http://web.mit.edu/tdqm/www/papers/94/94-03.html

(Analyzes various modeling techniques in terms of their ability to satisfy information quality requirements.)

23 Andreas L. Opdahl, Associate Professor, Department of Information Science, University of Bergen. Towards a Faceted Modeling Language in *Proceedings of the Fifth European Conference on Information Systems*, 353–366, Cork, Cork Publishing Ltd, 1997.

24 National Committee for Information Technology Standards, Technical Committee H7. Object Model Features Matrix (document number X3H7-93-007v12b) May 25, 1997 at http://www.objs.com/x3h7/omfm12b.doc

(Describes the object management group core meta model. And compares it with various other metamodels and standards, such as Eiffel and CORBA. Information about Technical Committee H7 may be found at http://www.objs.com/x3h7/h7home.htm.) *(Knowledge Reuse Projects.)*

25 John Kingston, AIAI, University of Edinburgh. Merging Top Level Ontologies for Scientific Knowledge Management (ref. EDI-INF-RR-0171) in Proceedings of the AAAI Workshop on Ontologies and the Semantic Web, AAAI-02 Conference, Edmonton, Canada, July 29, 2002 at http://www.inf.ed.ac.uk/publications/report/0171.html

(Lists, describes and compares various major knowledge reuse and ontology projects.)

26 List of some key domain specific and cross industry software component reuse projects with links to each at http://marexpo.balport.com/Project-Navigator/project_navigator.htm and http://marexpo.balport.com/Project-Navigator/matrix12.htm

27 Peter Clark of Boeing. Some Ongoing KBS/Ontology Projects and Groups, compiled at http://www.cs.utexas.edu/users/mfkb/related.html

28 Institut für Angewandte Informatik und Formale Beschreibungsverfahren links to world-wide ontology and knowledge use projects and researchers at http://www.aifb.uni-karlsruhe.de/ or http://www.aifb.uni-karlsruhe.de/Projekte/

29 COMMET & KREST. Knowledge Reusability and Configurability Projects at http://arti.vub.ac.be/www/krest/information/commet-krest.html

30 TOVE (Toronto Virtual Enterprise) Knowledge Reuse Project (as of February 18, 2002) at http://www.eil.utoronto.ca/comsen.html

31 TOVE ontologies at http://www.eil.utoronto.ca/tove/toveont.html

32 KACTUS Reusable Knowledge Modeling, 1995 at http://www.swi.psy.uva.nl/projects/Kactus

33 Wielinga, Schreiber and others. Project 8145 (The project partners were Cap Gemini Innovation (January 1994–September 1995), Integral Solutions Limited (September 1995–September 1996), CAP Programator, DELOS S.p.A, FINCANTIERI, IBERDROLA, LABEIN, Lloyd's Register, RPK Universität Karlsruhe, STATOIL, SINTEF Automatic Control, University of Amsterdam: KACTUS ESPRIT). Modeling Knowledge about Complex Technical Systems for Multiple Use, several papers at http://hcs.science.uva.nl/projects/Kactus/Papers.html

34 PROJECTXML™ (Year: 2000) by Project.Net San Diego, CA. The firm's website is http://www.project.net/scripts/SaISAPI.dll/website/products/ProjectXML.jsp

35 The One World Information System (OWIS) General Enterprise Management (GEM), Engineering, and Improvement Framework at http://one-world-is.com/rer/owis/emeif.htm

36 Rational Corporation: Rational Requirements Framework, Net Market Edition, 2000.

37 Simon Cox of Dublin Core Metadata Initiative. DCMI Box Encoding Scheme: Specification of the Spatial Limits of a Place, and Methods for Encoding This in a Text String, July 28, 2000 at http://dublincore.org/documents/2000/07/28/dcmi-box/

38 Simon Cox of Dublin Core Metadata Initiative. DCMI Period Encoding Scheme: Specification of the Limits of a Time Interval, and Methods for Encoding This in a Text String, July 28, 2000 at http://dublincore.org/documents/2000/07/28/dcmi-period/

39 Simon Cox of Dublin Core Metadata Initiative. DCMI Point Encoding Scheme: A Point Location in Space, and Methods for Encoding This in a Text String, July 28, 2000 at http://dublincore.org/documents/2000/07/28/dcmi-point/

40 Tim Menzies, Department of Artificial Intelligence, University of New South Wales. KBS Methodologies: KADS and Others, technical report TR95-28, Department of Software Development, Monash University, 1995.

41 KADS: A Development Methodology for Knowledge-based Systems at http://www.mdx.ac.uk/www/ai/samples/ke/53-kads.htm

42 Philippe Martin, University of Adelaide (Australia) – Computer Sciences Department. KADS Top-level Ontology of Concept Types and Relations Types at http://meganesia.int.gu.edu.au/~phmartin/WebKB/kb/topLevelOntology.html, http://meganesia.int.gu.edu.au/~phmartin/WebKB/interface/hierarchyBrowser.html?objectKind = concept+type&top = Thing&relation = Subtype&minDepth = 0&openNodes = Entity+Situation+Spatial_entity+Information_entity, http://meganesia.int.gu.edu.au/~phmartin/WebKB/interface/hierarchyBrowser.html?objectKind = relation+type&top = BinaryRel&relation = Subtype&minDepth = 0&openNodes = BinaryRel_from_a_situation+BinaryRel_from_a_Process

43 A. Th. Schreiber, J. M. Ackkermans, A. A. Anjewierden, R. de Hoog, N. R. Shadbolt, W. Van de Velde, and B. J. Wielinga. Knowledge Engineering and Management: The Common KADS Methodology at http://www.commonkads.uva.nl/frameset-commonkads.html and http://www.commonkads.uva.nl/frameset-commonkads.html

44 Kieron O'Hara, Artificial Intelligence Group, University of Nottingham, UK. A Representation of KADS-I Interpretation Models Using a Decompositional Approach in Proceedings of 3rd KADS Meeting, 1993 at http://eprints.ecs.soton.ac.uk/4164/

45 John Kingston AIAI, University of Edinburgh. Common KADS: Overview of Knowledge Engineering Methods at http://www.aiai.ed.ac.uk/~jkk/kadspubs.html

46 I. Laresgoiti and A. Bernarasl of LABEIN, Spain; A. Anjewierden, A. TH. Schreiber and B. J. Wielinga of University of Amsterdam, Department of Social Science Informatics, The Netherlands; and J. Corera of IBERDROLA, Spain. Ontologies as Vehicles for Reuse: A Mini-experiment, 1996 at http://ksi.cpsc.ucalgary.ca/KAW/KAW96/laresgoiti/k.html

47 Knowledge Interchange Format (KIF) draft of proposed American National Standard (dpANS) NCITS.T2/98-004: A Framework for Comparing Methods for Using or Reusing Multiple Ontologies in an Application, 1998 at http://logic.stanford.edu/kif/dpans.html

48 Rational Corporation. Rational Reusable Asset Specification (1999 technical report: major contributors Grady Booch, Catapulse, CTO Peter Eeles, Rational, RSO UK; Luan Doan-Minh, Rational, SSO US; Kelli Houston, Rational, A&AF Senior Architecture Specialist; Ivar Jacobson, Rational, VP of Business Engineering; Wojtek Kozaczynski, Rational, Director of A&AF; Philippe Kruchten, Rational Fellow; Grant Larsen, Catapulse, Senior Architecture Specialist; Jon Lawrence, Rational, A&AF Product Manager; Davyd Norris, Rational Software, RSO Australia,

Jim Rumbaugh, Rational Fellow, Bran Selic, Rational, Methodologist, Jim Thario, Rational, A&AF Senior Software Engineer).

Unified Modeling Language (UML)

49 Mike Lee, of Project Technology Inc. Object Oriented Analysis in the Real World, 1992.
50 Rational Software Corporation. UML Quick Reference for Rational Rose, 2001 at http://www.rational.com/uml/resources/quick/index.jsp

UML General Purpose Concepts
UML Class Diagram
UML Class Diagram Relationships
UML Collaboration Diagram
UML Component Diagrams
UML Class Visibility Notation
UML State Transition Diagrams
UML Sequence Diagram

Extended modeling language (XML)

51 XML Information Set W3C Working Draft March 16, 2001 and XML Information Set (second edition) W3C Recommendation, February 4, 2004 at http://www.w3.org/TR/xml-infoset/#infoitem.element
52 XML Schema Part 1: Structures, W3C Candidate Recommendation October 24, 2000 of the World Wide Web Consortium and XML Schema Part 1: Structures (second edition), W3C Recommendation, October 28, 2004 at http://www.w3.org/TR/xmlschema-1/
53 XML Schema Part 0: Primer, W3C Proposed Recommendation, March 16, 2000; editor David C. Fallside (IBM) at http://www.w3.org/TR/2001/PR-xmlschema-0-20010316/primer.html (World Wide Web Consortium, Massachusetts Institute of Technology, Institut National de Recherche en Informatique et en Automatique, Keio University.)
54 XML Core Metamodel at http–www.omg.org-cgi-bin-docad-01-02-03.txt and ftp://ftp.omg.org/pub/docs/ad/01-02-03.txt
55 XML 1.0 (second edition) W3C (MIT, INRIA, Keio), 2000 and XML 1.0 (third edition) W3C Recommendation February 4, 2004 at http://www.w3.org/TR/REC-xml

Process/task/schedule management and models

56 Veryard Projects. Process Management Workflow, Workload, Work Control, 1995–2001 at http://www.users.globalnet.co.uk/~rxv/sebpc/workflow.htm
57 M. C. Tanuan, Software Engineering Manager of Waterloo EAServer QA, eBusiness Division, Sybase, Inc. An Introduction to Workflow and Business Process Modeling, December 2, 1997 at http://se.uwaterloo.ca/~mctanuan/cs645/IntroBPMWF.htm
58 Stephen Russell Jernigan, M. S. E. and K. S. Barber, The University of Texas at Austin. Distributed Search Method for Scheduling Flow Through a Factory Floor, 1966 at http://www-lips.ece.utexas.edu/~stevej/papers/thesis/masters.html

59 Various practitioners and academics. Diverse manufacturing process summaries of Refereed Conference Papers of ICME 2000, 8th International Conference on Manufacturing Engineering in Sydney, August 27–30, 2000 at http://www.unisa.edu.au/ame/pubs/2000.asp
60 S. R. Jernigan, S. Ramaswamy, and K. S. Barber, The Laboratory for Intelligent Processes and Systems, The Department of Electrical and Computer Engineering, The University of Texas at Austin. A Distributed Search and Simulation Method for Job Flow Scheduling November 30, 1995 at http://www-lips.ece.utexas.edu/~stevej/papers/simulation/simulation.html
61 Project Management Institute (PMI). A Guide to Project Management Body of Knowledge: PMBOK Guide 2000 edition. PMI web page is at http://www.pmi.org/info/default.asp
62 Jürgen Sauer and L. Jain, Intelligent Techniques in Industry, In *Knowledge-Based Scheduling Techniques in Industry*, CRC Press, 1998 (excerpts available at http://www-is.informatik.uni-oldenburg.de/~sauer/paper/scheduling.html).

Process algebras and techniques

63 Assaf Arkin of Intalio Inc. from the Business Process Management Initiative (BPMI) Consortium. Business Process Markup Language (BPML), Working Draft 0.4 3/8/2001 (the BPMI site is http://www.bpmi.org).

("Business Process Modeling Language (BPML) is a meta-language for the modeling of business processes, just as XML is a meta-language for the modeling of business data. BPML provides a . . . model for collaborative & transactional business processes based on a . . . finite-state machine.")

64 Vitria Technology, Inc. Executive Overview: Value Chain Markup Language™-VCML™. A Collaborative E-Business Vocabulary, 2001 (Vitria Technology Inc. home page at http://www.vitria.com).
65 Vitria Technology, Inc. Downloads: Value Chain Markup Language™-VCML™. A Collaborative E-Business Vocabulary, 2001.

(Lists transactions in different industries – indicator of functions that are similar and different across industries. Visitors may download sample schemas and documentation.)

66 Alexander James Cowie, School of Computer and Information Science, University of South Australia. The Modeling of Temporal Properties in a Process Algebra Framework, 1999 at http://www.cis.unisa.edu.au/~cisajc/thesis.pdf

(For the mathematically inclined reader, a comprehensive review of process algebras, their meaning, operation, utilization and properties.)

67 Deepa Pandalai, Honeywell Technology Center, Honeywell Inc. Minneapolis, USA and Lawrence Holloway, Center for Robotics and Manufacturing Systems, University of Kentucky, Lexington, KY, USA. Template Languages for Fault Monitoring of Concurrent and Non-Concurrent Discrete Event Processes, March 1997 at http://citeseer.ist.psu.edu/74396.html

(An algebra that deals with the rules of single and multiple interleaved instances of identical concurrent processes.)

68 Rajeev Alur and David Dill, Computer Science Department, Stanford University, CA, USA. A Theory of Timed Automata, 1994. Abstract available at http://www.cis.upenn.edu/~alur/Icalp90.html and http://citeseer.ist.psu.edu/alur94theory.html
69 Petri Nets at http://pdv.cs.tu-berlin.de/~azi/petri.html#pnresearch (Maintained by Armin Zimmermann Dr.-Ing., research assistant, Technische Universität Berlin.)
70 Graph Theory: Color Petrinet at http://markun.cs.shinshu-u.ac.jp/learn/graph/cn7/colorPetrinet.html (Maintained by Shinshu University, Japan.)
71 Graph Theory: CO Petrinet at http://markun.cs.shinshu-u.ac.jp/learn/graph/cn7/coPetrinet.html
72 Srinivasan Ramaswamy, University of Southwestern Louisiana. Ph.D. thesis, Hierarchical Time-Extended Petri Nets (H-EPNs) for Integrated Control and Diagnostics of Multilevel Systems, 1994.

(For the mathematically inclined reader this is an excellent dissertation on the properties of processes, as expressed by petrinets.)

73 Vijay K. Garg, University of California, Berkely, CA, USA and M. T. Raghunath of the University of Texas, Austin, TX, USA. Concurrent Regular Expressions and their Relationship to Petri Nets, 1992 at http://citeseer.ist.psu.edu/garg92concurrent.html

(For mathematically inclined readers only! Flexible way of specifying concurrent processes, also deals with interleaving, interleaving closure, synchronous composition and renaming of processes.)

74 C. Ramchandani, Timed Petri nets, Technical Report 120, Project MAC, Massachusetts Institute Technology. A Study of Asynchronous Concurrent Systems, February 1974.

(An old but interesting paper. Project MAC [MAC was an acronym for "Man and Computer"] was one of the first visionary attempts to program common sense in the form of business rules into automation.)

75 Calculi for Mobile Processes at http://lamp.epfl.ch/mobility/

(A set of links to research papers on pi-calculus. You may need permission from LAMP Programming Methods Laboratory, Institute of Core Computing Science, School of Computer and Communication Sciences, Swiss Institute of Technology, Lausanne at http://lamp.epfl.ch to access the site.)

76 Marcus Lumpe at the University of Berne, Germany. A Pi-Calculus Based Approach to Software Composition, an inaugural dissertation, January 21, 1999 at http://www.iam.unibe.ch/~scg/Archive/PhD/lumpe-phd.pdf
77 Lucian Wischik, University of Bologna. New Directions in Implementing Pi Calculus, August 30, 2002 at http://www.newcastle.research.ec.org/cabernet/workshops/radicals/2002/Papers/Bertinoro/18.pdf

(A succinct, if mathematical description of pi-calculus.)

78 J.-P. Courtiat, C. A. S. Santos, C. Lohr, and B. Outtaj of LAAS-CNRS, France and Ecole Mohamedia d'Ingénieurs, Rabat, Maroc. Experience with RT-LOTOS: A Temporal Extension of the LOTOS Formal Description Technique, in *Computer Communications*, 23, 1104–1123, 2000

at http://www.laas.fr/RT-LOTOS/doc-src/CompCom99/CompCom99/ and http://www.laas.fr/~courtiat/PAPERS/ComCom00.pdf

(Real-Time LOTOS Process Algebra.)

79 David Harel, On Visual Formalisms, *Communications of the ACM*, 31(5), 521–523, 527, May 1988.

(An algebra to simplify state representations in real world finite state automata.)

80 Alan M. Davis, A Comparison of Techniques for the Specification of External System Behavior, *Communications of the ACM*, 31(9), 1105, 1988.

81 David Harel and A. Pnueli, Department of Applied Mathematics, The Weizmann Institute of Science, Rehovot, Israel. *On Development of Reactive Systems*, 8–10, NATO ASI Series F, Vol. 13, SpringerVerlag, 1985.

82 David Harel, On Visual Formalisms, *Communications of the ACM*, 31(5), 519, 1988.

83 Robert L. Jones III of Langley Research Center, VA, USA. NASA Technical Paper 3491, Design Tool for Multiprocessor Scheduling and Evaluation of Iterative Data Flow Algorithms, August 1998 at http://www.iis.sinica.edu.tw/JISE/2000/200005_07.pdf

(Although this paper focuses on distributed computer capacity and process efficiency, several concepts are also germane to real-world, distributed business processes.)

84 Sergio Bandinelli, Alfonso Fuggetta, and Sandro Grigolli, Process Modeling in-the-large with SLANG at Proceedings of the Second International Conference on the Software Process, 1993.

(Deals with evolution of large process models and high level petrinets.)

85 Mark E. Pitstick and William L. Garrison, Path Research Report UCB-ITS-PRR-91–7, Institute for Transportation Studies, University of California, Berkely, USA. Restructuring the Automobile Highway System for Lean Vehicles: The Scaled Precedence Activity Network (SPAN) Approach, April 1991.

(SPAN Process diagrams.)

86 Klaus Neumann and Welf G. Schneider, Dokumenteserver der Universitätsbibliothek Karlsruhe, in a 1997 Technical Report, Heuristic Algorithms for Job Shop Scheduling Problems with Stochastic Precedence Constraints, at http://www.ubka.uni-karlsruhe.de/cgi-bin/psview?document = /1997/ wiwi/6&search = /1997/wiwi/6

(Describes GERT concepts.)

87 Bin-Shiang Liang and Feng-Jiang Wang, Institute of Computer Science and Information Engineering, National Chiao Tung University, Taiwan and Jenn-Nan Chen of Samar Electronics Corporation Ltd, Taiwan. A Project Model for Software Development, *Journal of Science and Engineering*, 16, 423–446, 2000 at http://www.iis.sinica.edu.tw/JISE/2000/200005_07.pdf

(SPREM Process Algebra.)

88 Workshop on Design of Algorithms, Dresden, Module 17 Design Algorithms Channel. A Matrix Calculus for the Analysis and Generation of Binary Relations: Generalizations and Applications, part 1, 1996 at http://marvin.sn.schule.de/~inftreff/modul17/task17_e.htm

89 Oyvind Forsbak, Department of Informatics, University of Oslo. Graduate thesis, A Critical Review of Aggregation in Object Models, and a Proposal for New Aggregation Concepts in UML.
90 William Paul Rogers, Senior Engineering Manager and Application Architect, Lutris Technologies. Reveal the Magic Behind Subtype Polymorphism, *Java World*, April 2001 at http://www.javaworld.com/javaworld/jw-04-2001/jw-0413- polymorph_p.html
91 Eric Allen, Ph.D. graduate student, Programming Language Technology Group, Rice University. Behold The Power Of Parametric Polymorphism *Java World*, February 2000 at http://www.javaworld.com/javaworld/jw-02-2000/jw-02-jsr_p.html
92 R. Armstrong, D. Gannon, A. Geist, K. Keahey, S. Kohn, L. McInnes, S. Parker, and B. Smolinski at CCA Forum. Towards a Common Component Architecture for High-Performance Scientific Computing, 1999.
93 Grady Booch, Rational Software Corp.; Magnus Christerson, Rational Software Corp.; Matthew Fuchs, Commerce One Inc.; Jari Koistinen, Commerce One Inc. at the UML resource Center. UML for XML Schema Mapping Specification, August 12, 1999.
94 Dr. James Rumbaugh, OMT Papers, September 1995.

(A collection of papers on OMT, objects and patterns.)

95 P. H. Aiken, Reverse Engineering of Data IBM, 1998 at http://www.research.ibm.com/journal/sj/372/aiken.txt

Demand and supply chains and standards

96 Bill Hakanson, Executive Director Supply Chain Council (SCC). Supply Chain Management: Where Today's Businesses Compete, December 2, 1997 at http://www.ascet.com/documents.asp?d_ID = 228

(Supply chain meaning and process overview.)

97 Gaps in Common Knowledge Between Professions, Kenneth Kmack Associates August 2000 at http://web.archive.org/web/20030602070819/kenkmack.com/august_2000_topic_gaps_in_common_knowledge_between_professions.html

(Analysis of gaps between standard business process models and initiatives such as SCOR, CFPR, ARIS, XML, and others.)

98 Gordon Stewart, Supply Chain Operations Reference Model (SCOR): The First Cross Industry Framework for Integrated Supply Chain Management, *Logistics Information Management*, 10(2), 62–67, 1997.
99 QPR Software: SCOR Supply Chain Model from the Supply-Chain Council at http://www.qprportal.com/pg/scor_eng/

(You may have to access this publication via http://www.qpronline.com/)

100 S95 Standard May 2001, Enterprise – Control System Integration Standard, ANSI / ISA S95.00.01–2000, May 2001 at http://www.pera.net/Standards/Stds_S95.html
101 Dennis Brandl Director, Enterprise Initiative Sequencia Corporation, NC USA; Peter Owen, Eli Lilly & Co. A Tutorial on the SP95 Enterprise/Control Integration Standard at http://www.iee.org/oncomms/sector/manufacturing/Articles/Download/EF5A16ED-3D3C-466F-BB4FBD56A2FB90CB

102 Keith Unger of EnteGreat Inc. Integrate ERP with Control Systems Using the S95 model, Mountain Systems Inc, a presentation at the Mountain Systems Conference, May 13–17, 2002 at http://www.entegreat.com/eg_downloads_presentations_mountainsystems2002.htm

(An overview of S95 standard and its object model. You may have to access the site via www.entegreat.com)

103 Controls Definition and MES to Controls Data Flow Possibilities, MESA International White Paper Number 3, MESA International, Pittsburgh, PA, USA at http://www.mesa.org

104 Paul Sawyer, PES Associates. CAPE Tools for the Design and Operation of Batch Processes presented on 13 August 2001 at CAPE-21 (a conference on Computer Aided Process Engineering Tools and Techniques for the twenty-first Century).

(Information on CAPE-21 is at http://cape-21.ucl.org.uk/ and http://cape-alliance.ucl.org.uk/)

105 Brian A. Johnson, managing partner for strategy, research and thought leadership in the Cross Financial Services Solutions group, Accenture Corp. Fault Lines in CRM: New E-Commerce Business Models and Channel Integration Challenges, White Paper, January 15, 1999 at http://www.crmproject.com/documents.asp?d_ID = 706

106 Jagdish Sheth, Professor of Marketing, Goizueta Business School and Dr. Rajendra Sisodia Trustee Professor of Marketing, Bentley College Waltham MA. Marketing's Final Frontier – The Automation of Consumption, 2000 at http://www.jagsheth.net/pubs_articlesbytype.html and http://www.crmproject.com/documents.asp?grID = 187&d_ID = 709

107 Osman Turan, Department of Industrial and Systems Engineering, Virginia Tech. Introduction to Supply Chain Management, 2000 at http://hokies.ise.vt.edu/oturan/SCM/introduction.html

108 QPR Software, Introduction to Supply Chain Management, 2000 at http://www.qpronline.com/supplychainmanagement/supplychain_intro.html

109 Cyber M@rketing Services. Demand Chain Management: New Strategies for E-Business, IMA, 1998, 1999, 2000.

(Cyber M@rketing Services, Teaneck, NJ, USA may be found at http://www.elsnet.org/orgs/1697.html, Information Management Associates, Inc. (IMA), Irvine, CA, USA may be found at http://www.elsnet.org/orgs/0770.html)

110 Jan Holmström, and Tiina Tissari, The ECOMLOG Research Program, Department of Industrial Engineering and Management, Helsinki University of Technology. IT Value Capture: Creating an Effective Demand–Supply Chain for IT Solutions, presented at Logistics Research Network (LRN) 5th Annual Conference, Cardiff, UK, September 2000 at http://www.tuta.hut.fi/logistics/publications.html

(Information Technology Value Chain.)

111 REM Associates of Princeton, Inc. Supply Chain Move Over, It's Time for Demand Chain (March 2000), 1999, 2000 at http://www.remassoc.com/news/demandchain.asp.

112 Jim Noller, project consultant, Renaissance Worldwide. Integrating the Demand Chain and the Supply Chain: Technology and Trends, 1999 at http://www.afsmi.org/journal/jun99/jun-003.htm. Please register at register at: http://www.afsmi.org/ to access this site.

(There is an inherent difference between enterprise resource planning (ERP) systems and customer management systems. ERP systems measure financial transactions. Customer management systems measure customer contact events. However, to maximize the demand-chain value, processes and events need to be defined and systems implemented to reduce the demand cycle time.)

113 Jan Holmstrom, William E Hoover Jr., Perttu Louhiluoto, and Antti Vasara, Mckinsey & Co. The Other End of the Supply Chain, *McKinsey Quarterly*, (1), 62–71, 2000.

114 Roberto Michel, Manufacturing Systems. Why Best Practices Make Perfect: CFAR and SCOR Initiatives Aim to Improve Supply Chain Operations, *Manufacturing Business Technology*, 1997 at http://www.mbtmag.com/Default.asp

115 Elgar Fleisch and Hubert Oesterle, Institute of Information Management, University of St. Gallen, Switzerland. A Process Oriented Approach to Business Networking, *Virtual Organization Net*, 2(2), 2000 at http://verdi.unisg.ch/org/iwi/iwi_pub.nsf/wwwPublYearEng/9A080ADF5173DC38C1256FC600471E98 or http://www.ve-forum.org/Projects/264/Issues/eJOV%20Vol2/Fleisch%20-%202000%20-%20eJOV2,2-1%20-%20A%20Process- oriented%20Approach%20to%20Business%20Networking.pdf (access through http://verdi.unisg.ch/org/iwi/iwi_pub.nsf/wwwPublRecentEng/9A080ADF5173DC38C1256FC600471E98)

(Maps supply and demand chain models to collaborative business networking models.)

116 Kevin Crowston, School of Business Administration, MIT Sloan Center for Coordination Science University of Michigan, Michigan, USA. A Taxonomy of Organizational Dependencies and Coordination Mechanisms, May 1999 at http://ccs.mit.edu/papers/CCSWP174.html

(Task and resource coordination models.)

117 R. Alexander Milowski, XML Architect and Ray Waldin, Senior XML Engineer Lexica LLC. iLingo – The Language of Insurance e-Business at http://xml.coverpages.org/ilingowhitepaper19991218.html

(Insurance Supply Chain.)

118 Prof. A. W. Scheer, Instutut Fur Wirtschaftinformatik der universitat des saarlander. ARIS Business Process Model at http://www.iwi.uni-sb.de/teaching/ARIS/aris-i/aris-e-i/index.htm and http://www.iwi.uni-sb.de/teaching/ARIS/aris-i/aris-e-i/

119 Hau L. Lee and Corey Billington, Managing Supply Chain Inventory: Pitfalls and Opportunities, *Sloan Management Review*, Cambridge 33(3), 1992.

120 Ranier Alt, Elgar Fleissch, and Hubert Osterle, Institute of Information Management, University of St. Gallen, Switzerland. Electronic Commerce and Supply Chain Management at ETA Fabriques d' Ebauches SA, 2000 at http://www.csulb.edu/web/journals/jecr/issues/20002

(Maps complementary relationship between the intensively collaborative processes that support electronic commerce and traditional supply chain process models such as SCOR.)

121 Voluntary Interindustry Commerce Standards (VICS) Association. The CPFR Process Model at http://www.cpfr.org/ProcessModel.html and http://www.cpfr.org/Images/5.htm

122 Voluntary Interindustry Commerce Standards (VICS) Association. The CPFR Data Model, 2002 at http://www.cpfr.org/Images/AppendixH.HTM or http://havinghadlunch.com:8080/tamikin/GLS/matter/CPFR_Tabs_061802.pdf (See also links to current papers at http://www.vics.org/committees/cpfr/)

123 Collaborative Practices Research Initiative Sponsored by The Neeley Supply and Value Chain Center, Texas Christian University. November 15, 2004 at http://www.vics.org/committees/cpfr/academic_papers/academic_papers

124 VICS, Process and Results Metrics. Measuring the Success of a Process-Driven Value Chain at http://www.cpfr.org/Process-Results%20.html

(See also http://havinghadlunch.com:8080/tamikin/GLS/matter/CPFR_Tabs_061802.pdf)

125 VICS CPFR XML Messaging Model standard draft, January 17, 2001 for public comment at http://www.cpfr.org/XMLMessageModel.doc

126 ICS/CPFR IDEF0 Format Model at http://www.cpfr.org/AppendixI.html. See also http://havinghadlunch.com:8080/tamikin/GLS/matter/CPFR_Tabs_061802.pdf

127 Rosettanet Standards at http://www.rosettanet.org/rosettanet/Rooms/DisplayPages/LayoutInitial?container = com.webridge.entity.Entity%5BOID%5B5F6606C8AD2BD411841-F00C04F689339%5D%5D&expanded = com.webridge.entity.Entity%5BOID%5B5F6606C8-AD2BD411841F00C04F689339%5D%5D

(You may have to access the site via www.rosettanet.org)

128 Rosettanet PIP Directory at http://www.rosettanet.org/rosettanet/Rooms/DisplayPages/LayoutInitial?Container = com.webridge.entity.Entity%5BOID%5B9A6EEA233C5CD411843C00C04F689339%5D%5D

(PIP is an acronym for Partner Interface Processes. The site has a list of standard rosettanet PIPs – transactions exchanged by trading partners in a supply chain. You may have to access the site via www.rosettanet.org)

129 Rosettanet PIPs at http://www.rosettanet.org/rosettanet/Rooms/DisplayPages/LayoutInitial?Container = com.webridge.entity.Entity%5BOID%5B279B86B8022CD411841F00C04F689339%5D%5D

(PIP is an acronym for Partner Interface Processes. The site classifies rosettanet PIPs – transactions exchanged by trading partners in a supply chain. You may have to access the site via http://www.rosettanet.org)

130 Rosettanet Fundamental Business Data Entities at http://www.rosettanet.org/rosettanet/Rooms/DisplayPages/LayoutInitial?Container = com.webridge.entity.Entity%5BOID%5B07C504EE1A96D411BD89009027E33DD8%5D%5D

(You may have to access the site via www.rosettanet.org)

131 Rosettanet Business Data Entities at http://www.rosettanet.org/rosettanet/Rooms/DisplayPages/LayoutInitial?Container = com.webridge.entity.Entity%5BOID%5BF7C104EE1A96D411BD89009027E33DD8%5D%5D

(You may have to access the site via www.rosettanet.org)

132 Rosettanet Business Properties at http://www.rosettanet.org/rosettanet/Rooms/DisplayPages/LayoutInitial?Container = com.webridge.entity.Entity%5BOID%5B62C104EE1A96D411BD89009027E33DD8%5D%5D

(You may have to access the site via www.rosettanet.org)

133 David Sprott, Open Market Components, A CBDi Forum Report, January 2000 at http://www.componentsource.com/services/cbdiopen_market.asp

(Analyzes the market and emerging supply chain standards in terms of how components must be defined.)

Financial accounting

134 AccountingSTUDY.com[SM]. Accounting Study Guide, 1999–2002 at http://accountinginfo.com/study/index.html

(Succinctly describes the key principles used in financial accounting.)

135 AccountingSTUDY.com[SM]. Accrual Basis vs. Cash Basis Accounting, 1999–2002 at http://accountinginfo.com/study/accrual-01.htm

(A succinct description of accrual and cash basis accounting with examples.)

136 AccountingSTUDY.com[SM]. Introduction to Adjusting Journal Entries, 1999–2002 at http://accountinginfo.com/study/aje-01.htm

(Describes reasons for accounting adjustment transactions, with examples.)

137 Wikipedia. US Generally Accepted Accounting Principles at http://en.wikipedia.org/wiki/U.S._generally_accepted_accounting_principles

(Brief description of generally accepted accounting principles and related standards.)

138 AccountingSTUDY.com[SM]. FASB Statements, Financial Accounting Standards Board, ARB, APB Opinions, American Institute of Certified Public Accountants, Generally Accepted Accounting Principles in the United States Index, 1999–2002 at http://cpaclass.com/gaap/gaap-us-01a.htm

(A comprehensive source of U.S. GAAP information.)

139 BookkeepersList.com, 1999–2003 at http://bookkeeperlist.com/gaap.shtml

(A succinct description of the principles that guide financial accounting practices.)

140 CPAclass.com. Ratios for Financial Statement Analysis Web Site, Financial Ratios: Summary, 1999–2002 at http://cpaclass.com/fsa/ratio-01a.htm

(Succinct definitions of key ratios used for financial analysis and evaluation.)

141 CPAclass.com. Ratios for Financial Statement Analysis Web Site, Financial Ratios: Index, 1999–2002 at http://cpaclass.com/fsa/ratio-01.htm

(List of common ratios used for financial analysis.)

142 CPAclass.com. Annual Report Project Resources, 1999–2001 at http://www.cpaclass.com/arp/

(A comprehensive source of information related to developing a corporate annual report.)

Software process

143 David Chappell of Chappell & Associates. The Next Wave: Component Software Enters the Mainstream, April 1997 at http://www.mc.edu/campus/users/gwiggins/syllabi/csc320/papers/dynamic-3.html

144 Philippe Kruchten of Rational Software Corp., Canada. The 4+1 View Model of Architecture, *IEEE Software*, 12 (6), 42–50, 1995.

145 D. E. Perry and A. L. Wolf. Foundations for the Study of Software Architecture, *ACM Software Engineering Notes*, 40–52, October 1992.

146 Capability Maturity Model for Software (version 1.1), Publication TR 25 from Software Engineering Institute (SEI)
147 Carnegie Mellon University. CMMI Models, 2002 at http://www.sei.cmu.edu/cmmi/models/models.html

(You may have to access the site through http://www.sei.cmu.edu/cmmi/)

148 Michael Paulk, Carnegie Mellon University. A History of the Capability Maturity Model for Software at http://www.dfw-asee.org/archive/cmm-history.pdf

(An overview of how the CMM was sponsored, how it evolved, the other models it absorbed in the process, and its continuing evolution. You may have to access the site through http://www.sei.cmu.edu/cmmi/)

149 Carnegie Mellon University. Concept of Operations for the CMMI, 2002 at http://www.sei.cmu.edu/cmmi/background/conops.html

(Background and introduction to the Capability Maturity Model Integration project. You may have to access the site through http://www.sei.cmu.edu/cmmi/)

150 Bob Rassa of Raytheon Corporation and Clyde Chittister of the Software Engineering Institute. State of the CMMI: Improving Processes for Better Products, Carnegie Mellon University, 2002 at http://www.raytheon.com/feature/stellent/groups/public/documents/legacy_site/cms01_042355.pdf
151 Sarah A. Sheard of the Software Productivity Consortium. The Frameworks Quagmire: A Brief Look, at http://www.software.org/quagmire/frampapr/frampapr.html

(A brief descriptions of several quality and process maturity frameworks and standards, and their relationships with each other.)

152 S. Bandinelli, A. Fugetta, and S. Ghezzi, Software Processes as Real Time Systems: A Case Study Using High Level Petri Nets, in Proceedings of the International Phoenix Conference on Computers and Communications, Phoenix, AZ, April 1992, pp. 231–242.
153 Giancarlo Succi, University of Calgary, Canada and Luigi Benedicenti, Paolo Predonzani and Tullio Vernazza of University Di Genova, Italy. Standardizing the Reuse of Software Processes at http://portal.acm.org/citation.cfm?id = 260564

(Develops a model for reuse of processes and contains some excellent references to other research in the area.)

User interface standards

154 Microsoft Inductive User Interface Guidelines at http://msdn.microsoft.com/library/default.asp?url = /library/en-us/ dnwui/html/iuiguidelines.asp
155 CSS2 Specification: Cascading Style Sheets, level 2 W3C Recommendation, May 12, 1998 at http://www.w3.org/TR/REC-CSS2/

Agile processes and adaptive software

156 Peter Norvig and David Cohn of Harlequin Incorporated. Adaptive Software at http://www.norvig.com/adapaper-pcai.html

157 Laura M. Meade of Automation and Robotics Research Institute's Enterprise Engineering Program, The University of Texas. Agile Process Design at http://arri.uta.edu/eif/lmmdis.html

158 Scott W. Ambler, Agile Software Development at http://www.agilemodeling.com/essays/agileSoftwareDevelopment.htm

159 Don Wells, Extreme Programming: A Gentle Introduction, 1999, 2000, 2001 at http://www.extremeprogramming.org/

160 Jim Dowling and Vinny Cahill of the Department of Computer Science, Trinity College, Dublin. K-Component Architecture Meta-model for Self-Adaptive Software at http://www.cs.tcd.ie/publications/tech-reports/reports.01/TCD-CS-2001-50.pdf

161 Howard Smith, chief technology officer (Europe) of Computer Sciences Corporation and co-chair of the Business Process Management Initiative, and Peter Fingar, executive partner with the Greystone Group. The Next Fifty Years, *Darwin*, December 2002 at http://www.darwinmag.com/read/120102/bizproc.html

(A discussion of how computers have been seen as record keeping machines for 50 years as opposed to adaptable management machines. The need is now to use computers to gain actionable insight. For this, the authors say, corporations must shift their focus from "systems of record" to "systems of process." Moreover, "data processing" must give way to "process processing." The basic unit of automated support will then become the process, not data or the application system. The concept of databases will thus give way to "process bases," which will record and track past, present, and future of business process structures because, in the words of the authors, "business processes are the business." The authors describe how business processes will be made the central focus and basic building blocks of all automation and business systems in support of agility and responsiveness, and assert that the manual development of supporting information systems will be eliminated.)

162 Peyman Oreizy, Ph.D candidate, University of California, Irvine; Michael Gorlick, Research Scientist, Aerospace Corporation; Richard Taylor, Professor, Department of Information and Computer Science, UCI and Director of the Irvine Research Unit in Software; Dennis Heinsbigner, Research Associate Professor, University of Colorado, Boulder; Gregory Johnson, Member of Technical Staff, Concept Shopping Inc.; Nenad Medvidevic, Assistant Professor, Computer Science Department, University of Southern California; Alex Quilici, Associate Professor of Electrical Engineering, University of Hawaii, Manoa; David Rosenblum, Associate Professor, Department of Computer Science, UCI; and Alexander Wolf, Associate Professor, Department of Computer Science, University of Colorado, Boulder. An Architecture Based Approach to Self-Adaptive Software at http://ftp.ics.uci.edu/pub/c2/papers/ieee-is99.pdf

163 Paul Robertson of Dynamic Language Labs, Andover, MA. Self-Adaptive Software, a white paper for the Workshop on New Visions for Software Design and Productivity at http://www.hpcc.gov/iwg/sdp/vanderbilt/position_papers/paul_robertson_self_adaptive_software.pdf

164 Karyl Scott, Computer, Heal Thyself, *InformationWeek*, April 1, 2002 at http://www.informationweek.com/story/IWK20020329S0005 or http://www.informationweek.com/story/IWK20020329S0005

Mathematical foundations: set theory, number theory, category theory, theory of functions, lambda calculus, spaces and their properties, borel sets and tensors

165 Mathematical and Logical Vocabulary, Cycorp 1996, 1997, 1998. Cycorp is based in Austin, Texas. The cycorp home page is at http://www.cyc.com/cyc/company.

(Mathematical Sets, Categories, Topoii, Groups and Rings.)

166 Kyle Siegrist, Department of Mathematical Sciences, University of Alabama in Huntsville. Sets and Events, Virtual Laboratories in Probability and Statistics, 1997–2001. The tutorial is available at http://www.ds.unifi.it/VL/VL_EN/prob/prob2.html

(Describes Set theory and Sigma Algebra.)

167 Set theory at http://www.wikipedia.com/wiki/Set_theory (Describes the basic axioms of set.)

168 Basic set theory from Wikipedia at http://www.wikipedia.com/wiki/Basic+Set+Theory

169 Axiom of choice at http://www.wikipedia.com/wiki/Axiom_of_choice

(About creating sets by choosing elements from a collection of sets, even if they are sets with infinite members.)

170 Power set at http://www.wikipedia.com/wiki/Power_set

(The power set of any given set is the set of all possible subsets of the set.)

171 Axiom of regularity from Wikipedia at http://www.wikipedia.com/wiki/Axiom+of+regularity

(No set belongs to itself . . . otherwise [it] would violate the axiom of regularity.)

172 Mathematical class from Wikipedia at http://www.wikipedia.com/wiki/mathematical+class

(Describes the differences between classes and sets, and how the mathematical concept of *class* subsumes the mathematical concept of *set*: "A class is a collection of sets that can be unambiguously defined by a property that all its members share.")

173 Category theory from Wikipedia at http://www.wikipedia.com/wiki/category+theory

(A category attempts to capture the essence of a class of structures, instead of focusing on individual objects . . . the structure preserving maps between these objects are emphasized.)

174 John Baez, Professor of Mathematics, University of California, Riverside Categories, Quantization, and Much More, August 7, 1992 at http://math.ucr.edu/home/baez/categories.html

(Although it is written primarily for mathematical physicists, the paper is a good source of information on category theory, groups, and morphisms, including higher-order morphisms and categories, as well as their application in diverse areas.)

175 Chris Hillman, Ph.D., Mathematics, University of Washington. A Categorical Primer, a tutorial paper, July 2, 2001 at http://www.di.uminho.pt/~lsb/mmc_ap/Hilmann.pdf

(A reasonably simple mathematical introduction to category theory and Topoi.)

176 Goldblatt. Topoii: The Categorical Analysis of Logic at http://www.andrew.cmu.edu/~cebrown/notes/goldblatt.html

(An introduction to categories and Topoii, the need for them, and how categories and Topoii generalize the concept of set.)

177 John Baez, Professor of Mathematics, University of California, Riverside. This Week's Finds in Mathematical Physics (Week 68) October 29, 1995 at http://math.ucr.edu/home/baez/week68.html

(A relatively benign discussion of Topoii for beginners, and a non-mathematical description of how sub-objects emerge from commonalities based on the logic of Topoii.)

178 Steven Vickers, Department of Computing, Imperial College, London, UK. Topical Categories of Domains *Mathematical Structures in Computer Science*, Volume 11, Cambridge University Press, 1995 at http://mcs.open.ac.uk/sjv22/TopCat.ps.gz

(A geometric form of constructive mathematics . . . enables toposes as "generalized topological spaces" to be treated . . . in a . . . spatial way . . . it is quite in order to treat a topos as a "space" whose points are models of the theory and to treat a geometric morphism . . . as a transformation of points of one such space to points in another . . . a topos can be considered both as a "generalized topological space" and as a "generalized universe of sets.")

179 Heyting Algebra at http://publish.uwo.ca/~jbell/HEYTING.pdf

(A brief introduction to Heyting Algebra as a generalization of Boolean Algebra.)

180 Masao Mori, Department of Information Systems, Interdisciplinary Graduate school of Engineering Science, and Yasuo Kawahara of Research Institute of Fundamental Information Science, both of Kyushu University, Japan. Heyting Algebra at http://www.i.kyushu-u.ac.jp/~masa/fuzzy-graph/node2.html

(A mathematical, but brief introduction to Heyting Algebra, without proofs.)

181 Robert Goldblatt, Topoii, Categorical Analysis of Logic, *Studies in Logic and the Foundations of Mathematics*, Volume 98, New York, North Holland, 1984 at http://www.mcs.vuw.ac.nz/~rob/ or http://www.library.cornell.edu/math/digital-books.php#index

182 Andrew M. Pitts of Computer Laboratory, University of Cambridge, England. Non-trivial Power Types Can't be Subtypes of Polymorphic Types, a paper presented in Proceedings Fourth Annual IEEE Symposium on Logic in Computer Science, 6–13, Asilomar, CA, July 1989, IEEE Computer Society Press, 1989 at http://www.cl.cam.ac.uk/~amp12/papers

183 Mathematical topos from Wikipedia at http://www.wikipedia.com/wiki/mathematical+topos

("A topos (plural: Topoii) in mathematics is a type of category which allows to formulate all of mathematics inside it.")

184 John Baez, Professor of Mathematics, University of California, Riverside, January 3, 2001: Topos Theory in a Nutshell John Baez at http://math.ucr.edu/home/baez/topos.html

185 Law of excluded middle from Wikipedia at http://www.wikipedia.com/wiki/law+of+the+excluded+middle

("The law of excluded middle states that for any proposition, either it or its contradictory obtains; for any proposition P, either P or not-P." This law may not be true for all Topoii.)

186 Functor from Wikipedia at http://www.wikipedia.com/wiki/functor

(In category theory a functor is a mapping from one category to another which maps objects to objects and morphisms to morphisms in such a manner that the composition of morphisms and the identities are preserved.)

187 Monoid from Wikipedia at http://www.wikipedia.com/wiki/Monoid

(The set of all morphisms from this object to itself, with composition as the operation [is an example of a Monoid] . . . categories[are] generalizations of monoids.)

188 Mathematical Group from Wikipedia at http://www.wikipedia.com/wiki/mathematical+group

(Groups underlie other algebraic structures such as fields and vectors . . . also important . . . for studying symmetry.)

189 Semigroup from Wikipedia at http://www.wikipedia.com/wiki/semigroup
190 Subgroup from Wikipedia at http://www.wikipedia.com/wiki/subgroup

(The abstract mathematical theories that support the concept of subtyping by partitioning sets, and show that subsets are subtypes of supersets.)

191 Group Action from Wikipedia at http://www.wikipedia.com/wiki/group+action
192 Mathematical Ring from Wikipedia at http://www.wikipedia.com/wiki/Mathematical+ring

(A kind of mathematical group that generalizes commutative and associative operations.)

193 Fundamental Group from Wikipedia at http://www.wikipedia.com/wiki/fundamental+group

(Mathematical structures that convey information on loops and the one dimensional structure of space.)

194 Group representation Lie Algebra from Wikipedia at http://www.wikipedia.com/wiki/group+representation
195 Abelian group from Wikipedia at http://www.wikipedia.com/wiki/abelian+group

(The mathematics of commutative operators.)

196 Lie Group from Wikipedia at http://www.wikipedia.com/wiki/Lie+group
197 Lie Algebra from Wikipedia at http://www.wikipedia.com/wiki/Lie+algebra
198 Ring Ideal from Wikipedia at http://www.wikipedia.com/wiki/ring+ideal

(The mathematical theories behind "ideal," an abstraction and generalization of numbers.)

199 Integral domain from Wikipedia at http://www.wikipedia.com/wiki/Integral+domain
200 Field from Wikipedia at http://www.wikipedia.com/wiki/field
201 Finite Field from Wikipedia at http://www.wikipedia.com/wiki/Finite+field
202 Countable at http://www.wikipedia.com/wiki/Countable

(A set is countable if it is either finite or the same size as the set of positive integers, a set with infinite numbers of members.)

203 Countably infinite at http://www.wikipedia.com/wiki/Countably_infinite

(On countability in infinitely large sets.)

204 Continuum hypothesis at http://www.wikipedia.com/wiki/Continuum_hypothesis

(The set theoretic basis of a continuum based on the continuum of real numbers.)

205 Cantors Diagonal argument at http://www.wikipedia.com/wiki/Cantors_Diagonal_argument

(A logical argument that demonstrates that real numbers are not countably infinite.)

206 Cardinal number at http://www.wikipedia.com/wiki/Cardinal_number

(Gauges the relative sizes of sets, even sets with infinite members.)

207 Number from Wikipedia at http://www.wikipedia.com/wiki/number

(Describes numbers as abstract patterns and links to definitions of numbers of different kinds.)

208 Dense from Wikipedia at http://www.wikipedia.com/wiki/Dense

209 Jens Blanch, University of Gavle, Gavle, Sweden (1998): Domain representation of topological spaces at http://www.sm.luth.se/~jens/pdf/top.pdf

(Describes Scott–Ershov domains and their properties. Scott–Ershov domains can facilitate approximation of the infinite continuum of numbers in finite state machines.)

210 Pascal Hitzler, Universität Tübingen February 1998: Scott Domains, Generalized Ultrametric Spaces and Generalized Acyclic Logic Programs (now at University of Karlsruhe)

(Every object of interest can be arbitrarily closely approximated by [compact elements].)

211 Guy Davies, Decision Theory, Hösten 2000: Order and Value Assignment, a seminar series at ITE.

(A relatively benign discussion of ordinal value theory for those willing to brave it.)

212 Ordinal at http://www.wikipedia.com/wiki/Ordinal

(A set theoretic discussion of ordinalilty.)

213 Total Order at http://www.wikipedia.com/wiki/Total_order

(Mathematical basis of ordered sets and ordinal domains.)

214 Well-founded set at http://www.wikipedia.com/wiki/Well-founded_set

(The set theoretic basis of the origin in a coordinate system, especially in an ordinal domain.)

215 Well-order at http://www.wikipedia.com/wiki/Well-order

(A set theoretic discussion of lower bounds on ordinal domains.)

216 Ordered field at http://www.wikipedia.com/wiki/ordered+field

(Describes the set theoretic basis of the "natural zero" of a domain.)

217 Partial order at http://www.wikipedia.com/wiki/Partial_order

(Mathematical descriptions of subtyping and relationship to set theory, especially "posets.")

218 Lattice from Wikipedia at http://www.wikipedia.com/wiki/Lattice (Numbers, functions, and number theory.)

219 Natural number from Wikipedia at http://www.wikipedia.com/wiki/Natural_number

220 Rational number from Wikipedia at http://www.wikipedia.com/wiki/rational+number

221 Irrational number from Wikipedia at http://www.wikipedia.com/wiki/irrational+number

222 Real number from Wikipedia at http://www.wikipedia.com/wiki/Real+number

223 Complex number from Wikipedia at http://www.wikipedia.com/wiki/complex+number

224 Transcendental number at http://www.wikipedia.com/wiki/transcendental+number
225 Hyperreal numbers from Wikipedia at http://www.wikipedia.com/wiki/hyperreal+numbers
226 Hypercomplex numbers from Wikipedia at http://www.wikipedia.com/wiki/Hypercomplex+numbers
227 Octonions from Wikipedia at http://www.wikipedia.com/wiki/octonions
228 Quaternions from Wikipedia at http://www.wikipedia.com/wiki/quaternions
229 Sedenions from Wikipedia at http://www.wikipedia.com/wiki/sedenions
230 P-adic numbers from Wikipedia at http://www.wikipedia.com/wiki/p-adic+numbers
231 Surreal numbers from Wikipedia at http://www.wikipedia.com/wiki/Surreal_numbers
232 Functions and Random Variables at http://www.math.uah.edu/stat/

(Elementary introduction to the mathematical theory of functions.)

233 Function at http://www.wikipedia.com/wiki/Function

(Another easily readable introduction to the mathematical theory of functions.)

234 Injective, surjective and bijective functions from Wikipedia at http://www.wikipedia.com/wiki/Injective,+surjective+and+functions
235 Cartesian product at http://www.wikipedia.com/wiki/Cartesian_product
236 Direct Product from Wikipedia at http://www.wikipedia.com/wiki/direct+product

("In mathematics, one can often define a direct product of objects already known, giving a new [object]" – focuses on mathematical groups.)

237 Recursion definition at http://www.wikipedia.com/wiki/Recursion_definition
238 Transfinite induction at http://www.wikipedia.com/wiki/Transfinite_induction

(Transfinite Induction is a technique of proving that a property applies to all ordinals) (*Lambda Calculus, functional programming and semantics*.)

239 Luca Cardelli of AT&T Bell Laboratories, Murray Hill, NJ, USA and Peter Wegner, Department of Computer Science, Brown University, Providence, USA. On Understanding Types, Data Abstraction, and Polymorphism, *Computing Surveys*, 17(4): 471–522, December 1985 at http://research.microsoft.com/Users/luca/ Papers/OnUnderstanding.pdf

(A mathematical treatment of polymorphism and inheritance based on λ-calculus.)

240 Lambda calculus at http://www.wikipedia.com/wiki/Lambda_ calculus

(A brief informal discussion of λ-calculus, including emergence of functions, arithmetic operations and recursion, as well as a discussion of equivalence of rule expressions.)

241 The Lambda Calculus: A Brief description and history at http://www.kids.net.au/encyclopedia-wiki/la/Lambda_calculus#History
242 Jim Larson at the JPL Section 312. An Introduction to Lambda Calculus and Scheme, a talk in a Programming Lunchtime Seminar on 7/26/1996 at http://www.jetcafe.org/~jim/lambda.html

(Describes how polymorphism emerges from λ-calculus, and how λ-calculus is a universal model of computation. Also describes a programming language, scheme, which facilitates application of λ-calculus.)

243 Andrew Myers, Cornell University. Advanced Programming Languages at http://www.cs.cornell.edu/courses/cs611/2000fa/slides/lec09.pdf

(A brief presentation on normalizing rule expressions with Lambda Calculus. "Two functions are equal by *Extension* if they have the same meaning: they give the same result when applied to the same argument.")

244 H. Zhang of Iowa University: Lambda Calculus at http://www.cs.uiowa.edu/~hzhang/c123/Lecture5.pdf

(A simple but mathematical definition of lambda calculus and normal form with reduction algorithms and examples.)

245 M.-J. Dominus Church–Rosser Theorem, 1999 at http://perl.plover.com/yak/lambda/samples/slide014.html

(A brief presentation of the Church–Rosser theorem that reduces rule expressions to a normal form.)

246 Chris Clack, Senior Lecturer and MSc. CS Course Director, Department of Computer Science, UCLA. The Lambda Calculus: A Deeper Look at http://www.cs.ucl.ac.uk/teaching/3C11/HTML_Lectures/lecture3_3C11/sld011.htm

(Another good presentation on the essence of the Church–Rosser theorem.)

247 Stephen Fenner, Normal Forms and the Church–Rosser Theorem, 1996 at http://www.cs.usm.maine.edu/class/cos370/handouts/ lambda/node7.html

(Describes when rule expressions can and cannot be reduced to normal forms.)

248 P. Selinger, University of Pennsylvania. Functionality, Polymorphism, and Concurrency: A Mathematical Investigation of Programming Paradigms, Ph.D. thesis, 1997.

(Both formal and intuitive descriptions of the normal forms and the Church–Rosser theorem.)

249 Peter V. Homeier, US Department of Defense, Ph.D. in Computer Science, UCLA. A Proof of the Church–Rosser Theorem for Lambda Calculus in Higher Order Logic, 1995 at http://www.cis.upenn.edu/~hol/lamcr/lamcr.pdf

250 Entscheidungsproblem at http://www.wikipedia.com/wiki/Entscheidungsproblem

("Entscheidungsproblem" is German for "the decision problem." In mathematics Entscheidungsproblem addresses the issue of the same rule being expressed in different ways. It specifically proves that there is no general algorithm that will show that algebraic expressions that consist of different terms are equivalent.)

251 First-order predicate calculus at http://www.wikipedia.com/wiki/First-order_predicate_calculus

(Deals with symbolic logic that is the basis of set theory, values, relationships, arithmetic and logical operators.)

252 Manfred Kanka, A paper on Semantics, Manfred Krifka, Institut für deutsche Sprache und Linguistik, HU Berlin, WS 2000/2001at http://amor.rz.hu-berlin.de/~h2816i3x/SemanticsI-07.pdf

253 Anthony J. Roy, Department of Computer Science, University of Keele, UK. A Comparison of Rough Sets, Fuzzy Sets and Non-monotonic Logic Technical Report TR99–11, June 1999 at http://pages.britishlibrary.net/aroy/ant/revigis/Comparisonpdf.pdf
254 Functional programming at http://www.wikipedia.com/wiki/Functional_programming

(Functional programming expresses logic by combining functions instead of focusing on execution of computer commands. Arguments as well as results of functions can be functions.) (*Spaces and their properties*.)

255 Eric W. Weisstein, Space from Eric Weissensteinn's Treasure Trove of Science at http://hades.ph.tn.tudelft.nl/Internal/PHServices/Documentation/ MathWorld/math/math/s/s513.htm

(Succinct but very abstract mathematical definitions of various spaces including metric spaces and state spaces.)

256 Tensor from Wikipedia at http://www.wikipedia.com/wiki/Tensor
257 Tensor/Old from Wikipedia at http://www.wikipedia.com/wiki/Tensor/Old

("Tensors are quantities that describe a transformation between coordinate systems. in such a way that the physical laws [are described in a way that is] . . . independent of the coordinate system chosen . . . tensors were introduced as specific representations of the group of all changes of coordinate systems.")

258 Tensor at http://hades.ph.tn.tudelft.nl/Internal/PHServices/ Documentation/MathWorld/math/math/t/t078.htm
259 Metric Tensor at http://hades.ph.tn.tudelft.nl/Internal/PHServices/ Documentation/MathWorld/math/math/m/m217.htm
260 Vector space from Wikipedia at http://www.wikipedia.com/wiki/vector+space
261 Normed vector space from Wikipedia at http://www.wikipedia.com/wiki/normed+vector+space

(Description of mathematical "norm" and isometry.)

262 Topology from Wikipedia at http://www.wikipedia.com/wiki/Topology
263 Pointless Topology at http://www.wikipedia.com/wiki/Pointless+topology
264 Topology Glossary from Wikipedia at http://www.wikipedia.com/wiki/Topology+Glossary
265 Dr. Paul Cairns (principal Investigator) and Jeremy Gow (researcher) of Interaction Design Centre at the School of Computing Science, Middlesex University, UK. The definition of a metric space based on lecture notes by Peter Collins in Elements of Euclidean and Metric Topology of the Interactive, available at the Mathematical Proofs research (IMP) project (January 2001 to June 2002), funded by The Engineering and Physical Sciences Research Council (EPSRC), UK at http://www.uclic.ucl.ac.uk/imp/

(A simple definition of metric spaces.)

266 Spaces with richer structures especially metric spaces in The Mathematical Atlas at http://www.math.niu.edu/~rusin/known-math/index/54EXX.html

(A minimally mathematical definition of metric spaces and discussion of a metric as a generalized concept of topological distance.)

267 Bruce MacLennan, Computer Science Department University of Tennessee, Knoxville. Discrete Metric Space, 1996.

268 Manifold from Wikipedia at http://www.wikipedia.com/wiki/Manifold

("A manifold, in mathematics, can be thought of as a 'curved' surface or space which locally looks like Euclidean space and therefore admits the introduction of local charts or coordinate systems . . . Every manifold has a dimension, the number of coordinates needed in local coordinate systems.")

269 Hausdorff space from Wikipedia at http://www.wikipedia.com/wiki/Hausdorff+space

("A Hausdorff space is a topological space in which any two distinct points have disjoint neighbourhoods.")

270 Tychonoff space from Wikipedia at http://www.wikipedia.com/wiki/Tychonoff+space

("A Hausdorff space X is called a Tychonoff space if, for every nonempty closed subset C and every x in the complement of C, there is a continuous function f: X –> [0,1] such that f(x) = 0 and f(C) = {1}" – i.e. a tychonoff space is a space of distinct points that may be partitioned into two mutually exclusive sets of points. This is the mathematical theory that supports partitioning objects and state spaces.)

271 Dimensional Analysis from Wikipedia at http://www.wikipedia.com/wiki/Dimensional+analysis

(A description of how other physical domains emerge from fundamental physical domains and the use of this information in engineering sciences.)

272 Fundamental Dimensions from Wikipedia at http://www.wikipedia.com/wiki/Fundamental+dimension

(A description of fundamental physical domains of this book, called "dimensions" in this publication.)

273 Hausdorff Dimension from Wikipedia at http://www.wikipedia.com/wiki/Hausdorff+dimension

(A mathematical description of the dimensionality of complex metric spaces that subsumes the "normal" Euclidean concept of dimension.)

274 Infimum from Wikipedia at http://www.wikipedia.com/wiki/Infimum

(A type of lower bound. The Hausdorff dimension is related to this concept.)

275 Hamel dimension from Wikipedia at http://www.wikipedia.com/wiki/Dimension+of+a+vector+space

(A mathematical description of the dimensionality of vector spaces that subsumes the "normal" Euclidean concept of dimension and accounts for the cardinality – see cardinal number – of the space.)

276 Connectedness from Wikipedia at http://www.wikipedia.com/wiki/connectedness

(Mathematically describes the concept of points in space being connected to points in their neighborhood, as well as the weirder concept of points being isolated from others.)

277 Simply Connected from Wikipedia at http://www.wikipedia.com/wiki/simply+connected

(A mathematical description of paths and connections in abstract spaces.)

278 Eric W. Weisstein, Distance at http://hades.ph.tn.tudelft.nl/Internal/PHServices/Documentation/MathWorld/math/math/d/d325.htm

(A succinct description of a generalized concept of distance in a manifold.)

279 Eric W. Weisstein, Metric at http://hades.ph.tn.tudelft.nl/Internal/PHServices/Documentation/MathWorld/math/math/m/m213.htm

(A succinct description of a metric as a generalized concept of distance.)

280 Measure at http://www.wikipedia.com/wiki/Measure

(Gauges the relative sizes of sets.)

281 Thierry Coquand, How to Define Measures of Borel Sets at http://www.cs.chalmers.se/~coquand/riesz.pdf

(A complex mathematical discussion of Borel sets and Cantor spaces.)

282 Borel Measure at http://www.wikipedia.com/wiki/Borel_measure

("The Borel Measure is the measure on the smallest set algebra containing the intervals which gives to the interval [a,b] the measure b-a.")

283 Koji Tsuda, Machine Understanding Division, Electrotechnical Laboratory, Japan. Subspace Classifier in Hilbert Space, *Pattern Recognition Letters*, 20(5), 513–519, May 1999.

(Using Hilbert spaces to automate creation of classes and subtypes based on similarities between objects. The paper is a sophisticated mathematical discussion of how objects might be classified from a large number of samples using statistical methods.)

284 Prof. C-I Tan Department of Physics, Brown University: Notes on Hilbert Space at http://www.chem.brown.edu/chem277/Tan_on_Hilbert_Space.html

285 Norman D. Megill, Hilbert Space Explorer (GPL), 2000 at http://us.metamath.org/mpegif/mmhil.html

(A set of definitions, theorems and explanations about Hilbert space.)

286 Jack Sarfatti, A Semi-Pop Non-Mathematical Tutorial on Hilbert Space in Quantum Mechanics at http://www.qedcorp.com/pcr/pcr/hilberts.html

(This paper focuses on representing quantum mechanical states with the help of Hilbert spaces: "Hilbert space contains infinite dimensions, but these are not geometric. Rather, each dimension represents a state of possible existence for a quantum system. All possible states coexist."

The book you are reading is about business systems, not quantum states, and the metamodel in this book focuses on purely deterministic systems. In contrast, the state of a quantum system is unknown, and merely querying it can change its state. However, mathematically astute readers will find interesting analogs that can be extended to describe the states of non-deterministic business systems in Sarfatti's paper – especially those that might change state by merely querying the information in them. This can happen in all real-world systems, but is beyond the scope of this book. We can safely ignore Hilbert space in this book.)

Buckingham's pi theorem – about the independence of physical laws from their units of measure

287 Harald Hanche-Olsen, Department of Mathematical Sciences, Norwegian University of Science and Technology (NTNU), Trondheim, Norway. Buckingham's Pi Theorem (Version 2001-09-15, 1998).

(Describes Buckingham's pi theorem with illustrative examples of its use in finding solutions to physical problems. Includes a mathematical discussion of values, measurement, and units of measure – non-mathematicians beware!)

288 Buckingham's Pi Theorem, *Academic Press Dictionary of Science*, Editor Christopher Morris, Elsevier Science & Technology Books, December 1991.

(Concise description of Buckingham's pi theorem.)

289 Eric Weisstein, Buckingham's Pi Theorem, Eric Weisstein's World of Physics at http://scienceworld.wolfram.com/physics/ BuckinghamsPiTheorem.html

(Description and mathematical proof of Buckingham's pi theorem.)

Information theory, chaos theory and miscellaneous publications

290 Information Theory from Wikipedia at http://wikipedia.com/wiki/information+theory

(A brief introduction to Shannon's information theory and measures of information.)

291 Sapir–Whorf Hypothesis from Wikipedia at http://wikipedia.com/wiki/Sapir-Whorf+hypothesis

(An overview of the impact of language on meaning and perception)

292 Chaos theory from Wikipedia at http://www.wikipedia.com/wiki/Chaos+theory

(A succinct introduction to the theory of chaos.)

293 Takashi Kanamaru, Department of Electrical and Electronic Engineering, Tokyo University of Agriculture and Technology, Japan and J. Michael T. Thompson, Department of Applied Mathematics and Theoretical Physics, Cambridge. Introduction to Chaos and Nonlinear Dynamics, September 1997 at http://brain.cc.kogakuin.ac.jp/~kanamaru/Chaos/e/. See Time Series of Logistic map at http://brain.cc.kogakuin.ac.jp/~kanamaru/Chaos/e/Logits/

(An interactive site that can give the reader a hands-on experience in chaotic systems.)

Books

294 Ronald G. Ross, *The Business Rule Book: Classifying, Defining and Modeling Rules*, Database Research Group Inc, 1997.

295 Michael Hammer and James Champy, *Reengineering The Corporation*, Harper Collins, 1993.
296 Peter Herzum and Oliver Sims, *Business Component Factory: A Comprehensive Overview of Component-Based Development for the Enterprise*, John Wiley & Sons, 1999.
297 G. M. Nijssen and T. A. Halpin, *Conceptual Schema and Relational Database Design: A Fact Oriented Approach*, Prentice Hall, 1989.
298 Paul Harmon and David King, *Artificial Intelligence in Business*, John Wiley & Sons, 1985, Chapter 4: Representing Knowledge, Chapter 5: Drawing Inferences.
299 Martin Fowler, *Analysis Patterns: Reusable Object Models*, Addison Wesley Longman, 1997.
300 August-Wilhelm Scheer, *ARIS, Business Process Frameworks*, Springer-Verlag, August 1999.
301 August-Wilhelm Scheer, *ARIS, Business Process Modeling*, Springer-Verlag, November 1996.
302 James Rumbaugh, Michael Blaha, William Premerlani, Frederick Eddy, and William Lorensen at the General Electric Research and Development Center. *Object-Oriented Modeling and Design*, Prentice Hall, 1991.
303 *Structured Systems Analysis and Design Methodology* (version 4) from the SSADM College Ltd, 1996 at http://www.comp.glam.ac.uk/pages/staff/tdhutchings/chapter4.html.
304 Candace C. Fleming and Barbara von Halle, *Handbook of Relational Database Design*, Addisson Wesley, 1989.
305 Maisell and Gnugnoli of Science Research Associates. *Simulation of Discrete Stochastic Systems*, Published in 1972.
306 Raphael Finkel, *Functional Programming*, Addison-Wesley, 1996: Chapter 4 at ftp://ftp.aw.com/cseng/authors/finkel/apld/finkel04.pdf
307 Paul Taylor, *Practical Foundations of Mathematics*, section 2.3 Sums, Products and Function-Types, Cambridge University Press, 1999, at http://www.dcs.qmul.ac.uk/~pt/Practical_Foundations/html/s23.html.

(A simple, physical explanation of lambda calculus and the need for it in addressing practical real world problems: "[We] discussed how functions act, but they must also be considered as entities in themselves. Early . . . problems arose in which the unknown was a function as a whole, rather than its value at particular or even all points: the Sun's light takes that path through the variable density of the atmosphere which minimizes the time of travel; the motion of a stretched string depends on its initial displacement along its whole length."

308 Daniel Finkbeiner II, *Kenyon College. Matrices and Linear Transformations*, W. H. Freeman & Co., 1966, Chapter 1.

(Includes a mathematical discussion of sets, set operations, functions, mapping between sets, relationships and domains.)

309 George Thomas and Ross Finney, *Calculus and Analytical Geometry*, Addison-Wesley, 1996.

(Third edition published in 1960). (Has simple mathematical descriptions of functions, domains, ranges, existence and continuity.)

310 Emanuel Parzen, Stanford University. *Modern Probability Theory and Its Applications*, John Wiley, 1960, 1992.
311 Sidney Siegel, Research Professor of Psychology, The Pennsylvania State University. *Nonparametric Statistics for the Behavioral Sciences*, McGraw-Hill, Kogakusha Ltd, 1956, 1988.

312 M. Harvey Wagner of Yale University. *Principles of Operations Research with Applications to Managerial Decisions*, Prentice Hall, 1969, 1975.

313 Billy E. Gillett, Professor of Computer Science, University of Missouri-Rolla. *Introduction to Operations Research: A Computer Oriented Algorithmic Approach*, McGraw-Hill, 1976.

314 Erwin Kreysig, Professor of Mathematics, Ohio State University. *Advanced Engineering Mathematics*, John Wiley & Sons, 1967, 1999.

315 A. W. Goodman, University of South Florida. *Modern Calculus with Analytic Geometry*, New York: MacMillan, 1967, 1974.

316 William R. Durell, *The Complete Guide To Data Modeling*, Data Administration Inc.

317 W. Durell, *Data Administration: A Practical Guide to Data Management*, New York: McGraw-Hill, 1985.

318 Anastasia Pagnoni, *Project Engineering: Computer-Oriented Planning and Operational Decision Making*, Berlin: Springer-Verlag, 1999.

(Describes various techniques for modeling and managing tasks, including complex stochastic, models of repetitive processes using techniques such as GERT and petrinets.)

319 Kenneth M. Dymond, *A Guide to the CMM: Understanding the Capability Maturity Model for Software*, Process Transition International Inc., 1995.

(Describes the dynamics of Best Practices and processes needed to institutionalize change based on the System Engineering Institute's [SEI] Capability Maturity Model [CMM].)

320 Jeanie Daniel Duck, Senior Vice President, the Boston Consulting Group. *The Change Monster*, Random House, 2001.

(Excellent and very readable work on the social, emotional and organizational dynamics of change.)

321 Geoffrey Moore, *Crossing the Chasm*, New York: HarperCollins, 2002.

(On the acceptance of technological innovation in the market place.)

322 W. H. Inmon, *A Brief History of Data Base Design*, John Wiley, 1999.

(Describes some changes in business environments and assumptions that have disrupted legacy systems.)

323 Tom Mullins, Department of Physics, University of Oxford; David Holton, Department of Hydrogeology, Harwell Laboratory; Robert May, Department of Zoology, University of Oxford; J. M. T. Thompson, Center for Nonlinear Dynamics and Applications, University College of London; Peter L. Read, Department of Physics, University of Oxford; M. S. Child, Department of Chemistry, University of Oxford; and Jonathan Keating, Department of Mathematics, University of Manchester. *The Nature of Chaos*, Oxford University Press, 1993.

324 G. J. Chamberlin and D. G. Chamberlin, *Colour: Its Measurement, Computation and Application*, Heyden & Sons, 1980.

325 V. Pekelis, *Cybernetics A to Z*, English translation, Moscow: Mir, 1974.

326 Maurice Aburdene, Bucknell University. *Computer Simulation of Dynamic Systems*, Dubuque, Iowa: William C. Brown, 1988.

327 Averill Law, President, Simulation Modeling and Analysis Company, Tucson, Arizona, USA and Professor of Decision Sciences, University of Arizona and W. Kelton, Associate Professor of Operations

Research and Management Science, University of Minnesota. *Simulation Modeling and Analysis*, second edition, McGraw Hill, 1991.

328 Bertrand Meyer, Object-Oriented Software Construction Interactive, Software Engineering, Inc., 1993–2001. Key extracts are available at http://www.eiffel.com/doc/manuals/technology/oosc/inheritance-design/ section_05.html

329 Martin Fowler and Kendall Scott, *UML Distilled: Applying the Standard Object Modeling Language*, Addison-Wesley Longman Inc., 1999.

330 Bruce Powel Douglass, *Real-Time UML: Developing Efficient Objects for Embedded Systems*, Second Edition, Addison-Wesley Longman Inc., 1997.

331 James Rumbaugh, Ivar Jacobson, and Grady Booch, *The Unified Modeling Language Reference Manual*, Addison-Wesley Longman Inc., 2004.

332 Grady Booch, James Rumbaugh, and Ivar Jacobson, *The Unified Modeling Language User Guide*, Addison-Wesley Longman Inc., 1998.

333 Pierre-Alain Muller, *Instant UML*, Birmingham, UK: Wrox Press Ltd, 1997.

334 Eugene Blanchard (Edited by Joshua Drake, Bill Randolph, and Phuong Ma), Introduction to Networking and Data Communications, Commandprompt, Inc and Copyright, 2001 at http://www.linuxports.com/howto/intro_to_networking/book1.htm

335 Howard Smith and Peter Fingar, *Business Process Management: The Third Wave*, Meghan-Kiffer Press, 2003.

336 *The New Encyclopedia Britannica*, 15th Edition, Encyclopedia Britannica Inc., 1988.

337 Amit Mitra and Amar Gupta, *Agile Systems with Reusable Patterns of Business Knowledge: A Component Based Approach*, Artech House, 2005.

338 James Martin, *After the Internet: Alien Intelligence*, Regenery Publishing Inc., 2000.

339 Amit Mitra and Amar Gupta, *Knowledge Reuse and Agile Processes: Catalysts for Innovation*, The Idea Group, 2006.

Index

P9-CMA-356

696 SILLY SCHOOL JOKES & RIDDLES

by **JOSEPH ROSENBLOOM**

illustrated by
Dennis Kendrick

Sterling Publishing Co., Inc. New York

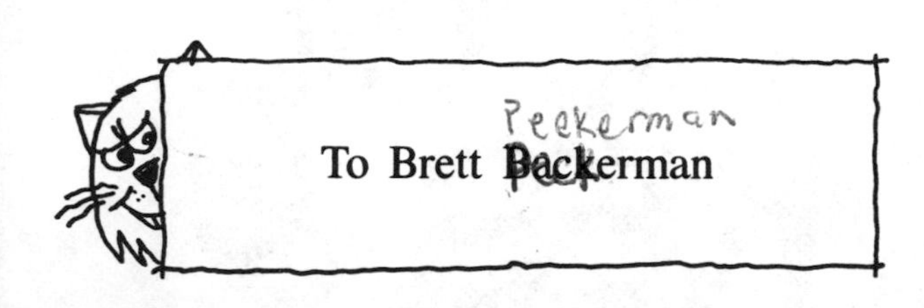

Library of Congress Cataloging-in-Publication Data

Rosenbloom, Joseph.
696 silly school jokes & riddles.

Includes index.
Summary: A collection of 696 jokes, riddles, and tongue twisters on the subject of school.
1. Education—Juvenile humor. 2. Wit and humor, Juvenile. [1. Schools—Wit and humor. 2. Jokes. 3. Riddles. 4. Tongue twisters] I. Kendrick, Dennis, ill. II. Title. III. Title: Six hundred ninety-six silly school jokes and riddles.
PN6231.S3R67 1986 818′.5402 85-27855
ISBN 0-8069-4726-8
ISBN 0-8069-4727-6 (lib. bdg.)

ISBN 0-8069-6392-1 (paper)

Published by Sterling Publishing Co., Inc.
Two Park Avenue, New York, N.Y. 10016
Distributed in Australia by Capricorn Book Co. Pty. Ltd.
Unit 5C1 Lincoln St., Lane Cove, N.S.W. 2066
Distributed in the United Kingdom by Blandford Press
Link House, West Street, Poole, Dorset BH15 1LL, England
Distributed in Canada by Oak Tree Press, Ltd.
℅ Canadian Manda Group, P.O. Box 920, Station U
Toronto, Ontario, Canada M8Z 5P9
Manufactured in the United States of America

Contents

Books by Joseph Rosenbloom

Biggest Riddle Book in the World
Daffy Definitions
Doctor Knock-Knock's Official Knock-Knock Dictionary
Funniest Joke Book Ever!
Funniest Riddle Book Ever!
Funny Insults & Snappy Put-Downs
Gigantic Joke Book
Knock-Knock Who's There
Looniest Limerick Book in the World
Mad Scientist
Monster Madness
Nutty Knock Knocks
Official Wild West Joke Book
Ridiculous Nicholas Haunted House Riddles
Ridiculous Nicholas Pet Riddles
Ridiculous Nicholas Riddle Book
Silly Verse (and Even Worse)
Wacky Insults and Terrible Jokes
Zaniest Riddle Book in the World

1. FIRST THINGS FIRST

What is the first thing a little snake learns in school?
Hiss-tory.

What is the first thing a little gorilla learns in school?
The Ape B C's.

What do little astronauts get when they do their homework?
Gold stars.

Why did the little vampires stay up all night?
They were studying for a blood test.

TEACHER: Alice, name four members of the cat family.
ALICE: Mother, father, sister and brother.

TEACHER: Pablo, name six wild animals.
PABLO: Two lions and four tigers.

WHY DID THEY DO WELL IN SCHOOL?

—The firefly?
It was so bright.

—The duck?
It was a wise quacker.

—The two-headed monster?
Two heads are better than one.

—The elephant?
It had a lot of grey matter.

—The balloon?
It went to the top of the class.

TEACHER: How old were you on your last birthday?
STUDENT: Seven.
TEACHER: How old will you be on your next birthday?
STUDENT: Nine.
TEACHER: That's impossible.
STUDENT: No, it isn't, teacher. I'm eight today.

WHY DID THEY FLUNK OUT?

—The little witches?
They couldn't spell.

—The skeleton?
His heart wasn't in it.

—The owl?
It didn't give a hoot.

Where do monsters study?
In ghoul school.

Who sits in front of the class in ghoul school?
The creature teacher.

"Teacher, may I leave the room?"

"Well, you certainly can't take it with you."

TEACHER: George, go to the map and find North America.

GEORGE: Here it is!

TEACHER: Correct. Now, class, who discovered America?

CLASS: George!

Where do you find prehistoric cows?

In a moo-seum.

TEACHER: Willy, name one important thing we have today that we didn't have ten years ago.

WILLY: Me!

FIRST DAY OF SCHOOL

Sally came home from her first day at school.

"It was all right," she told her mother, "except for some lady named Teacher who kept spoiling our fun."

Larry came home from his first day at school.

"I'm not going back," he said.

"Why not?" asked his father.

"Because my teacher doesn't know anything," said Larry. "All she ever does is ask questions."

Elroy came home from his first day at school.

"Nothing much happened," he told his mother. "Some lady didn't know how to spell 'cat.' I told her."

TEACHER: Do you know "London Bridge Is Falling Down?"

SARA: No, but I hope no one gets hurt.

SUZIE: I won a prize in kindergarten today. The teacher asked me how many legs a hippopotamus had. I said three.

FATHER: Three? How on earth did you win the prize?

SUZIE: I came the closest.

PRINCIPAL: What are you going to be when you get out of school?

HARVEY: An old man.

Marvin came into his kindergarten class with a squirming worm.

"What are you doing with that disgusting worm?" asked his teacher.

"We were playing outside," said Harvey, "and I thought I'd show him my kindergarten."

TEACHER SAYS

Say each sentence three times quickly:

- Rubber buggy bumpers.
- Bobby Blue blows big blue bubbles.
- The big beautiful blue balloon burst.

How does a skeleton study for tests?
It bones up.

With tears in his eyes, the little boy told his kindergarten teacher that only one pair of boots was left in the classroom and they weren't his.

The teacher searched and searched, but she couldn't find any other boots. "Are you sure these boots aren't yours?" she asked.

"I'm sure," the little boy sobbed. "Mine had snow on them."

TEACHER: Why do traffic lights turn red?
DOLLY: You would too if you had to stop and go in the middle of the street.

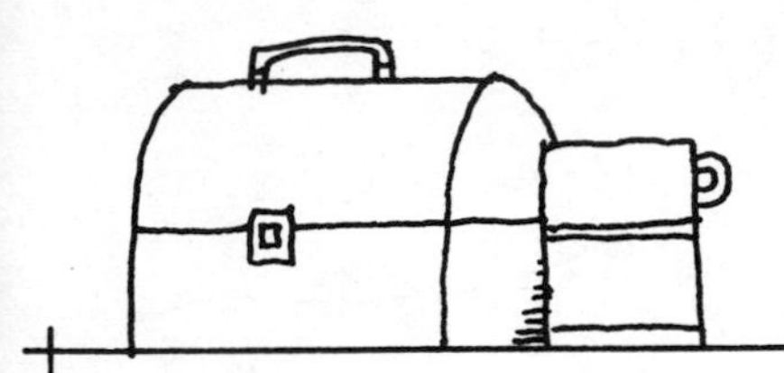

WHAT DID YOU LEARN IN SCHOOL TODAY?

MOTHER: What did you learn in school today?
AUDREY: Not enough. I have to go back tomorrow.

MOTHER: What did you learn in school today?
LOUELLA: How to talk without moving my lips.

FATHER: What did you learn in school today?
LOUIE: My teacher taught us how to write.
FATHER: What did you write?
LOUIE: I don't know, Dad. She didn't teach us how to read yet.

TEACHER: Goodness, Mildred, haven't you finished washing that blackboard yet? You've been at it for an hour.
MILDRED: I know, but the more I wash it, the blacker it gets.

WATSON: What school did you go to, Holmes?
SHERLOCK: Elementary, my dear Watson!

It was the first day of school. As the principal made his rounds, he heard a terrible commotion coming from one of the classrooms. He rushed in and spotted one boy, taller than the others, who seemed to be making the most noise. He seized the lad, dragged him into the hall, and told him to wait there until he was excused.

Returning to the classroom, the principal restored order and lectured the class for half an hour about the importance of good behavior.

"Now," he said, "are there any questions?"

One girl stood up timidly. "Please, sir," she asked, "may we have our teacher back?"

SUBSTITUTE TEACHER: Are you chewing gum?
BILLY: No, I'm Billy Anderson.

What is big and yellow and comes in the morning to brighten Mother's day?

The school bus.

Mrs. Jones brought her son Elmer to register at school. However, Elmer was only five, and the required age was six.

"I think," said Mrs. Jones to the principal, "that he can pass the six-year-old test."

"We'll see," replied the principal. "Elmer, say the first thing that comes to your mind."

"Do you want logically connected sentences," said Elmer, "or purely irrelevant words?"

What's one and one?
Two.
What's four minus two?
Two.
Who wrote Tom Sawyer?
Twain.
Now say all the answers together.
Two, two, twain.
Have a nice twip!

2. HOME ROOM HI-JINKS

TEACHER: Alfred, how can one person make so many stupid mistakes in one day?
ALFRED: I get up early.

GRACE: What time do you wake up in the morning?
ACE: About an hour and a half after I get to school.

NAN: Let's play school.
DAN: Okay, let's play I'm absent.

What is yellow, has wheels and lies on its back?
A dead school bus.

How do bees get to school?
By school buzz.

What would happen if you took the school bus home?
The police would make you bring it back.

What is the difference between a school bus driver and a cold?

One knows the stops and the other stops the nose.

A class has a top and a bottom. What lies between?

The student body.

MOTHER: How do you like your new teacher?
JUNIOR: I don't. She told me to sit up front for the present and then she didn't give me one.

SUE: Does your teacher like you?
LEW: *Like* me! She *loves* me! Look at all the kisses she puts on my papers!

"My little sister is so smart! She's only in nursery school and she can spell her name backwards and forwards."

"Really? What's her name?"

"Anna."

SON: I'm glad you named me Timothy, Dad.
FATHER: Why?
SON: Because that's what the kids in school call me.

HENRY: What was my name in first grade?
TEACHER: Henry.
HENRY: What was my name in second grade?
TEACHER: Henry.
HENRY: Knock-knock.
TEACHER: Who's there?
HENRY: Henry.
TEACHER: Henry who?
HENRY: Don't tell me you've forgotten me already!

Knock-knock.
Who's there?
Quiet Tina.
Quiet Tina who?
Quiet Tina classroom, monkey wants to speak!

FLO: How did you find school today?
JOE: Oh, I just got off the bus—and there it was!

TEACHER: Freddie, are you the youngest member of your family?
FREDDIE: No, my puppy is.

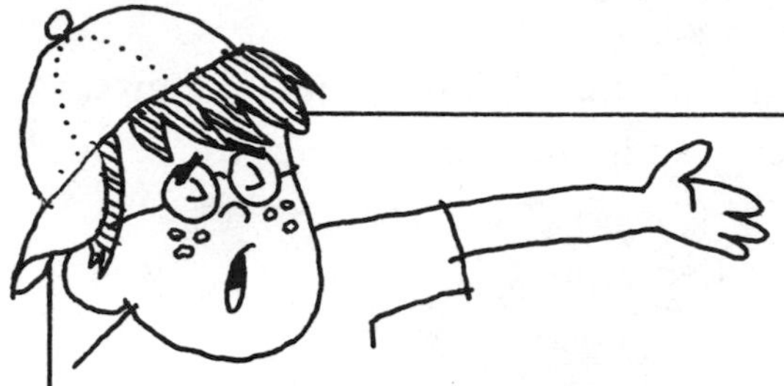

ONE OF THE KIDS IN MY CLASS IS A REALLY BIG PEST.

HOW BIG A PEST IS HE?

He's such a pest, he gives aspirins a headache.

He's such a pest, he has more crust than Betty Crocker.

He's such a pest, people throw parties just not to have him.

He's such a pest, when he goes to the zoo, the monkeys throw peanuts at him.

He's such a pest, if he throws a boomerang, it doesn't come back.

He's such a pest, even echoes don't answer him.

TEACHER: If you don't stop making so much noise, I'll go crazy.
DUFFY: Too late, teacher. We stopped an hour ago.

TEACHER: Didn't you promise to behave?
STUDENT: Yes, sir.
TEACHER: And didn't I promise to punish you if you didn't?
STUDENT: Yes, sir, but since I broke my promise, you don't have to keep yours.

TEACHER: Who was older, David or Goliath?
STUDENT: David must have been, because he rocked Goliath to sleep.

TEACHER: What is the definition of ignorance?
ED: I don't know.
TEACHER: Correct!

TEACHER: Define the word "disease."
CARMEN: "Disease" is de grade you get below de B's.

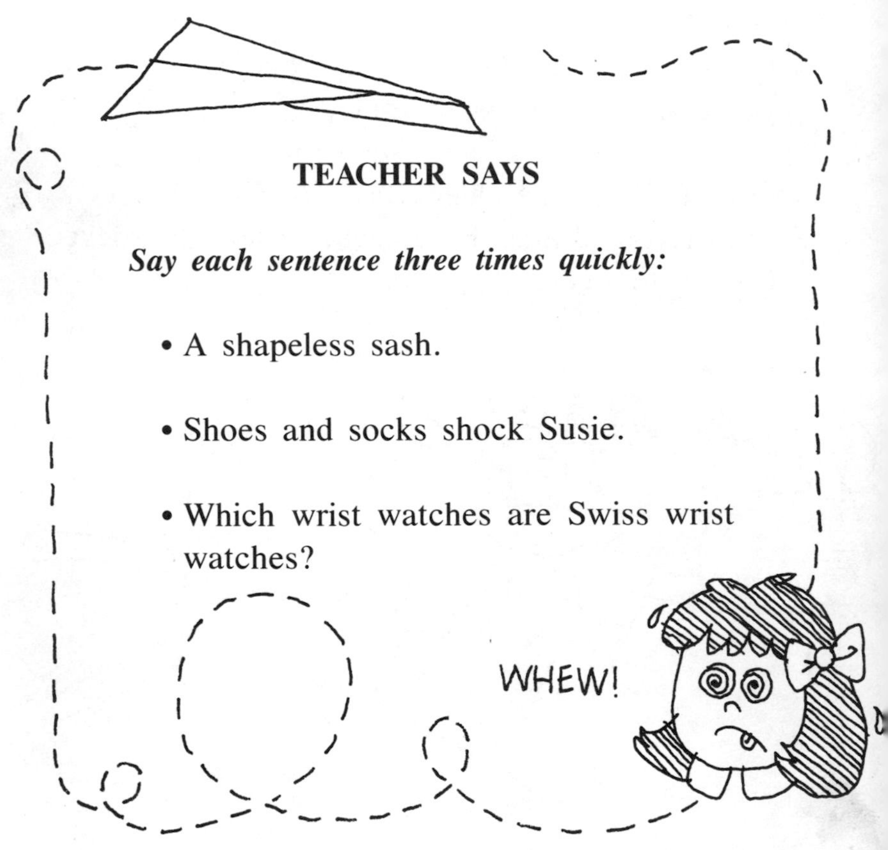

TEACHER SAYS

Say each sentence three times quickly:

- A shapeless sash.
- Shoes and socks shock Susie.
- Which wrist watches are Swiss wrist watches?

TEACHER: Tommy, why do you always get so dirty?
TOMMY: Well, I'm a lot closer to the ground than you are.

HOMEWORK

HAROLD: Teacher, would you punish me for something I didn't do?

TEACHER: Of course not.

HAROLD: Good, because I didn't do my homework.

TEACHER: Did you do your homework?

ARTHUR: No, teacher.

TEACHER: Do you have an excuse?

ARTHUR: Yes. It's all my mother's fault.

TEACHER: She kept you from doing it?

ARTHUR: No, she didn't nag me enough.

TEACHER: This homework looks like your father's writing.

DWIGHT: Sure, I used his pen.

SON: Dad, I'm tired of doing homework.

FATHER: Now, son, hard work never killed anyone yet.

SON: I know, Dad, but I don't want to be the first.

TEACHER: How do you like doing homework?

PUPIL: I like doing nothing better.

IT'S LATER THAN YOU THINK

TEACHER: Why are you late?
WEBSTER: Because of the sign.
TEACHER: What sign?
WEBSTER: The one that says, "School Ahead, Go Slow." That's what I did.

TEACHER: Why are you crawling into my classroom?
SONYA: Because you always say, "Don't anyone dare walk in late!"

TEACHER: Are you late to school again?
WENDELL: Yes, Miss Jones, but didn't you tell us it's never too late to learn?

Harry came home from school, very unhappy.

"I'm not going back tomorrow," he said.

"Why not, dear?" asked his mother.

"Well, I can't read and I can't write, and they won't let me talk, so what's the use?"

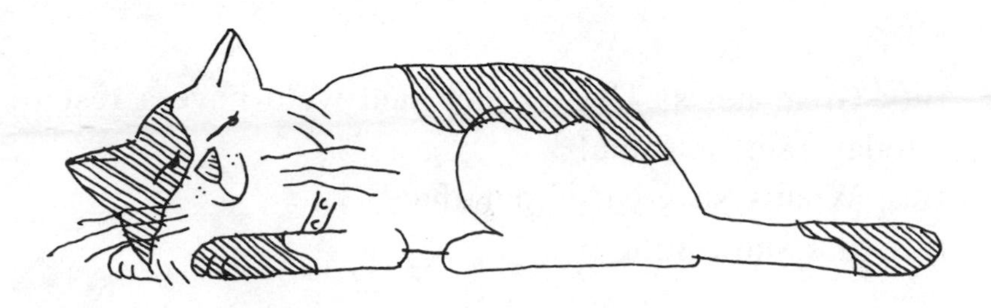

3. JEST TESTING

THE TEST PRAYER

Now I lay me down to rest,
I pray to pass tomorrow's test.
If I should die before I wake,
That's one less test I'll have to take.

TEACHER: I hope I didn't see you looking at Don's paper.
RON: I hope you didn't either.

TEACHER: Seymour, you copied from Elmo's paper, didn't you?
SEYMOUR: How did you find out?
TEACHER: Elmo's paper says, "I don't know," and yours says, "Me neither."

TED: Great news! The teacher said we'd have a test today rain or shine!

RED: What's so great about that?

TED: It's snowing!

What would you get if you crossed a vampire and a teacher?

Lots of blood tests.

What kinds of tests do they give witches?

Hex-aminations.

FATHER: What did the teacher think of your idea?

JUNIOR: She took it like a lamb.

FATHER: Really? What did she say?

JUNIOR: Baa!

FATHER: How were the exam questions?
SON: Easy.
FATHER: Then why do you look so unhappy?
SON: The questions didn't give me any trouble—just the answers.

FATHER: How did the exams go?
SON: I got nearly 100 in every subject.
FATHER: What do you mean—nearly 100?
SON: Well—I got the naughts.

GARY: I don't think I deserve a zero on this test.
TEACHER: I agree, but it's the lowest mark I can give you.

MOTHER: Why did you get such a low mark on that test?
JUNIOR: Because of absence.
MOTHER: You mean you were absent on the day of the test?
JUNIOR: No, but the kid who sits next to me was.

FATHER: It says here that you're at the bottom in a class of 20. That's terrible.
JUNIOR: It could be worse.
FATHER: I don't see how.
JUNIOR: It could have been a bigger class.

FATHER: Aren't you first in anything at school?
JUNIOR: Sure, Dad. I'm first out when the bell rings!

FATHER: Why are your marks so low?

SON: Because I sit in the last row of the class.

FATHER: What difference does that make?

SON: Well, when the teacher hands out marks, there just aren't enough good ones left for the people in the back.

MOTHER: Why have your grades been so low since the holidays?

BARBARA: Well, Mother, you know how everything gets marked down after Christmas.

TEACHER: Do you know why you have such poor grades?

STUDENT: I can't think.

TEACHER: Exactly!

TEACHER: Have you had your eyes checked lately?

DIGBY: No, they've always been plain brown.

REPORT CARDS

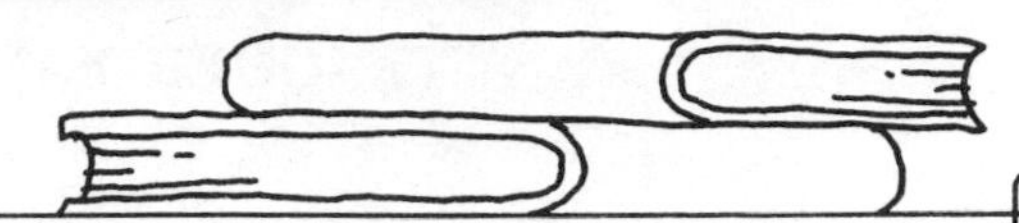

STANLEY (*after the teacher handed out the report cards*): I don't want to scare you, teacher, but my father said that if I didn't bring home a good report card, *somebody* was going to get spanked!

FATHER: Look at these bills! Taxes, rent, telephone, clothes, food! The cost of everything keeps going up. I'd like to see just one thing going down.

SON: Dad, here's my report card.

FATHER: Junior, what does this "C" on your report card mean?

JUNIOR: Colossal!

SYLVIA: Dad, can you write in the dark?

FATHER: I think so. What do you want me to write?

SYLVIA: Your name on this report card.

FATHER: What a terrible report card! What's the matter with you?

MORRIS: I'm not sure, sir, if it's heredity or environment.

TEACHER SAYS

Say each sentence three times quickly:

- Dick kicks sticky bricks.
- Shave a single shingle thin.
- Stick six thick sticks there.

Did you hear about the little kid who copied from his friend's arithmetic test paper by using a mirror? He got all his answers backwards. His friend got a grade of 93 and he got 39.

TEACHER: Young man, are you the teacher of this class?
STUDENT: No, ma'am.
TEACHER: Then don't talk like an idiot!

SECOND-GRADER: I really liked being in your class, Miss Jones. I'm sorry you're not smart enough to teach us next year.

4. SHOW & TELL

TEACHER SAYS

Name the picture:

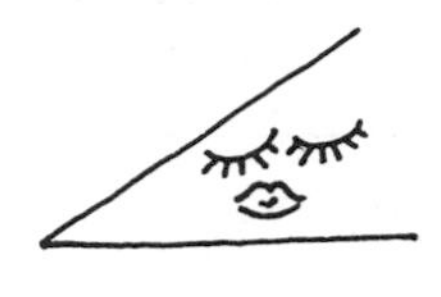

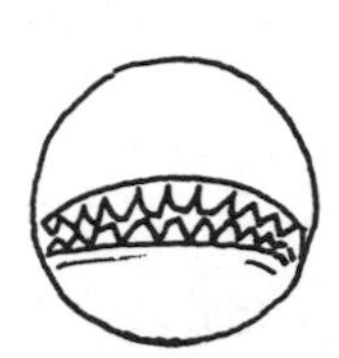

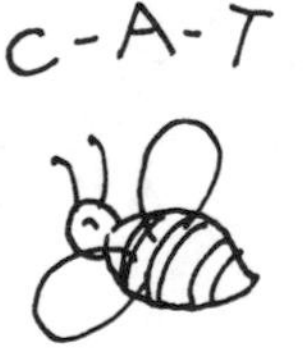

Answers on the next page

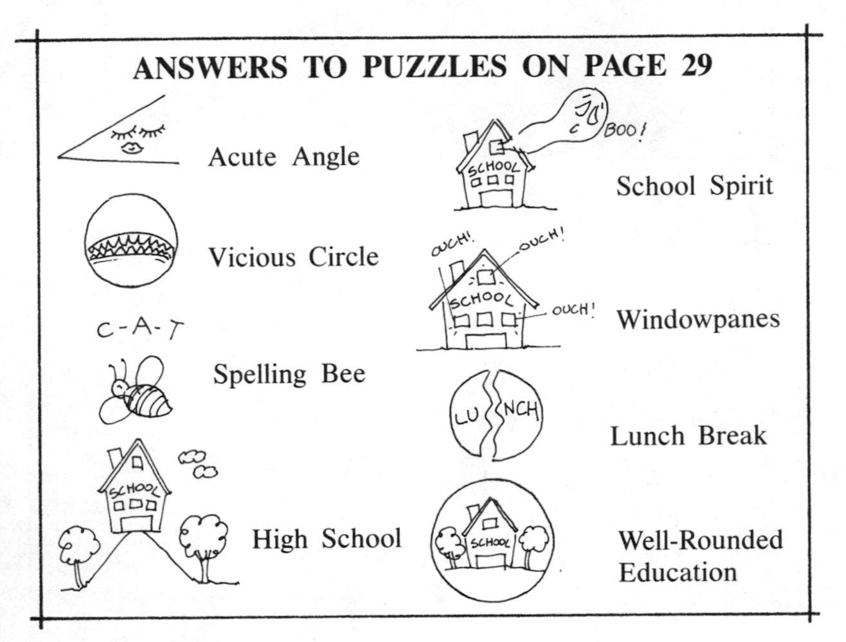

TEACHER: Where do bugs go in winter?

HERBERT: Search me.

TEACHER: No, thanks, I just wondered if you knew.

LIBRARIAN: Sssshhh! The people next to you can't read.

SECOND-GRADER: What a shame! I've been reading since last year.

ROSES ARE RED

Roses are red,
Violets are blue,
I copied your paper
And I flunked, too.

TEACHER: Laura, why are you laughing?
LAURA: I'm sorry—I was just thinking of something.
TEACHER: Once and for all, Laura, remember that during school hours you're not supposed to think!

TEACHER: Sheldon, what are you doing? Are you learning anything?
SHELDON: No, teacher, I'm listening to you.

"I've changed my mind."
"Good, let's hope this one works better."

ELWOOD: May I bring my pet hen to school?
TEACHER: No, I've heard enough fowl language.

TEACHER: Where do blue eggs come from?
PATSY: From sad chickens.

TEACHER: Why do chickens lay eggs?
RAMONA: Because if they dropped them, they would break.

VERN: Why was the baby chicken thrown out of school?
FERN: It was caught peeping during a test.

NAN: How do you treat an injured bird?
DAN: Give it first-aid tweet-ment.

LENORE: How do they grade chickens?
TEACHER: They give them eggs-ams (exams).

TEACHER: Emma, spell mouse.
EMMA: M-O-U-S.
TEACHER: Yes—and what's on the end of it?
EMMA: A tail?

What would happen if an elephant sat in front of you in class?

You'd never see the blackboard.

SAL: Where are you taking that skunk?
VAL: To school.
SAL: What about the smell?
VAL: Oh, he'll get used to it.

What do you get if you cross a bear and a skunk?

Winnie-the-Phew (Pooh)!

TEACHER: Wendy, why do you look over your eye-glasses instead of through them?
WENDY: So I won't wear them out.

During Show and Tell, Miss Johnson showed pictures of different birds.

"George," she said, "what kind of bird do you like best?"

George thought for a while. "Fried chicken," he replied.

ANDREW: I'm teacher's pet.
MOTHER: How come?
ANDREW: She can't afford a dog.

MARK: Is your teacher strict?
TIM: I don't know. I'm too scared to ask.

(At party) "Will you pass the nuts, teacher?"
"No, I think I'll flunk them."

"I will now illustrate what I have in mind," said the teacher as she erased the blackboard.

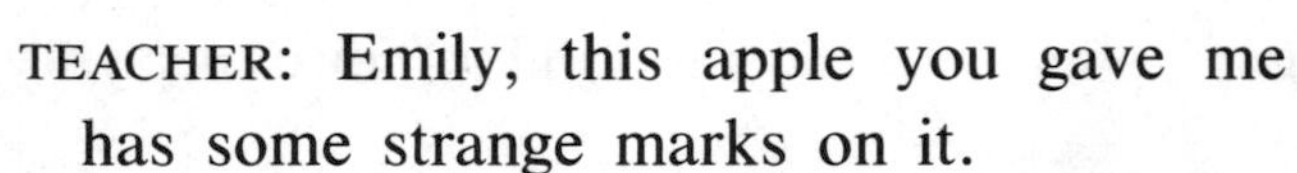

TEACHER: Emily, this apple you gave me has some strange marks on it.
EMILY: Well, so does the report card you gave me.

SON: Dad, I know how you can save money.

FATHER: That's fine, Son. How?

SON: You remember you promised me $5 if I got passing grades?

FATHER: Yes.

SON: Well, you don't have to pay me.

SON: Good news, Dad!

FATHER: What do you mean?

SON: You won't have to buy me any new books next year. I'm taking all of this year's work over again.

TEACHER: Well, at least there's one thing I can say about your son.

FATHER: What's that?

TEACHER: With grades like these, he couldn't be cheating.

DON'T WORRY

Don't worry if your grades are low
And your rewards are few.
Remember that the mighty oak
Was once a nut like you.

TEACHER: Where do they weigh whales?
RAY: At a whale weigh station.

TEACHER: Where do they weigh fish?
STEPHANIE: On their scales.

TEACHER: Where do fish sleep?
KEVIN: In water beds.

TEACHER: Where do fish wash?
SEAN: In river basins.

TEACHER: In this box, I have a 10-foot snake.
SAMMY: You can't fool me, teacher. Snakes don't have feet.

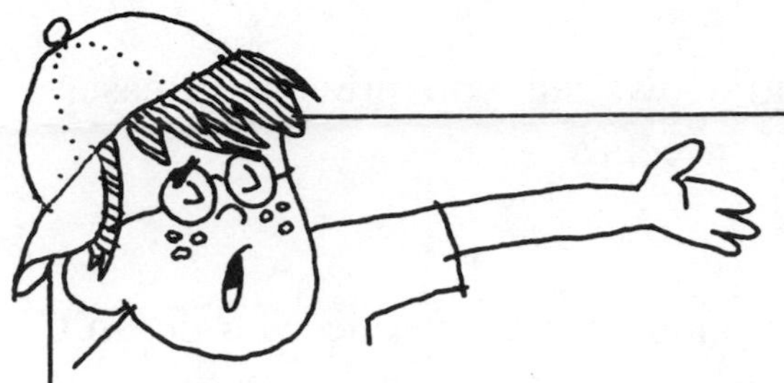

ONE OF THE KIDS IN MY SCHOOL IS REALLY MEAN.

HOW MEAN IS HE?

He's so mean, he makes the teacher stay after school.

He's so mean, his parents ran away from home.

He's so mean, when he comes to school, the teacher plays hookey.

He's so mean, when he tried to join the human race, he was turned down.

He's so mean, when he graduated from high school, they gave him a no-class ring.

He's so mean, they named a cake after him—crumb!

HYGIENE TEACHER: How can you prevent diseases caused by biting insects?

JOSÉ: Don't bite any.

TEACHER: An anonymous person is one who doesn't wish to be known.

STUDENT: What a stupid definition!

TEACHER: Who said that?

STUDENT: An anonymous person.

TEACHER: Johnny, why did you kick Bobby in the stomach?

JOHNNY: It was his own fault, teacher. He turned around.

"Some kids are bad, but you're an exception."

"Really?"

"Yes, exceptionally bad."

BOB: I love to tinker around in the workshop.

ROB: That doesn't surprise me. You're the biggest tinker I know.

5. READING & WRITING & RIDDLING

TEACHER: What are you writing, Tommy?
TOMMY: A letter to myself.
TEACHER: What does it say?
TOMMY: I don't know. I won't get it till tomorrow.

TEACHER: Where is your pencil, Harmon?
HARMON: I ain't got none.
TEACHER: How many times have I told you not to say that, Harmon? Now listen: I do not have a pencil. You do not have a pencil. They do not have a pencil. Now, do you understand?
HARMON: Not really. What happened to all the pencils?

NIT: Want to hear the story about the broken pencil?
WIT: No, thanks, I'm sure it has no point.

TEACHER: Why do they say the pen is mightier than the sword?
CLASS COMEDIAN: Because no one has yet invented a ballpoint sword.

TEACHER: Dorothy, what did you write your report on?

DOROTHY: A piece of paper.

Sammy did a report about the phone book. He wrote: "This book hasn't got much of a plot, but boy, what a cast!"

Mrs. Johnson asked the class to write a composition about what they would do if they had a million dollars.

Everyone except Fannie began to write. Fannie twiddled her thumbs and looked out the window.

When Mrs. Johnson collected the papers, Fannie's sheet was blank.

"Fannie," said Mrs. Johnson, "everyone has written two pages or more, but you've done *nothing*. Why is that?"

"Nothing is what I'd do," replied Fannie, "if I had a million dollars."

TEACHER (*holding book report*): Elwood, your ideas are like diamonds.

ELWOOD: You mean they're so valuable?

TEACHER: No, I mean they're so rare.

TEACHER: Patricia, the story you handed in called "Our Dog," is exactly like your brother's.

PATRICIA: Of course. It's the same dog.

TEACHER: Howard, your poem is the worst in the class. It's not only ungrammatical, it's rude and in bad taste. I'm going to send your father a note about it.

HOWARD: I don't think that would help, teacher. He wrote it.

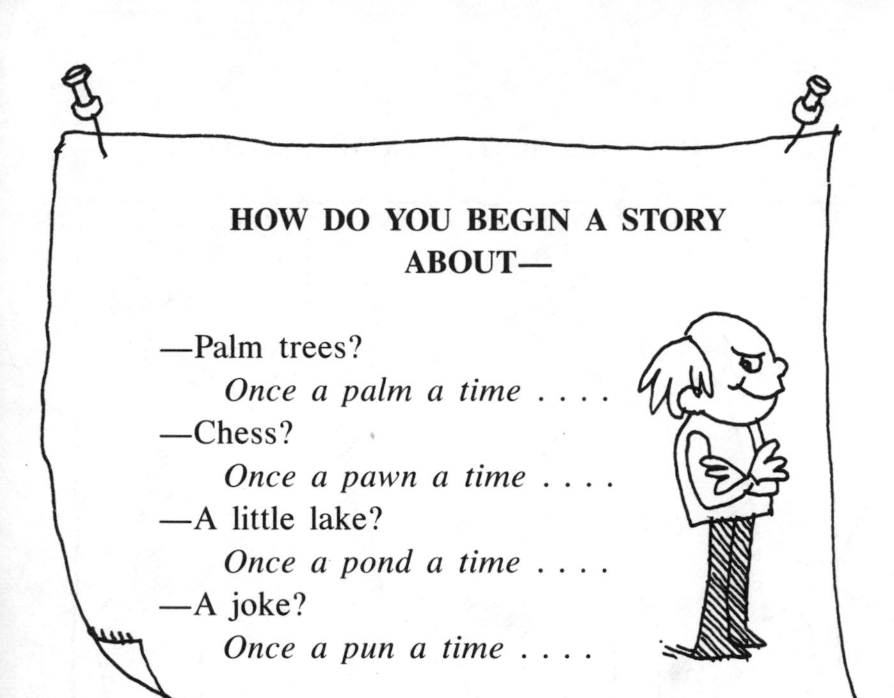

TEACHER: What does "coincidence" mean?
TRACY: Funny, I was just going to ask you that.

TEACHER: Define "procrastination."
PAM: May I answer that question tomorrow?

TEACHER: How nice that you have your new glasses, William. Now you'll be able to read everything.
WILLIAM: You mean, I don't have to come to school anymore?

TEACHER: Rhonda, please explain the difference between sufficient and enough.
RHONDA: If my mother helps me to cake, I get sufficient. If I help myself, I get enough.

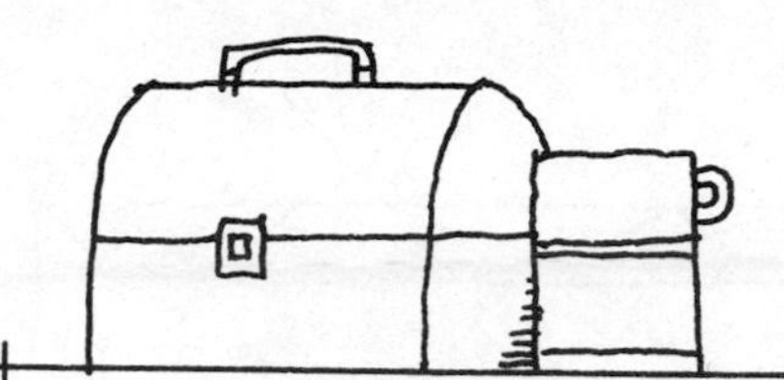

STUPID SPELLING

TEACHER: Your spelling is much better, Ronald. Only five mistakes that time.

RONALD: Thank you, Miss Smith.

TEACHER: Now let's go on to the next word.

TEACHER: George, how many "i"s do you use to spell Mississippi?

GEORGE: None. I can do it blindfolded.

TEACHER: Mort, how do you spell Mississippi?

MORT: The river or the state?

TEACHER: Whitney, spell "rain."

WHITNEY: R-A-N-E.

TEACHER: That's the worst spell of rain we've had around here in a long time!

TEACHER: Carlos, how do you spell "imbecile?"

CARLOS: I-M-B-U-S-L.

TEACHER: The dictionary spells it "I-M-B-E-C-I-L-E."

CARLOS: Yes, teacher, but you asked me how *I* spelled it.

SILLY SENTENCES

The teacher asked for sentences using the word "beans."

"My father grows beans," said a girl.

"My mother cooks beans," said a boy.

Then a third child spoke up, "We're all human beans (beings)," he said.

TEACHER: Charles, use the word "knockwurst" in a sentence.

CHARLES: A chicken joke is bad; an elephant joke is worse, but I'd rate a knock-knockwurst (a knock-knock worst).

TEACHER: Ellen, give me a sentence starting with "I."

ELLEN: I is . . . .

TEACHER: No, Ellen. Always say "I am."

ELLEN: All right. "I am the ninth letter of the alphabet."

TEACHER: Max, use "defeat," "defense" and "detail" in a sentence.

MAX: The rabbit cut across the field, and defeat went over defense before detail.

AWFUL ALPHABETS

TEACHER: Alvin, how many letters are there in the alphabet?
ALVIN: 18.
TEACHER: Wrong, there are 26.
ALVIN: No, teacher, there used to be 26, but ET went home in a UFO and the CIA went after them.

TEACHER: Max, how many letters are there in the alphabet?
SANCHO: Eleven.
TEACHER: Eleven!
SANCHO: T-H-E A-L-P-H-A-B-E-T—11!

How do they say good-bye using the alphabet?
A B C'ing you!

How do they say good-bye in England?
B B C'ing you!

How do they say good-bye on the Johnny Carson Show?
N B C'ing you!

GARBLED GRAMMAR

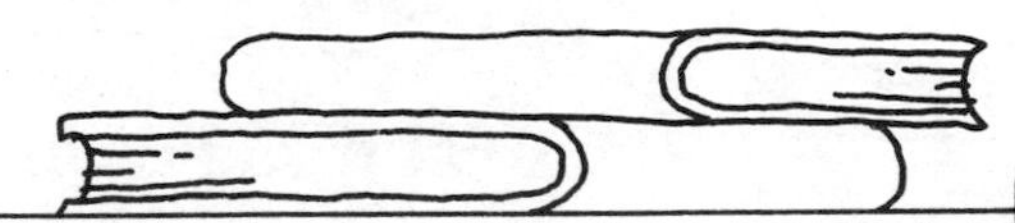

TEACHER: If "can't" is short for "cannot," what is "don't" short for?
NATALIE: Doughnut.

JOHNNY: Him and me helped clean up the yard.
TEACHER: Now, Johnny, don't you mean he and I helped?
JOHNNY: No, Mr. Smith, you weren't even there.

TEACHER: Sylvia, what are subordinate clauses?
SYLVIA: Santa's helpers.

AND MORE GARBLED GRAMMAR

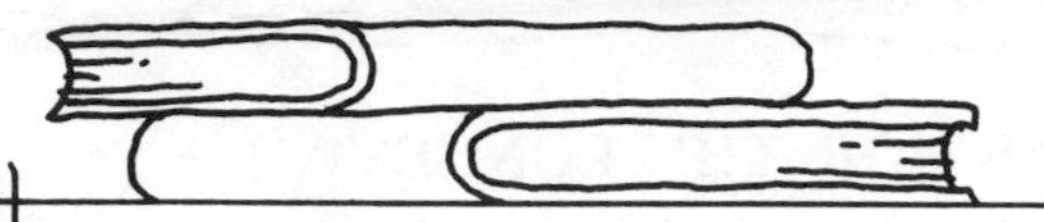

TEACHER: Rudolph, describe a synonym.
RUDOLPH: A word you use when you can't spell the other word.

TEACHER: Herman, name two pronouns.
HERMAN: Who, me?
TEACHER: Correct!

TEACHER: Wade, give me an example of a double negative.
WADE: I don't know none.
TEACHER: Excellent!

BURT: I ain't going.
TEACHER: That is not correct. Listen: I am not going. We are not going. You are not going. They are not going. Now do you understand?
BURT: Sure, teacher. Nobody ain't going.

MRS. JONES (*to the class*): Can anyone tell me the imperative of the verb "to go?" (*No reply*.)
MRS. JONES: Go, class, go!
CLASS: Thanks, Mrs. Jones! See you tomorrow!

CANDIDATES FOR THE LONGEST WORD IN THE ENGLISH LANGUAGE

Smiles
Because there's a mile between the first and last letter.

Rubber
Because it stretches.

Post Office
Because it has the most letters.

Equator
Because it circles the globe.

How is an English teacher like a judge?
They both hand out sentences.

TEACHER: Toby, what are you doing under your desk?
TOBY: Didn't you tell us to read Dr. Jekyll and Hyde (hide)?

What is an autobiography?
A car's life story.

REQUIRED READING LIST

Baker's Men by Pat E. Cake

The Making of a Hot Dog by Frank Furter

How to Raise Lambs by Shep Hurd

A History of Valentines by Bea Mine

How to Make an Igloo by S. K. Moe

Bell-Ringing by Paula D. Rope

The Nasty Kid by Enid A. Spanking

Basic Math by Adam Upp

Grade School Is Easy by Ella Mann Tree

How to Keep Things Oiled by Russ T. Gates

All About Weeds by Dan D. Lyons

Playing in the School Band by Clara Nette

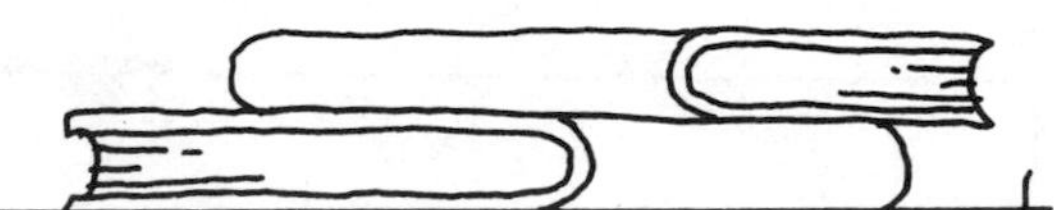

"Please hush," said the librarian to some noisy children. "The people around you can't read."

"Really?" asked one little girl. "Then why are they here?"

Melvin took a book from the library because the cover read "How to Hug."

It turned out to be Volume VII of an encyclopedia.

Why do snobs like books?

Because they have titles.

What famous book is about young cats?

"A Tale of Two Kitties" (Cities).

Where is the best place to find books about trees?

In a branch library.

What reference book lists famous owls?

"Whoo's Who."

What would you get if you crossed a book of nursery rhymes and an orange?

Mother Juice (Goose).

MORE LIBRARY SKILLS

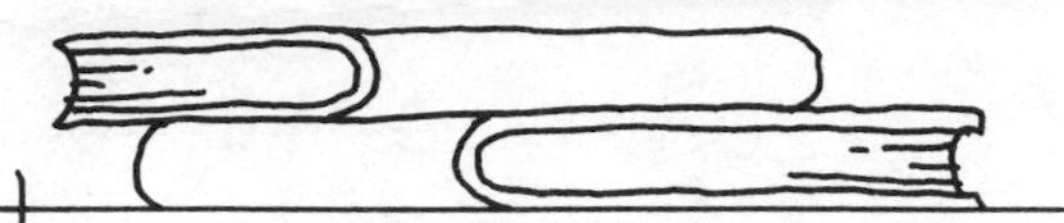

Have you read Shakespeare?

No.

Have you read Poe?

No.

Have you read Longfellow?

No.

I give up. What have you read?

I have red suspenders.

What part of a book is like a fish?

The fin-ish.

How does a book about zombies begin?

With a dead-ication.

How do you begin a book about ducks?

With an intro-duck-tion.

SAID A BOY

Said a boy to his teacher one day,
"Wright has not written 'rite' right, I say."
So the teacher replied,
As the error she eyed,
"Right. Wright, write 'rite' right right away."

NIT: Did you hear about the riot in the library?
WIT: No, what happened?
NIT: Someone found "dynamite" in the dictionary.

LIBRARIAN: Will you two please stop passing notes!
STUDENT: We're not passing notes. We're playing cards.

6. IS THIS ASSEMBLY NECESSARY?

The principal was annoyed by the noise during the assembly program.

"There seem to be several idiots in the auditorium this morning," he snapped. "Wouldn't it be better to hear one at a time?"

A voice shouted, "Okay—you start."

The person who said all things must end never heard my principal talk.

My principal is so boring, when he gives a speech even your feet fall asleep.

Until I heard my principal talk I thought my butter knife was dull.

Until I heard my principal talk, I thought water was colorless.

DIT: Where is the principal?
DOT: He's round in front.
DIT: I know what he looks like. I was just wondering where he went.

What do you get if you cross one principal with another principal?

Don't do it. Principals don't like to be crossed.

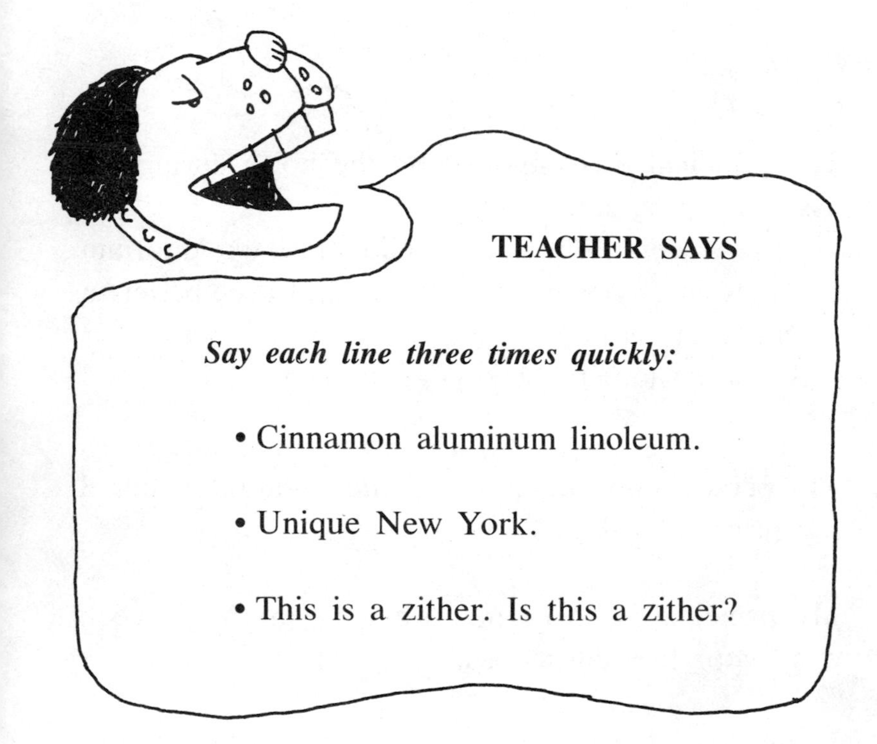

"Wise men hesitate," said the principal. "Only fools are certain."

"Is that true?"

"I'm certain."

HAPPY HOLIDAY

How many months have 28 days?
All of them.

NIT: What comes before March?
WIT: Forward!

How many seconds are there in a year?
Twelve. January 2nd, February 2nd, March 2nd . . . .

NICK: What is there in December that isn't in any other month?
VIC: The letter "D"!

TEACHER: Why do bells ring at Christmas?
PETER: Because someone pulls the rope.

TEACHER: What does the Christmas tree stand for?
SKEETER: It would take too much room lying down.

TEACHER: What nationality is Santa Claus?
DENNIS: North Polish.

TEACHER: Why is there a Mother's Day and a Father's Day, but no Son's Day?

PUPIL: Because there is a Sunday (Son-day) every week.

Before Thanksgiving a first-grade teacher asked her pupils to tell what they were thankful for.

"I'm thankful," said one small boy, "that I'm not a turkey."

"What did you think of the ventriloquist?" the teacher asked one of her first-graders after the show in the school auditorium.

"He wasn't very good," replied the first-grader. "But the little guy on his knee was terrific."

TEACHER: Doris, what are you going to do in the school talent show?
DORIS: Imitations.
TEACHER: Let's hear them.
DORIS: "I love you—ouch! I love you—ouch!"
TEACHER: I give up—what are you imitating?
DORIS: Two porcupines kissing.

TEACHER: Boris, what are you going to do in the school talent show?
BORIS: Bird imitations.
TEACHER: Are you going to warble?
BORIS: No, I'm going to eat worms.

What musical key do cows sing in?
Beef-flat (B♭).

TEACHER: Norris, what are you going to do in the school talent show?
NORRIS: I'm going to sing "Old Lady River."
TEACHER: Don't you mean "Old Man River"?
NORRIS: No, I'm singing about a lady river—Mississippi!

DELORES (*after singing a song horribly*): How did you like my execution?
MUSIC TEACHER: I'm all in favor of it.

MUSIC TEACHER (TO VOICE STUDENT)

I didn't say your voice was out of this world—I said it was unearthly.

You have a fine voice. Don't spoil it by singing.

I like the song you sang. One day you should put it to music.

You sing like a bird—a screech owl!

I've heard better sounds coming from a leaking balloon.

Every time I tear a rag, it reminds me of your voice.

You couldn't carry a tune if it had a handle.

Of course your voice is pure. You strain it every time you sing.

You're a natural musician. Your tongue is sharp and your head is flat.

VOICE TEACHER: Now, please sing the scale.
STUDENT: Do-re-mi-fa-sol-la-do.
VOICE TEACHER: You left out the "ti."
STUDENT: I know—every time I try to hit a high note, my voice sinks.
VOICE TEACHER: Again you left out the "t."

TEACHER: What are your favorite songs?
GORDON: I have five of them—"Three Blind Mice" and "Tea for Two."

(*Overheard in the school auditorium at a glee club performance*):

"Is that a popular song?"

"It was before today."

"That tune has been running through my head all day."

"Of course—there's nothing there to get in its way."

PUPIL (*at concert*): What is the book the orchestra leader is looking at?
TEACHER: That's the score.
PUPIL: Really? Who's winning?

Why did George bring a ladder to the assembly program?

The music teacher asked him to sing higher.

STUDENT (*after playing the piano*): I've never had a lesson in my life—and I can prove it.
MUSIC TEACHER: Never mind, you just did.

MUSIC TEACHER: Why did you put that vegetable on the piano?
STUDENT: You told me my playing would improve if I had a beet (beat)!

BAND STUDENT: Our high school band played Beethoven yesterday.
ATHLETE: Who won?

The class laughed when I sat down at the piano—no stool.

The class laughed when I sat down at the piano with my hands tied behind my back. They didn't know I played by ear.

How do you clean a tuba?
With a tuba toothpaste!

"Look at how gracefully that girl eats her corn on the cob," said Mrs. Jones to her son, Harry, at the restaurant.

"Of course," Harry replied, "she plays the flute in the school band."

7. WAY OUT TO LUNCH

What is the worst thing you're likely to find in the school cafeteria?

The food.

The food in our school cafeteria is perfect, if you happen to be a termite.

The food in our school cafeteria is so bad, you get a prescription with every meal.

The food in our school cafeteria is so bad, flies go there to commit suicide.

"What is this on my plate in case I have to describe it to my doctor?"

"Is that school food spicy?"

"No, smoke always comes out of my ears."

What did the computer do in the lunchroom?
It had a byte.

LITTLE MONSTER: Mother, I hate my teacher.
MOTHER MONSTER: Then just eat your salad, dear.

MOTHER: Why on earth did you swallow the money I gave you?
JUNIOR: You said it was my lunch money.

TEACHER: What is a mushroom?
MAXWELL: The place they store the school food.

TEACHER SAYS

Say each sentence three times quickly:

- A cupcake cook's cap.
- For cheap sheep soup, shoot sheep.
- Cheeky chimps chomp cheap chop suey.

TEACHER: When do astronauts eat?
SCOTT: At launch time.

What do astronauts eat from?
Flying saucers.

What gets served at science fairs?
Fish 'n' chips (fission chips).

What kind of food do math teachers eat?
Square meals.

Where do math teachers go to eat?
To the lunch counter.

Where do smart frankfurters end up?
On the honor roll.

LITTLE BOY (*opening lunch bag*): Not again! Day after day, the same old thing—cheese sandwiches on white bread. I'm sick and tired of them.

TEACHER: Why don't you ask your mother to make a different kind of sandwich for you?

LITTLE BOY: I can't.

TEACHER: Why not?

LITTLE BOY: Because my mother doesn't make them—I do.

TEACHER: What does an 800-pound gorilla eat?

PUPIL: Anything it wants.

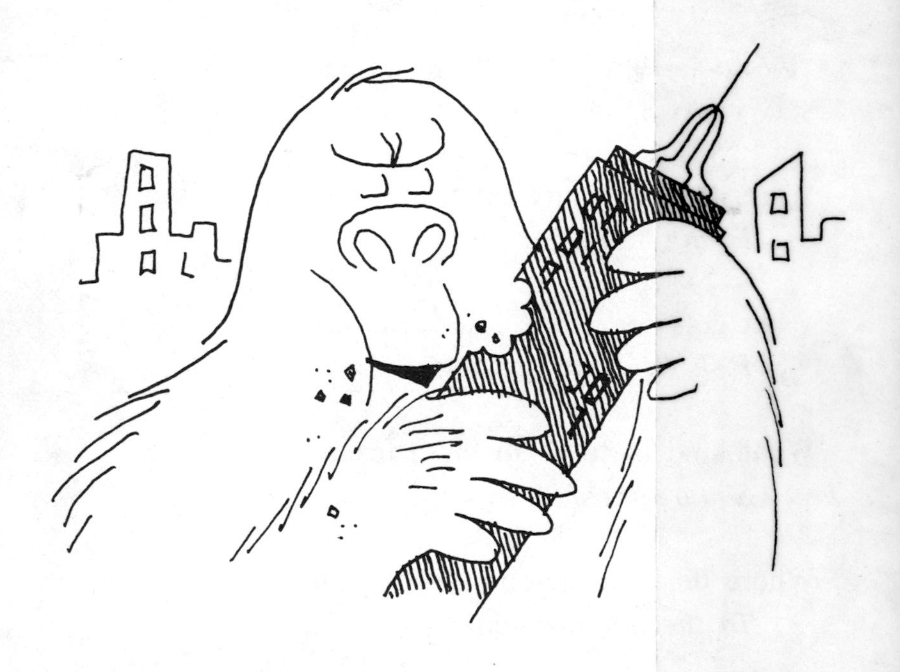

BARRY: Ugh—this plate is wet.

CARY: That's the soup.

My school cafeteria is a place where they serve soup to nuts.

"I thought this was barley soup—not barely soup!"

NIT: I thought this was supposed to be pea soup, but it tastes like soap!
WIT: Oh, it must be tomato soup. Pea soup tastes like gasoline.

"Why do you have alphabet soup every day for lunch?"
"So I can eat and practise reading at the same time."

What happened to the bad egg in the lunchroom?
It got eggs-pelled.

NIT: Do you feel like a doughnut?
WIT: Of course not, do I look like one?

What is the difference between a teacher and a doughnut?
You can't dunk a teacher in a glass of milk.

TEACHER: There is a general belief that fish is brain food.
KIRK: Yes, teacher, I eat it all the time.
TEACHER: Oh, well, there goes another scientific theory.

TEACHER: Order, children! Order!
CLASS COMEDIAN: I'll have a burger with French fries.

I feel like a sandwich.
Funny, you look like a pork chop.

I feel like a piece of chocolate.
Well, stick around. If I get hungry, I'll bite you.

IKE: There's a fly in this ice cream.
SPIKE: Serves him right. Let him freeze!

LEM: Is this peach or apple pie?
CLEM: If you can't tell by the taste, what difference does it make?

ART: That crust on the apple pie was tough.
BART: That wasn't the crust. That was the paper plate.

"If I cut an apple in two, what would I get?"
"Two pieces."
"If I cut a pear in four, what would I get?"
"Four pieces."
"If I cut a banana in eight, what would I get?"
"Eight pieces."
"Now, if all the pieces were added together, what would I get?"
"Fruit salad!"

TEACHER: How do you make Mexican chili?
PUPIL: Take him to the North Pole.

TEACHER: How do you make meat loaf?
PUPIL: Send it on a vacation.

TEACHER: Why is that pickle behind your ear?
PUPIL: Gosh! I must have eaten my pencil!

FLOYD: Is a chicken big enough to eat when it's two weeks old?

TEACHER: Of course not!

FLOYD: Then how does it manage to live?

ONE OF THE GIRLS IN MY SCHOOL IS TERRIBLY CONCEITED.

HOW CONCEITED IS SHE?

She's so conceited, she goes out in the garden so the flowers can smell her.

She holds her nose so high in the air, there's an inch of snow on it.

She holds her nose so high in the air that every time she sneezes, she wets the ceiling.

She holds her nose so high in the air that every time she hiccups, she blows her hat off.

8. IT'S A FUNNY WORLD

"It's clear," said the teacher, "that you haven't studied your geography. What's your excuse?"

"Well—my dad says the world is changing every day. So I decided to wait a little while until it settles down."

TEACHER: Josh, what can you tell us about the Dead Sea?

JOSH: I didn't even know it was sick!

TEACHER: What are the small rivers that run into the Nile?

GRACE: The Juve-niles!

TEACHER: Why is the Mississippi such an unusual river?

SYLVESTER: Because it has four eyes and can't see.

TEACHER: What are the Great Plains?

PIP: The 747s.

TEACHER: Where is the English Channel?
ALEX: I don't know. My television set doesn't pick it up.

GERTRUDE: My teacher was mad because I didn't know where the Andes were.
MOTHER: Well, dear, next time remember where you put things.

Why does the Statue of Liberty stand in New York Harbor?

Because it can't sit down.

TEACHER: Is Lapland heavily populated?
BRUCE: No, there are not many Lapps to the mile.

TEACHER: Name an animal that lives in Lapland.
PUPIL: A reindeer.
TEACHER: Good. Now name another.
PUPIL: Another reindeer.

TEACHER: Rachel, can you tell us where elephants are found?
RACHEL: We don't have to find elephants. They're so big, they don't get lost.

TEACHER: Rupert, what fur do we get from the leopard?
RUPERT: As fur as possible.

ERNIE: I only got 35 in Arithmetic and 50 in Spelling, but I knocked them cold in Geography.
BERNIE: What did you get?
ERNIE: Zero.

TEACHER: What birds are found in Portugal?
ALVIN: Portu-geese.

TEACHER: Name three famous Poles.
CLASS COMEDIAN: North, South and tad.

What do we do with crude oil?
Teach it manners.

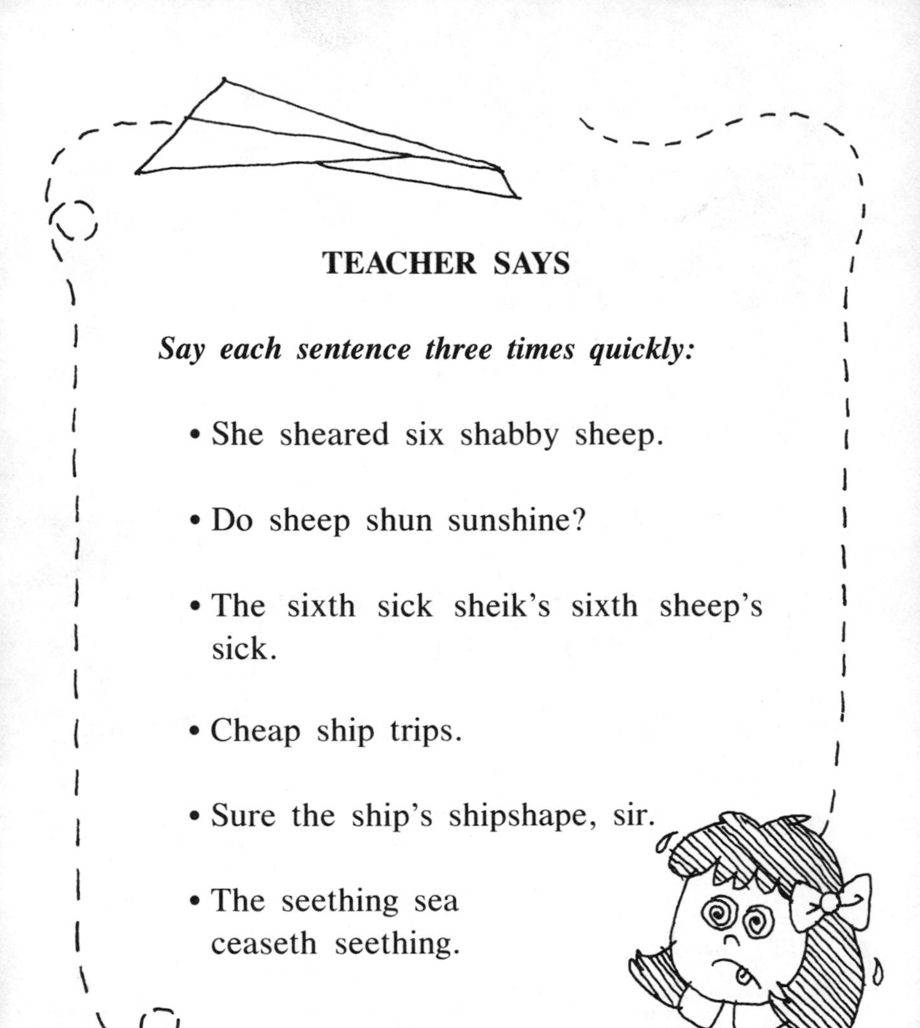

TEACHER SAYS

Say each sentence three times quickly:

- She sheared six shabby sheep.
- Do sheep shun sunshine?
- The sixth sick sheik's sixth sheep's sick.
- Cheap ship trips.
- Sure the ship's shipshape, sir.
- The seething sea ceaseth seething.

Herman's teacher always rewarded good work by putting a gold star at the top of her students' homework. One day Herman came home with a big zero at the top of his paper.

"Herman, what does this mean?" asked his mother.

"Oh," Herman explained, "my teacher ran out of stars, so she gave me a moon."

What do frogs like to sit on?
Toadstools.

What do frogs wear on their feet?
Open-toed (toad) shoes.

How do you get nuts from a squirrel?
Walk up to the squirrel and say, "This is a stickup!"

TEACHER: Do sailors go on safaris?
PUPIL: Not safaris (so far as) I know.

TEACHER: What is an island?
RUTHIE: An island is a piece of land surrounded by water except in one place.
TEACHER: What place is that?
RUTHIE: On top.

Knock-knock.
  Who's there?
Oscar and Greta.
  Oscar and Greta who?
Oscar foolish question,
Greta foolish answer.

Knock-knock.
  Who's there?
Summertime.
  Summertime who?
Summertime I get the right answer
and summertime I don't.

TEACHER (*correcting a pupil*): When I asked you what shape the world was in, I meant "round" or "flat"—not "rotten."

TEACHER: Danny, give me three reasons why you know the earth is round.
DANNY: Because my mother says so, my father says so—and you say so.

"Dorrie, do you know a girl named Louise?"
  "Yes, Mom, she sleeps next to me in Geography."

9. MATH-A-RAMA

FLIP: I failed every subject except algebra.
FLOP: How did you keep from failing that?
FLIP: I didn't take algebra.

TEACHER: Are you good at math?
PUPIL: Yes and no.
TEACHER: What does that mean?
PUPIL: Yes, I'm no good in math.

"Pop, will you help me find the lowest common denominator in this problem?"

"Good heavens, don't tell me they haven't found it yet! They were looking for it when I was a boy!"

TEACHER: Lisa, did your father help you with these math problems?
LISA: No, teacher. I got them wrong all by myself.

"Teacher, I can't do this problem."

"Any five-year-old should be able to do that problem."

"No wonder I can't do it! I'm almost ten!"

COUNTING TO 10

The teacher was reviewing counting with her first-grade class.

"Pauline," she asked, "can you count to 10 without mistakes?"

"Yes," said Pauline, and she did.

"Now, Philo," said the teacher, "can you count from 10 to 20?"

"That depends," said Philo, "with or without mistakes?"

TEACHER: Can you count to 10?

SUZANNE: Yes, teacher. (*counting on her fingers at waist level*) One, two, three, four, five, six, seven, eight, nine, ten.

TEACHER: Good. Now can you count higher?

SUZANNE: Yes, teacher. (*She puts her hands over her head and counts on her fingers.*) One, two, three, four, five, six, seven, eight, nine, ten.

TEACHER: Can you count to 10?

ERIC: Yes, teacher—one, two, three, four, five, six, seven, eight, nine, ten.

TEACHER: Now go on from there.

ERIC: Jack, Queen, King.

TEACHER: What is two and two?
HUGH: Four.
TEACHER: That's good.
HUGH: Good? It's perfect!

TEACHER: If one and one make two, and two and two make four, how much do four and four make?
ANNIE: That's not fair, Teacher. You answer the easy ones yourself and leave the hard ones for us.

TEACHER: How much is half of eight?
WENDELL: Up and down or across?
TEACHER: What do you mean?
WENDELL: Up and down it's 3 and across it's 0.

TEACHER: Now, class, whatever I ask, I want you to answer at once. Amy, how much is eight and eight?
AMY: At once!

TEACHER: If there were 10 cats in a boat, and one jumped out, how many would be left?
MARYLOU: None, because they were all copycats.

TEACHER: If you received $10 from 10 people, what would you get?
SASHA: A new bike.

Why was the math book unhappy?

It had too many problems.

TEACHER: If I lay one egg here and another there, how many eggs will there be?

CLARK: None!

TEACHER (*surprised*): Why not?

CLARK: Because you can't lay eggs!

TEACHER: Stella, take 932 from 1,439. What is the difference?

STELLA: That's what I say—what's the difference?

Who invented fractions?

Henry the Eighth.

"Our teacher has a bad memory. For three days she asked us how much is two and two. We told her it was four. But she still doesn't know. Today she asked us again."

MYSTERY

Birds on the mountain,
Fish in the sea,
How you passed math
Is a mystery to me.

HECTOR: I've added these figures ten times.
TEACHER: Good work!
HECTOR: And here are my ten answers.

Why is a dog with a lame leg like adding 6 and 7?
He puts down the three and carries the one.

Why is six afraid of seven?
Because seven eight (ate) nine.

What animal is best at math?
Rabbits—they multiply fastest.

DIT: My dog is great at math.
DOT: Really?
DIT: Ask him how much is two minus two.
DOT: But two minus two is nothing!
DIT: That's what he'll answer—nothing!

TEACHER: If I gave you three rabbits today and five rabbits tomorrow, how many rabbits would you have?

WENDY: Nine.

TEACHER: Sorry Wendy, you'd have eight.

WENDY: No, Teacher, I'd have nine. I already have one rabbit at home.

TEACHER: If you add 3452 and 3096, then divide the answer by 4 and multiply by 6, what would you get?

LILY: The wrong answer.

FATHER: How are you doing in arithmetic?

DIRK: I've learned to add up the zeros, but the numbers are still giving me trouble.

What makes arithmetic hard work?

All those numerals you have to carry.

JASON: I got 100 in school today.

MOTHER: Wonderful. What did you get 100 in?

JASON: Two things: I got 50 in Spelling and 50 in History.

MOTHER (*sighing*): Well, at least you can add.

PAMELA: I got 100 in an arithmetic test and still didn't pass.

FATHER: Why not, for goodness sake?

PAMELA: Because the answer was 200.

Why are misers good math teachers?

They know how to make every penny count.

What kind of tree does a math teacher climb?

Geometry.

What do you have to know to get top grades in geometry?

All the angles.

What kind of pliers do you use in arithmetic?

Multipliers.

TEACHER: Vincent, if you had one dollar and you asked your father for another, how many dollars would you have?

VINCENT: One dollar.

TEACHER (*sadly*): You don't know your arithmetic.

VINCENT (*sadly*): You don't know my father.

Lucille stood quietly as her father examined her report card.

"What is this 45 in math?" asked her father.

"I think that's the size of the class," she said quickly.

FATHER: If I had five coconuts and I gave you three, how many would I have left?

FRANKIE: I don't know.

FATHER: Why not?

FRANKIE: In our school we do all our arithmetic in apples and oranges.

TEACHER: If I had seven oranges in one hand and eight oranges in the other, what would I have?

CLASS COMEDIAN: Big hands!

The teacher was giving her first-grade class a quiz on counting. Naomi got things started by counting from 1 to 10.

"Now, Charles," said the teacher, "you take over, beginning with 11."

"11, 14, 23, 42, 26," said Charles.

"What kind of counting is that?" asked the teacher.

"Who's counting?" replied Charles. "I'm calling signals."

10. HISTORY HEE-HAWS

SUZI: I wish I'd been born 1000 years ago.
MOTHER: Why is that, dear?
SUZI: Just think of all the history I wouldn't have to study.

My teacher reminds me of history—she's always repeating herself.

What do history teachers make when they want to get together?
Dates.

What do they talk about?
The good old days.

Who was the biggest thief in history?
Atlas, because he held up the whole world.

What was Noah's profession?
Ark-itect.

What kind of illumination did Noah use on the ark?
Floodlights.

The first-grade teacher brought her little pupils to the museum. They stood in front of a mummy case. At the bottom of the case were the words "1286 BC."

"Does anyone know what that number means?" asked the teacher.

One little kid spoke up, "That must be the license of the car that hit him."

Did they play tennis in ancient Egypt?
Yes, the Bible tells how Joseph served in Pharaoh's court.

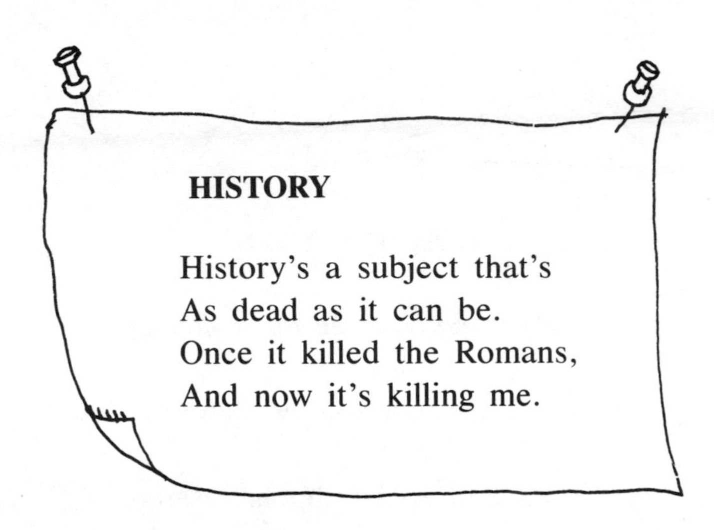

TEACHER: Where did knights learn to kill dragons?
VICTOR: In knight (night) school.

TEACHER: How did Vikings communicate with each other?
HESTER: By Norse code.

What is a forum?
Two-um plus two-um.

How did Columbus's men sleep on their ships?
With their eyes shut.

What did Napoleon become when he was 41 years old?
42 years old.

How do we know Napoleon loved spicy food?
First he mustard (mustered) his army, and then he salted (assaulted) the city.

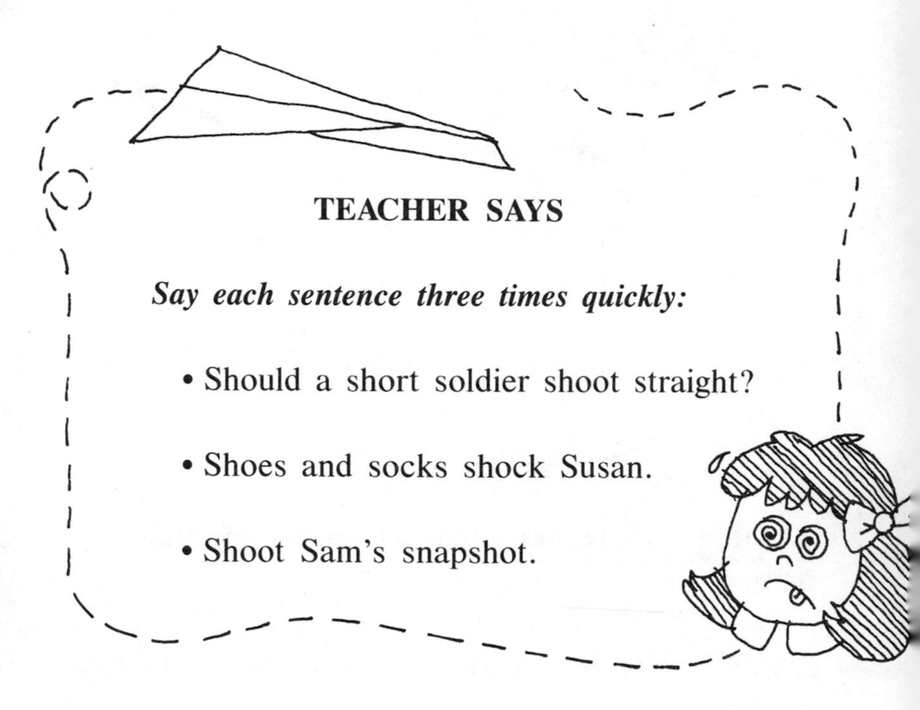

TEACHER SAYS

Say each sentence three times quickly:

- Should a short soldier shoot straight?
- Shoes and socks shock Susan.
- Shoot Sam's snapshot.

TEACHER: The Spanish explorers travelled around the world on a galleon.

CLASS COMEDIAN: How many galleons did they get to a mile?

"Class, you've been so bad, you're all going to have to stay after school," said the teacher.

"Give me liberty or give me death!" came a voice from the back of the room.

"Who said that?" snapped the teacher.

"Patrick Henry!" said the class.

TEACHER: What did Paul Revere say when he got on his horse?

HELGA: "Giddy-up!"

TEACHER: Andrea, what did they do at the Boston Tea Party?

ANDREA: I don't know, teacher, I wasn't invited.

TEACHER: Helen, what did they wear at the Boston Tea Party?

HELEN: T-shirts?

TEACHER: Where was the Declaration of Independence signed?

STANLEY: At the bottom.

TEACHER: Did the Indians hunt bear?

DONALD: Not in winter.

What is red, white and blue and says, "Ouch"?

Betsy Ross, sewing the flag, without her eyeglasses.

What did Betsy Ross say when they asked if the flag was ready?

"Give me a Minute, Man!"

TEACHER: Why did George Washington chop down the cherry tree?
PUPIL: I'm stumped.

TEACHER: When crossing the Delaware River, why did George Washington stand up in the boat?
CLASS COMEDIAN: He was afraid if he sat down someone would hand him an oar.

TEACHER: Why was George Washington buried at Mount Vernon?
PETER: Because he was dead.

FATHER: How did you do in your tests, Elvis?
ELVIS: I did what George Washington did.
FATHER: What's that?
ELVIS: I went down in history.

TEACHER: Carol, do you know the 20th President of the United States?
CAROL: No, we were never introduced.

"Abraham Lincoln had a hard childhood," explained the first-grade teacher. "He had to walk nearly seven miles to school every day."

"It was his own fault," said Norman. "Why couldn't he get up and catch the school bus like everybody else?"

A MYSTERY

He asked me *when?* I could not tell.
He asked me *who?* Again I fell.
He named a man, to me a stranger,
I could see myself in danger.
What was this plight—this mystery?
Oh—just my course in history.

TEACHER: Who gave the Liberty Bell to Philadelphia?
DANNY: It must have been a duck family.
TEACHER: A duck family!
DANNY: Didn't you tell us there was a quack in it?

What is the biggest telephone company in space?
ET & T (AT & T).

TEACHER: Why did the pioneers cross the country in covered wagons?
CLASS COMEDIAN: Because they didn't want to wait 40 years for a train.

FATHER: Ingrid, I see from your report card that you're not doing well in history. Why not?
INGRID: Because my teacher is always asking me about things that happened way before I was born.

FATHER: I see you've flunked history again, Junior.
JUNIOR: Yes, dad. You always told me it's best to let bygones be bygones.

WHEN I DIE

When I die, bury me deep.
Bury my history book at my feet.
Tell the teacher I've gone to rest
And won't be back for the history test.

11. STUDY HALL

PRINCIPAL: If you study hard, you'll get ahead.
SAL: No, thanks, I already have a head.

Knock-knock.
  Who's there?
Don Juan.
  Don Juan who?
Don Juan to study today.

WHY STUDY?

The more we study, the more we know,
The more we know, the more we forget,
The more we forget, the less we know,
So, why study?

PRINCIPAL: Do you believe in clubs for children?
TEACHER: If all else fails.

10 REASONS WHY I DON'T HAVE MY HOMEWORK

1. My little sister ate it.
2. I was mugged on the way to school, and the mugger took everything I had.
3. Our puppy toilet-trained on it.
4. Some creatures from outer space borrowed it so they could study how the human brain worked.
5. I put it in a safe, but lost the combination.
6. I loaned it to a friend, but he suddenly moved away.
7. Our furnace stopped working, and we had to burn it to keep from freezing.
8. I left it in my shirt, and my mother put it in the washing machine.
9. I didn't do it because I didn't want to add to your already heavy workload.
10. I lost it fighting with a kid who said you weren't the best teacher in the school.

MOTHER: Our son's teacher says he ought to have an encyclopedia.

FATHER: Why? Let him walk to school like I did.

What does an elf do when it gets home from school?
Gnomework (Homework).

FRANK: Day after day, the boy and his dog went to school together. Then the day came when they had to part.

HANK: What happened?

FRANK: The dog graduated.

My dog is so bad,
he was expelled from
obedience school.

TEACHER SAYS

Say each sentence three times quickly:

- Big B-52 bombers.
- The back brake block broke.
- Peggy Babcock blushes.

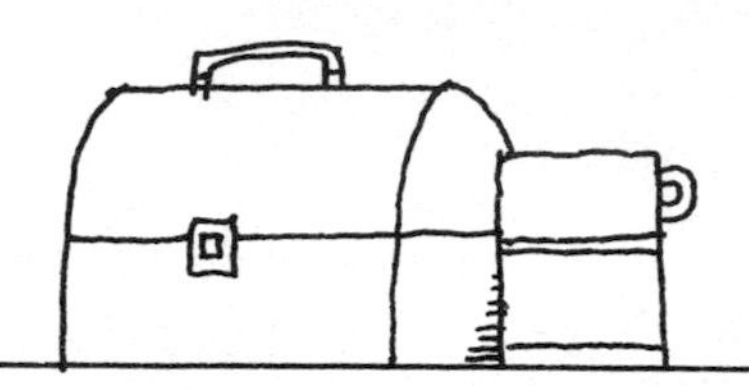

LATE AGAIN?

TEACHER: Why are you late?

LORISSA: It rained last night. The road was so wet and slippery that for every step I made forward, I slipped backwards two steps.

TEACHER: Well, if it rains again tomorrow, start walking in the opposite direction.

TEACHER: Why are you late?

AMOS: I lost my quarter.

TEACHER: And why are you late, Oliver?

OLIVER: I was standing on it.

TEACHER: Why are you late?

BARNEY: I sprained my ankle.

TEACHER: What a lame excuse!

TEACHER: Why are you late?

TIMMY: I was riding my bike and I ran into a tree.

TEACHER: Well, well, that's the first time I ever heard of sap running *into* a tree.

TEACHER: What's the idea of coming to school two hours late?

BASIL (*in bandages*): But, teacher, I was run over.

TEACHER: It doesn't take two hours to get run over.

TEACHER: Do clocks tell time?
CASSIE: No, Teacher, you have to look at them.

Why was the school clock punished?
Because it tocked (talked) too much.

TEACHER: Frieda, how long is a minute?
FRIEDA: Which kind of minute do you mean—a real minute—or "wait-a-minute"?

MOTHER (*to sleeping son*): Aldo, it's time to get up. It's twenty to eight.
ALDO: In whose favor?

THEODORE: My father beats me up every morning.
TEACHER: How terrible!
THEODORE: It's not too bad. He gets up at 7 and I get up at 8.

SOMETHING GOOD

Go on to college, continue your knowledge,
Be smart, be brave, be true.
If they make penicillin from moldy cheese,
They can make something good out of you.

ONE OF THE KIDS IN MY SCHOOL IS VERY DUMB.

HOW DUMB IS HE?

He's so dumb, the nearest he ever came to a brainstorm is a light drizzle.

He's so dumb, he cut up the calendar because he wanted to take some time off from school.

He's so dumb, he flunked recess.

He's so dumb, the only kind of poetry he can make up is blank verse.

He's so dumb, he trips over his IQ.

He's so dumb, if he were twice as smart, he'd still be a half-wit.

ONE OF THE KIDS IN MY SCHOOL IS VERY DUMB.

HOW DUMB IS HE?

He's so dumb he thinks noodle soup is brain food.

He's so dumb he pals around with morons so he can have someone to look up to.

He's so dumb he tries to blow out light bulbs.

He's so dumb, he put a rubber band around his head to stretch his mind.

He's so dumb, the only time he has something on his mind is when he wears a hat.

SCHOOL DAZE

From what school do you have to drop out in order to graduate?

Parachute school.

What school has a sign on it that says, "Please don't knock."

Karate school.

At which college can you learn how to drive tanks?

Tank U.

What do they teach vampires in business courses?

How to type blood.

TEACHER: How can you be such a perfect idiot?
EDWIN: I practise a lot.

What happened when the teacher wrote, "Please wash," on the blackboard?

The school janitor took a bath.

Why did the teacher marry the school janitor?

He swept her off her feet.

What is the difference between a teacher and a train engineer?

One trains the mind, the other minds the train.

What's the difference between a train and a teacher?

One says, "Choo-choo," and the other says, "Take the gum out of your mouth."

TEACHER: Did you take a bath today?
KERMIT: Why, is one missing?

ARNOLD: Mom, my teacher told me not to take any more baths.
MOTHER: Are you sure that's what she said?
ARNOLD: Well, she told me to stay out of hot water—or else.

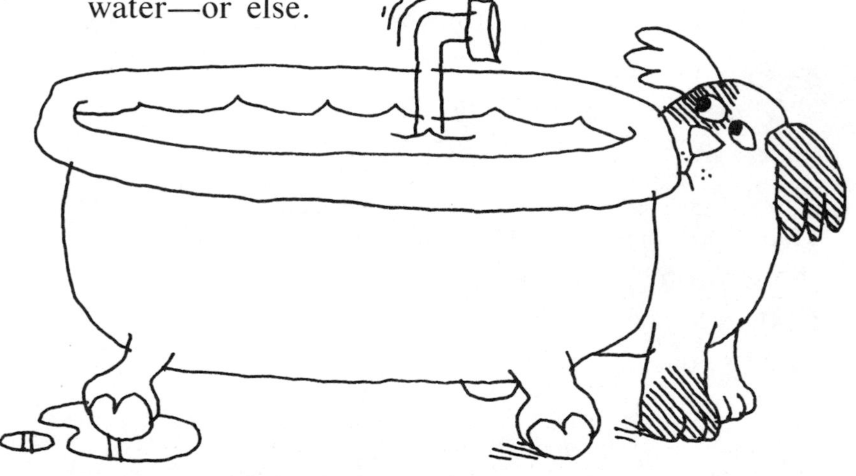

TEACHER: Edna, I drove by your house yesterday and I saw your family wash in the back yard.
EDNA: You must have had the wrong house. We all wash in the bathroom.

What is green and wet and teaches school?

The Teacher from the Black Lagoon.

"Emily," said the teacher, "I don't know what I'm going to do with you. Everything goes in one ear and out the other."

"Of course," said Emily, "isn't that why we have two ears?"

FATHER: Stuart, your teacher tells me you're at the bottom of your class.

STUART: So what, Dad? We learn the same thing at both ends.

MOTHER CANNIBAL: Junior was sent home from school today.

FATHER CANNIBAL: What happened?

MOTHER CANNIBAL: They caught him buttering up the teacher.

10 MORE REASONS WHY I DON'T HAVE MY HOMEWORK

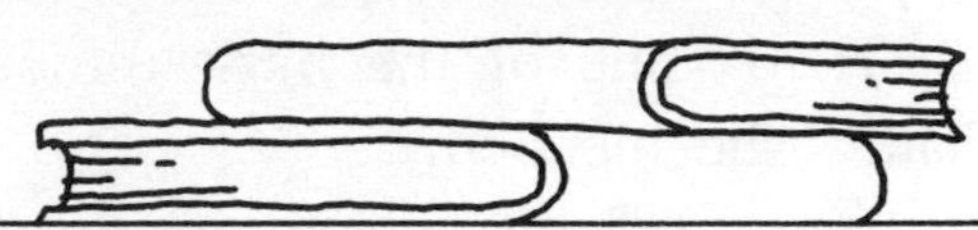

1. A sudden gust of wind blew it out of my hands, and I never saw it again.
2. I was kidnapped by terrorists and they just let me go, so I didn't have time to do it.
3. The lights in our house went out, and I had to burn it to get enough light to see the fuse box.
4. A kid fell in the lake, and I jumped in to rescue him. My homework drowned.
5. I used it to fill a hole in my shoe.
6. My father had a nervous breakdown, and he cut it up to make paper dolls.
7. My pet gerbils had babies, and they used it to make a nest.
8. I didn't do it, because I didn't want the other kids in class to look bad.
9. I made a paper airplane out of it and it got hijacked.
10. ET took it home.

The teacher was complaining about one of her pupils.

"He is one of the most difficult students I ever had," she moaned.

"How difficult can a nine-year-old possibly be?" she was asked.

"To give you an idea," replied the teacher, "his mother comes to PTA meetings in disguise."

Who belongs to the Monster PTA?

Mummies and deadies.

LESTER: Dad, there's a small PTA meeting tomorrow that you have to come to.

FATHER: If it's a small one, do I have to be there?

LESTER: I'm afraid so, Dad. It's just between you, me and my teacher.

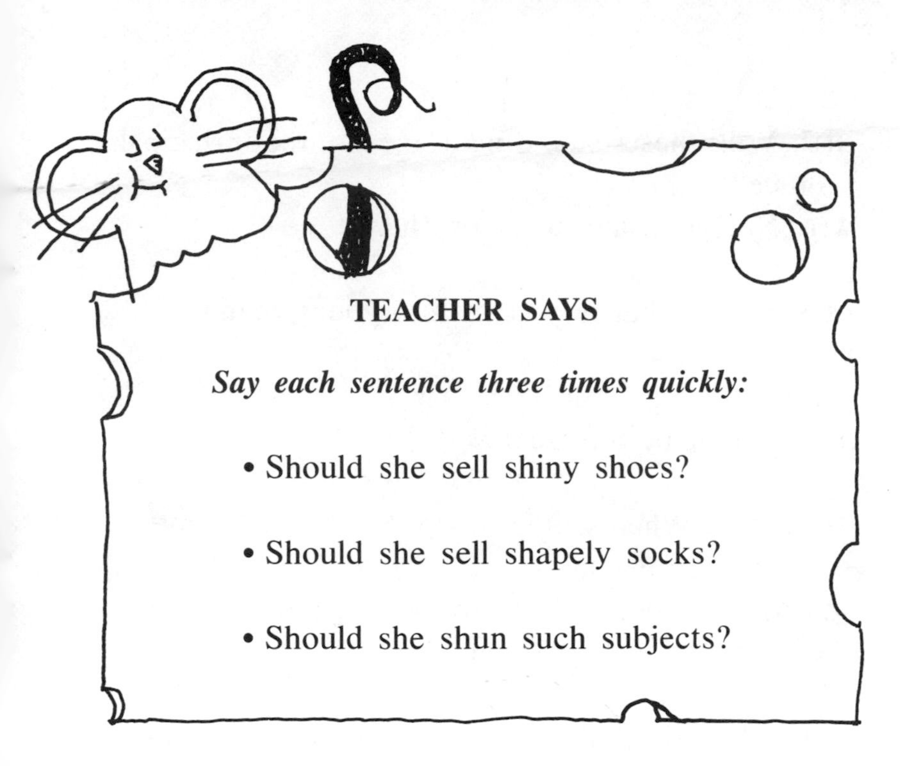

"No one likes me at school," said the son to his mother. "The kids don't and the teachers don't. I want to stay home."

"You have to go, son," insisted his mother. "You're not sick, and you have a lot to learn. Besides, you're 45 years old. You're the principal and you have to go to school!"

DORA: All the criticism of the American school system in newspapers and magazines is absolutely justified.

FATHER: Do you think so?

DORA: Yes, I do. And if you want proof of how bad it is, just look at the terrible marks on my report card.

NIT: Will these stairs take me to the principal's office?
WIT: No, you have to climb them.

PRINCIPAL: What is your name, young man?
BOY: Henry.
PRINCIPAL: Say "Sir."
BOY: All right. Sir Henry!

PRINCIPAL: When you grow up, Seymour, I want you to be a gentleman.
SEYMOUR: I don't want to be a gentleman, sir. I want to be just like you.

"Isn't the principal a dummy!" said a boy to a girl.
"Say, do you know who I am?" asked the girl.
"No."
"I'm the principal's daughter."
"And do you know who I am?" asked the boy.
"No," she replied.
"Thank goodness!"

Old principals never die,
they just lose their faculties.

Old teachers never retire,
they just lose their class.

12. RECESS RIOT

My teacher told me to exercise with dumbbells. Will you join me in the gym?

TEACHER: Why are you taking your math book to the gym?
LEONARD: I have to reduce some fractions.

TEACHER: It is well known that exercise kills germs.
CLASS COMEDIAN: But how do you get the germs to exercise?

CLASS COMEDIAN: I heard a new joke the other day. I wonder if I told it to you.
TEACHER: Was it funny?
CLASS COMEDIAN: Yes.
TEACHER: You didn't.

TEACHER: Did you make up that joke all by yourself?
CLASS COMEDIAN: Yes, out of my head.
TEACHER: You must be!

Let's talk again when your brain gets back from recess.

MORE MESSAGES FOR THE CLASS COMEDIAN

You have a ready wit. Please let us know when it's ready.

Jokes like that will make humor illegal.

You think you're a wit? Well, you're half right.

If Adam came back to earth, the only thing he would recognize are your jokes.

That joke was so bad, you'd need a microscope to see the point.

That joke was so corny, it could feed a chicken for five years.

"What kind of marks did you get in physical education?"

"I didn't get any marks—only a few bruises."

QUENTIN: My brother is in a fight in the school yard.

PRINCIPAL (*rushing into the hall*): How long has it been going on?

QUENTIN: Half an hour, sir.

PRINCIPAL: And you're just telling me now?

QUENTIN: Well, up to now he was winning.

LITTLE FRED: Show me a tough guy and I'll show you a coward.

BIG BULLY: Well, I'm a tough guy.

LITTLE FRED: Well, I'm a coward.

PRINCIPAL: Why are you running in the hall?

SHELDON: I'm running to stop a fight.

PRINCIPAL: That's good. Between whom?

SHELDON: Between me and the guy who's chasing me!

Omar's parents were shocked by the note from his teacher. She wanted a written excuse for his presence.

The French teacher received the following note:

Please excuse my daughter from class today. Her throat is so sore, she can barely speak English.

"I think I have a cold or something in my head."
"It must be a cold."

What is the difference between a dressmaker and the school nurse?

One cuts the dresses and the other dresses the cuts.

SCHOOL NURSE: Can I take your pulse?
NIGEL: Why? Haven't you got one of your own?

Why is the school yard larger at recess?

Because there are more feet in it.

"I have the body of an athlete."
"Better give it back. You're getting it out of shape."

What athlete is never promoted?

The left back.

TEACHER SAYS

Say each sentence three times quickly:

- Surely the sun shall shine soon.
- Sascha slashes sheets slightly.
- She says she shall sew a sheet shut.

What three R's do cheerleaders learn at school?
"Rah, rah, rah!"

Why do soccer players do well in school?
Because they use their heads.

Who was the fastest runner of all time?
Adam. He was first in the human race.

What famous runner had the most peculiar trainer?
Cinderella—she had a pumpkin for a coach.

What subjects do runners like best?
Jog-raphy (geography).

ANGUS: Mom, we played baseball in school today and I stole second base.

MOTHER: Well, you march right over to school and put it back.

AARON: I went out for the football team, Dad.

FATHER: Did you make it?

AARON: I think so. The coach looked at me and said, "This is the end."

TEACHER: The national sport in Spain is bullfighting and in England it's cricket.

PERCY: I'd rather play in England.

TEACHER: Why is that?

PERCY: It's easier to fight crickets.

When do boxers start wearing gloves?

When it gets cold.

MESSAGES FROM THE COACH

Just because the world is out of shape doesn't mean you have to be, too.

You're in such bad shape, if you threw yourself on the floor you'd miss.

You're in such bad shape, you get winded playing checkers.

You're in such bad shape, you get winded when your stocking runs.

You're in such bad shape, if you tried to whip cream, the cream would win.

You're in such bad shape, you better not try to lick an envelope.

Your muscles are like potatoes—mashed potatoes.

The only regular exercise you ever get is reaching for seconds.

You'd make a great football player. Even your breath is offensive.

Geoffrey sent his father a letter from college:

Dear Dad:

No money. Not funny.

Sonny.

The father wrote back:

Dear Son:

How sad. Too bad.

Dad

NIT: I have a chance on the school soccer team.

WIT: I didn't know they were raffling it off.

TEACHER: Ralph, how many times have I told you not to speak without permission?

RALPH: I didn't know I had to keep score.

MOTHER: Julius, what did you learn in school today?

JULIUS: I learned to say "No, ma'am"; "Yes, sir"; and "Yes, ma'am."

MOTHER: You did?

JULIUS: Yup!

AL: My father wants me to have everything he didn't have when he was a boy.

SAL: What didn't he have?

AL: A's on his report card.

13. LAUGHTER FROM THE LAB

FRED: When I die, I'm going to leave my brain to science.

NED: That's nice. Every little bit helps.

TEACHER: Why did the germ cross the microscope?

SYLVIA: To get to the other slide.

SCIENCE QUIZ

1. What does it mean when the barometer falls?
2. How do you fix a short circuit?
3. What should you do with a dead battery?
4. How do you charge a battery?
5. Who invented spaghetti?
6. What is barium?
7. What is camphor?
8. What happens when you swallow uranium?
9. What is the difference between lightning and electricity?
10. Explain "mean temperature."

"C"

BATTERY

Answers on next page.

ANSWERS TO SCIENCE QUIZ

1. That whoever nailed it up didn't do such a good job.
2. Lengthen it.
3. Bury it.
4. With a credit card.
5. A fellow who used his noodle.
6. What you do to the dead.
7. For having fun in the summer.
8. You get an atomic (stomach) ache.
9. We have to pay for electricity.
10. Ten degrees below zero when you're not wearing long johns.

TEACHER: Who can give me a definition of claustrophobia?
SIGMUND: An unnatural fear of Santa Claus.

TEACHER: Kenneth, can you tell me what death is?
KENNETH: Patrick Henry's second choice.

TEACHER: What is stucco?
STUDENT: What you get when you sit on gum-mo?

TEACHER: How can you tell the difference between a boy moose and a girl moose?
ANITA: Er—by his mous-tache?

FAILING STUDENT: If a person's brain stops working, does he die?
TEACHER: You're alive, aren't you?

TEACHER: Name a conductor of electricity.

JOSIE: Why—er—

TEACHER: Wire is right. Name a unit of electrical power.

JOSIE: What?

TEACHER: The watt is absolutely correct.

TEACHER: How did Edison's invention of electricity affect society?

SHEP: If it weren't for him, we'd have to watch television by candlelight!

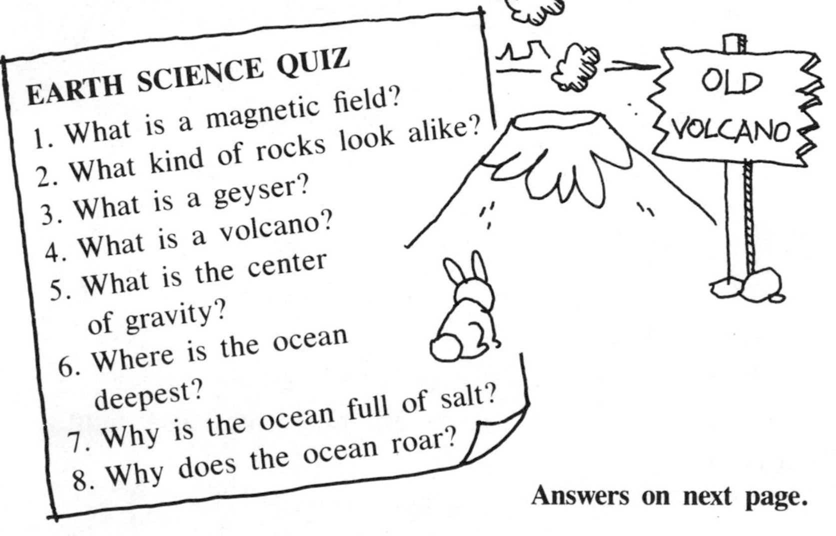

Answers on next page.

What is the difference between ammonia and pneumonia?

Ammonia comes in bottles, pneumonia comes in chests.

TEACHER: Linda, what is a vacuum?

LINDA: I can't think of it just now, but I've got it in my head.

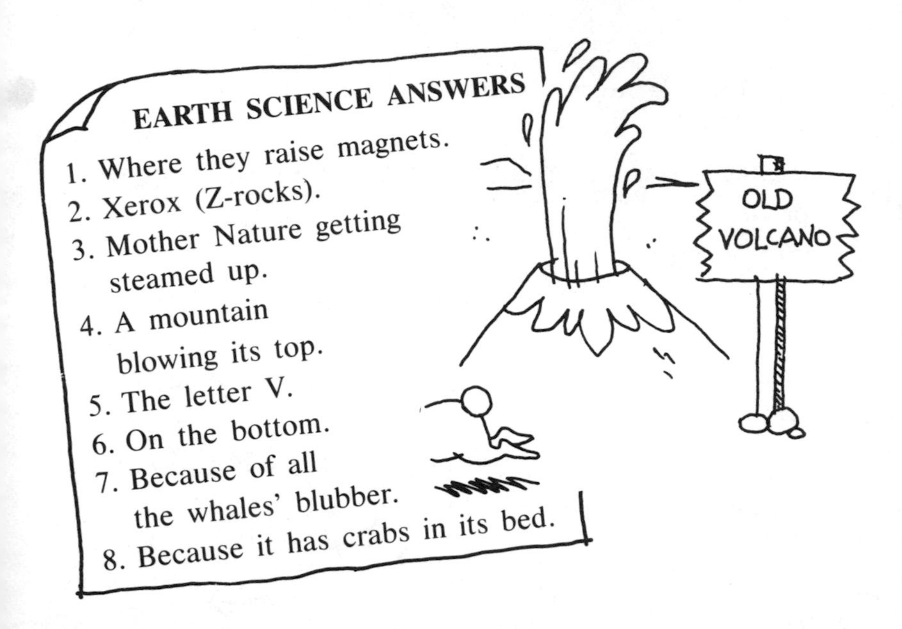

Did you hear about the deer who took a test? He did so well, the teacher passed the buck.

TEACHER: How fast does light travel?
MARCEL (*yawning*): I don't know, teacher, but it gets here too early in the morning.

TEACHER: Eric, will you tell me how fast light travels?
ERIC: The same way slow light travels.

TEACHER: Do you realize that light travels at the rate of 186,000 miles per second?
GWEN: Sure, it's downhill all the way.

TEACHER: Which is faster, hot or cold?
HAL: Hot. You can always catch cold.

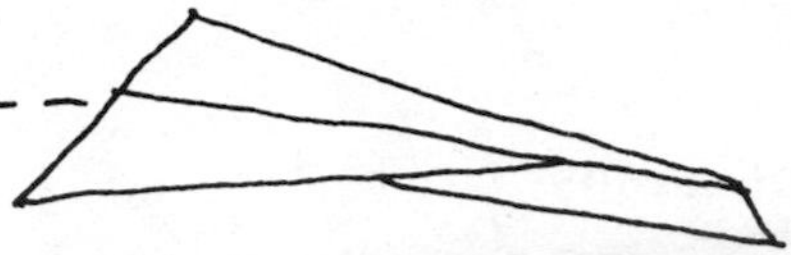

TEACHER SAYS

Say each sentence three times quickly:

- Three shy thrushes.
- Thick thinkers tinker.
- Free thugs set free thugs free.

ZOOLOGY QUIZ

1. Why do lions eat raw meat?
2. How do chimpanzees communicate?
3. Why do polar bears have fur?
4. How are elephants and hippopotamuses alike?
5. Why do elephants have trunks?
6. Why do elephants live alone?
7. What can a bird do that a man cannot do?
8. What animals spend most of the day in the principal's office?
9. How do you keep an angry rhinoceros from charging?
10. How do you fix a broken ape?

Answers on next page.

TEACHER: To which family does the lion belong?

GREGORY: I don't know, teacher. No family in our neighborhood owns one.

How do scientists count atoms?

They atom (add them) up.

What do atomic scientists do when they have time off?

They go fission (fishin').

LEWIS: It's a good thing it was adults who split the atom.

ELVIS: Why?

LEWIS: Well, if one of us kids did it, they'd make us put it back together again.

ZOOLOGY QUIZ ANSWERS

1. Because they don't know how to cook.
2. They speak Chimpaneese.
3. They'd look funny in plastic raincoats.
4. Neither can play tennis.
5. They'd have trouble carrying suitcases.
6. Because two's a crowd.
7. Take a bath in a saucer.
8. Cheetahs.
9. Take away his credit cards.
10. With a monkey wrench.

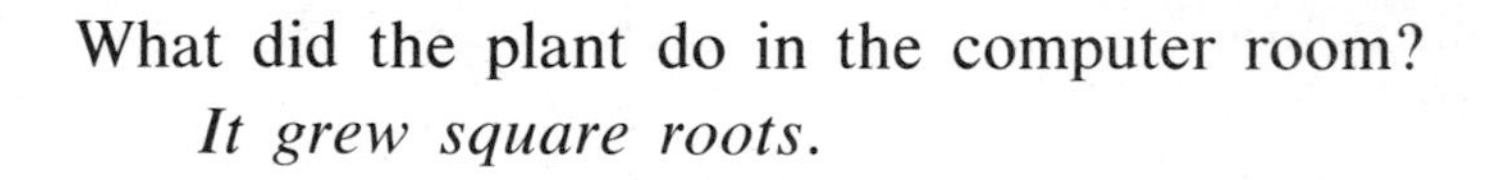

What did the plant do in the computer room?

It grew square roots.

TEACHER (*to unruly class*): This afternoon, I want to tell you about the hippopotamus. Please pay attention, all of you! If you don't look at me, you'll never know what a hippopotamus is like!

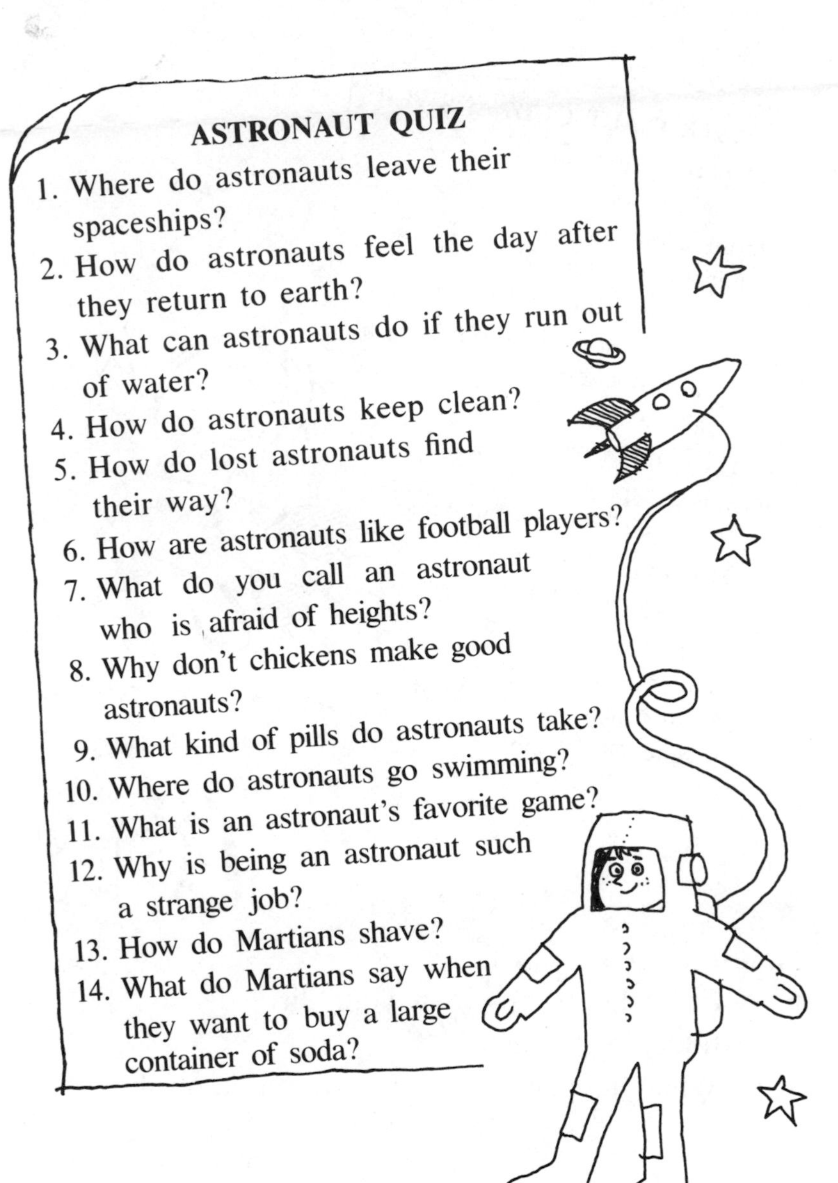

ASTRONAUT QUIZ

1. Where do astronauts leave their spaceships?
2. How do astronauts feel the day after they return to earth?
3. What can astronauts do if they run out of water?
4. How do astronauts keep clean?
5. How do lost astronauts find their way?
6. How are astronauts like football players?
7. What do you call an astronaut who is afraid of heights?
8. Why don't chickens make good astronauts?
9. What kind of pills do astronauts take?
10. Where do astronauts go swimming?
11. What is an astronaut's favorite game?
12. Why is being an astronaut such a strange job?
13. How do Martians shave?
14. What do Martians say when they want to buy a large container of soda?

Answers on next page

TEACHER: Why did the cow jump over the moon?
MELODY: Because the farmer had cold hands.

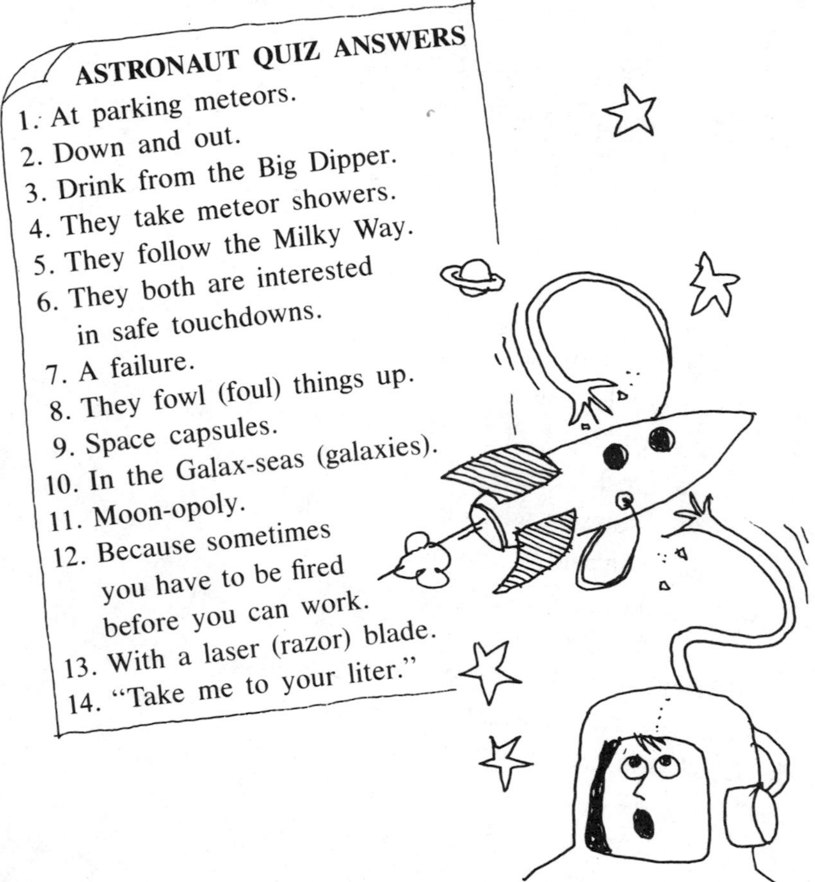

TEACHER: What is zinc?

ROY: Er—where you wash your dirty dishes?

TEACHER: No, try again.

ROY: What happens when you don't know how to zwim.

TEACHER: I'm happy to be able to give you a 70 in Science.

FRED: Why don't you really enjoy yourself and give me 100?

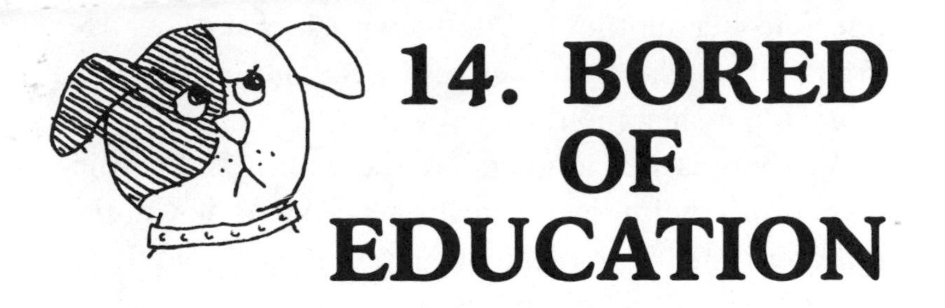

14. BORED OF EDUCATION

TEACHER: When you yawn, you're supposed to put your hand to your mouth.
WILLY: What? And get bitten?

TEACHER: Emil, you aren't paying attention to me. Are you having trouble hearing?
EMIL: No, teacher, I'm having trouble listening.

It was a long, boring lecture, but Mary thought she had to say something nice to the teacher.

"I'd like to apologize for dozing off," she explained, "but I want you to know, I didn't miss a thing."

TEACHER: Why are you late?
CYNTHIA: Sorry, teacher, I overslept.
TEACHER: You mean, you sleep home, too?

TEACHER: Class, we will only have half a day of school this morning.
CLASS: Yippee! Hooray!
TEACHER: We will have the other half this afternoon.

Little Alice wanted to take a day off from school. She told her mother she felt feverish and achy. Her mother brought her a cup of tea and put a thermometer in her mouth.

As soon as her mother left the room, Little Alice dipped the thermometer in the tea. When her mother returned and read the thermometer, she told Little Alice to get dressed and go to school.

"But my fever is so high!" gasped Little Alice.

"You're right, dear," her mother agreed. "It's way up to 140. That means you're dead. So you might as well go to school."

GLADYS: I can't go to school today.
MOTHER: Why not?
GLADYS: I don't feel well.
MOTHER: Where don't you feel well?
GLADYS: In school.

TEACHER: That's quite a cough you have, Michael. What are you taking for it?
MICHAEL: I don't know, teacher. What will you give me?

TEACHER: You missed school yesterday, didn't you?
ROD: Not very much.

FATHER: I hear you played hookey from school to play baseball.
JUNIOR: No, Dad, and I have the fish to prove it.

VOICE (*on telephone*): My son has a bad cold and won't be able to attend school today.
ASSISTANT PRINCIPAL: Who is this?
VOICE: This is my father speaking.

A WORD

TO THE WISE GUY

Playing hookey from school is like a credit card—fun now, pay later.

Laugh and the class laughs with you, but you stay after school alone.

Some people drink at the fountain of knowledge—some just gargle.

"The brain is a wonderful thing."
"Why do you say that?"
"Because it starts working the minute you get up in the morning and never stops until you're called on in class."

KELLY: I didn't do my homework because I lost my memory.
TEACHER: How long has this been going on?
KELLY: How long has what been going on?

TEACHER'S COMMENTS ON YOUR COMPOSITION

You have nothing to say, but that doesn't stop you from saying it.

Some people can write on any subject. You don't need a subject.

Your ideas are so corny, you should dish them up with butter and salt.

I've heard brighter ideas from my parrot.

FATHER: When I was your age, I thought nothing about walking 10 miles to school.
JUNIOR: I agree, Dad. I don't think much of it myself.

"How do you like going to school?" asked Billy's aunt.

"I like the going fine," said Billy. "I also enjoy the coming home. But I don't care much for the time in between."

TEACHER: Now, Terry, be sure you go straight home.
TERRY: I can't, teacher.
TEACHER: Why not?
TERRY: Because I live around the corner.

The teacher was annoyed with her students, who kept checking the clock on the wall. She covered it with a sign that read, "Time will pass. Will you?"

When the teacher came into the classroom, she noticed a girl sitting with her feet in the aisle and chewing gum.

"Eloise," she said, "take that gum out of your mouth and put your feet in this instant!"

Knock-knock.
Who's there?
Ike, Anne, Howard, Lee, Wyatt, Tillie.
Ike, Anne, Howard, Lee, Wyatt, Tillie who?
Ike, Anne, Howard, Lee, Wyatt, Tillie (I can hardly wait till) it's three o'clock!

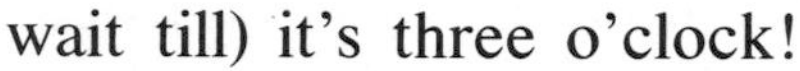

HOMEWORK

I love to do my homework,
It makes me feel so good.
I love to do exactly
As my teacher says I should.

I love to do my homework,
I never miss a day.
I love the little men in white
Who're taking me away.

TEACHER: Henry, I'd like to go through one whole day without having to scold you!
HENRY: You have my permission.

MY TEACHER

My teacher loves me,
Thinks I'm dear.
She's kept me for
The fourth straight year.

TEACHER: Alfred, why don't you answer me?
ALFRED: I did, teacher. I shook my head.
TEACHER: You don't expect me to hear it rattle from up here, do you?

FLO: Our teacher talks to herself. Does yours?
JOE: Yes, but she doesn't realize it. She thinks we're listening.

That last joke about the teacher was so bad, we put it at the end of this book.

Did you hear my last joke?
I sure hope so!

Index

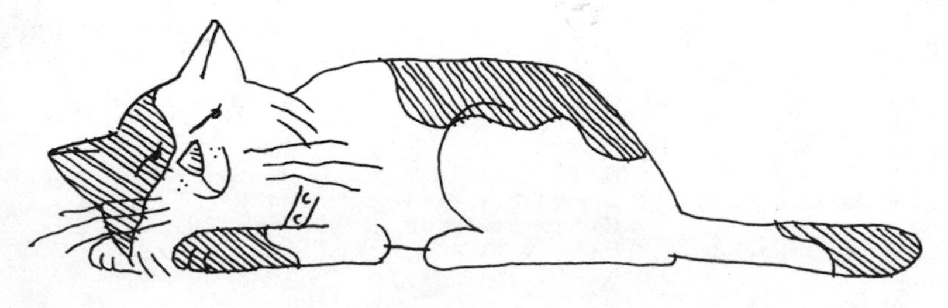